Lecture Notes in Networks and Systems

Volume 1314

Series Editor

Janusz Kacprzyk, Systems Research Institute, Polish Academy of Sciences, Warsaw, Poland

Advisory Editors

Fernando Gomide, Department of Computer Engineering and Automation—DCA, School of Electrical and Computer Engineering—FEEC, University of Campinas—UNICAMP, São Paulo, Brazil

Okyay Kaynak, Department of Electrical and Electronic Engineering, Bogazici University, Istanbul, Türkiye

Derong Liu, Department of Electrical and Computer Engineering, University of Illinois at Chicago, Chicago, USA
Institute of Automation, Chinese Academy of Sciences, Beijing, China

Witold Pedrycz, Department of Electrical and Computer Engineering, University of Alberta, Alberta, Canada
Systems Research Institute, Polish Academy of Sciences, Warsaw, Poland

Marios M. Polycarpou, Department of Electrical and Computer Engineering, KIOS Research Center for Intelligent Systems and Networks, University of Cyprus, Nicosia, Cyprus

Imre J. Rudas, Óbuda University, Budapest, Hungary

Jun Wang, Department of Computer Science, City University of Hong Kong, Kowloon, Hong Kong

AF793633

The series "Lecture Notes in Networks and Systems" publishes the latest developments in Networks and Systems—quickly, informally and with high quality. Original research reported in proceedings and post-proceedings represents the core of LNNS.

Volumes published in LNNS embrace all aspects and subfields of, as well as new challenges in, Networks and Systems.

The series contains proceedings and edited volumes in systems and networks, spanning the areas of Cyber-Physical Systems, Autonomous Systems, Sensor Networks, Control Systems, Energy Systems, Automotive Systems, Biological Systems, Vehicular Networking and Connected Vehicles, Aerospace Systems, Automation, Manufacturing, Smart Grids, Nonlinear Systems, Power Systems, Robotics, Social Systems, Economic Systems and other. Of particular value to both the contributors and the readership are the short publication timeframe and the world-wide distribution and exposure which enable both a wide and rapid dissemination of research output.

The series covers the theory, applications, and perspectives on the state of the art and future developments relevant to systems and networks, decision making, control, complex processes and related areas, as embedded in the fields of interdisciplinary and applied sciences, engineering, computer science, physics, economics, social, and life sciences, as well as the paradigms and methodologies behind them.

Indexed by SCOPUS, EI Compendex, INSPEC, WTI Frankfurt eG, zbMATH, SCImago.

All books published in the series are submitted for consideration in Web of Science.

For proposals from Asia please contact Aninda Bose (aninda.bose@springer.com).

Siba K. Udgata · Srinivas Sethi · George Ghinea ·
Sanjay Kumar Kuanar
Editors

Intelligent Systems

Proceedings of 4th International Conference on Machine Learning, IoT and Big Data (ICMIB 2024), Volume 1

Springer

Editors
Siba K. Udgata
University of Hyderabad
Hyderabad, Telangana, India

George Ghinea
Department of Computer Science
Brunel University London
Uxbridge, UK

Srinivas Sethi
Department of Computer Science
Engineering and Applications
Indira Gandhi Institute of Technology
Dhenkanal, Odisha, India

Sanjay Kumar Kuanar
School of Engineering and Technology
GIET University
Gunupur, Odisha, India

ISSN 2367-3370 ISSN 2367-3389 (electronic)
Lecture Notes in Networks and Systems
ISBN 978-981-96-3796-6 ISBN 978-981-96-3797-3 (eBook)
https://doi.org/10.1007/978-981-96-3797-3

© The Editor(s) (if applicable) and The Author(s), under exclusive license to Springer Nature Singapore Pte Ltd. 2025

This work is subject to copyright. All rights are solely and exclusively licensed by the Publisher, whether the whole or part of the material is concerned, specifically the rights of translation, reprinting, reuse of illustrations, recitation, broadcasting, reproduction on microfilms or in any other physical way, and transmission or information storage and retrieval, electronic adaptation, computer software, or by similar or dissimilar methodology now known or hereafter developed.
The use of general descriptive names, registered names, trademarks, service marks, etc. in this publication does not imply, even in the absence of a specific statement, that such names are exempt from the relevant protective laws and regulations and therefore free for general use.
The publisher, the authors and the editors are safe to assume that the advice and information in this book are believed to be true and accurate at the date of publication. Neither the publisher nor the authors or the editors give a warranty, expressed or implied, with respect to the material contained herein or for any errors or omissions that may have been made. The publisher remains neutral with regard to jurisdictional claims in published maps and institutional affiliations.

This Springer imprint is published by the registered company Springer Nature Singapore Pte Ltd.
The registered company address is: 152 Beach Road, #21-01/04 Gateway East, Singapore 189721, Singapore

If disposing of this product, please recycle the paper.

Preface

It is really a pleasure to pen down a message on the eve of the International Conference on Machine learning, Internet of Things and Big data (ICMIB-2024) being organized by the prestigious GIET University, Gunupur, Odisha, India, during March 29–31, 2024. The conference draws some excellent technical keynote talks and papers. We received 368 full paper submissions and we accepted only 75 papers. The contributing authors are from different parts of the globe that includes India. All the papers are reviewed by at least three independent reviewers and in some cases by as many as four reviewers. All the papers are also checked for plagiarism and similarity score. It was really a tough job for us to select the best papers out of so many good papers for presentation in the conference. We had to do this unpleasant task, keeping the Springer guidelines and approval conditions in view. We take this opportunity to thank all the authors for their excellent work and contributions and also the reviewers who have done an excellent job.

On behalf of the program chair, we are indebted to Prof. L. M. Patnaik, General Chair of the Conference, for his timely and valuable advice. We cannot imagine the conference without his active support at all the crossroads of decision-making process. We would also like to place on record our thanks to all the keynote speakers, tutorial speakers, reviewers, session chairs, authors, technical program committee members, various chairs to handle finance, accommodation, hospitality, publication and publicity, and above all to several volunteers.

We are also thankful to Springer Nature publication house for agreeing to publish the accepted papers in their Lecture Notes in Networks and Systems (LNNS) series.

The organizing committee made all arrangements to make the conference enjoyable and a good learning experience for everybody. Hope all the delegates attending the virtual sessions also enjoy the presentations and also use the opportunity to exchange their ideas.

Gunupur, India

Prof. Siba K. Udgata
Prof. Srinivas Sethi
Prof. Sanjay Kumar Kuanar
Prof. George Ghinea
Program Chairs, ICMIB-2024

Reviewers and Committees

Reviewers

Alekha Mishra
Alok Prusty
Alok Tripathy
Anitha A.
Anup Maharana
Arunima Pattanayak
Ashima Rout
Ashis Das
Atish Nanda
Bhargavi B.
Bibhudatta Sahoo
Bibudhendu Pati
Bichitra Mandal
Bichitra Sethi
Bunil Balabantaray
Chhabi Rani Panigrahi
Chinmayee Rout
Chiranji Lal Chowdhary
Debabrata Dansana
Debasis Mohapatra
Debendra Muduli
Deepak Gupta
Deepak Tosh
Devesh Bandil
Diptendu Singha Roy
P. K. Panigrahi
Rabindra Behera
Srinivas Sethi

Abhinav Tomar
Bikramaditya Das
Krishnaveni
Sangharatna Godboley
Santosh Majhi
Suvendu Narayan Mishra
Gopal Behera
Hemant Apat
Indrajeet Gupta
J. Chandrkanta Badajena
Janmenjoy Nayak
Jitendra Kumar
Jitendra Rout
K. Hemant Reddy
Kali Rath
Kalyan Kumar Jena
Kaushik Mishra
Krishna Kumar
Kshira Sahoo
Lakhmi Das
Lalatendu Muduli
Lalitha Krishnasamy
Madhu Panda
Malti Nagle
Manoj Kumar Patra
Niranjan Panigrahi
Om Prakash Jena
Pavan Mishra
Pradumn Kumar
Prahallad Sahu
Prajna Paramita Nanda
Prasant Dhal
Pratap Puhan
Preeti Chandrakar
Pritee Parwekar
Priyadarsan Parida
Priyanka Sen
Punyaban Patel
Purna Sethi
Pushkar Kishore
Raghunandan Swain
Raghvendra Kumar
Rajendra Nayak
Rajiv Senapati
Rakesh Chandra Balabantaray

Rakesh Swain
Ramesh Sahoo
Ranumayee Sing
Rashmi Sahoo
Rohit Kumar Bondugula
Rohit Sharma
S. Gopal Krishna Patro
Saifulla Md.
Salman Abdul Moiz
Sambit Kumar Mishra
Sampa Sahoo
Sandhya Sahoo
Sangita Pal
Sanjay Kuanr
Sanjaya Panda
Sanjib Nayak
Santanu Dash
Saroj Panigrahy
Sashi Nayak
Sasmita Acharya
Sasmita Behera
Satyapada Das
Satyasundara Mahapatra
Sibarama Panigrahi
Sipra Swain
Sohan Pande
Sourav Bhoi
Sreyasee rout
Srichandan Sobhanayak
Subasish Mohapatra
Subhashish Nayak
Subhasish Pani
Sudhansu Patra
Suman Das
Sumit Kar
Sunita Dhalbisoi
Suraj Sharma
Surya Das
Susil Mohanty
Suvendra Jayasingh
Swarupananda Bissoyi
Tirimula Rao Benala
Umakanta Samantsinghar

Umashankar Ghugar
Umesh Sahu
Yatindra Gopal

Committees

Chief Patrons

Prof (Dr.) Satya Prakash Panda, President, GIET University, India
Prof (Dr.) Chandra Dhwaj Panda, Vice President, GIET University, India
Prof (Dr.) Jagadish Panda, Director General, GIET University, India

Patrons

Dr. A. V. N. L. Sharma, Vice Chancellor, GIET University, India
Dr. N. V. Jagannadha Rao, Registrar, GIET University, India

General Chair

Prof. Lalit Mohan Patnaik, National Institute of Advanced Studies and Indian Institute of Science, Bengaluru

Program Chair

Prof. Siba K. Udgata, University of Hyderabad, India

Program Co-chairs

Prof. Srinivas Sethi, IGIT Sarang
Prof. George Ghinea, Computer Science, Brunel University
Prof. Sanjay Kumar Kuanar, GIET University, India

Organizing Chairs

Prof. S. N. Mishra, IGIT Sarang
Dr. Raghvendra Kumar, GIET University, India

Convenors

Prof. Srinivas Sethi, IGIT Sarang
Dr. Sanjay Kumar Kuanar, GIET University, India

Publicity Chairs

Dr. K. Murali Gopal, GIET University, India
Dr. Neelamadhab Padhy, GIET University, India
Dr. A. V. S. Pavan Kumar, GIET University, India
Dr. Sachikanta Dash, GIET University, India
Prof. B. P. Panigrahi, IGIT Sarang
Prof. Ashima Rout, IGIT Sarang

Finance Chairs

Dr. Nilambar Sethi, GIET University, India
Dr. Bidush Kumar Sahoo, GIET University, India

Hospitality Chairs

Dr. Premansu Sekhara Rath, GIET University, India
Dr. Jyotirmaya Mishra, GIET University, India
Mrs. Rasmita Panigrahi, GIET University, India
Mr. K. Siva Krishna, GIET University, India

Logistic Chairs

Prof. Satya Narayan Das, GIET University, India
Mr. Ranjit Patnaik, GIET University, India
Mr. Ranjeet Panigrahi, GIET University, India
Mr. Santosh Kumar Panda, GIET University, India

Accommodation Chairs

Mr. Sibo Prasad Patro, GIET University, India
Mr. P. Sudheer Babu, GIET University, India
Mr. Chandrakanta Mahanty, GIET University, India

Organising Committee

Mr. Bhavani Sankar Panda, GIET University, India
Ms. Jemarani Jaypuria, GIET University, India
Mr. G. V. S. Narayana, GIET University, India
Mr. Jagadish Sahoo, GIET University, India

Technical Committee Members

George Ghinea, Computer Science, Brunel University
Kumar Padmanabh, EBTIC (A Research Lab of British Telecom)
Deepak Tosh, Computer Science, University of Texas at El Paso
Haesik Kim, VTT Technical Research Centre, Finland
Soroj Meher, ISI, Bengaluru
Manas Ranjan Patra, Berhampur University
Siba K. Udgata, University of Hyderabad
O. B. V. Ramanaiah, OBV, JNTU Hyderabad
G. Suvarna Kumar, MVGRCE, Vizianagaram
G. Sandhya MVGRCE, Vizianagaram
R. Hemalatha, University College of Engineering, Osmania University
Amit Kumar Mishra, University of Cape Town, South Africa
R. Thangarajan, Kongu Engineering College, Tamil Nadu
Birendra Biswal, Gayatri Vidya Parishad College of Engineering, Vishakhapatnam
Somanath Tripathy, IIT Patna
A. K. Turuk, NIT Rourkela

B. D. Sahoo, NIT Rourkela
D. P. Mohapatra, NIT Rourkela
P. M. Khilar, NIT Rourkela
P. G. Sapna, CIT, Coimbatore
R. K. Dash, NIC, Mizoram
B. K. Tripathy, VIT, Vellore
Moumita Patra IIT, Guahati
S. N. Das, GIET University, Gunupur, Odisha
Ram Kumar Dhurkari, IIM, Sirmaur, Himachal Pradesh
Chitta Ranja Hota, BITS Pilani, Hyderabad
A. Kavitha, JNTU, Hyderabad
Subasish Mohapatra, CET Bhubaneswar
Sanjaya Kumar Panda, NIT Warangal
S. Mini, NIT, Goa
S. Nagender Kumar, University of Hyderabad
Lalit Garg, University of Malta, Malta
Lalitha Krishna, Kongu Engineering College, Tamil Nadu
C. Poongodi, Kongu Engineering College, Tamil Nadu
Sumanth Yenduri, Kennesaw University, USA
Shaik Shakeel Ahamad, Majmaah University, Saudi Arabia
K. Srujan Raju, CMR Technical Campus, Hyderabad
R. Hemalatha, University College of Engineering, Osmania University, Hyderabad
P. Sakthivel, Anna University
Pavan Kumar Mishra, NIT, Raipur
Tapan Kumar Gandhi, Department of Electrical Engineering, IIT, Delhi
Annappa B., NIT Surathkal
Prafulla Kumar Behera, Utkal University, Bhubaneswar, Odisha
Nekuri Naveen, School of Computer and Information Sciences, University of Hyderabad
Ch. Venkaiah, School of Computer and Information Sciences, University of Hyderabad
Rajendra Lal, School of Computer and Information Sciences, University of Hyderabad
Dillip Singh Sisodia, Department of Computer Science and Engineering, NIT, Raipur
Pradeep Singh, Department of Computer Science and Engineering, NIT, Raipur
Jay Bagga, Ball State University, USA
Sumagna Patnaik, JB Institute of Engineering and Technology, Hyderabad
Ajit K. Sahoo, University of Hyderabad
Atluri Rahul, Neurolus Systems, Hyderabad
Samrat L. Sabat, Center for Advanced Studies in Electronic Science and Technology (CASEST), University of Hyderabad
Nihar Satapathy, Sambalpur University
Susil Kumar Mohanty, Computer Science and Engineering, IIT, Patna
Kagita Venkat, NIT, Warangal
Sanjay Kumar Kuanar, GIET University, Gunupur, Odisha

Bhabendra Biswal, College of Engineering, Bhubaneswar
Padmalaya Nayak, GR Institute of Engineering and Technology, Hyderabad
Bhibudendu Pati, R. D Women's University, Bhubaneswar
Chabi Rani Panigrahi, R. D. Women's University, Bhubaneswar
Rajesh Verma, Infosys Ltd., Hyderabad
Arun Avinash Chauhan, School of Computer and Information Sciences, University of Hyderabad
Khusbu Pahwa, Delhi Technological University, New Delhi
Soumen Roy, DRDL, Hyderabad
Satyajit Acharya, Tech Mahindra, Hyderabad
Subhrakanta Panda, BITs Pilani, Hyderabad
Vineet P. Nair, School of Computer and Information Sciences, University of Hyderabad
Subash Yadav, Department of Computer Science, Central University of Jharkhand, Ranchi
Layak Ali, Central University Karnataka, Gulbarga
Deepak Kumar, NIT Meghalaya
Bunil Balabantaray, NIT, Meghalaya
Sumanta Pyne, NIT Rourkela
Asis Tripathy, VIT, Vellore
Mousumi Saha, NIT Durgapur
Abhijit Sharma, NIT Durgapur
Mayukh Sarkar, MNNIT, Allahabad
Oishila Bandyopadhyay, IIIT, Kalayani
Subrat Kumar Mohanty, IIIT, Bhubaneswar
Ramesh Chandra Mishra, IIIT Manipur
Hirak Maity, Kolaghat Engineering College
Sandeep Kumar Panda, ICFAI, Hyderabad
Ashim Rout, IGIT Sarang
Srinivas Sethi, IGIT Sarang
S. N. Mishra, IGIT Sarang
Sangita Pal, IGIT Sarang
Sanjaya Patra, IGIT Sarang
Biswanath Sethi, IGIT Sarang
Niroj Pani, IGIT Sarang
Dillip Kumar Swain, IGIT Sarang
B. P. Panigrahy, IGIT Sarang
Rabindra Behera, IGIT Sarang
L. N. Tripathy, OUTR Bhubaneswar
B. B. Choudhary, IGIT Sarang
Dhiren Behera, IGIT Sarang
Anand Gupta, IGIT Sarang
Ayaskanta Swain, NIT Rourkela
Gayadhr Panda, NIT Meghalaya
S. K. Tripathy, IGIT Sarang

Devi Acharya, VIT, Vellore
Saurov Bhoi, PMEC Berhampur
Kalyan Kumar Jena, PMEC Berhampur
Alok Ranjan Prusty, Skill Development, Delhi
Babita Majhi, GGU, Chhattisgarh
Prof. Subhrashu Dash, GCE, Keonjhar
Puspalata Pujahari, GGU, Chhattisgarh
Tirimula Rao, JNTU Kakinada
Kshirsagar Sahoo, VIT-AP University
Niranjan Panigrahy, PMEC Berhampur
Trilochon Rout, PMEC Berhampur
P. K. Panigrahy, GIET, India
Mihir Kumar Sutar, UCE Burla
Gopal Behera, GCEK, Bhawanipatna
Rajendra Prasad Nayak, GCEK, Bhawanipatna
Kaliprasan Sethi, GCEK, Bhawanipatna
Jitendra Kumar Rout, NIT Raipur

Contents

Optimizing Lung Cancer Risk Prediction: A Comparative Analysis of Machine Learning Models with Emphasis on Logistic Model Tree Accuracy

Satyaprakash Swain, Binod Kumar Pattanayak, Mihir Narayan Mohanty, Kumar Janrdan Patra, Suvendra Kumar Jayasingh, and Tapas Kumar Swain

Abstract In this research, using a dataset of 1000 patient records that we have taken from Kaggle dataset repository, on which we have performed a thorough analysis to forecast the risk levels of lung cancer, which were divided into three categories: low, medium and high. We attempted to train and test the data for optimal predictive accuracy using sophisticated machine learning models, specifically Naive Bayes (NB), Logistic Model Tree (LMT), Random Tree (RT) and Decision Stump (DS). The Logistic Model Tree (LMT) was found to be the best performer following a thorough evaluation based on key performance indicators like accuracy, Root Absolute Error (RAE), Mean Absolute Error (MAE), Root Relative Square Error (RRSE)

S. Swain (✉) · B. K. Pattanayak
Department of Computer Science and Engineering, Institute of Technical Education and Research (ITER), Siksha 'O' Anusandhan (Deemed to be University), Bhubaneswar, Odisha, India
e-mail: satyaimit@gmail.com

B. K. Pattanayak
e-mail: binodpattanayak@soa.ac.in

M. N. Mohanty
Department of Electronics and Communication Engineering, Institute of Technical Education and Research (ITER), Siksha 'O' Anusandhan (Deemed to be University), Bhubaneswar, Odisha, India
e-mail: mihirmohanty@soa.ac.in

K. J. Patra
Department of Computer Science and Enginerring, OUTR, Bhubaneswar, Odisha, India
e-mail: janardanpatra1997@gmail.com

S. Swain · S. K. Jayasingh · T. K. Swain
Department of Computer Science and Enginerring, Institute of Management and Information Technology(IMIT), Cuttack, Odisha, India
e-mail: sjayasingh@gmail.com

T. K. Swain
e-mail: tapaskumarr00@gmail.com

© The Author(s), under exclusive license to Springer Nature Singapore Pte Ltd. 2025
S. K. Udgata et al. (eds.), *Intelligent Systems*, Lecture Notes in Networks and Systems 1314, https://doi.org/10.1007/978-981-96-3797-3_1

and Root Mean Square Error (RMSE). Notably, the LMT model outperformed the other three models with an astounding accuracy rate of 98%. To further support its effectiveness, the Random Tree (RT) model also showed minimal errors across all error metrics. The high accuracy attained by the LMT model demonstrates its exceptional performance, which forms the basis of the recommendation to use it for lung cancer risk prediction. Not only did this model perform better than the others, but its low error rates across multiple metrics also demonstrated its superior predictive power. Accurately estimating the risk of lung cancer has significant potential clinical implications, providing doctors with a useful tool for early intervention and individualized patient care. The results of this study highlight how crucial it is to use advanced machine learning methods in healthcare analytics, especially when it comes to cancer risk assessment. We believe that this research will advance predictive analytics for healthcare and lead to improved patient outcomes and a proactive approach to lung cancer management, as we recommend the use of the LMT model in lung cancer risk prediction.

Keywords ML · LMT · Clustering · Error · Accuracy

1 Introduction

Lung cancer is still a major global health concern that requires novel strategies for risk assessment and early detection. In this work, we utilized a Kaggle dataset repository containing the medical records of one thousand patients to create a predictive model that classifies lung cancer risk into three categories like low, medium and high. We investigated the capabilities of four different models: Naive Bayes (NB), Logistic Model Tree (LMT), Random Tree (RT) and Decision Stump (DS), realizing the crucial role that machine learning plays in healthcare analytics. Using critical performance metrics like accuracy, Root Absolute Error (RAE), Mean Absolute Error (MAE), Root Relative Square Error (RRSE) and Root Mean Square Error (RMSE), the main goal was to thoroughly assess these models.

It is impossible to exaggerate the significance of precise and timely risk assessment given the steadily increasing incidence of lung cancer. Large datasets can be sorted through using machine learning models, which can help uncover patterns and insights that conventional diagnostic techniques might miss. Our study explores the comparative analysis of these models, paying particular attention to how well they predict outcomes and reduces errors in risk classification.

In our analysis, the Logistic Model Tree (LMT) showed the highest accuracy rate, i.e., 98%, making it the clear winner. Not only did this model outperform the other three in terms of accuracy, but it also showed minimal errors in terms of RAE, MAE, RRSE and RMSE, among other error metrics. In light of the potential therapeutic benefits of precise lung cancer risk assessment, our research suggests that clinicians utilize the LMT model as a valuable resource. Beyond predictive analytics, the implications offer a proactive lung cancer management strategy that may greatly enhance patient outcomes through prompt interventions.

Our research, which aims to advance predictive analytics for lung cancer risk assessment, makes use of a Kaggle dataset repository that includes comprehensive data from one thousand patients. The dataset comprises 19 parameters that provide distinct insights into different aspects of an individual's health and 1 event that are essential for predicting the risk levels of lung cancer, which can be classified as high, medium or low. The parameters cover a broad range of variables, including lifestyle decisions, exposure to the environment and particular health indicators, in addition to demographic data like age and gender.

In order to create a reliable predictive model and identify complex patterns in this dataset, this study aims to leverage machine learning. The 18 parameters of dataset collectively contribute to a thorough understanding of the complex nature of lung cancer risk, encompassing variables like genetic risk, smoking habits, chronic lung disease and air pollution, among others.

We have carefully separated the dataset into four subsets to improve the effectiveness of our machine learning models and to guarantee a thorough and representative training procedure. Our approach is based on the use of cutting-edge machine learning models [1, 2], such as Naive Bayes (NB), Random Tree (RT), Decision Stump (DS) and the top performer Logistic Model Tree (LMT).

With an astounding accuracy rate of 98%, the results of our analysis demonstrate the superiority of the Logistic Model Tree. This superiority over Naive Bayes, Random Tree and Decision Stump models highlights the capacity of advanced machine learning methods to decipher intricate relationships found in healthcare datasets. In order to guarantee a thorough assessment, our research goes beyond accuracy [3–6] and examines critical performance metrics like mean absolute error (MAE), root mean square error (RMSE), root absolute error (RAE) and root relative square error (RRSE).

We support the use of the Logistic Model Tree as the best option for estimating the risk of lung cancer in clinical settings in light of these findings. This model's nuanced insights have the potential to revolutionize the field of lung cancer risk assessment by giving medical professionals a valuable tool for early intervention and individualized patient care. Our overall objective as we work through the complexities of this dataset is to advance the field of predictive analytics in healthcare, with a particular emphasis on improving our comprehension and management of lung cancer risk.

2 Existing Work

In order to predict the likelihood that a patient may have a stroke, Swain et al. [7] compared the Neural Basis Model (NBM), Random Forest (RF), Logistic Model Tree (LMT) and Decision Tree (DT) using data from 25,000 heart disease patients. They found that the Decision Tree (DT) performs the best after analyzing the various models. In order to diagnose heart diseases accurately, they proposed a model that will aid in prediction with lower prediction error and higher accuracy.

Six deep learning models were developed by Vani Rajasekar et al. [8] to facilitate the accurate diagnosis of lung cancer. Their CNN, CNN GD, Inception V3, Resnet-50, VGG-16 and VGG-19 machine learning techniques are used to detect lung cancer. The histopathological and CT scan images served as the basis for the experimental analyses. Based on factors like recognition accuracy, $F1$-Score, precision, sensitivity and specificity, among others, the suggested approach's performance was evaluated. Owing to the inherent benefit of the suggested methodology, lung cancer can be effectively detected using this method, which will help those in need. With an accuracy of 97.86%, the CNN GD model demonstrates the improved performance.

Elias Dritsas et al. [9] have created models for identifying people whose lung cancer manifests based on a variety of features and symptoms by utilizing supervised learning. AUC, F-Measure, recall, accuracy, precision and KNN were assessed for a number of machine learning models, including AdaBoostM1, NB, BayesNet, SGD, SVM, LR, ANN, KNN, J48, LMT, RF, RT, RepTree and RotF. With an accuracy, precision, recall and F-Measure of 97.1% and an AUC of 99.3%, the RotF outperformed the other models based on the experiment results and after applying SMOTE with tenfold cross-validation.

Several models that use a chest X-ray or CT scan as an input to identify a specific disease have been used by Deepapriya et al. [10]. They conduct research to find the most effective deep learning methods for predicting lung disease. Many performance metrics, including precision, recall, accuracy and Jacquard index, are used to assess the method's performance. The outcome showed that, when compared to other models taken into consideration for experimental analysis, MSD-NET produces superior results.

Risk assessment models are presented by Nemlander et al. [11] and have the potential to be developed into clinical RATs that assist clinicians in determining a patient's risk of lung cancer.

A CNN neural network method has been demonstrated by Bharathy et al. [12] for early cancer detection. The CNN algorithm's high computational speed and accuracy make it a good choice for lung cancer screening decision-making. Information can be communicated more effectively with video than with still images. It is beneficial to detect lung cancer early because treatment can begin immediately to prevent the disease's harmful effects.

Timor et al. [13] have provided an overview of the primary methods for classifying nodules and predicting lung cancer from CT imaging data. Convolutional Neural Networks (CNNs) trained using Deep Learning techniques achieve the current state-of-the-art with classification performance in the low nineties for AUC points. There is enough training data available. When assessing system performance, it is critical to understand the limitations related to the training and validation datasets used.

The ability of machine learning to precisely forecast the Length of Stay (LOS) for ICU-based hospitalization in patients with lung cancer was shown by Belal Alsinglawi et al. [14]. They assessed appropriate class balancing strategies for clinical Electronic Health Records (EHR) in order to address the imbalanced class problem that is common in clinical health records data, especially in predictive modeling tasks with skewed class distributions. By introducing the LOS research framework—the first of its kind for predicting days that patients with lung cancer would spend in the intensive care unit, they gave clinical oncologists, hospital bed managers and other healthcare professionals a useful tool. The Random Forest ensemble classifier was found to be resilient in a range of feature selection processes (RFE or clinical significance) among the numerous machine learning techniques, feature selection and class balancing strategies investigated.

Cruz et al. [15] performed a thorough evaluation of the different machine learning methods used in cancer prognosis and prediction in their review. They noticed trends in the way machine learning was applied, including the kinds of methods, training sets, endpoint predictions, kinds of cancer and overall effectiveness. Artificial Neural Networks (ANNs) continue to be widely used, but different alternatives are being used more often to predict different outcomes for different types of cancer. In terms of predictive accuracy, machine learning approaches typically perform better than conventional statistical or expert-based systems. Although they acknowledge well-designed studies with sufficient validation, they also call for greater focus on the design and conduct of experiments, particularly in light of the volume and caliber of biological data. Enhancing the generality, repeatability and general quality of machine-based classifiers is considered to require advances in experimental design and biological validation. With ongoing advancements in research quality, the authors predict a rise in the use of machine learning classifiers in clinical and hospital settings.

The existing literature on predicting lung cancer risk reveals significant advancements in utilizing various machine learning techniques and models, particularly focusing on deep learning methods like Convolutional Neural Networks (CNNs) and ensemble classifiers such as Random Forests. However, despite these strides, there remains a noticeable research gap in the integration of diverse risk factors beyond traditional biomarkers and imaging features. While studies have extensively explored genetic predispositions, environmental exposures and imaging-based markers, there's a limited emphasis on incorporating socio-economic factors, lifestyle choices and co-morbidities such as chronic obstructive pulmonary disease (COPD) into predictive models. Furthermore, there is a lack of standardization and validation across different datasets and populations, hindering the scalability and robustness of predictive algorithms. Thus, there is a clear need for research that comprehensively integrates diverse risk factors and addresses the challenges of standardization and validation to develop more accurate and clinically applicable predictive models for lung cancer risk assessment.

3 Methodology

This study used a dataset from Kaggle that included records from 1000 patients across 19 parameters in order to predict lung cancer risk levels using a methodical and thorough methodology. The dataset included demographic data that is essential

Fig. 1 Flowchart of this work

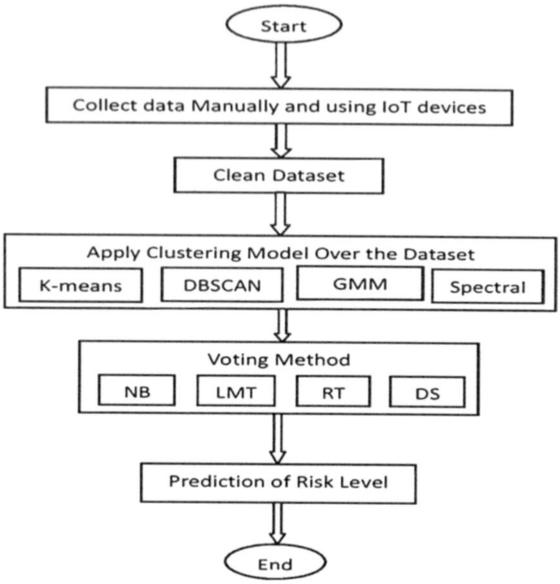

for comprehending the intricate relationship between lung cancer risk and lifestyle factors, environmental exposures and particular health indicators. The dataset was carefully split into four subsets to improve the resilience of machine learning models and enable a thorough training and testing procedure. Our work is organized as shown in Fig. 1.

After obtaining the dataset from Kaggle respiratory, preprocessing the data was an essential step in our study to guarantee its accuracy and dependability. Popular clustering techniques including K-means, DBSCAN, Gaussian Mixture Model (GMM) and spectral clustering were applied to this process. Each technique had a unique function in the dataset's organization and cleaning, helping to produce a polished and analytically sound dataset. To help find possible outliers and anomalies, K-means clustering was first used to divide the dataset into different clusters based on similarity. Then, dense regions in the data were found using DBSCAN (Density-Based Spatial Clustering of Applications with Noise), which successfully distinguished between noise and outliers. The Gaussian Mixture Model provided information about the dataset's underlying probability distributions and helped to understand its structure. Furthermore, by capturing complex relationships between data points, spectral clustering which makes use of the spectral decomposition of the similarity matrix further refined the dataset. To evaluate how well each clustering technique cleaned and ready the dataset for further analysis, the results were carefully compared based on the Silhoutte Score. Table 1 displays each clustering technique's with their silhouette values.

Notably, DBSCAN stood out as the most promising technique, exhibiting superior performance as illustrated in Fig. 2 in terms of handling outliers, identifying noise and maintaining the data's natural structure. A thorough analysis of DBSCAN's results in comparison with the other clustering techniques led to the decision to give it priority over the other methods for cleaning datasets.

With this methodology, you can be sure that the dataset used for further analyses is clean, well-organized and free of artifacts that could jeopardize the accuracy and dependability of the research findings [16–19]. Multiple clustering techniques are applied methodically, demonstrating a thorough approach to data cleaning and the significance of choosing the best technique to improve dataset quality. Naive Bayes (NB), Logistic Model Tree (LMT), Random Tree (RT) and Decision Stump (DS) are the four different machine learning models (Voting Models) that were chosen for examination. By using these models, the dataset's patterns were to be found and predictive models for precise lung cancer risk classification were to be established. A portion of the dataset was used to train the models and their effectiveness was then assessed through testing. For an extensive evaluation, key performance metrics such as accuracy, mean absolute error (MAE), root mean square error (RMSE), root absolute error (RAE) and root relative square error (RRSE) were used.

The model with the highest potential, the Logistic Model Tree, outperformed the other three models with an astounding accuracy rate of 98%. As illustrated in Figs. 3 and 4, the LMT model is the best option for estimating the risk of lung cancer in clinical settings due to its higher accuracy and low errors across a variety of metrics. This study's methodology, which makes use of cutting-edge machine learning techniques to extract complex insights into the intricacies of lung cancer risk prediction, reflects a methodical and rigorous approach. The results are reliable and valid due to the division of the dataset and the careful assessment of model performance. This highlights the importance of implementing advanced machine learning techniques in healthcare research.

Table 1 Silhoutte score of clustering models

Heading level	Example
K-Means	0.25
DBSCAN	0.64
GMM	0.11
Sprectral	0.14

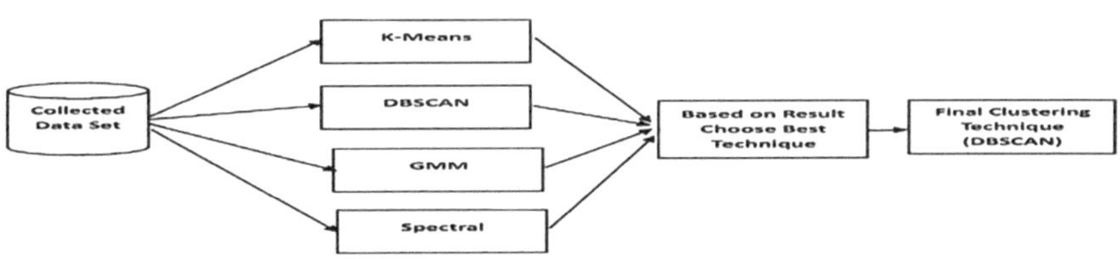

Fig. 2 Implementation of clustering technique

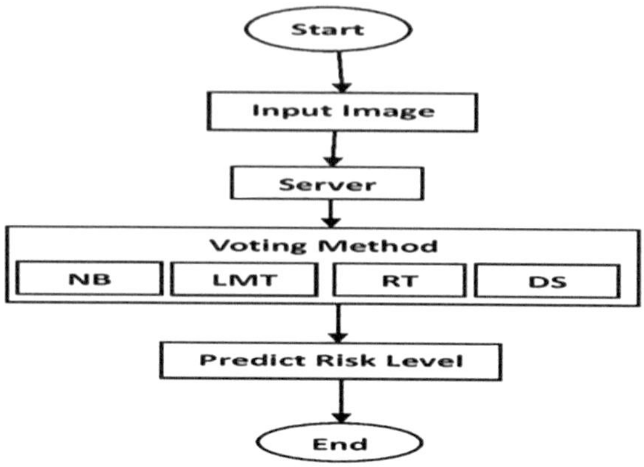

Fig. 3 Implementation of voting models

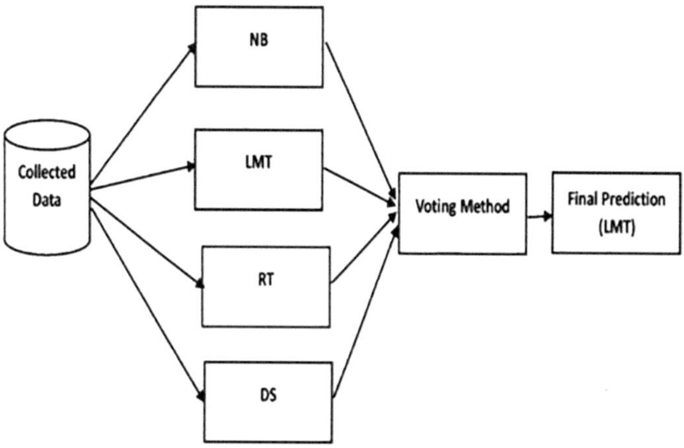

Fig. 4 Selection of proposed voting model

3.1 Naive Bayes (NB)

This algorithm is generative and probabilistic. Its foundation is the Bayes theorem, which states that given some observed evidence, a hypothesis has a probability. Text classification, spam filtering and other tasks where the independence assumption holds reasonably well are among the common uses for it.

3.2 Logistic Model Tree (LMT)

It is an ensemble model that blends logistic regression and decision trees. To determine the ultimate probability, it constructs a decision tree for every class and uses logistic regression on each leaf node. It can handle both numerical and categorical data and captures non-linear relationships between features and the target variable. It works well for classification tasks that require modeling intricate relationships between features.

3.3 Random Tree (RT)

It is a decision tree-based ensemble model. Every decision tree in the Random Forest ensemble is trained using a random subset of features and a random subset of training data at each split. It gives feature importance information, boosts robustness and lessens over fitting. It is frequently used to enhance decision trees' performance and is useful for a variety of classification tasks.

3.4 Decision Stump (DS)

It is a decision tree in its most basic form, having just one level. A decision stump bases their choice on the importance of a single characteristic. It functions as a one-level decision tree, basically. In ensemble techniques like AdaBoost, it is frequently employed as a building block. A boosting algorithm can produce a strong classifier by combining multiple decision stumps, whereas a single stump is weak.

4 Simulation and Result Analysis

The data used in the suggested model is taken from the Kaggle dataset repository. In order to provide the best possible care for the human body, the collected data are stored in the cloud [20] and analyzed using various machine learning techniques [21–25], such as NB, LMT, RT and DS models, to predict the lung cancer risk beforehand. In order to predict diseases more accurately and with less error, parameters are chosen based on the needs of various machine learning models. Numbers ranging from 2 to 6 in the Tables 2, 3, 4, 5 and 6 indicate the accuracy values and the values gathered from various errors. Additionally, the graphical representation of Tables 2, 3, 4, 5 and 6 is shown in Figs. 5, 6, 7, 8 and 9.

Based on Tables 2, 3, 4, 5 and 6 values and Figs. 5, 6, 7, 8 and 9, we got LMT model performs best with accuracy 98% and minimal different errors as compared to other models. The high accuracy attained by the LMT model demonstrates its exceptional performance, which forms the basis of the recommendation to use it for lung cancer risk prediction. Not only did this model perform better than the others, but its low error rates across multiple metrics also demonstrated its superior predictive power.

Table 2 Comparison of NB, LMT, RT, DS models based on its correct accuracy values

Accuracy				
Input	NB	LMT	RT	DS
Dataset 1	87.6	98	77.2	62.8
Dataset 2	88.8	96.8	74	64.4
Dataset 3	89.2	95.6	80.4	67.2
Dataset 4	90.8	95.6	75.2	70.4

Table 3 Comparison of NB, LMT, RT, DS models based on MAE values

MAE				
Input	NB	LMT	RT	DS
Dataset 1	0.0885	0.0174	0.1778	0.2829
Dataset 2	0.0789	0.0424	0.2035	0.2887
Dataset 3	0.0745	0.0416	0.1616	0.2755
Dataset 4	0.0666	0.0374	0.1962	0.2327

Table 4 Comparison of NB, LMT, RT and DS models on the based on RMSE values

RMSE				
Input	NB	LMT	RT	DS
Dataset 1	0.282	0.1059	0.3051	0.3861
Dataset 2	0.2699	0.1407	0.3312	0.382
Dataset 3	0.2668	0.1556	0.2869	0.3725
Dataset 4	0.2441	0.1533	0.3181	0.3419

Table 5 Comparison of NB, LMT, RT, DS models based on RAE values

RAE				
Input	NB	LMT	RT	DS
Dataset 1	20.0344	3.9387	40.2499	64.0387
Dataset 2	17.969	9.6524	46.3344	65.718
Dataset 3	17.1038	9.5603	37.0848	63.2442
Dataset 4	15.501	8.7118	45.7004	54.2052

Table 6 Comparison of NB, LMT, RT, DS models on the basis of its RRSE values

RRSE				
Input	NB	LMT	RT	DS
Dataset 1	60.0065	22.5408	64.9199	82.1541
Dataset 2	57.5821	30.028	70.6627	81.5181
Dataset 3	57.1761	33.3334	61.47	79.8175
Dataset 4	52.6984	33.0879	68.6727	73.7938

Fig. 5 Graphical comparison NB, LMT, RT and DS models based on its correct accuracy

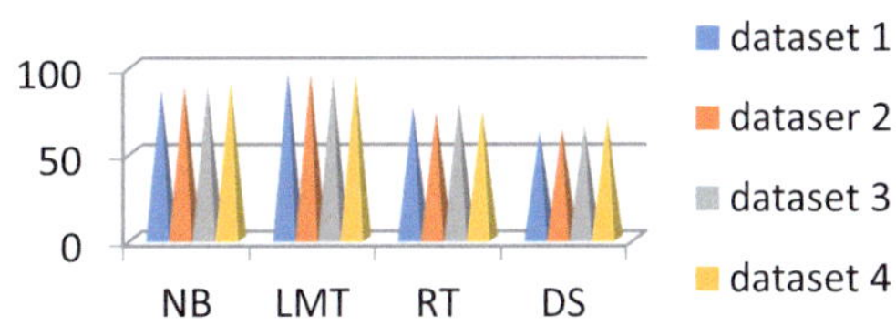

Fig. 6 Graphical comparison NB, LMT, RT and DS models based on MAE values

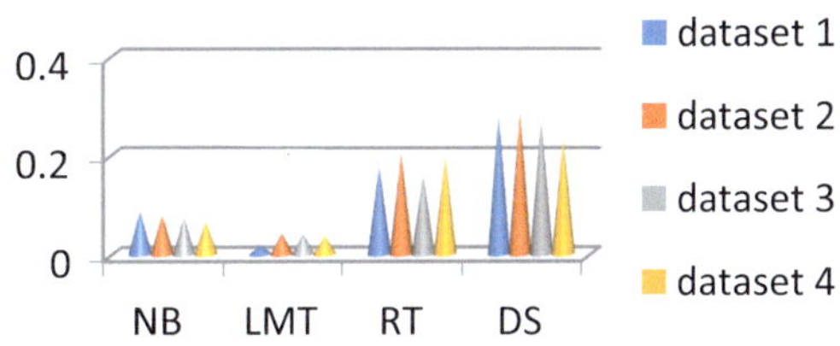

Fig. 7 Graphical comparison NB, LMT, RT and DS based on RMSE values

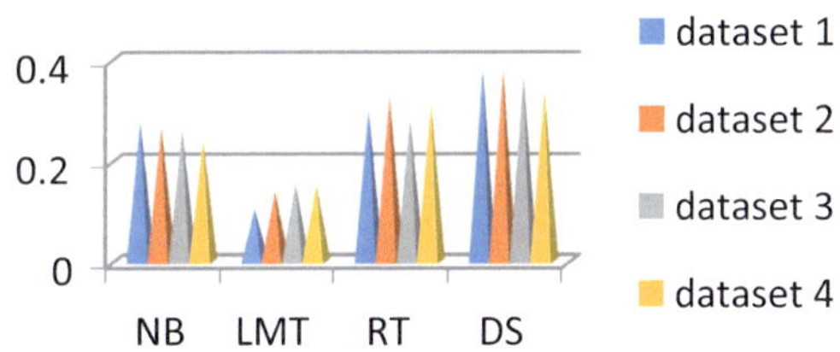

Fig. 8 Graphical comparison NB, LMT, RT and DS based on RAE values

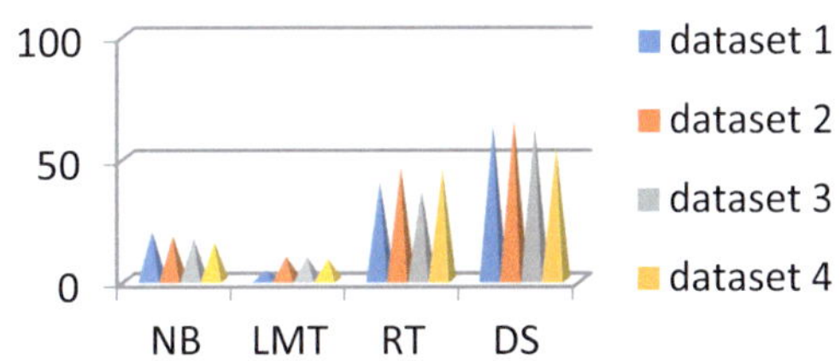

Fig. 9 Graphical comparison NB, LMT, RT and DS on the basis of RRSE values

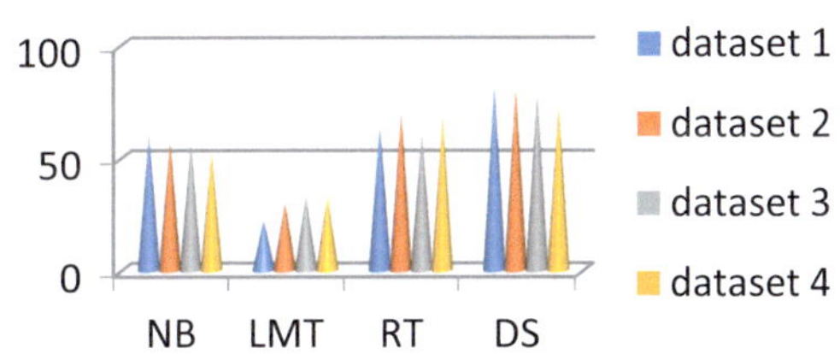

5 Conclusion

Using a dataset of 1000 patient records from Kaggle respiratory, our research attempts to address the urgent need for precise lung cancer risk prediction. The Logistic Model Tree (LMT) has been determined to be the most promising candidate after a thorough comparative analysis of machine learning models, including Naive Bayes, Logistic Model Tree, Random Tree and Decision Stump. The robust predictive abilities of the LMT model were demonstrated with an impressive accuracy rate of 98% and minimal errors across key metrics, including RAE, MAE, RRSE and RMSE, outperforming the competing models. It is impossible to overstate the potential significance of such precise risk assessment in lung cancer, providing doctors with an invaluable tool for early intervention and individualized patient care. We anticipate a revolutionary effect on the treatment of lung cancer and the dawning of a new era of targeted and preemptive healthcare approaches as we push for the implementation of the LMT model in clinical settings. This work advances the field of predictive analytics in healthcare and emphasizes the value of utilizing cutting-edge machine learning methods to improve patient outcomes and address lung cancer's challenges in a more efficient manner. Our findings support additional research and integration of machine learning models for enhanced diagnostic and prognostic capabilities in the field of cancer risk assessment, as technology continues to play a critical role in healthcare.

References

1. Anisa S, Devaki P, Rachitha P, Gracewell JJ (2022) Automatic identification of cancer affect in lungs using machine learning algorithm. In: 2022 fourth international conference on emerging research in electronics, computer science and technology (ICERECT), Mandya, pp 1–5. https://doi.org/10.1109/ICERECT56837.2022.10059997
2. Agarwal S, Thakur S, Chaudhary A (2022) Prediction of lung cancer using machine learning techniques and their comparative analysis. In: 2022 10th international conference on reliability. Infocom Technologies and Optimization (Trends and Future Directions) (ICRITO), Noida, pp 1–5. https://doi.org/10.1109/ICRITO56286.2022.9965052
3. Puneet, Chauhan A (2020) Detection of lung cancer using machine learning techniques based on routine blood indices. In: 2020 IEEE international conference for innovation in technology (INOCON), Bengaluru, pp 1–6. https://doi.org/10.1109/INOCON50539.2020.9298407
4. Al-Tawalbeh J, Alshargawi B, Alquran H, Al-Azzawi W, Mustafa WA, Alkhayyat A (2022) Classification of lung cancer by using machine learning algorithms. In: 2022 5th international conference on engineering technology and its applications (IICETA), Al-Najaf, pp 528–531. https://doi.org/10.1109/IICETA54559.2022.9888332
5. Rehman A, Kashif M, Abunadi I, Ayesha N (2021) Lung cancer detection and classification from chest CT scans using machine learning techniques. In: 2021 1st international conference on artificial intelligence and data analytics (CAIDA), Riyadh, pp 101–104. https://doi.org/10.1109/CAIDA51941.2021.9425269
6. Saini S, Maithani A, Dhiman D, Bisht A (2021) Analysis of different machine learning algorithms used for identification of lung cancer disease. In: 2021 9th international conference on reliability, Infocom technologies and optimization (trends and future directions) (ICRITO), Noida, pp 1–5. https://doi.org/10.1109/ICRITO51393.2021.9596308

7. Swain S, Behera N, Swain AK, Jayasingh SK, Patra KJ, Pattanayak BK, Mohanty MN, Naik KD, Rout SS (2023) Application of IoT framework for prediction of heart disease using machine learning. Int J Recent Innov Trends Comput Commun 11(10s):168–176. https://doi.org/10.17762/ijritcc.v11i10s.7616
8. Rajasekar V, Vaishnnave MP, Premkumar S, Sarveshwaran V, Rangaraaj V (2023) Lung cancer disease prediction with CT scan and histopathological images feature analysis using deep learning techniques. Results Eng 18:101111. ISSN 2590-1230. https://doi.org/10.1016/j.rineng.2023.101111
9. Dritsas E, Trigka M (2022) Lung cancer risk prediction with machine learning models. Big Data Cogn Comput 6:139. https://doi.org/10.3390/bdcc6040139
10. Deepapriya BS, Kumar P, Nandakumar G et al (2023) Performance evaluation of deep learning techniques for lung cancer prediction. Soft Comput 27:9191–9198. https://doi.org/10.1007/s00500-023-08313-7
11. Nemlander E, Rosenblad A, Abedi E, Ekman S, Hasselström J, Eriksson LE et al (2022) Lung cancer prediction using machine learning on data from a symptom e-questionnaire for never smokers, formers smokers and current smokers. PLoS ONE 17(10):e0276703. https://doi.org/10.1371/journal.pone.0276703
12. Bharathy S, Pavithra R, Akshaya B (2022) Lung cancer detection using machine learning. In: 2022 international conference on applied artificial intelligence and computing (ICAAIC), Salem, pp 539–543. https://doi.org/10.1109/ICAAIC53929.2022.9793061
13. Kadir T, Gleeson F (2018) Lung cancer prediction using machine learning and advanced imaging techniques. Transl Lung Cancer Res 7(3)
14. Alsinglawi B, Alshari O, Alorjani M et al (2022) An explainable machine learning framework for lung cancer hospital length of stay prediction. Sci Rep 12:607. https://doi.org/10.1038/s41598-021-04608-7
15. Cruz JA, Wishart DS. Applications of Machine Learning in Cancer Prediction and Prognosis. Cancer Informatics. 2006;2. doi:10.1177/117693510600200030
16. Omar AA-C, Nassif AB (2023) Lung cancer prediction using machine learning based feature selection: a comparative study. In: 2023 advances in science and engineering technology international conferences (ASET), Dubai, pp 1–6. https://doi.org/10.1109/ASET56582.2023.10180436
17. Verma A, Shah CK, Kaur V, Shah S, Kumar P (2022) Cancer detection and analysis using machine learning. In: 2022 second international conference on computer science, engineering and applications (ICCSEA), Gunupur, pp 1–5. https://doi.org/10.1109/ICCSEA54677.2022.9936457
18. Rao BD, Arshad M (2023) Early detection of lung cancer using machine learning technique. In: 2023 international conference on computer communication and informatics (ICCCI), Coimbatore, pp 1–5. https://doi.org/10.1109/ICCCI56745.2023.10128389
19. Karthick K, Rajkumar S, Selvanathan N, Saravanan UKB, Murali M, Dhiyanesh B (2021) Analysis of lung cancer detection based on the machine learning algorithm and IOT. In: 2021 6th international conference on communication and electronics systems (ICCES), Coimbatore, pp 1–8. https://doi.org/10.1109/ICCES51350.2021.9489084
20. Swain SP, Swain PK (2019) Performance analysis of a priority based buffer-aided relaying. Int J Comput Sci Eng 7(6):1006–1011
21. Qureshi R, Alam T, Wu J (2022) Drug response prediction for lung cancer patients using biophysical simulation and machine learning. In: 2022 IEEE international conference on bioinformatics and biomedicine (BIBM), Las Vegas, NV, pp 3867–3869. https://doi.org/10.1109/BIBM55620.2022.9995622
22. Degadwala S, Vyas D, Kothari A, Khunt U (2023) Cancer death cases forecasting using supervised machine learning. In: 2023 4th international conference on electronics and sustainable communication systems (ICESC), Coimbatore, pp 903–907. https://doi.org/10.1109/ICESC57686.2023.10193685

23. Ingle K, Chaskar U, Rathod S (2021) Lung cancer types prediction using machine learning approach. In: 2021 IEEE international conference on electronics, computing and communication technologies (CONECCT), Bangalore, pp 01–06. https://doi.org/10.1109/CONECCT52877.2021.9622568
24. Sruthi G, Ram CL, Sai MK, Singh BP, Majhotra N, Sharma N (2022) Cancer prediction using machine learning. In: 2022 2nd international conference on innovative practices in technology and management (ICIPTM), Gautam Buddha Nagar, pp 217–221. https://doi.org/10.1109/ICIPTM54933.2022.9754059
25. Rajaguru H, SCSR, Chidambaram S (2022) Gaussian mixture model based hybrid machine learning for lung cancer classification using symptoms. In: 2022 smart technologies, communication and robotics (STCR), Sathyamangalam, pp 1–4. https://doi.org/10.1109/STCR55312.2022.10009440

Drug Recommendation System for Humans Using NLP and Machine Learning Approach

Priyadarshan Dhabe, Yash Munde, Nitin Choudhary, Muaz Sayyed, Ayush Vidhale, and Netal Zanwar

Abstract Drug recommendation systems have gained importance in recent years due to the increased need for personalized healthcare and efficient management of medical conditions. The research paper presents a drug recommendation system that utilizes natural language processing (NLP) and machine learning (ML) approaches. The objective of this study is to develop a robust model that can accurately predict medical conditions and recommend the most suitable drugs based on various factors such as predicted medical conditions, top reviews, and useful review counts. To achieve this, the TF-IDF technique is used to preprocess the textual data, effectively representing the drugs and reviews in a numerical format. Furthermore, it utilizes a diverse dataset containing information about drug names, medical conditions, user reviews, ratings, and useful counts. By using this dataset, a training and testing framework is created to train and evaluate the performance of machine-learning models. The research delves into the implementation and comparison of various algorithms, including the passive aggressive classifier and Multinomial NB, to identify the most accurate and efficient model for drug recommendation. The results of the experiments demonstrate the effectiveness of the proposed drug recommendation system, highlighting its ability to accurately predict medical conditions and provide tailored drug suggestions. This research contributes to the growing body of knowledge in the field of healthcare and medicine by offering a comprehensive analysis of a drug recommendation system using NLP and ML techniques. The findings of this study provide valuable insights for healthcare professionals, pharmaceutical companies, and researchers alike, advancing the field of personalized medicine and enhancing patient care.

Keywords Drug · Healthcare · Recommendation system · NLP · TF-IDF · Machine learning · Naive Bayes · Passive aggressive classifier

P. Dhabe · Y. Munde · N. Choudhary · M. Sayyed · A. Vidhale (✉) · N. Zanwar
Department of Information Technology, Vishwakarma Institute of Technology, Pune, Maharashtra, India
e-mail: ayush.vidhale211@vit.edu

© The Author(s), under exclusive license to Springer Nature Singapore Pte Ltd. 2025
S. K. Udgata et al. (eds.), *Intelligent Systems*, Lecture Notes in Networks and Systems 1314, https://doi.org/10.1007/978-981-96-3797-3_2

1 Introduction

In recent years, the field of personalized healthcare has gained significant attention for its potential to revolutionize patient care and improve treatment outcomes. One area of particular interest is the development of drug recommendation systems, which utilize cutting-edge technologies such as "natural language processing (NLP) and machine learning (ML)" to provide tailored drug suggestions based on individual medical conditions and patient preferences. These systems have the potential to greatly enhance the efficiency and effectiveness of medical treatments, enabling healthcare professionals to make informed decisions regarding drug prescriptions. While several studies have explored the development of drug recommendation systems using NLP and ML approaches, there are still gaps in understanding the most accurate and efficient methodologies. Existing research often fails to consider crucial factors such as top reviews and useful review counts in the recommendation process. Additionally, the lack of comprehensive studies comparing multiple ML algorithms limits our understanding of the optimal approach for drug recommendation. The research paper aims to address these gaps by presenting a comprehensive study on a drug recommendation system that utilizes NLP and ML methods. The primary objective is to develop a robust model that can accurately predict medical conditions and recommend suitable drugs based on factors like predicted medical conditions, top reviews, and useful review counts. By incorporating these previously overlooked elements, the proposed system aims to provide more accurate and personalized drug recommendations, ultimately improving patient outcomes. Specifically, the project will assess the performance of various ML algorithms, including the TF-IDF with passive aggressive classifier and TF-IDF with multinomial NB, in the drug recommendation process. By comparing their accuracy, efficiency, and stability, the aim is to identify the most effective algorithm for this particular application. Additionally, the study will examine the impact of incorporating top reviews and useful review counts in the recommendation system, analyzing their influence on the accuracy of drug suggestions. The significance and relevance of this research lie in its potential to advance the field of personalized medicine and enhance the quality of patient care. By developing a more accurate and efficient drug recommendation system, healthcare professionals can make informed decisions, ensuring that patients receive the most suitable medications for their specific medical conditions. This research also contributes to the understanding of ML algorithms in healthcare applications and highlights the importance of incorporating various factors in the recommendation process. It is important to acknowledge the scope and limitations of this study. While the research provides valuable insights into the development of a drug recommendation system using NLP and ML approaches, the generalizability of the findings may be limited to the specific dataset and algorithms utilized. Additionally, the accuracy of the recommendations may depend on the quality and availability of data. Despite these limitations, the research offers valuable contributions to the field and is a foundation for further investigations into personalized healthcare and drug recommendation systems. Overall, the project aims to fill the gaps in existing knowledge by developing a comprehensive drug recommendation system using NLP and ML techniques. By enhancing the accuracy and efficiency of drug suggestions, this study strives to improve patient care, advance the field of personalized medicine, and contribute to the growing body of knowledge in healthcare informatics.

2 Literature Review

Many researchers have been working on drug recommendation and here are some of the projects related to it [1]. The paper proposes a drug recommendation system that utilizes sentiment analysis of drug reviews to provide personalized medication suggestions. The system employs various text vectorization techniques, including bag-of-words (BoW), term frequency–inverse document frequency (TF-IDF), and Word2Vec, to transform drug reviews into numerical representations suitable for machine learning algorithms. The sentiment of each review is then classified as positive, negative, or neutral using various classification models, such as support vector machines (SVM), random forest, and Naive Bayes. The system aggregates the sentiment scores of reviews for each drug to generate a sentiment-based drug recommendation [2]. The paper introduces a drug-drug interaction (DDI) extraction system that utilizes natural language processing (NLP) techniques to identify and extract potential interactions between drugs from unstructured text sources. The system employs a pipeline of NLP tasks, including drug-named entity recognition (DNER), relation extraction, and classification, to extract DDIs from medical literature and clinical documents. The system's performance is evaluated on a benchmark dataset, demonstrating its effectiveness in accurately identifying and classifying DDIs [3]. The paper proposes a blockchain and machine learning-based drug supply chain management and recommendation system (DSCMR) to address the challenges of drug counterfeiting and inefficient drug recommendation in the pharmaceutical industry. The DSCMR system utilizes Hyperledger Fabric, a permissioned blockchain framework, to track and verify the authenticity of drugs throughout the supply chain, ensuring transparency and tamper-proof records. Additionally, the system employs machine learning algorithms to analyze drug reviews and provide personalized drug recommendations to consumers [4]. By merging the knowledge graph (KG) and recommendation system, the research provides a unified framework for drug-target interaction (DTI) prediction. The framework develops a low-dimensional representation for each entity in the KG and merges multimodal data using a neural factorization machine (NFM). The proposed methodology is tested in three actual circumstances and achieves accurate and robust predictions on four benchmark datasets, particularly in the cold start scenario for proteins [5]. The research paper presents a drug recommender system based on a stacked artificial neural network (ANN) model to improve the fairness and safety of infectious disease treatment. The proposed method recommends medications based on a patient's previous health profile, lifestyle, and habits and achieves 97.5% accurate results. The report emphasizes the need to suggest safe drugs, especially during health emergencies [6]. The paper presents a system that uses the R package "recommenderlab" to help locate the best suitable medicine based on conditions. The study proposed a method that provides the best drug accessible based on user ratings in a database. The system uses recommendation systems (RS) to recommend the appropriate medicine to the appropriate user [7]. The research paper covers the application of machine learning-based recommender systems in medicine. It emphasizes the significance of such systems in making accurate, precise, and trustworthy clinical predictions while using fewer resources. The study additionally examines how these systems can be used to propose safe medications to patients based on their symptoms, blood pressure, diabetes, temperature, and other factors [8]. Benjamin Stark et al. did a literature survey on existing solutions for medical recommender systems. They compared these systems based on diseases, data storage, interface, data collecting, data preparation, platform/technology, algorithm, and future work. The purpose of this paper is to show existing methods for healthcare providers to improve the medicine selection process and choose an appropriate drug for patients [9]. Based on collaborative filtering and clustering techniques, the paper proposes a drug recommendation system for diabetic patients. For dimensionality reduction and patient clustering, the system employs unsupervised learning techniques. A user-based collaborative filtering approach is used to generate drug predictions, which creates a patient profile that can be compared to the profiles of other patients with similar characteristics. Based on collaborative filtering and clustering techniques, the system achieved good results in the quality of predictions with a mean squared error metric of 0.51 and accuracy in the quality of recommendations with a mean squared error metric of 0.61, which is acceptable [10]. GalenOWL, a semantic-enhanced online framework, is presented in the study to assist professionals in finding information about the drugs. The study presents a framework that implies medications for a patient with consideration of their sensitivity, infections, and drug interactions. Clinical data and terminology were first translated into ontological terms using international standards like ICD-10 and UNII to enhance GalenOWL. Then, the ontological terms were correctly combined with the clinical data.

3 Methodology

The methodology for developing an NLP and machine learning-driven drug recommendation system involves several key phases, as depicted in the project workflow diagram (Fig. 1). The process begins with data acquisition from an established open repository, focusing on attributes such as drug names, conditions, user ratings, and reviews. The raw data undergoes preprocessing and cleaning, including formatting standardization, redundancy removal, and terminology normalization, to optimize input for subsequent algorithms. Techniques such as TF-IDF are then applied to transform the text into vector representations, capturing underlying semantic signals for classification. By utilizing TF-IDF bigram weighting schemes, meaningful dimensionality reduction is achieved, producing inputs for training predictive models like passive aggressive classifiers. These models learn patterns linking user reviews to

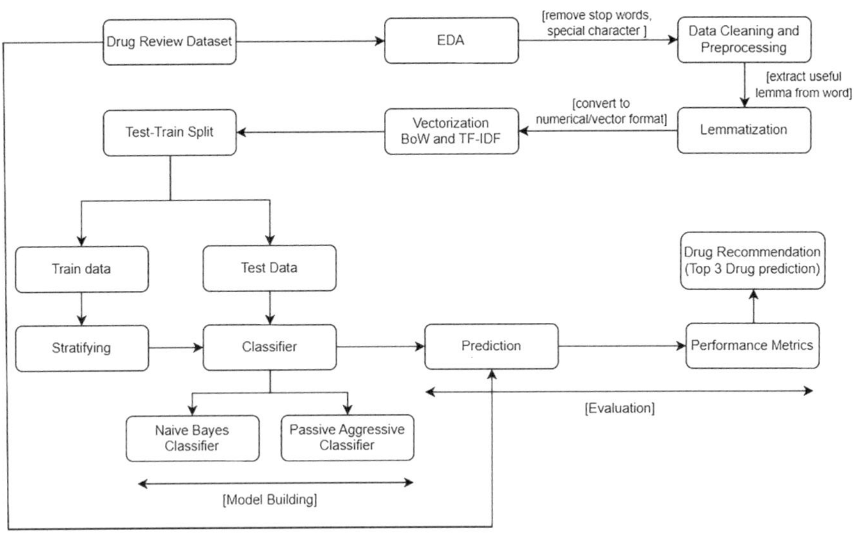

Fig. 1 Project workflow

medical conditions. Comprehensive model evaluation, including metrics such as accuracy, is conducted on a separate test set to demonstrate the framework's reliability in mapping user data to predicted medical conditions while maximizing performance.

1. ***Dataset***

In this research "Drug Review Dataset (Drugs.com)" from the "UCI ML Repository" is used. The dataset has 161,297 reviews of the users with seven features. The dataset consists of the following "features: 'uniqueID,' 'drugName,' 'condition,' 'review,' 'rating,' 'date,' 'usefulCount.'" From these features, in this research, five attributes will be used. The attributes drugName, condition, review, rating, and useful counts are used for further processing and classification of disease. A sample of the dataset is shown in Fig. 2. Also, the conditions into consideration can be seen in Fig. 3.

2. ***Data Cleaning and Preprocessing***

The textual data in the dataset needs to be preprocessed to represent it in numerical format to apply the NLP techniques. This involves checking and removing things from the textual data, such as stop words (like "the," "is," "are," "for," "in," "to," etc.), special characters, removing HTML tags, punctuations, quotes, URLs, etc. (Fig. 4).

	Unnamed: 0	drugName	condition	review	rating	date	usefulCount
0	206461	Valsartan	Left Ventricular Dysfunction	"It has no side effect, I take it in combinati...	9.0	May 20, 2012	27
1	95260	Guanfacine	ADHD	"My son is halfway through his fourth week of ...	8.0	April 27, 2010	192
2	92703	Lybrel	Birth Control	"I used to take another oral contraceptive, wh...	5.0	December 14, 2009	17
3	138000	Ortho Evra	Birth Control	"This is my first time using any form of birth...	8.0	November 3, 2015	10
4	35696	Buprenorphine / naloxone	Opiate Dependence	"Suboxone has completely turned my life around...	9.0	November 27, 2016	37

Fig. 2 Sample of the dataset

Fig. 3 Distribution of medical conditions was taken into account

```
Birth Control            38436
Depression               12164
Pain                      8245
Anxiety                   7812
Insomnia                  4904
Diabetes, Type 2          3362
High Blood Pressure       3104
Smoking Cessation         2440
Anxiety and Stress        2236
Cough                     1224
Name: condition, dtype: int64
```

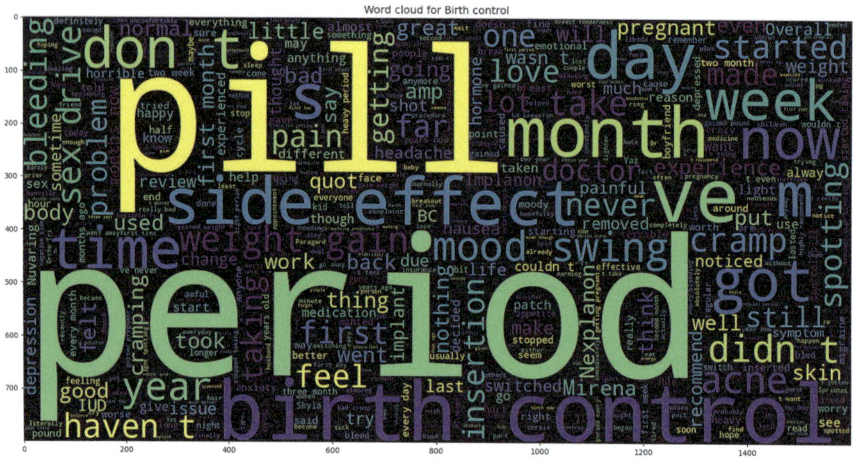

Fig. 4 Word cloud plot of birth control

3. ***Lemmatization***

A technique that involves reducing words to their base or root form is known as the "lemma," e.g., the lemma of "running" is "run," and the lemma of "better" is "good." This was done to extract meaningful information, as can be seen in Fig. 5.

4. ***Feature Extraction and Vectorization***

After the preprocessing part is done, the data needs to be in proper form more specifically in vectorized form, "Word Embedding" is one such technique where text can be represented using vectors so that machine learning algorithms can easily analyze and process it. So this research uses two techniques for this: "Bag-of-words (BoW) and term frequency–inverse document frequency (TF-IDF)."

(1) **BoW**: Bag-of-words model is a foundational technique in NLP that transforms text into a numerical format or vector form. This model has some drawbacks

	condition	review	review_clean
2	Birth Control	I used to take another oral contraceptive, which had 21 pill cycle, and was very happy- very light periods, max 5 days, no other side effects. But it contained hormone gestodene, which is not available in US, so I switched to Lybrel, because the ingredients are similar. When my other pills ended, I started Lybrel immediately, on my first day of period, as the instructions said. And the period lasted for two weeks. When taking the second pack- same two weeks. And now, with third pack things got even worse- my third period lasted for two weeks and now it's the end of the third week- I still have daily brown discharge.\r\nThe positive side is that I didn't have any other side effects. The idea of being period free was so tempting... Alas.	used take another oral contraceptive pill cycle happy light period max day side effect contained hormone gestodene available u switched lybrel ingredient similar pill ended started lybrel immediately first day period instruction said period lasted two week taking second pack two week third pack thing got even worse third period lasted two week end third week still daily brown discharge positive side side effect idea period free tempting ala
3	Birth Control	This is my first time using any form of birth control. I'm glad I went with the patch, I have been on it for 8 months. At first It decreased my libido but that subsided. The only downside is that it made my periods longer (5-6 days to be exact) I used to only have periods for 3-4 days max also made my cramps intense for the first two days of my period, I never had cramps before using birth control. Other than that in happy with the patch	first time using form birth control glad went patch month first decreased libido subsided downside made period longer day exact used period day max also made cramp intense first two day period never cramp using birth control happy patch
9	Birth Control	I had been on the pill for many years. When my doctor changed my RX to chateal, it was as effective. It really did help me by completely clearing my acne, this takes about 6 months though. I did not gain extra weight, or develop any emotional health issues. I stopped taking it bc I started using a more natural method of birth control, but started to take it bc I hate that my acne came back at age 28. I really hope symptoms like depression, or weight gain do not begin to affect me as I am older now. I'm also naturally moody, so this may worsen things. I was in a negative mental rut today. Also I hope this doesn't push me over the edge, as I believe I am depressed. Hopefully it'll be just like when I was younger.	pill many year doctor changed rx chateal effective really help completely clearing acne take month though gain extra weight develop emotional health issue stopped taking bc started using natural method birth control started take bc hate acne came back age really hope symptom like depression weight gain begin affect older also naturally moody may worsen thing negative mental rut today also hope push edge believe depressed hopefully like younger
11	Depression	I have taken anti-depressants for years, with some improvement but mostly moderate to severe side affects, which makes me go off them.\r\n\r\nI only take Cymbalta now mostly for pain.\r\n\r\nWhen I began Deplin, I noticed a major improvement overnight. More energy, better disposition, and no sinking to the low lows of major depression. I have been taking it for about 3 months now and feel like a normal person for the first time ever. Best thing, no side effects.	taken anti depressant year improvement mostly moderate severe side affect make go take cymbalta mostly pain began deplin noticed major improvement overnight energy better disposition sinking low low major depression taking month feel like normal person first time ever best thing side effect
14	Birth Control	Started Nexplanon 2 months ago because I have a minimal amount of contraception's I can take due to my inability to take the hormone that is used in most birth controls. I'm trying to give it time because it is one of my only options right now. But honestly if I had options I'd get it removed.\r\nI've never had acne problems in my life, and immediately broke out after getting it implanted. Sex drive is completely gone, and I used to have sex with my boyfriend a few days a week, now its completely forced and not even fun for me anymore. I mean I'm on birth control because I like having sex but don't want to get pregnant, why take a birth control that takes away sex? Very unhappy and hope that I get it back with time or I'm getting it removed.	started nexplanon month ago minimal amount contraception take due inability take hormone used birth control trying give time one option right honestly option get removed never acne problem life immediately broke getting implanted sex drive completely gone used sex boyfriend day week completely forced even fun anymore mean birth control like sex want get pregnant take birth control take away sex unhappy hope get back time getting removed

Fig. 5 After removing stop words and performing lemmatization

like its vector form is sparse, if there are too many words then vector size will increase and it will be computationally intensive.

(2) **TF-IDF**: "Term frequency–inverse document frequency, is a numerical statistic that is intended to reflect how important a word is to a document in a collection or corpus." It contains information about less important and more important words as well. Term frequency (TF) can be calculated as

$$tf_{(t,d)} = \frac{n_{td}}{\text{No. of terms in the document}} \tag{1}$$

Here, n is the frequency of the term "t" appearing in the document "d." Thus, each sentence and term would have its own TF value.

Finally, IDF is calculated as

$$idf_t = \log\left(\frac{\text{no. of documents}}{\text{no. of documents with term } t}\right) \tag{2}$$

So finally *tf–idf* is a product Eqs. (1) and (2), i.e., $tf_{(t,d)} * idf_t$.

5. ***Test Train Split***

The dataset is split into 80–20 split and stratification is done to ensure that the distribution of the important variable is maintained in the split.

6. ***Classifiers***

Two machine learning classifiers were used in this project to classify the condition of the patient:

1. BoW with multinomial "Naive Bayes" classifier and "passive aggressive classifier."
2. "TF-IDF" with multinomial "Naive Bayes classifier and passive aggressive classifier."

Then TF-IDF was tested using n-grams to ensure better results.

7. ***Performance Metrics***

The predicted results were evaluated using four metrics, precision, recall, accuracy, and $F1$ score. Precision, recall, accuracy, and $F1$ score were calculated using,

$$\text{Precision} = \frac{\text{true positive}}{\text{true positives} + \text{false positives}}$$

$$\text{Recall} = \frac{\text{true positive}}{\text{true positives} + \text{false negatives}}$$

$$\text{Accuracy} = \frac{(\text{true positive} + \text{true negative})}{(\text{true positive} + \text{true negative} + \text{false positive} \ + \text{false negative})}$$

$$F1\text{score} = 2 * \frac{\text{precision} * \text{recall}}{\text{precision} + \text{recall}}$$

4 Results and Discussion

The study evaluates the performance of various machine learning models using various feature extraction techniques, like bag-of-words (BoW) and term frequency–inverse document frequency (TF-IDF), along with different n-gram approaches (unigram, bigram, and trigram), as can be seen in Table 1.

BoW versus TF-IDF

The table compares the performance metrics (precision, recall, $F1$ score, and accuracy) of the models using BoW and TF-IDF vectorization techniques with different machine learning algorithms. TF-IDF generally performs better than BoW, indicating that TF-IDF captures more nuanced information from the textual data.

Algorithm Comparison

Among the algorithms tested (multinomial Naive Bayes and passive aggressive classifier), the passive aggressive classifier consistently outperforms multinomial Naive Bayes across all vectorization techniques and n-gram approaches. For instance, the TF-IDF bigrams with passive aggressive classifier achieve the highest accuracy of

Table 1 Vectorization with machine learning models

Model	Precision	Recall	$F1$ score	Accuracy
BoW with mutinomial Naive Bayes	0.89	0.90	0.88	0.89
BoW with passive aggressive classifier	0.92	0.94	0.92	0.92
TF-IDF with mutinomial Naive Bayes	0.76	0.65	0.58	0.63
TF-IDF with passive aggressive classifier	0.95	0.96	0.93	0.95
TF-IDF bigram with mutinomial Naive Bayes	0.72	0.62	0.57	0.66
TF-IDF bigram with passive aggressive classifier	0.93	0.97	0.95	0.97
TF-IDF trigram with mutinomial Naive Bayes	0.69	0.58	0.54	0.59
TF-IDF trigram with passive aggressive classifier	0.95	0.98	0.97	0.98

95%, demonstrating its superiority in predicting medical conditions and suggesting appropriate drugs.

Figures 6 and 7 depict the graphical user interface (GUI) through which the predicted medical condition and three recommended medicines are displayed. This interface serves as a user-friendly tool for healthcare professionals and patients to access personalized drug recommendations based on the input medical condition.

Interpretation of Results

- **Significance of TF-IDF Approach**

  The notable superiority of the TF-IDF approach over BoW signifies its efficacy in capturing the semantic relevance of terms within the textual data. This enhanced representation enables more accurate predictions and recommendations within the drug recommendation system.

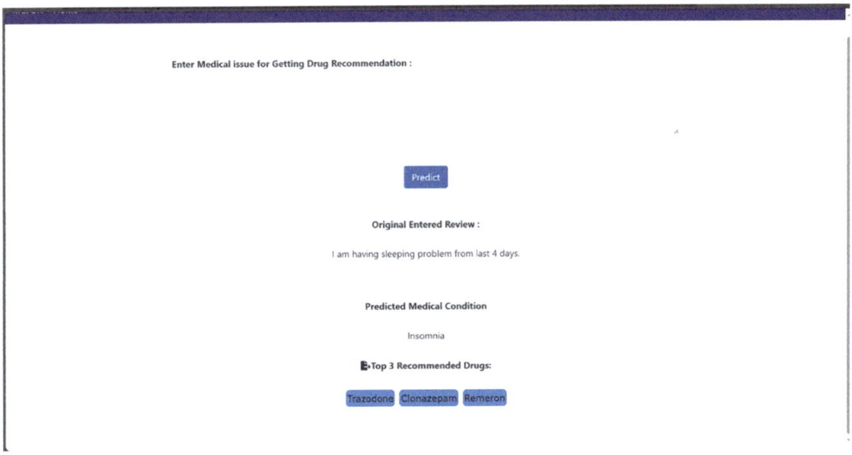

Fig. 6 GUI in which condition and three medicine recommendations are displayed

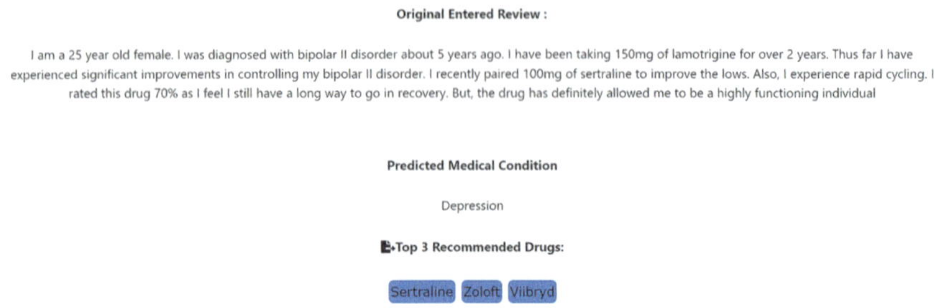

Fig. 7 Detailed medical conditions and recommendation of drugs are shown

- **Insight into Algorithm Performance**

  The performance comparison between the multinomial Naive Bayes and passive aggressive classifier algorithms offers valuable insights into their respective capabilities. The consistent outperformance of the passive aggressive classifier underscores its suitability for handling the complexities inherent in medical data, thereby enhancing the accuracy and reliability of the drug recommendation system.
- **Implications for Practical Implementation**

  The experimental results validate the effectiveness of the proposed methodology in predicting medical conditions and suggesting appropriate drugs. The high accuracy and robust performance achieved across various vectorization techniques and machine learning algorithms affirm the viability of the developed model in real-world healthcare applications.

5 Future Scope

1. The dataset could be expanded to include more medical conditions, drugs, user demographics, and other supplementary data. This would allow for training more robust models and evaluating generalizability.
2. More advanced neural network architectures like LSTMs and transformers could be explored for text classification and recommendation tasks. These deep learning models may improve accuracy.
3. The system could incorporate patient medical history, lab test results, and genomic data to provide personalized recommendations tailored to an individual.
4. Model interpretability should be evaluated to ensure recommendations are clinically sound and providers can understand suggestions.
5. User testing by healthcare professionals and patient groups would provide valuable feedback on the usability and utility of the system.

6 Conclusion

The paper presents a drug recommendation system using NLP and machine learning to predict medical conditions accurately and suggest three suitable drugs. The system effectively classified conditions and recommended drugs, with TF-IDF bigram + passive aggressive classifier achieving the best accuracy of 95.00%. Comparing multiple models identified the optimal ML approach for this application. Incorporating supplementary data like top reviews and useful counts improved recommendation accuracy and relevance. The system has the potential to significantly advance personalized medicine by providing tailored drug suggestions to inform better decisions. The scope was limited to the specific dataset and algorithms used. More diverse data and models are needed to generalize findings. Valuable contributions made toward applying NLP/ML in healthcare. This establishes a foundation for personalized medicine solutions. Future work should address limitations, experiment with advanced methodologies, increase personalization, evaluate interpretability, and test real-world viability. Overall, the paper presented a robust drug recommendation system with promising capability. Limitations exist but research sets a strong baseline for future enhancements toward clinical implementation. In summary, the key highlights are: effectiveness demonstrated, optimal ML approach identified, supplementary data improved performance, potential for personalized medicine, limitations exist, and future work defined to make system production-ready.

References

1. Garg S (2021) Drug recommendation system based on sentiment analysis of drug reviews using machine learning. In: 2021 11th international conference on cloud computing, data science & engineering (confluence) 175–181
2. Machado J, Rodrigues C, Sousa R, Gomes LM (2023) Drug–drug interaction extraction-based system: a natural language processing approach. Exp Syst e13303. https://doi.org/10.1111/exsy.13303
3. Abbas K, Afaq M, Ahmed Khan T, Song W-C (2020) A blockchain and machine learning-based drug supply chain management and recommendation system for smart pharmaceutical industry. Electronics 9(5):852. https://doi.org/10.3390/electronics9050852
4. Ye Q, Hsieh CY, Yang Z et al (2021) A unified drug–target interaction prediction framework based on knowledge graph and recommendation system. Nat Commun 12:6775. https://doi.org/10.1038/s41467-021-27137-3
5. Bhimavarapu U, Chintalapudi N, Battineni G (2022) A fair and safe usage drug recommendation system in medical emergencies by a stacked ANN. Algorithms 15(6):186. https://doi.org/10.3390/a15060186
6. Abhishek A, Bindal D, Yadav, Abhishek, Bindal AK, Dharminder Y (2022) Medicine recommender system: a machine learning approach 50015
7. Kretika T, Singh DK (2022) Machine learning-based recommendation system for disease-drug material and adverse drug reaction: comparative review. Mater Today Proc 51(Part 1):304–313. ISSN 2214-7853. https://doi.org/10.1016/j.matpr.2021.05.404
8. Stark B et al (2019) A literature review on medicine recommender systems. Int J Adv Comput Sci Appl 10(8)

9. Granda Morales LF, Valdiviezo-Diaz P, Reátegui R, Barba-Guaman L (2022) Drug recommendation system for diabetes using a collaborative filtering and clustering approach: development and performance evaluation. J Med Internet Res 24(7):e37233. https://doi.org/10.2196/37233. PMID: 35838763, PMCID: 9338420
10. Doulaverakis C, Nikolaidis G, Kleontas A et al (2012) GalenOWL: ontology-based drug recommendations discovery. J Biomed Semant 3:14. https://doi.org/10.1186/2041-1480-3-14

Assessment of Tooth Wear in a Profile Modified Spur Geared System Using Vibration Signal Analysis

Akash Mishra, Jawaz Alam, and Sumanta Panda

Abstract Gears are one of the most significant machines elements for transmitting torque. Its application can be found in various machines. Examination of gear teeth interaction has been considered as one of the most cumbersome tasks. The variation in working conditions like rise in load, temperature, decrease in dynamic viscosity result diminution in *EHL* and decrease of characteristics of lubricant thereby initiating various types of wears on gear tooth surfaces viz. spalling, scoring, pitting and micro-pitting etc., these wears effects on change in vibration signature. This article shows the outcomes of experimental examination performed to monitor wear in standard and profile shifted spur gears drives under static loading conditions. The article considered the vibration signal analysis along with empirical mode decomposition (*EMD*) approach to extract the wear related characteristics from the vibration signatures attained from the gear drive. The outcomes highlight the significance of *EMD* approach for efficient estimation of faults in gear drive.

Keywords Spur gear · Profile shift · Fault diagnosis · *EMD*

1 Introduction

The gears are crucial components of machines that are continuously working under various environments. Any breakdown in the gear set has the potential to destroy the entire machine and create catastrophic mishaps. Rotating machinery's vibration signature includes a wealth of useful data that quickly reflects the equipment's operational status. In a study, Samanta [1] had proposed the measures known to be sensitive to varying degrees of errors, including mean, root means square, variance, skewness, and kurtosis. Tan et al. [2] have carried out an experimental investigation on a spur gear drive where the rate of variation of three parameters, Fe concentration, acoustic emission, and vibration root mean square value (RMS) with regard to gear box working duration rises with rising applied torque. Furthermore, in an

A. Mishra · J. Alam · S. Panda (✉)
Department of Mechanical Engineering, VSSUT, Burla, Odisha, India
e-mail: sumanta.panda@gmail.com

© The Author(s), under exclusive license to Springer Nature Singapore Pte Ltd. 2025
S. K. Udgata et al. (eds.), *Intelligent Systems*, Lecture Notes in Networks and Systems 1314, https://doi.org/10.1007/978-981-96-3797-3_3

important research, Parey et al. [3] have found that RMS measurements on gear sets with defects show an increase in vibration as the defect size rises. The wear diagnostic data was improved by the adoption of statistical characteristics including crest factor and kurtosis. Statistical factors such impulse factor, RMS, kurtosis, crest factor and skewness values to offer a better wear investigative information in spur gear utilising the suggested methodologies, boosting the geared drive's capacity for condition monitoring [4]. Further, Hakan [5] has found that information regarding wear and surface roughness may be obtained using the statistical metrics kurtosis, crest factor, and RMS. The kurtosis value and RMS similarly rise as the wear does; however, the crest factor shows erratic changes.

The widespread literature survey reveals that very few literatures have been proposed related to observation of fault initiation and propagation in standard and profile modified gear set. The raw vibration signature and decompose vibration signature, i.e. EMD, are used to precise examination of gear fault identification. The rest of the article are organised as follows: Sect. 2 presents the experimental set up developed for the testing of the gear set. In Sect. 3, the result and discussion have been presented. The study has been concluded in Sect. 4.

2 Experimental Setup

The experimental setup of the standard back-to-back gearbox with oil lubrication used throughout this study is shown in Fig. 1. It consists of one D.C motor (1 hp) connected with two parallelly connected shaft (one driving and another driven shaft), each shaft was supported by two bearing and input load is given by a pully that is connected to a band break spring balance system on the driver shaft that contain the gear. The driving shaft is connected to motor by a coupling and the motor is connected to variable frequency drive (VFD), to maintain the speed of motor. In this experiment, the gear box was operated at 800 RPM and a torque of 5000 and 7500 N mm was given on the shaft connected to gear for a certain period of time for efficient study. Multi-grade lubricant (20W40) has been used to lubricate the gear drives.

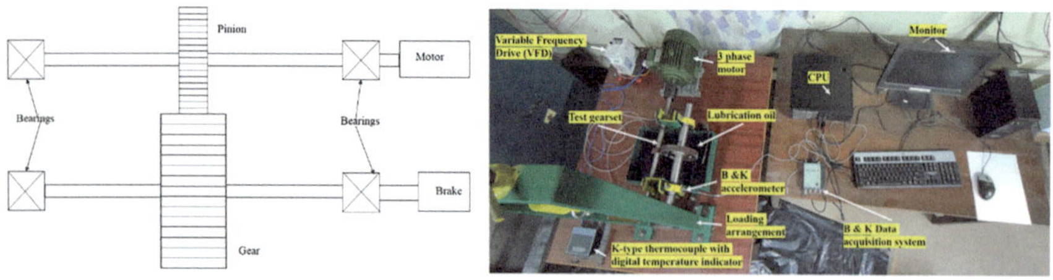

Fig. 1 Experimental setup

Table 1 Gear test rig specification

Parameter	Specification
Pressure angle	25°
m	2 mm
Diameter of gear shaft	36.4 mm
Diameter of pinion shaft	20 mm
Gear material	EN24
Spring capacity	1000 N
Belt width and thickness	5 cm and 2 cm
Gear oil	20 W40
Pinion shaft speed	800 RPM
Thermocouple	K-type with digital temperature indicator
Number of pinion teeth	18
Transmission ratio	4:1

Analytical methods can be employed to examine standard and modified spur gear efficiency prior to testing. But the researchers were focussed on testing the functioning of the spur gear set. Earlier articles based their standards for examining of faults by several statistical variables of raw vibration signal and EMD for the fault diagnostics information. So, in this article, the authors have experimentally examined the resistance of wear of the optimised gear drive by estimating the above-mentioned parameters for all the spur gear set. The performance of standard (set 1) and profile modified (set 2) spur gear set with module (2 mm) and gear ratio (4:1) was examined and compared with the standard gear in this study. Description of this experimental set up is reported in Table 1.

2.1 Instruments and Sensors

The vibration signature is examined at regular interval. B&K RT pro Dactron data acquisition system was used for gear wear analysis using vibration signal. B&K accelerometer was mounted in bearing housing. Further, K-type thermocouple has been used to examine the temperature of the oil bath during operation.

2.2 Experimental Procedure

The gear box was kept running for over 40 h after that the experiment is carried out with different sets of gear. The gear sets 90 standard (set 1) and 90 modified (set 2) were run for 2.5×10^5 cycles in a day with a load of 5000 N mm (test condition 1) and 7500 N mm (test condition 2) which was given on the pulley by the help of spring balance and the operating speed is given as 800RPM. The lubricant temperature was measured at regular intervals of time (15 min). The severe wear of gear samples is expected to begin after 5×10^5 cycles based on this operating condition. Each gear sets are run up-to 10^5 cycles. Experiments were paused after every 2.5×10^5 cycle to examine the vibration signature of the gear set.

2.3 Signal Processing Technique

Since damaged gears are clear obstacle in rotating equipment, condition monitoring has drawn a lot of attention for a long time. It is well acknowledged that monitoring vibration signals can help identify early defects in rotating machine components. Vibration/acoustic signal monitoring now plays a significant part in the condition monitoring of crucial components in complex rotating machines because of the advancements in signal processing methods and computer power. In this analysis, the authors have initially observed the raw vibration signal for the analysis of initiation and propagation of wear in gear set. Afterwards, the decompose signal techniques like empirical mode decomposition (EMD) have been adopted for the precise measurement of the faults in the gear sets.

3 Results and Discussion

In this paper, experimental investigations are conducted under two separate test situations to evaluate wear in set 1 and set 2 gear drives. It has been observed that the rate of increase of *RMS* acceleration (mm/s^2) in set 1 gear set in the both test condition, i.e. 5000 N mm (test condition 1) and 7500 N mm (test condition 2) is more as compared to profile shifted gear set, i.e. set 2 gear set. So, in this study, the authors have considered the statistical parameters of raw vibration signal for comparison of gear sets vibration signal for detection of initiation and rate of propagation of wear in gear set. The time domain characteristics of vibration signal have been calculated of two gear trains, i.e. 90 standard (set 1 gear set) and 90 modified (set 2 gear set). The statistical parameters of raw vibration signals were recorded in every 2.5×10^5 cycles under two test conditions, i.e. test condition 1 and test condition 2 which also covers skewness, the root means square, the crest factor, and other statistical variables.

It was found in comparative study that the rate of wear in set 1 gear set was more than set 2 gear set and decrease in specific film thickness is sharper in set 1 gear set. Vibration analysis shows better performance in set 2 gear set because profile modification reduces the contact stress and results in low rate of wear of modified gear set as compared to standard gear set [6].

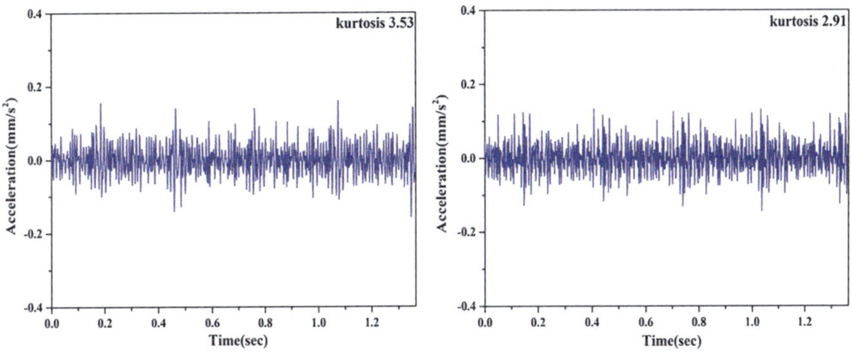

Fig. 2 Raw vibration signal in time domain of set 1 gear set in test condition 1

3.1 Statistical Parameters of Raw Vibration

Signal analysis is a meek and conventional technique, for the analysis of wear characteristics of a gear drive. The time domain data were used to calculate the statistical parameters. The examination of the spatial distribution of the vibration amplitudes included in a time domain signal was made possible by the values of kurtosis, crest factor, RMS, and skewness. The kurtosis study has been carried out in this analysis. Due to its great sensitivity to the erroneous effects of noises and high frequency vibration signatures, these kurtosis values revealed an uneven pattern in the vibration signature. Initial stages of wear were marked by a rise in signal, but later stages of wear did not significantly increase in kurtosis values in accordance with this pattern. As it can be observed, set 2 gear operating hours increased more steadily than set 1 gear operating hours as evidenced by the kurtosis value.

An increase in RMS acceleration (mm/s^2) signature is observed in both test condition 1 and 2. However, the rate of increase of RMS acceleration (mm/s^2) signature in set 2 gear is less as compared to set 1 gear. As in previous literatures [3–5]. it has been reported that in defected gear set, RMS measurements reveal that the vibration threshold rises as the size of the defect grows. Hence due to low contact stress in modified gear set, i.e. set 2 gear, the defect is less as compared to set 1 gear and the propagation of defect is also low. Here, we can observe that wear severity and propagation on set 1 gear set are superior to set 2 gear set. Figures 2, 3, 4, and 5 depict initial and final vibration signals reading of gear sets from 40 to 80 h at two different load conditions (test conditions 1 and 2), respectively.

3.2 Empirical Mode Decomposition (EMD)

To overcome the limitation of vibration raw signal wear analysis, EMD is an approach used to extract six intrinsic mode functions (IMFs), i.e. IMF1 to IMF6; therefore, six

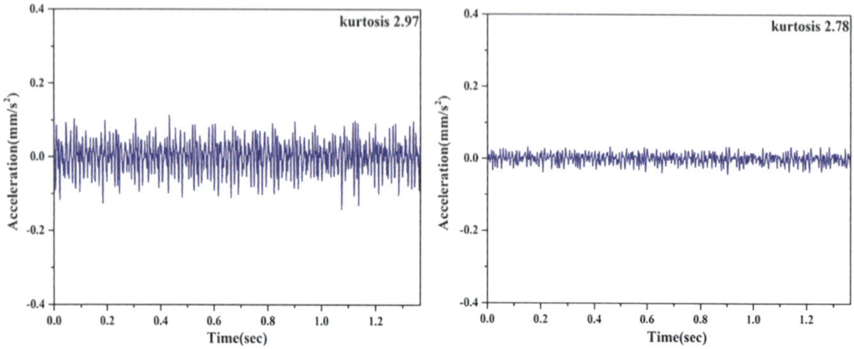

Fig. 3 Raw vibration signal in time domain of set 2 gear set in test condition 1

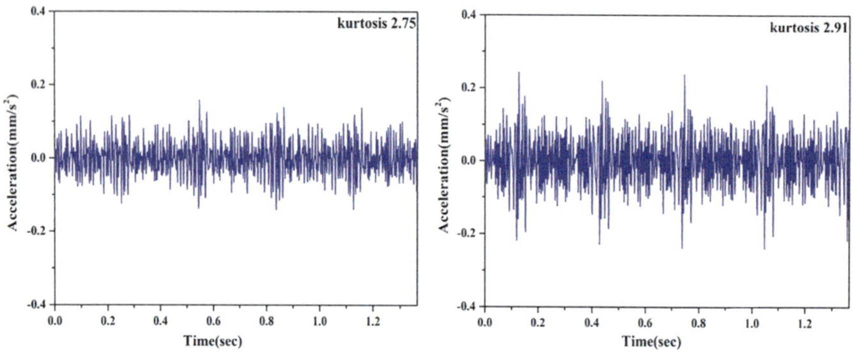

Fig. 4 Raw vibration signal in time domain of set 1 gear set in test condition 2

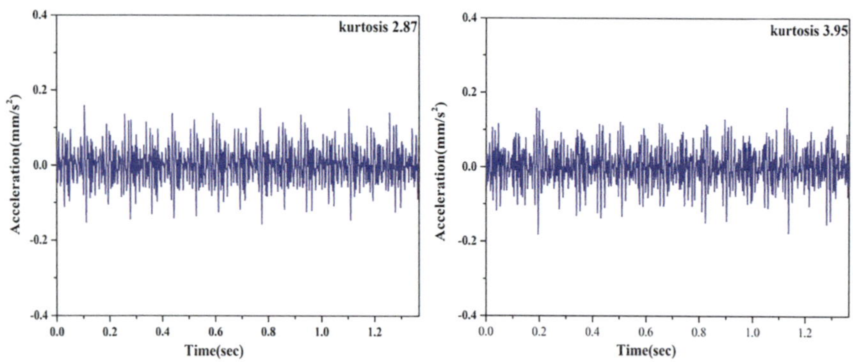

Fig. 5 Raw vibration signal in time domain of set 2 gear set in test condition 2

IMFs have been evaluated and the corresponding signal processing characteristics magnitudes were taken to obtain analytic information. These signatures got divided up into several frequencies, i.e. IMF1 (c_1) to IMF6 (c_6) as in Figs. 6, 7, 8, and 9 of set 1 and set 2 gear sets, where IMF6 is a residual signal.

The residual signal (IMF6) is the residue value of the previous IMF signal (in this case IMF5) which lacks maxima and minima and hence cannot be further decomposed. Clearly in Figs. 6, 7, 8, and 9, second IMF(c_2) shows better transient characteristics than other non-stationary IMFs and is viable for analysis. Kurtosis, crest factor, RMS, and skewness magnitudes of initial six IMF (IMF1–IMF6) of vibration signatures examined from healthy gear drive attained at two different loading conditions according to applied torque of 5000 and 7500 N mm over 40 h of time duration for set 1 and set 2 gear sets.

3.3 Comparative Study of Vibration Signals

A comparison of raw signal and decompose signal EMD methods is made in Fig. 10. In this analysis, the comparison of kurtosis values attained from raw signal and decomposed vibration signature based on EMD techniques are presented. Figure 10 clearly shows a higher kurtosis value of decompose signal as compared to raw vibration signal.

The limitations of raw vibration signal are discussed earlier; hence, to improve the investigative information, EMD approach is used on the same vibration signature. Due to the presence of higher harmonics of gear mesh frequencies in the first two IMFs, it can be seen that the trend for these IMFs is increasing. The EMD exhibits a tendency of progressive rise as seen in Fig. 10 for its second intrinsic mode function (c_2) of the EMD. Because of this, the second intrinsic mode function, c_2, has been taken into consideration for the identification and diagnostics of gear teeth that are wearing out. Additionally, the kurtosis values of vibrational signals and kurtosis values based on EMD are taken into consideration for comparison. These results, together with the lubricant specific film thickness (λ), are compared and correlated in Figs. 11 and 12. Since it has been noted that the reduction closely relates to the degree of surface wear, and a gain in surface wear on the teeth causes an upsurge in vibration signatures [7]. In Figs. 11 and 12, for test condition 1 and 2 of set 1 and 2 gear set respectively, the upper and lower bounds raw vibration signal kurtosis readings as well as kurtosis data retrieved from the c_2 of the decomposed signals were both covered. Plotted values decline with time; this decline in λ is brought on by a rise in temperature and ensuing variations in viscosity.

Further, it is evident from Figs. 11 and 12 that in case of set 1 gear, the kurtosis value is high as compared to gear set 2 for both the test conditions. So, there is initiation of wear in set 1 gear, which is supported by the fact that the decrease in λ is sharper than set 2 gear set. Also, it is noticeable that the particular film thickness steadily decreases from the state of no damage to the probable damage condition. In particular,

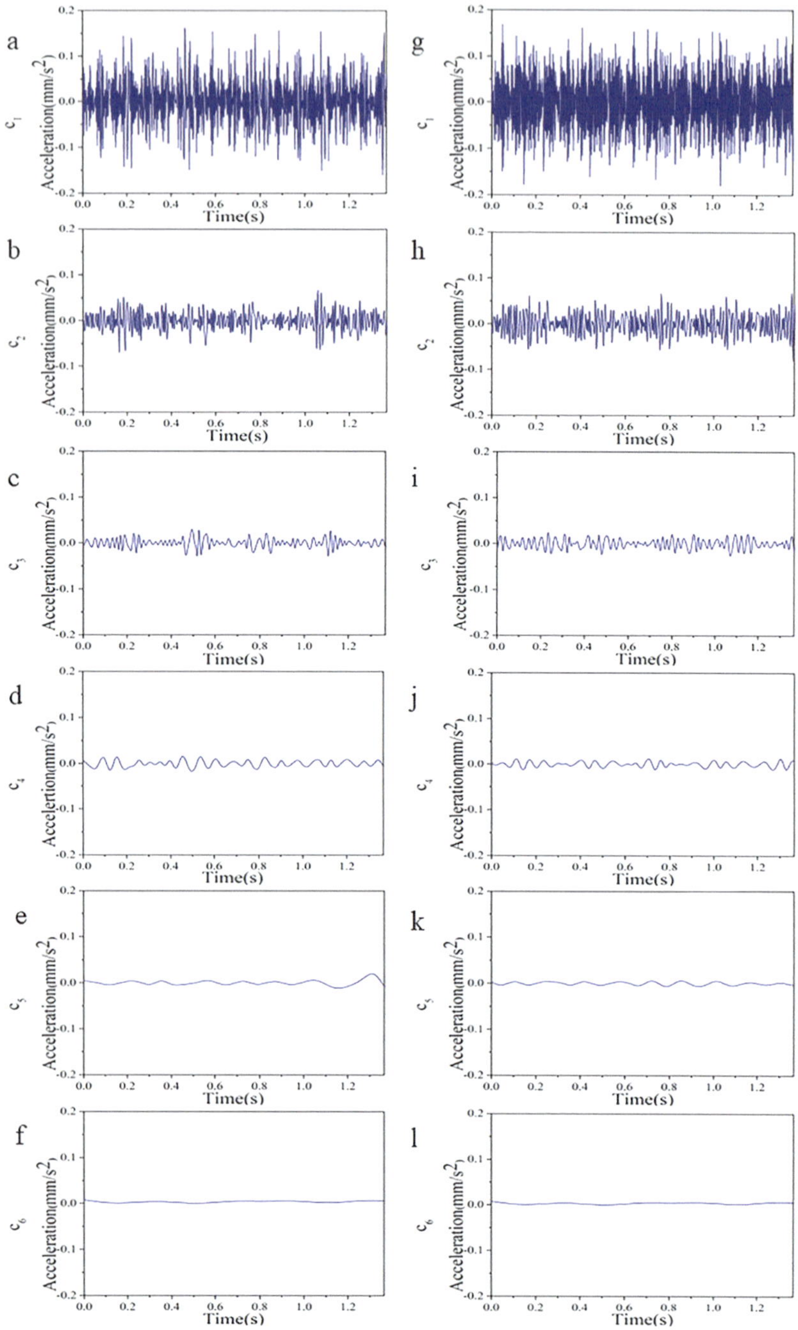

Fig. 6 IMFs of EMD of vibration signal of set 1 gear in test condition 1: **a–f** IMFs after 40 h and **g–l** IMFs after 60 h

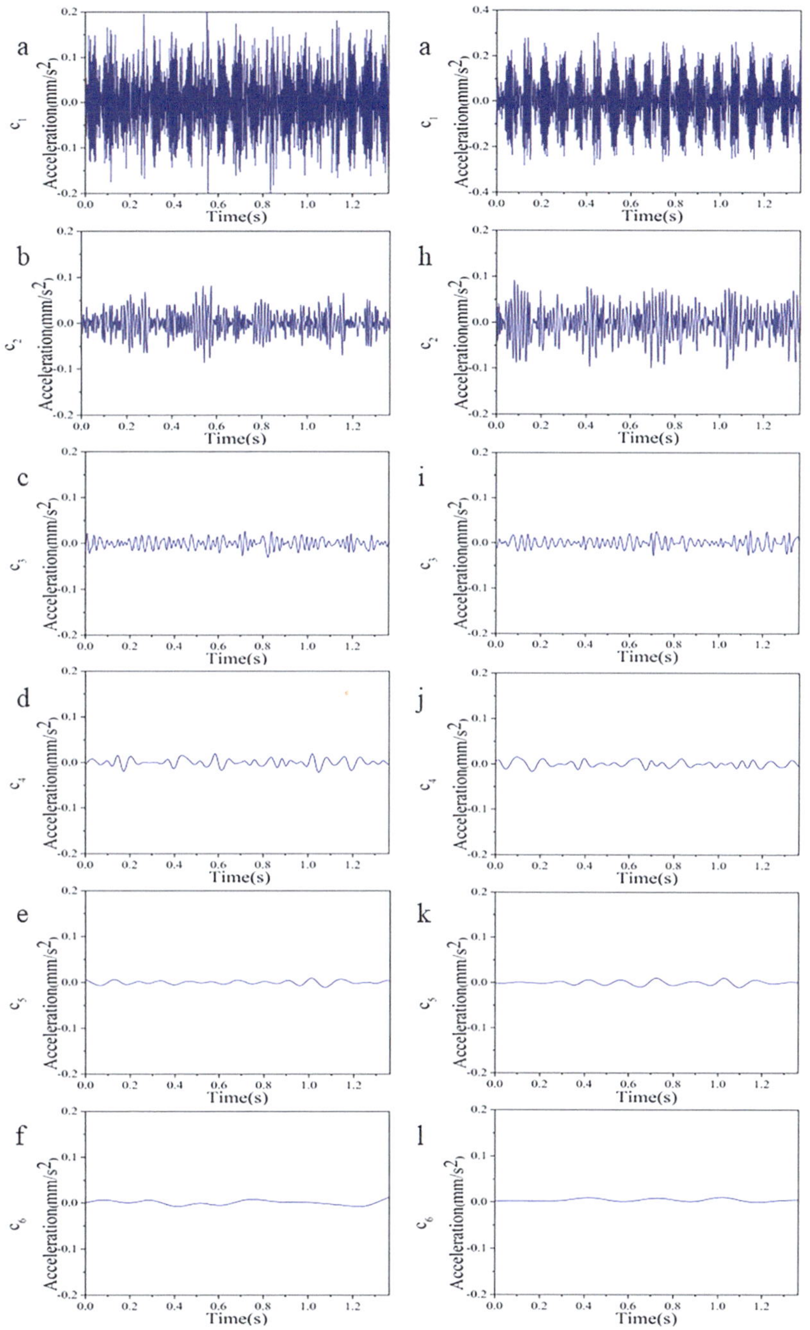

Fig. 7 IMFs of EMD of vibration signal of set 1 gear in test condition 2: **a–f** IMFs after 60 h and **g–l** IMFs after 80 h

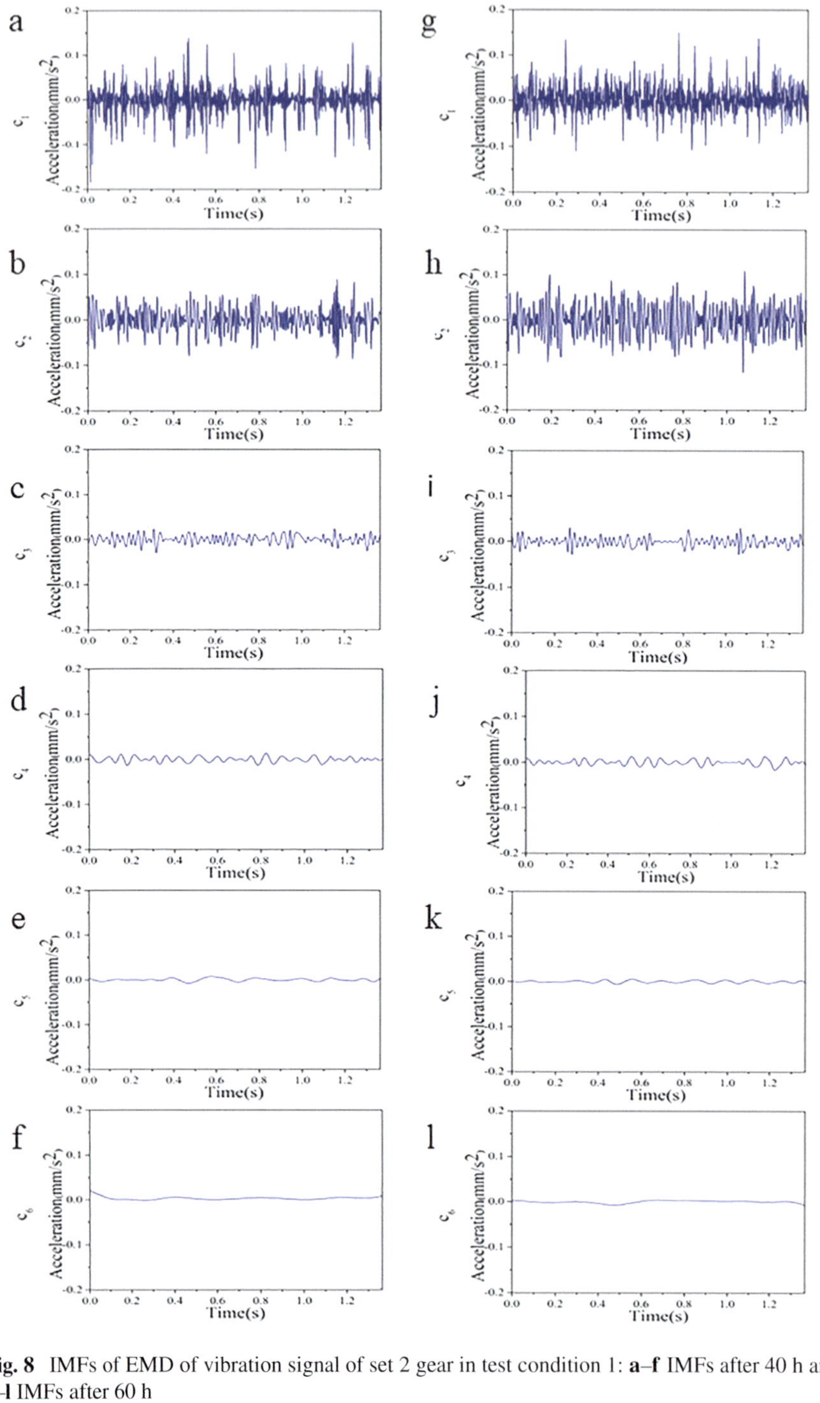

Fig. 8 IMFs of EMD of vibration signal of set 2 gear in test condition 1: **a–f** IMFs after 40 h and **g–l** IMFs after 60 h

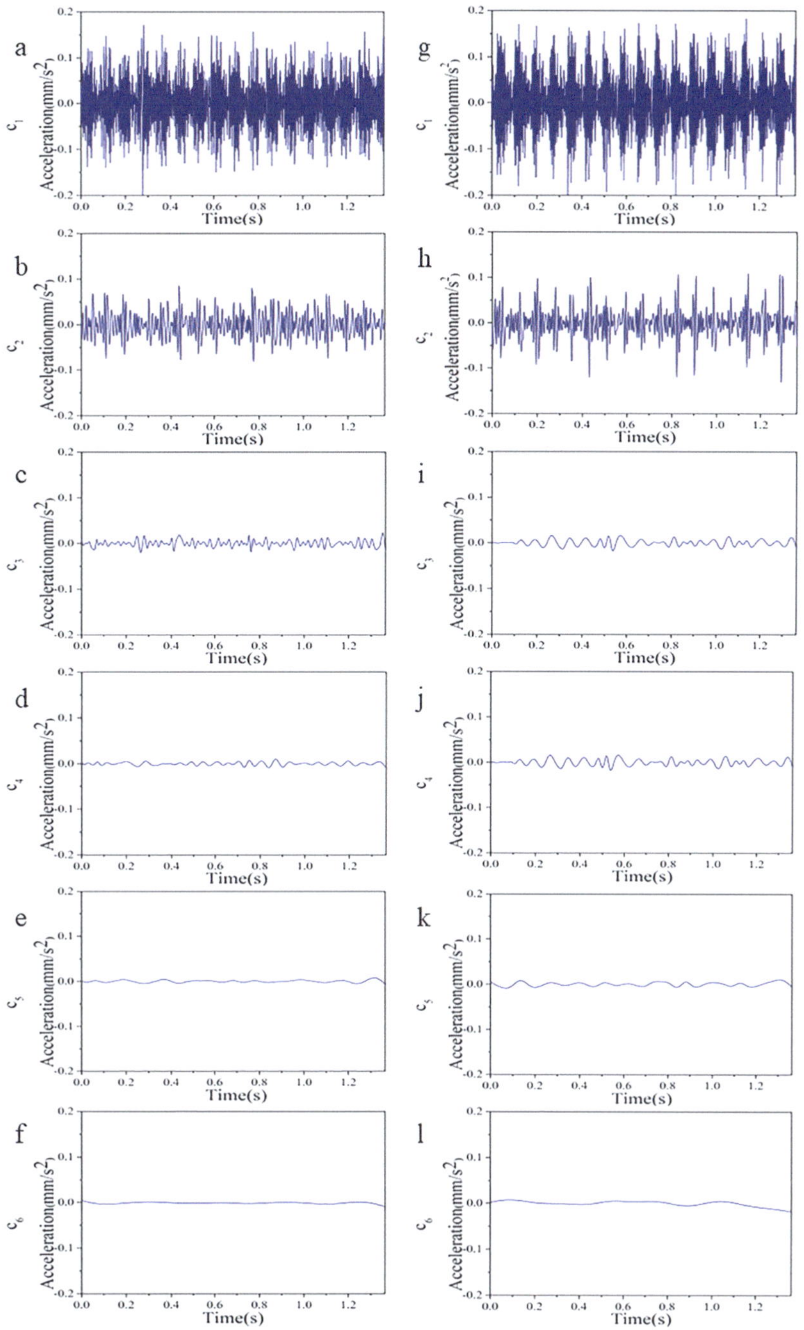

Fig. 9 IMFs of EMD of vibration signal of set 2 gear in test condition 2: **a–f** IMFs after 60 h and **g–l** IMFs after 80 h

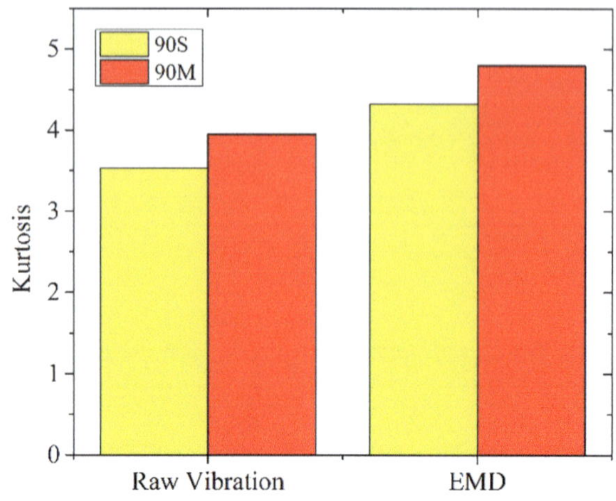

Fig. 10 Comparison of kurtosis values

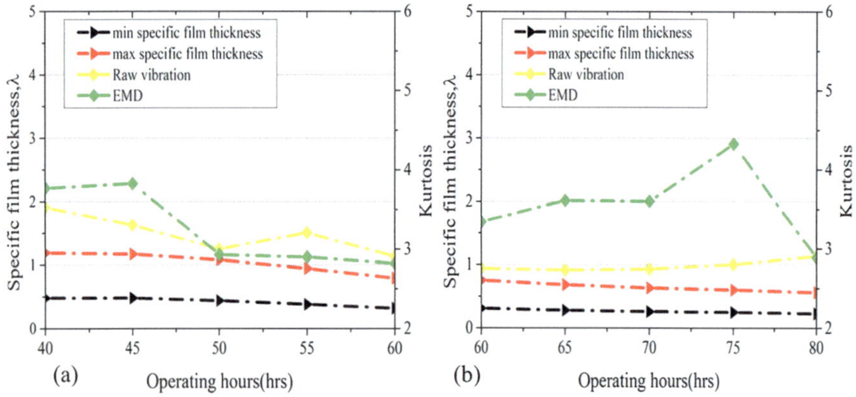

Fig. 11 Kurtosis values of raw vibration signal, second IMFs of EMD and λ with respect to operating time of set 1 gear set in **a** test condition 1 and **b** test condition 2

it was discovered that λ was less than 1 for a considerable amount of time under test condition 2. Due to accumulated damage from greater loads, there is a significant drop at this load. The tip contact, the region near the pitch line, and the pitch line itself are the only three portions of the tooth profile that are affected by this operating situation, which propagates dispersed defects on gear teeth. The regions in the current work had pitting. In a severely loaded geared gearbox system, the lubricant coating is crucial in preventing the spread of tooth damage. Lubrication performance is influenced by a number of factors, including environmental, rheological, thermodynamic, and chemical characteristics. These characteristics guarantee little gear tooth contact wear. When power is first transmitted through a gear drive system, minor wear manifests as mirror-like tooth surfaces, which improves gear performance. After

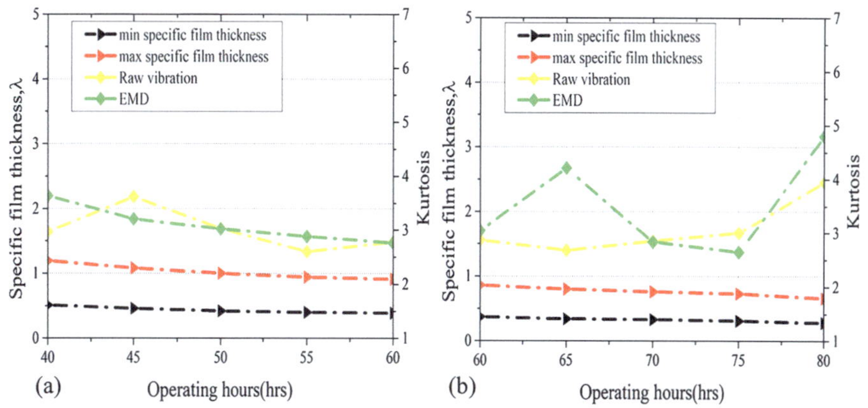

Fig. 12 Kurtosis values of raw vibration signal, second IMFs of EMD, and λ with respect to operating time of set 2 gear set in **a** test condition 1 and **b** test condition 2

55 h, modest wear and micro-pitting are also visible, and a steady decrease in λ causes micro-pitting to spread. The critical temperature where the lubricant coating breaks and micro-welding among the asperities occurs causes the scuffing depicted in Fig. 13.

EMD-based kurtosis magnitudes present superior rise in vibration signature, this may be because of the sensitivity of these techniques, it is possible to use improved kurtosis magnitudes to detect the onset and spread of wear at an early stage. Thus, kurtosis values acquired from decomposed signals (EMD technique) in combination with film thickness are used to evaluate the degree of wear on spur gears. This information has trustworthy diagnostic data that is helpful in determining the wear growth on gear teeth.

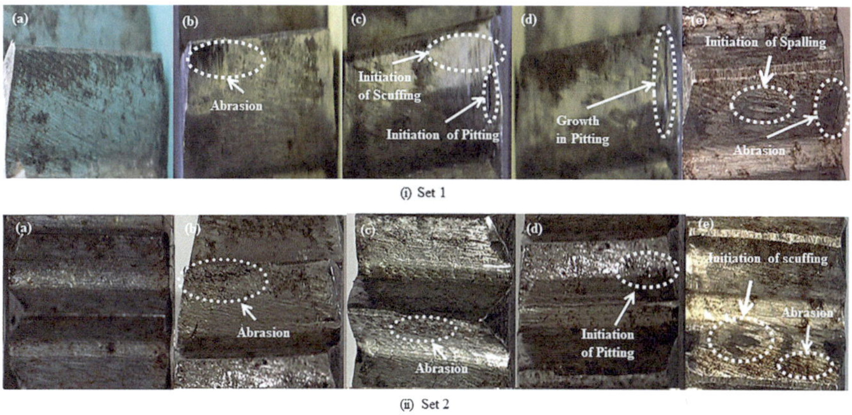

Fig. 13 Wear mechanism of set 1 and set 2 gear sets on cycle **a** 0, **b** 2.5×10^5, **c** 5×10^5, **d** 7.5×10^5 and **e** 10^6

4 Conclusions

Spur gear due to its simplicity in design is widely used and is most widely used in power transmission. Profile modification in spur gear enables low contact stress as compared to standard gear set. Statistical parameters of raw time domain signal, EMD decomposition enabled a vivid study. The following conclusions were made from the study are (i) It is observed from the experimental study that the performance of profile modified gear set is better than standard gear set in terms of efficiency and wear life. (ii) To compare the conditions of gear sets, statistical characteristics like kurtosis values of time domain vibration signature are evaluated. The values are useful in diagnostic information, which shows set 2 gear set showing better characteristics data than set 1 gear set. (iii) Second intrinsic mode function showed fresh transient characteristics than other decomposed non-stationary signature of EMD process and hence was used for further comparison and analysis. (iv) The decrease in specific film thickness, wears like abrasion, pitting, spalling, scuffing proliferated with the number of cycles. The decrease in film thickness has been confirmed through an increasing trend in the kurtosis values of the transient signal.

References

1. Samanta B (2004) Gear fault detection using artificial neural networks and support vector machines with genetic algorithms. Mech Syst Signal Process 18:625–644. https://doi.org/10.1016/S0888-3270(03)00020-7
2. Tan CK, Irving P, Mba D (2007) A comparative experimental study on the diagnostic and prognostic capabilities of acoustics emission, vibration and spectrometric oil analysis for spur gears. Mech Syst Signal Process 21:208–233. https://doi.org/10.1016/j.ymssp.2005.09.015
3. Parey A, Tandon N (2007) Impact velocity modelling and signal processing of spur gear vibration for the estimation of defect size. Mech Syst Signal Process 21:234–243. https://doi.org/10.1016/j.ymssp.2005.12.011
4. Amarnath M, Krishna IRP (2013) Detection and diagnosis of surface wear failure in a spur geared system using EEMD based vibration signalanalysis. Tribol Int 61:224–234. https://doi.org/10.1016/j.triboint.2013.01.001
5. Arslan H, Er AO, Orhan S, Aslan E (2016) Tool condition monitoring in turning using statistical parameters of vibration signal. Int J Acoust Vib 21:371–378. https://doi.org/10.20855/ijav.2016.21.4432
6. Alam J, Panda S (2021) A comprehensive study on multi-objective design optimization of spur gear. Mech Based Des Struct Mach. https://doi.org/10.1080/15397734.2021.1996246
7. Amarnath M, Sujatha C, Swarnamani S (2009) Experimental studies on the effects of reduction in gear tooth stiffness and lubricant film thickness in a spur geared system. Tribol Int 42:340–352. https://doi.org/10.1016/j.triboint.2008.07.008

SSTSM: A Black Box Test Suite Minimization Algorithm Based on Fault-Exposing Potential and Squirrel Search

Aliazar Deneke Deferisha, Seifu Detso Bejo, Sudhir Kumar Mohapatra, Jukka Heikkonen, and Rajeev Kanth

Abstract The domain of software engineering involves applying engineering principles and methods to create software systems, focusing on producing high-quality software efficiently and cost-effectively. The implementation of software testing is necessary in order to actualize these commitments. The comparative analysis between manual testing and automated testing reveals that the latter, employing automatically generated test cases, surpasses the former in terms of efficiency and effectiveness. As the size of test suites increases, the execution of all test cases within the suite becomes increasingly laborious and time-consuming, rendering it impractical to execute every single test case. Several different methodologies have been suggested to tackle the challenge of minimizing test suites. However, the no-free-lunch (NFL) theorem and its NP-completeness reveal that no single methodology can ensure an optimal-sized collection of test suites. To address redundant test cases within test suites, the authors introduce SSTSM, a novel algorithm for test suite minimization. This algorithm, outlined in the study, is a black box technique leveraging the fault-exposing potential of test cases alongside a nature-inspired metaheuristic called Squirrel Search (SSTSM). Additionally, three other reduction techniques are explored. The study

A. D. Deferisha
Arba Minch Institute of Technology, Arba Minch University, Arba Minch, Ethiopia
e-mail: aliazar.deneke@amu.edu.et

S. D. Bejo
Wolkite University, Wolkite, Ethiopia
e-mail: seifu.detso@wku.edu.et

S. K. Mohapatra (✉)
Sri Sri University, Cuttack, India
e-mail: sudhir.mohapatra@srisriuniversity.edu.in

J. Heikkonen
University of Turku, Turku, Finland
e-mail: jukhei@utu.fi

R. Kanth
Savonia University of Applied Sciences, Kuopio, Finland
e-mail: rajeev.kanth@savonia.fi

© The Author(s), under exclusive license to Springer Nature Singapore Pte Ltd. 2025
S. K. Udgata et al. (eds.), *Intelligent Systems*, Lecture Notes in Networks and Systems 1314, https://doi.org/10.1007/978-981-96-3797-3_4

compares the performance of Particle Swarm Optimization (PSO), Bat Search Algorithm (BA or BAT), and Genetic Algorithm (GA) against the proposed approach. The comparison is conducted using the same input dataset, and the evaluation criteria include the size of the reduced set (RS), reduction factor (RF), and execution cost. The experimental findings indicate that SSTSM outperforms PSO, BA or BAT, and GA in all comparison projects.

Keywords Software testing · Regression testing · Test suite minimization · Squirrel search · Set covering · Software engineering · Software testing · Test suite minimization · Automated testing · Metaheuristic algorithms

1 Introduction

Software engineering applies engineering principles to create high-quality software at a reasonable cost. Testing is a crucial, complex, essential, and costly part of the software development cycle (SDLC), reportedly consuming around half of the total development expenses [1, 2]. Consequently, it receives considerable attention from both software companies and academia. In software development, engineers heavily rely on testing to detect bugs. Test cases are executed on the software during testing to identify defects. A collection of these test cases that satisfy all requirements forms a test suite, expanding as the program developed. Hence, it stands as a pivotal aspect that captivates the interest of software enterprises and the academic community. In the domain of software development, engineers heavily depend on the practice of testing in order to identify and rectify bugs. During software testing, the execution of test cases is carried out to identify and pinpoint any existing defects within the software. A test suite comprises all the test cases that successfully fulfill each and every requirement, and expands in size as the program undergoes development. Hence, the impracticality of executing all test cases arises as a result of the significant expenses associated with this approach. Consequently, the optimization problem of test suite minimization (TSM) arises as a significant endeavor that seeks to decrease the dimensions of the test suite while upholding its efficacy in identifying faults. The main goal of TSM is to decrease the effort and time required for executing test cases, particularly in regression testing scenarios. Rothermel et al. [3] emphasize the resource-intensive nature of conducting a thorough test suite for a system that consists of 20,000 lines of code, estimating a significant investment of seven weeks. Optimizing the test suite by eliminating redundant or overlapping test cases can result in substantial time and effort savings.

Researchers have conducted numerous comprehensive studies and surveys to investigate various strategies for addressing TSM fall into categories like search-based, clustering-based, coverage-based, and hybrid techniques [4]. Search-based techniques utilizing metaheuristic algorithms such as Genetic Algorithm (GA), Bat Search Algorithm (BA), Particle Swarm Optimization (PSO), and Firefly Algorithm

(FA) have shown promise in minimizing test suite sizes [5]. Even with comprehensive research efforts, TSM continues to be an NP-complete problem similar to the minimum set-covering problem, lacking an efficient solution [6, 7]. The absence of a universally optimal algorithm, as per the no-free-lunch theorem [8], necessitates further exploration.

Henceforth, the authors have introduced a new algorithm known as SSTSM. This innovative SSTSM algorithm is a black box regression testing technique that prioritizes the fault exposure potential of individual test cases. Drawing inspiration from the nature-inspired metaheuristic, squirrel search algorithm (SSA), this algorithm replicates the dynamic feeding behavior and efficient gliding locomotion observed in southern flying squirrels through mathematical simulations [9]. A comprehensive investigation is undertaken to evaluate the efficacy of SSA in comparison with other cutting-edge optimization algorithms, with a focus on convergence rate and optimization accuracy. The findings of the research indicate that the algorithm displays a notable rate of convergence and surpasses existing state-of-the-art methods in terms of accuracy [9]. Therefore, the primary factors that led to the selection of SSTSM for this study were its exceptional convergence and accuracy of the results.

This research study makes the following contributions:

- Proposed a Novel Nature-Inspired Metaheuristic: SSTSM introduces a pioneering algorithm that integrates fault-exposing potential and SSA, brings a fresh perspective to TSM. Emulating intelligent navigation and adaptation observed in biological systems, this approach addresses multi-objective optimization, crucial in TSM's varied objectives.
- Implemented SSTSM Prototype and Analysis: The authors created a prototype of SSTSM utilizing the Java programming language. They conducted thorough empirical and theoretical analyses to evaluate SSTSM's effectiveness in reducing test suite size and lowering execution costs.
- Outperforming State-of-the-Art Methods: Experimental results unequivocally showcase SSTSM's superiority over prevalent benchmark algorithms, notably TSM, in the realm of regression testing. SSTSM not only matches the requirement coverage of the original test suite but also significantly reduces suite size and lowers execution costs. This underscores the unparalleled effectiveness of SSTSM in comparison with current methodologies within TSM.

The subsequent sections of this study are structured as follows. Section 2 encompasses an extensive literature review. Section 3 concisely outlines the proposed methodology. The exploration of experimental results, along with a comparative analysis of SSTSM against other methodologies, is detailed in Sect. 4. Section 5 encapsulates the authors' findings, complemented by concluding remarks.

2 Literature Review

The section on literature review presents an exploration of methodologies and recent metaheuristic algorithms dedicated to the vital aspect of test suite minimization within the context of ensuring software system quality. A notable survey by Khan et al. [4] provides an overview of various techniques such as search-based, clustering-based, coverage-based, and hybrid methodologies are scrutinized for their contributions to reducing suite size while maintaining defect identification. Among these methods, search-based techniques stand out for their adeptness in utilizing optimization algorithms to efficiently navigate through diverse test case combinations, significantly minimizing suite size while retaining fault detection capabilities. However, a research gap remains regarding the scalability and adaptability of these approaches in addressing the escalating complexities inherent in contemporary software systems. This review sets the stage for further investigation by discussing the existing methodologies and algorithms, highlighting their strengths and limitations. Subsequently, the following section lists related works that complement and extend the understanding of test suite minimization techniques, paving the way for proposed improvements and novel approaches like the SSTM.

A. Deneke et al. introduced a set cover mapping of PSO for TSM and compared it to four heuristic techniques based on reduced test suite size and execution cost but their paper is limited to comparing the metaheuristic approach with heuristic techniques [10]. S. Mohanty et al. introduced an ant-colony optimization-based TSM approach, representing each test case within the suite as a node in a full graph. This method records the execution duration and corresponding test requirement in a matrix to identify a representative set (RS) meeting all requirements while minimizing runtime [11]. However, their approach's limitation lies in constructing a complete graph, which becomes computationally complex with larger test cases and requirements. A. Singh et al. introduced a Test Case Optimization (TCO) technique employing the Butterfly Algorithm, utilizing fault coverage information from the suite to improve fault detection performance [12].

D. Bala et al. proposed a TCO and prioritization approach utilizing a multi-objective GA. Their method considers various test case factors such as statement coverage data, requirements factors, risk exposure, and execution time [13]. On a similar note, S. Detso et al. introduced a method focusing on generating high-quality test data within the domain of mutation testing. They employ a hybrid optimization approach that emphasizes maximizing coverage metrics while simultaneously minimizing test suite size without compromising its quality. Their approach achieved a substantial test suite reduction ranging between 70% and 94% across nine Java benchmark programs [14].

Mohapatra et al. developed an intelligent local search algorithm focusing on test case minimization experimented across 8 projects of varying sizes, achieving an RS factor of 70.99 and an RF of 0 [15]. Radhika et al. recently introduced an AdaBoost Convolutional Neural Network model for TCM, evaluating it on the Defects4J dataset, attaining an accuracy of 99% [16]. Bajaj et al. employed BA or BAT in their proposed methodology, where their research indicated that BAT surpassed other algorithms in various aspects such as statistical ANOVA tests, Test Selection Percentage, Fault Detection Capability Loss Percentage, and Cost Reduction Percentage [17]. Nonetheless, in the context of this specific study, SSTSM demonstrated superior performance compared to BA.

Mochamad et al. [18] introduced a Test Cases Minimization approach based on Code Coverage Similarity Measurement using Machine Learning techniques. They compared Support Vector Machine, K-Nearest Neighbor, and Decision Tree for test suite minimization, reporting that Support Vector Machine exhibited the highest accuracy and lowest error rate among the three methods. However, their experiment was conducted on a relatively small dataset, without considering path and other coverage information. Emilio et al. [19] utilized Scalable Approaches from the FAST-R family for Test Suite Reduction. Their method resulted in a loss of fault detection compared to existing approaches but demonstrated superior efficiency in reducing time.

Raphael et al. [20] employed Greedy and HGS algorithms for TSM, achieving a reduction in the test suite size by over 50%. However, their approach resulted in 21% of faults or mutation coverage being missed, and it was limited only to statement coverage information. Mohapatra et al. proposed a test case minimization algorithm based on chemical reaction optimization [21]. Mansour et al. combined Genetic Algorithms with simulated annealing to control the test case size [22]. Additionally, Black et al. and Harrold et al. used the Greedy method for addressing test case reduction [23, 24]. Both methods are considered standard practices in the realm of test case reduction.

Drawing from the literature review, TSM has been categorized as an NP-complete problem, akin to the minimum set-covering problem [6, 7, 25, 26]. This classification suggests that devising an efficient solution algorithm for TSM presents a formidable challenge, aligning with the no-free-lunch theorem [8]. Current state-of-the-art techniques for TSM exhibit drawbacks such as sluggish convergence rates and suboptimal optimization accuracy. To surmount these challenges, this research's authors modeled and developed a prototype based on SSTSM. This model considers fault coverage, considering factors like the reduced test suite size, execution time, and cost. Notably, as far as the author is aware, no prior attempts have been made to tackle the TSM problem using SSTSM. The experimental analysis showcased that the proposed approach attains better convergence rates and optimization accuracy.

3 Methodology

This section outlines the problem formulation and the suggested solution approach to extract a representative subset of test cases from a test suite characterized by redundant test cases.

The core focus of the TSM problem formulation revolves around identifying the most condensed subset from a test suite, denoted as $Ti = \{tc^1, tc^2, tc^3 \ldots tc^n\}$, which effectively encompasses all test requirements (TR = $\{tr^1, tr^2, tr^3 \ldots tr^m\}$) present in the original test suite (TS). These test requirements must be executed for achieving comprehensive program coverage. Within the TS, subsets (TS = $\{T^1, T^2, T^3 \ldots.$ T^n) are specifically linked to individual requirements (tri). In this context, any test (tc^j) belonging to T^i satisfies the condition for tri. The essence of this problem is transformed into the TC matrix representation within TSM, where an m-row (representing test requirements) by n-column (representing test cases) matrix aims to find a subset of columns (test cases) that covers all rows (test requirements) at minimal cost (fitness value). The criterion for selection adheres to the rule $TC_{ij} = 1$ if test case i fulfills requirement j, and 0 otherwise.

Table 1 presents an illustration of generating a TC matrix based on provided test cases and requirements. In this matrix, the test cases are depicted in rows and the requirements in columns. The resulting TC matrix is a binary matrix measuring $m * n$, where 'm' signifies the test case count and 'n' represents the number of requirements. Next, TSM can be stated as a standard optimization problem:

$$\text{MinimizeTS}(t) = \sum_{i=1}^{n} c_i x_i \tag{1}$$

$$\text{Subject to} \sum_{i=0}^{n} a_{ij} x_i, i \geq 1, i = 1 \text{ to } m, X_j \in \{0, 1\} j = 1 \text{ to } n. \tag{2}$$

In Eq. (1), TS(t) represents the objective function, where ci stands for the execution cost related to covering the i-th column. The variable t signifies the test cases contained within the test suite, and x_i is the decision variable involved in this context. Equation (2) serves as a condition ensuring the coverage of each row by at least one column. Here, aij represents the matrix of constraint coefficients with dimensions $m \times n$, where x_i represents the integrality constraint taking values of either '1' or '0'. In this context, 'i' refers to test cases, and 'j' refers to test requirements.

The proposed SSTSM algorithm begins by setting input parameters and initializing test cases at random positions within a specified range. These cases' fitness is then evaluated and they are sorted in ascending order based on this fitness. Test cases

Table 1 Mapping of the test case to the requirement based on Eq. 1

TC	TR				
	TR_1	TR_2	TR_3	TR_4	TR_5
TC_1	0	1	0	0	1
TC_2	1	0	0	1	0
TC_3	1	0	1	1	0
TC_4	1	0	0	0	1
TC_5	0	1	1	1	0
TC_6	1	0	0	1	1

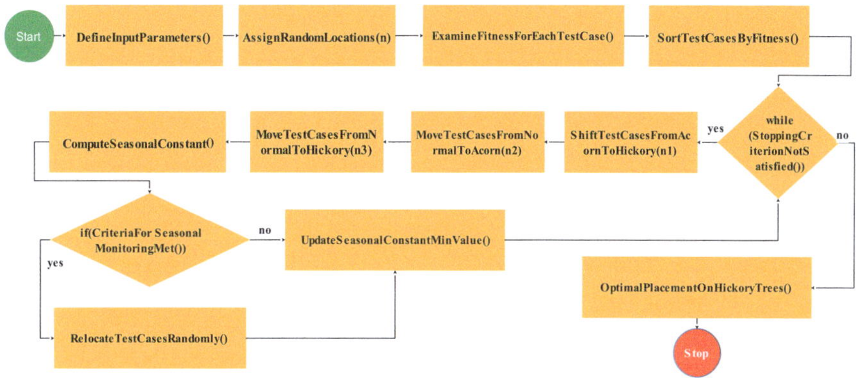

Fig. 1 Flowchart of SSTSM

are associated with three types of trees: hickory nut, normal, and acorn nut. A shifting mechanism is employed to move test cases from normal trees to either hickory nut or acorn nut trees based on their fitness levels, with this relocation process continuing iteratively until a predefined stopping criterion is met. This criterion involves reaching a predefined iteration limit or achieving near or exact optimal solutions. When shifting test cases from acorn to hickory nut trees, a probabilistic decision is made regarding whether to relocate the test case or choose a random location. Similarly, when test cases on normal trees move to acorn or hickory trees, a similar decision-making principle is applied. Additionally, seasonal monitoring is incorporated, allowing test cases to be randomly relocated based on a seasonal constant. This seasonal constant is regularly updated to ensure the optimized placement of test cases on hickory nut trees (Fig. 1).

The movement of the flying squirrels are governed by the following three cases.

Case 1: Test cases on acorn nut trees may be moved to hickory nut trees.

Case 2: To meet their daily energy needs, test cases on normal trees may migrate to acorn nut trees.

Case 3: Certain test cases that are currently on normal trees and have already consumed acorn nuts may migrate to hickory nut trees to stockpile hickory nuts for consumption during food insecurity.

4 Results and Discussion

The Software Artifact Infrastructure Library (SIR) provided the researchers with a single input dataset for the experiment. The dataset was obtained on November 30, 2023, from JMeter (http://jmeter.apache.org). The benchmark project test case matrix, similar to Table 1, is presented as a binary matrix in Excel format. The specific sizes of the test case matrices used in our experiments are detailed in Table 2. The authors executed the proposed SSTSM alongside three other methods (PSO, GA, and BA or BAT) for comparison. Java 1.8.0 was utilized for implementation, and Windows operating systems, i5-8300H CPUs running at 2.30 GHz, and 8 GB of RAM were installed on the PCs used for the experiments. All four algorithms in the study employed the same fitness evaluation function, as defined by Eq. 3, to evaluate the cost of execution for every test case in the test suite.

Moreover, a consistent number of iterations (1000) were employed for all algorithms. Metrics including the size of the reduced test suite, execution time, and execution costs were used to assess the studies. Additionally, the reduction set (RS) and reduction factor (RF) values were computed for each algorithm. Based on the information provided on Table 2 about the benchmark program the result obtained by the experiment is presented in Table 3.

Figure 2, shown above, provides a comparison of the performance of three algorithms in this area and a visual depiction of the test suite's execution cost. The reduced test suite's execution costs are represented by the y-axis, while the x-axis corresponds to the comparative algorithms. As the image illustrates, within the **Jemeter** program dataset, the cost of execution for SSTSM is 12145, PSO is 12351, GA is 13438, and BA is 12556. In execution cost also SSTSM performed better than PSO, GA, and BA.

In Fig. 3, the comparison is illustrated regarding what the four algorithms accomplished in terms of reducing the number of test cases in the test suite. The comparison algorithms are plotted on the x-axis, and the number of test cases in the condensed test suite is plotted on the y-axis. As the image illustrates, within the **Jemeter** program dataset, SSTSM decreased the quantity of test cases from 556 to 242, signifying a reduction of 56.74%. PSO reduced it to 246, resulting in a 55.75% reduction, GA to 271, a 51% reduction, and BA to 250, marking about a 55% reduction. These

Table 2 Detailed information of the input dataset (programs) used for experiment

Program	JMeter
Source file (LOC)	84,266
Test cases (T_c)	556
Test requirement (T_r)	167
TC matrix size	556 × 167

Table 3 Table presents the average of the outcomes from five separate runs of each algorithm, representing the experimental findings of the four methods

Program	JMeter			
	SSTSM	PSO	BAT	GA
No of test cases in minimized test suite	242	246	250	271
Execution cost (Fitness)	12,145	12,351	12,556	13,438

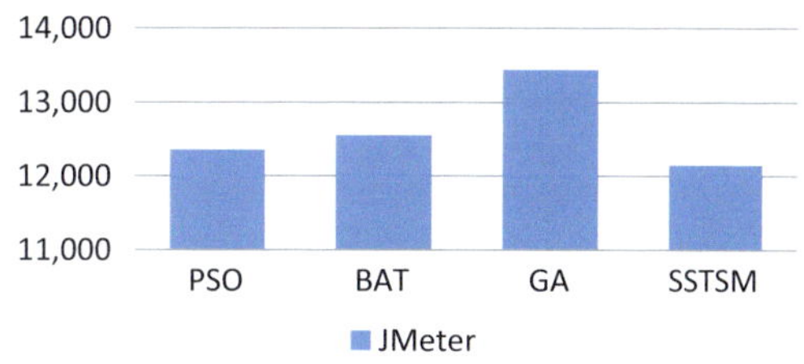

Fig. 2 Evaluation of the test suite's execution costs in relation to SSTSM, PSO, GA, and BA

results indicate that, when it comes to reducing the size of the test suite, SSTSM outperforms GA, BA, and PSO.

In evaluating the quality of the minimized test suite, the two metrics were presented by Zhang et al. [14], have become the standard for quality measurement following their publication. The **RS** metric assesses the quality of the TSM, whereas the **RF** metric gauges the fault-exposing quality of the test cases. The formulas provided denote the calculation of two metrics: Test suite size reduction (RS) and fault-detecting capability reduction (RF), expressed, respectively, as:

$$\text{RS} = \frac{|T| - |T_{\text{red}}|}{|T|} \times 100 \tag{3}$$

$$\text{RF} = \frac{|F| - |F_{\text{red}}|}{|F|} \times 100 \tag{4}$$

The RS metric Eq. (3) computes the reduction in the size of the test suite, where $|T|$ represents the total number of test cases in the original suite, and $|T_{\text{red}}|$ signifies the number of test cases in the minimized (reduced) suite. This metric's values indicate the percentage decrease in the number of test cases after optimization. Higher RS values indicate a smaller number of test cases in the reduced suite. Meanwhile, the RF metric Eq. (4) calculates the reduction in Fault-Detection Capability. Here, $|F|$ represents the total number of faults or test requirements in the original suite, while $|F_{\text{red}}|$ represents the number of faults or test requirements addressed by the minimized (reduced) suite. A RF value of 0 signifies that the test cases cover all faults, while non-zero values indicate the percentage decrease in fault coverage after optimization.

The results obtained for RS and RF metrics across the previously proposed meta-heuristic techniques exhibit less promising outcomes compared to the new proposed

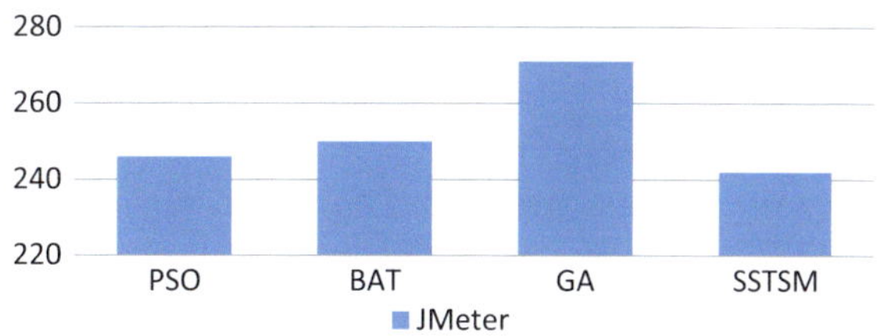

Fig. 3 Test suite size in relation to what SSTSM, PSO, GA, and BA accomplished

Table 4 RS, RF values for SSTSM, PSO, GA, and BAT on JMeter projects

Program	JMeter			
	SSTSM	PSO	BAT	GA
RS	56.5	55.8	55	51
RF	0	0	0	0

SSTSM technique for the JMeter benchmark project. These findings, elucidated in Table 4, demonstrate the impact and effectiveness of the SSTSM method in achieving better test case reduction factors for this specific project.

5 Conclusion

Nearly half of software development cost is considered to be spent on testing. Search-based software testing has emerged as an effective and cost-efficient approach for achieving optimized solutions, and as a result, the testing community now needs to maximize coverage metrics while simultaneously minimizing test suites, all without compromising the quality of testing. The SSTSM is a nature-inspired metaheuristic optimization approach that is inspired by southern flying squirrels' dynamic feeding habits and efficient glide mobility mechanisms. In this article, a novel SSTSM technique is employed to address the TSM problem. The fault coverage data was assessed using the test case to test requirement dataset retrieved from SIR. In terms of reduced set size, execution cost, SSTSM outperforms other nature-inspired algorithms PSO, GA, and BA, according to the findings of the studies. In terms of RS, RF, the SSTSM algorithm performance is far ahead for the benchmarked project than the PSO, GA, and BA algorithms. In the future, experiment will perform with other real-time project of different scale.

Declaration of Competing Interest The authors declare that they have no known competing financial interests or personal relationships that could have appeared to influence the work reported in this paper.

References

1. Pressman RS (2001) Software engineering: a practitioner's approach, 5th edn. McGraw Hill Publications, USA
2. Rothermel G, Untch RH, Chu C, Harrold MJ (2001) Prioritizing test cases for regression testing. IEEE Trans Softw Eng 27(10):929–948
3. Rothermel G, Untch RH, Chu C, Harrold MJ (2001) Prioritizing test cases for regression testing, IEEE Trans Softw Eng 27(10):929–948

4. Rehman Khan SU, Lee SP, Javaid N, Abdul W (2018) A systematic review on test suite reduction: approaches, experiment's quality evaluation, and guidelines. IEEE Access 6:11816–11841. https://doi.org/10.1109/access.2018.2809600
5. Sharma S, Rizvi SAM, Sharma VK (2019) Research on the use of nature-inspired algorithms in software testing. Regular Issue 8(11):3446–3452 (A novel nature-inspired algorithm for optimization: Squirrel search algorithm)
6. Garey DJMR (1979) Computers and intractability: a guide to the theory of NP completeness, Freeman
7. Cormen RRCSTH, Leiserson CE (2001) Introduction to algorithms. MIT Press
8. Wolpert DH, Macready WG (1997) No free lunch theorems for optimization. IEEE Trans Evol Comput 1(1):67–82. https://doi.org/10.1109/4235.585893
9. Jain M, Singh V, Rani A (2019) A novel nature-inspired algorithm for optimization: Squirrel search algorithm. Swarm Evol Comput 44:148–175
10. Deneke A, Gizachew Assefa B, Kumar Mohapatra S (2022) Test suite minimization using particle swarm optimization. Mater Today Proc
11. Mohanty S, Mohapatra SK, Meko SF (2020) Ant colony optimization (Aco-min) algorithm for test suite minimization. Adv Intell Syst Comput 55–63
12. Verma AS, Choudhary A, Tiwari S (2020) Test case optimization using butterfly optimization algorithm. In: 2020 10th international conference on cloud computing, data science & engineering (confluence), pp 704–709. https://doi.org/10.1109/Confluence47617.2020.9058334
13. Mishra DB, Mishra R, Acharya AA, Das KN (2018) Test case optimization and prioritization based on multi-objective genetic algorithm. In: Harmony search and nature inspired optimization algorithms, pp 371–381
14. Bejo SD, Assefa BG, Mohapatra SK (2021) BackIP: mutation based test data generation using hybrid approach. In: 2021 international conference on information and communication technology for development for Africa (ICT4DA), pp 178–183. https://doi.org/10.1109/ICT4DA53266.2021.9672216
15. Mohapatra SK, Mishra AK, Prasad S (2020) Intelligent local search for test case minimization. J Inst Eng India Ser B 101:585–595
16. Raamesh L, Jothi S, Radhika S (2022) Test case minimization and prioritization for regression testing using SBLA-based AdaBoost convolutional neural network. J Supercomput 78:18379–18403
17. Bajaj A, Sangwan OP, Abraham A (2022) Improved novel bat algorithm for test case prioritization and minimization. Soft Comput 26:12393–12419
18. Saputra MC, Katayama T (2020) Code coverage similarity measurement using machine learning for test cases minimization. In: 2020 IEEE 9th global conference on consumer electronics (GCCE)
19. Cruciani E, Miranda B, Verdecchia R, Bertolino A (2019) Scalable approaches for test suite reduction. In: 2019 IEEE/ACM 41st international conference on software engineering (ICSE)
20. Noemmer R, Haas R (2019) An evaluation of test suite minimization techniques. In: Software quality: quality intelligence in software and systems engineering, pp 51–66
21. Mohapatra SK, Prasad S (2017) Using chemical reaction optimization for test case minimization problem. Int J Softw Eng Technol Appl 2(1):22–40
22. Mansour N, El-Fakih K (1999) Simulated annealing and genetic algorithms for optimal regression testing. J Softw Maint 11(1):19–34
23. Black J, Melachrinoudis E, Kaeli D (2004) Bi-criteria models for all uses test suite reduction. In: Proceedings of 26th international conference on software engineering. IEEE Computer Society, Washington, DC, pp 106–115
24. Harrold MJ, Gupta R, Soffa ML (1990) A methodology for controlling the size of a test suite. In: Proceedings. Conference on software maintenance 1990, San Diego, CA, pp 302–310. https://doi.org/10.1109/ICSM.1990.131378

25. Mafi Z, Mirian-Hosseinabadi S-H (2024) Regression test selection in test-driven development. Autom Softw Eng 31(1):9
26. Khoshniat N et al (2024) Nature-inspired metaheuristic methods in software testing. Soft Comput 28(2):1503–1544

Doctor's Handwritten Prescription Recognition System Using Machine Learning

Kartik Patil and Smita Bhagwat

Abstract Handwriting serves as a means to articulate thoughts, concepts, and language. Throughout time, the scribbled cursive script often associated with medical practitioners has gained widespread acceptance despite its illegibility. The datasets utilized in this study comprise samples of cursive handwriting obtained from doctors and medical students across various clinics and hospitals. The paper introduces a Handwriting Recognition System utilizing a Deep Convolutional Recurrent Neural Network (CRNN) designed to decipher text from images of doctor-written prescriptions, presenting the converted readable text from the original handwriting. The CRNN model is trained using pre-processed input images for handwritten text recognition. The performance evaluation of CRNN, both with and without the Connectionist Temporal Classification (CTC) loss function, is detailed in this study. The findings indicate that the utilization of the CTC loss function resulted in a 92% accuracy rate during training, while without CTC, the model demonstrated an 87% accuracy rate. Validation revealed a rate of 52.5% with CTC and 45% without CTC. A total of 346 images were collected from doctors and medical students.

Keywords Doctor's handwriting · OCR · CRNN · LSTM · CTC · Zhang-Suen thinning algorithm

1 Introduction

Writing style varies from person to person. It also varies depending on their emotional status and with how much speed it was written. Doctor's handwritten prescriptions are difficult to understand. According to a study from Texas [1, 2], USA, pharmacists occasionally have trouble reading prescriptions, especially for individuals without medical backgrounds, who may not be familiar with medications and their correct

K. Patil (✉) · S. Bhagwat
E&TC Department, Vishwakarma Institute of Information Technology, Pune, India

S. Bhagwat
e-mail: smita.bhagwat@vit.edu

© The Author(s), under exclusive license to Springer Nature Singapore Pte Ltd. 2025

S. K. Udgata et al. (eds.), *Intelligent Systems*, Lecture Notes in Networks and Systems 1314, https://doi.org/10.1007/978-981-96-3797-3_5

dosage levels. Once, a cardiologist prescribed 10 mg of Nifedipine, yet a pharmacist misinterpreted it as 20 mg, leading to severe repercussions and, tragically, the patient's death. In a Philippine study [2], pre-written prescriptions from doctors were gathered. According to the results, it was determined that the assessors were having trouble reading the prescriptions, which caused them to make more mistakes. They allocated one set of assessors to each prescription who graded the legibility of the handwriting. Current technologies have been impacted by the difficulty of reading human handwriting [3]. One of the innovations brought forth by humans is the handwriting recognition system, which can recognize characters in text, photos, touch-screens, and other sources that will be translated into machine-encoded form [4]. Over the past few years, researchers have used a variety of techniques and methods to close the gap between human reading ability and recognition systems using various models [5]. Despite the abundance of studies on handwriting recognition, the challenge remains in accurately identifying cursive characters due to distortions, slants, varying sizes, diverse writing styles, incomplete strokes alongside connected characters, ligatures, and surrounding interference or noise. Since the stroke distinguishes cursive alphabets, one of the recognition mistakes that can occur is when the stroke resembles the curves of some alphabets [6, 7]. The goal of this study is to develop a model that can recognize the words and numbers in an input image of a doctor's cursive handwriting. Additionally, this study will make it easier for both medical professionals and non-professionals to read illegible handwriting. Creating a Deep Convolutional Recurrent Neural (CRNN) System for text-line detection of doctors' cursive handwriting with and without using the CTC loss function to analyze accuracy is the study's primary objective.

2 Literature Review

Recognizing handwriting can be challenging at times because each person has their own distinctive writing style [8]. It is a common stereotype that doctors have unreadable handwriting [2]. Numerous studies and real-world examples have shown that medical errors can be caused by a doctor's poor handwriting. According to a Reader's Digest article, doctors' hasty prescription writing is primarily due to the short amount of time they have to spend with each patient during consultations. The fact that some medicine names and abbreviations seem alike and might be confusing to readers if written illegibly is another factor that has made it harder to understand doctors' handwriting [2, 9, 10]. Digital image processing is used to enhance and extract recognition-related information from the stored bitmap image. A component of optical character recognition is the recognition of machine-printed characters [11]. The bitmaps of the template character and the read character will be examined by OCR to identify which character they would most closely resemble [12]. When data that is illegible by people and machines is needed and other data sources cannot be relied upon, OCR (Optical Character Recognition) is especially practical and useful. By carefully selecting the modalities of the optical character recognition system, script identification makes it easier to recognize text [13]. A data modeling technique that can capture and depict intricate input/output correlations is the neural network algorithm. The use of neural networks is motivated by the desire to create an artificial system that is capable of carrying out intellectual functions similar to those of the human brain [14]. Pre-processing, the first step in character recognition, is also known as a crucial stage to increase the recognition rate [15, 16]. Pre-processing is necessary in any OCR framework since it improves the image and reduces noise and incorrect letter interpretation [7]. Neural networks have emerged as the rapid and dependable instruments for classification in the off-line recognition system to achieve high efficiency [15]. Recurrent neural networks (RNNs) are a type of artificial neural network that are used to generate sequences in a variety of fields, including music, text, and motion capture data. By taking each step one at a time and anticipating the next, RNNs can be trained for sequence generation [17]. RNN systems do not have limits when it comes to handling long-term dependencies, in contrast to standard recognizers that employ HMM. Recurrent connections allow them to build up representations of past input events in the form of activations, allowing them to model complicated structures. They can also have numerous layers, which makes them highly useful for sequence modeling [18]. Compared to normal RNNs, the Long Short-Term Memory RNN architecture is designed to be more effective in storing and retrieving data. In various sequence processing tasks, including speech and handwriting recognition, LSTM has produced advanced results [19]. Compared to previous neural network architectures, the LSTM architecture is new and appears to solve several issues from those prior models. Recurrent neural networks are well renowned for their ability to analyze context-sensitive data and identify patterns in time series, however, typical RNNs do not perform as well as them in demanding applications like optical character recognition (OCR). The LSTM is a non-linear recurrent network with several 'gates' and additional feedback, and it was created to address this issue [20]. Building a program that can recognize handwritten characters more accurately [21] is a difficult undertaking because even humans sometimes make mistakes [22]. After all, it depends on the writer. Cursive handwriting recognition remains a challenge although numerous existing handwriting recognition systems have successfully used diverse algorithms like K-means and the Hidden Markov Model. Cursive characters were difficult to read due to the distortions and loops. The majority of handwriting recognition systems currently in use are only designed to recognize non-cursive characters and numerals. Currently, studies that have attempted to develop Cursive handwriting recognition algorithms have only managed to achieve a maximum accuracy rate of 90%. A study was done in 2019, on doctor's cursive handwriting recognition, which achieved 76% training accuracy rate and 35% validation accuracy (the model identified 17 images out of 48).

3 Design and Methodology

The proposed study which uses a four-stage methodology is depicted in Fig. 1.

3.1 Data Collection

Doctors and medical students wrote a preset prescription that included eight medicine names and their corresponding usage instructions. Given list of medications and usage instructions shown in Table 1 are used because they include frequently prescribed and misread medications with instructions based on numerous studies that are also mentioned in [2]. Hence, in total the dataset consists of 346 images collected from some clinical practitioner doctors and SMBT (Smt. Mathurabai Bhausaheb Thorat Sevabhavi Trust) Institute of Medical Sciences and Research Centre. All the prescriptions were written on one side of a page. The image of the pages was taken by a phone camera of 13 megapixels. The images were cropped as shown in Fig. 2 to isolate every prescription thus creating the prescription's dataset.

3.2 Pre-processing

Pre-processing is carried out in 4 steps as shown in Fig. 4. The images of each prescription were cleaned before using them for training and testing. To clean the images, they were converted to binary images then the Zhang-Suen thinning algorithm was used to filter out any noises. The Zhang-Suen thinning algorithm is used to thin the black-and-white images or binary images. The algorithm's goal is to create a 1-pixel-wide skeleton of the image while preserving its shape and structure.

The training process utilized skeletonized images as shown in Fig. 3 as inputs. A square-shaped sliding window scanned the text-line image, moving in the writing

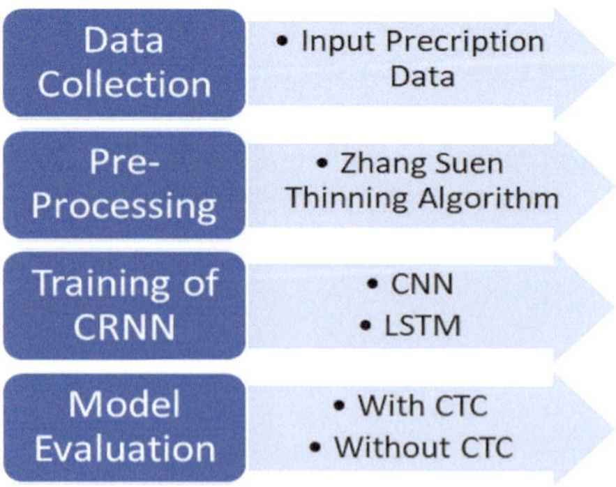

Fig. 1 Stages of methodology

Table 1 Catalog of medications alongside usage (instructions as described in the current prescription)

Prescription	Medicine
1	Amoxicillin: 300–500 mg/kg Per os OD
2	Ceftriaxone: 5 g IV q24h
3	Furosemide: 20–25 mg Per os 3 time a day
4	Warfarin: 5–15 mcg/kg/min
5	Clindamycin: 150–200 mg by IJ q24h
6	Omeprazole: 20 mg Per os 3 times a day
7	Ibuprofen: 200–400 g/kg Per os q6h
8	Atenolol: 5–20 mg per day qds

qds quater die sumendum (take 4 times daily), *q24h* quaque 24 horas (once every 24 h), *os* to be taken orally, *OD* once daily

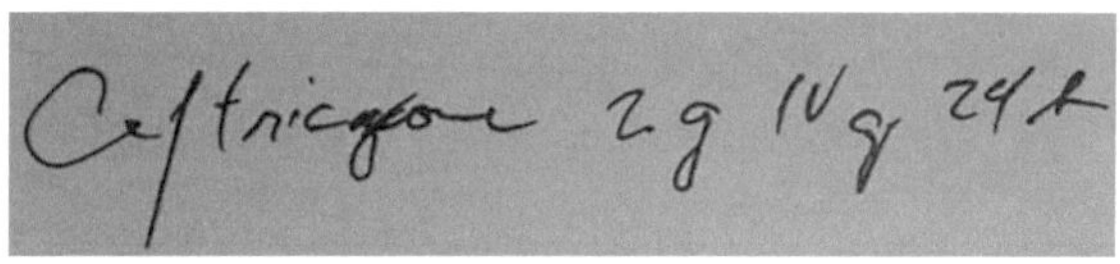

Fig. 2 Sample images of data set (original image)

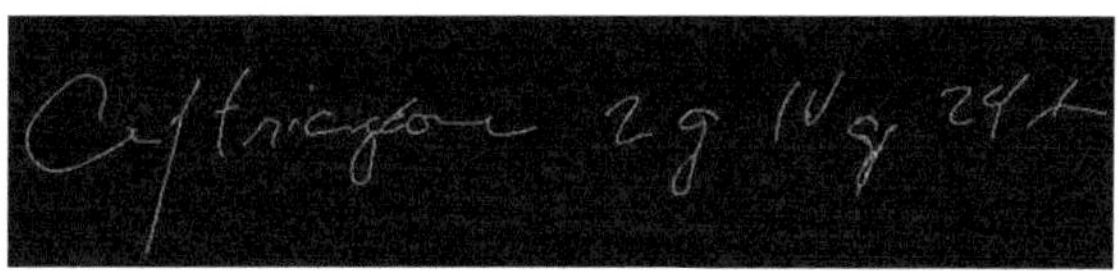

Fig. 3 Skeletonized image

direction. The window's height was adjusted based on the image size, and its pixel is normalized to 64 pixels. To facilitate convolution filter variations, a 3-pixel overlap was implemented between windows. An int64 data type has been employed to measure the sequence length, with the sequence capacity adapting dynamically according to the image's length.

Data augmentation was carried out to increase the number of images on which the model can train [1, 20]. A study demonstrated that employing data augmentation generates additional images for training the model, consequently yielding an increase in accuracy [23]. Data augmentation includes various transformations, such as vertical flip, horizontal flip, rotation, geometric transformation, vertical shift, and horizontal shift to the existing dataset (Fig. 4).

3.3 *Training of CRNN Model*

A combined deep learning algorithm was used to train the pre-processed data: Recurrent neural network and deep Convolutional Neural Network. From the dataset, 80% (276 images) were used to train data and 20% were for validation. The epoch size has

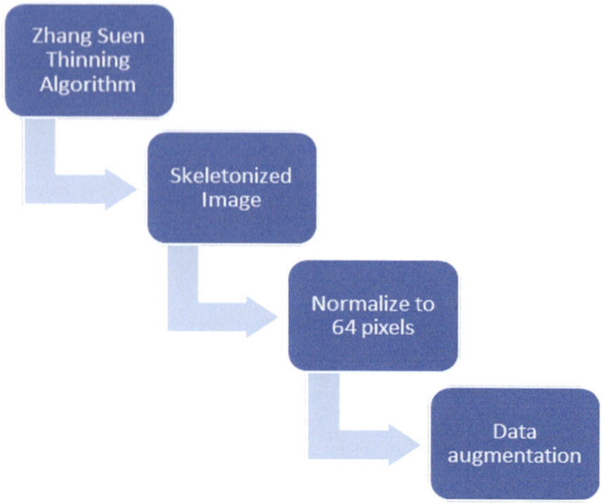

Fig. 4 Image pre-processing

been configured to 40, while the threshold has been established at 10. The testing.py was used to display the file of decoded text. A study demonstrated that the Rectified Linear Unit (ReLU) performs more effectively than the sigmoid activation function. According to a study, it has the benefit of not being impacted by the disappearing gradient problem and is easy to compute [24]. So for training the model of CRNN and Long Short-Term Memory (LSTM), the ReLU activation function is used. Based on a study, the ReLU activation function works better than the sigmoid or tach activation function [22]. The Convolutional Neural Network (CNN) extracts the features from the image by using a kernel of 3×3 matrix for images resized to 64×64 (width, height) pixel dimension. In this model, 20 convolutional layers with 3×3 filters and 1×1 stride was used. A dropout layer with 0.5 probability is used after every 10 convolutional layers are used. After dropout layer, max pooling was used.

The LSTM helps remember and store information obtained from processing the image for the long term. Bidirectional LSTM layers with 256 units and Rectified Linear Unit activation function was used due to its non-linearity. This model is trained with a cross-entropy loss function and Adam Optimizer with an initial learning rate of 0.001. Connectionist Temporal Classification (CTC) is applied with BLSTM as shown in Fig. 5. Batch Normalization was also used to lessen training time. After each layer, convolutional layer was used. The network was trained end-to-end way with Adam Optimizer and the initial learning rate, $n = 0.001$. Decoding was done by using a beam search as well as a token-passing algorithm.

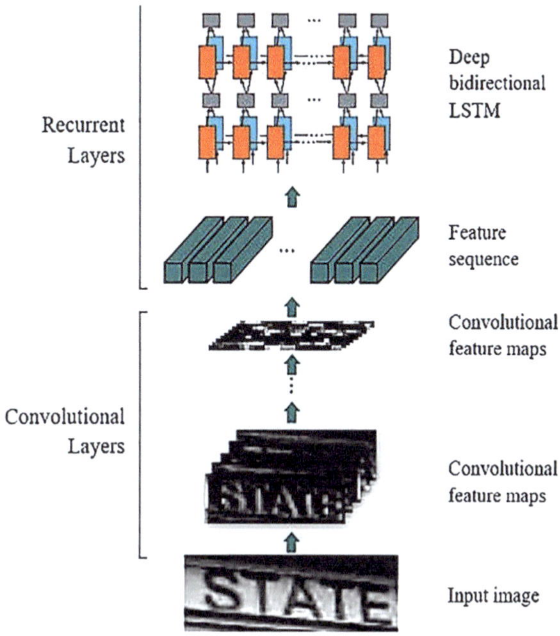

Fig. 5 CRNN model [25]

4 Result and Analysis

The prototype's performance was assessed using both the Accuracy percentage and $F1$ score metrics. The performance of CRNN and LSTM models was compared with and without CTC. Table 2 shows that the CRNN model with cross-entropy loss function gives an 87% accuracy rate and the CRNN model with CTC loss function gives 92% accuracy.

Precision measures the ratio of correctly predicted positive outcomes to the total predicted positives, showing the model's accuracy in predicting relevant instances. Whereas, recall measures the ratio of correctly predicted positive outcomes to the total actual positives, indicating the model's ability to identify all relevant instances.

The $F1$ score in machine learning is a metric that combines both precision and recall into a single value, providing a balanced assessment of a model's performance. It's calculated as the harmonic mean of precision and recall, and it helps gauge a

Table 2 Accuracy and evaluation of the two models

Model	Accuracy	$F1$ score	Precision	Recall
CRNN + LSTM	87	93	90	96
CRNN + LSTM + CTC	92	96	91	98

model's overall accuracy in classifying both positive and negative instances. Accuracy in machine learning represents the proportion of correctly classified instances (both true positives and true negatives) out of the total instances evaluated. It measures the overall correctness of the model's predictions.

Batch size is set to 10. The epoch was set to 40, but in the case of the use of the CTC loss function, it was observed that after the 32nd epoch, the accuracy remained the same so the training was stopped at the 33rd epoch while without the CTC loss function the training was stopped at 38th epoch.

Table 3 shows a detailed evaluation of the CRNN model with CTC, here the least accuracy is observed for 3rd prescription for which 27 datasets of images were available. The highest accuracy is achieved for the 1st prescription for which 29 datasets of images were available. For every prescription approx. 28 image datasets were available.

As shown in Table 4, the accuracy without using the CTC loss function is maximum for 2nd prescription and the least accuracy is observed for 3rd prescription. The training accuracy rate with and without CTC was 92% and 87%, respectively. During validation 22 out of 50 images were correctly recognized by the model using CTC which is 45%. For models without CTC, 26 out of 50 images were correctly recognized which would be 52.5% accuracy. A study was done in 2019, on doctor's cursive handwriting recognition, which achieved 76% training accuracy rate and 35% validation accuracy (the model identified 17 images out of 48) [20].

A batch was normalized to a binary image of 64 $\times$ 64 pixels to increase the speed of training and data augmentation helped in increasing accuracy for limited data. Using Zhang-Suen thinning algorithm lessened the noise from the image but it removed some considerable data like the dot of 'i' or 'j' sometimes making them more difficult to recognize.

Table 3 Result (%) of the proposed model with CTC

Prescription	Accuracy	*F*1 Score	Precision	Recall
P1	86	92	93	98
P2	95	99	95	99
P3	69	72	77	87
P4	92	98	94	99
P5	83	93	84	97
P6	88	94	98	87
P7	90	91	92	83
P8	80	81	78	96

Table 4 Evaluation of the model without using CTC

Prescription	Accuracy	$F1$ score	Precision	Recall
P1	85	90	98	96
P2	93	90	94	97
P3	68	68	72	85
P4	90	96	90	98
P5	79	89	79	99
P6	86	90	98	83
P7	88	84	89	80
P8	77	79	73	94

5 Conclusion

For a model with a CTC loss function, the repeating letters were considered only once example the word 'three' was read as 'thre' [1], and CTC is made for a model that needs a more robust sequence understanding during both input and output. For example, if it was the aim to recognize and reply to chats like 'Is it raining today?' and the reply should be 'Yes, it is raining today' the model should be able to recognize the sequence in which the word is performed from the letter and how a sentence is formed from a specific sequence of words, the sentence could be passive or active and how the output should be framed according to it. But in doctor's cursive handwriting, the sequence is not needed as much, there aren't many sequence-based sentences, and it's mostly about medicine names (letters) that are to be recognized. The little bit of sequential part of figuring out the letters in medicinal words could be done by LSTM and it can be improved by providing the model with a dictionary or spelling checker so it can get closer to accuracy. The accuracy could be improved by taking into consideration the SVM and related features like Pen length and time length.

Acknowledgements Recognition is attributed to the medical professionals who took part as respondents in this study. The study acknowledges the support given by all the individual clinic doctors and students of SMBT Institute of Medical Sciences and Research Centre.

References

1. Brits H, Botha A, Niksch L, Venter K, Terblanché R, Joubert G (2017) Illegible handwriting and other prescription errors on prescriptions at National District Hospital, Bloemfontein. Prof Nurs Today 21(2):53–56
2. Cerio AAP, Mallare NALB, Tolentino RMS (2015) Assessment of the legibility of the handwriting in medical prescriptions of doctors from public and private hospitals in Quezon City, Philippines. Proc Manuf 3:90–97
3. Lefakis L, Wiering MA (2007) Semi-supervised methods for handwritten character recognition using active learning

4. Fischer A, Bunke H (2011) Character prototype selection for handwriting recognition in historical documents. In: 2011 19th European signal processing conference. IEEE, pp 1435–1439
5. Hochuli AG, Oliveira LS, Britto AS, Sabourin R (2018) Segmentation-free approaches for handwritten numeral string recognition. In: 2018 international joint conference on neural networks (IJCNN). IEEE, pp 1–8
6. Ruiz-Pinales J, Jaime-Rivas R (2004) Cursive word recognition using a novel feature extraction method and a neural network. In: ICVGIP, pp 479–484
7. Choudhary U, Bhosale S, Bhise S, Chilveri P (2017) A survey: Cursive handwriting recognition techniques. In: 2017 2nd IEEE international conference on recent trends in electronics, information & communication technology (RTEICT). IEEE, pp 1712–1716
8. Ishak WHW, Rahman SA (1990) Neural network in handwritten recognition system: a survey. Wanhussain Com 1990:1–9
9. Roy V, Gupta P, Srivastava S (2005) Medication errors: causes & prevention. Health Adm 19(1):60–64
10. Merugu R, Mantoo A, Manna A, Meshram G, Patil PH (2023) Digitization, analysis & prediction of medical reports using deep learning. In: Analysis & prediction of medical reports using deep learning (May 26, 2023)
11. Bhushan A, Kalyani P, Supriya J (2001) Handwritten script recognition. J Comput Eng 30–33
12. Basu JK, Bhattacharyya D, Kim T (2010) Use of artificial neural network in pattern recognition. Int J Softw Eng Appl 4(2)
13. Garg N, Kumar M (2015) Clustering of multi scripts isolated characters using k-means algorithm. Int J Math Sci Comput 22–29
14. Rao NV, Sastry ASCS, Chakravarthy ASN, Kalyanchakravarthi P (2016) Optical character recognition technique algorithms. J Theor Appl Inf Technol 83(2)
15. Pradeep J, Srinivasan E, Himavathi S (2011) Diagonal based feature extraction for handwritten character recognition system using neural network. In: 2011 3rd international conference on electronics computer technology, vol 4. IEEE, pp 364–368
16. Dwivedi U, Rajput P, Sharma MK, Noida G (2017) Cursive handwriting recognition system using feature extraction and artificial neural network. Int Res J Eng Technol 4(03):2202–2206
17. Shen Y, He X, Gao J, Deng L, Mesnil G (2014) A latent semantic model with convolutional-pooling structure for information retrieval. In: Proceedings of the 23rd ACM international conference on conference on information and knowledge management, pp 101–110
18. Graves A, Schmidhuber J (2008) Offline handwriting recognition with multidimensional recurrent neural networks. Adv Neural Inf Process Syst 21
19. Goodfellow I, Bengio Y, Courville A (2016) Book review: deep learning. Deep Learn 22(4):351–354
20. Fajardo LJ, Sorillo NJ, Garlit J, Tomines CD, Abisado MB, Imperial JMR, Rodriguez RL, Fabito BS (2019) Doctor's cursive handwriting recognition system using deep learning. In: 2019 IEEE 11th international conference on humanoid, nanotechnology, information technology, communication and control, environment, and management (HNICEM). IEEE, pp 1–6
21. Saha S (2018) A comprehensive guide to convolutional neural networks—the ELI5 way. Towards Data Sci 15:15
22. Breuel TM, Ul-Hasan A, Al-Azawi MA, Shafait F (2013) High-performance OCR for printed English and Fraktur using LSTM networks. In: 2013 12th international conference on document analysis and recognition. IEEE, pp 683–687
23. Indian A, Bhatia K, Kumar K (2022) Offline handwritten Hindi character recognition using deep learning with augmented dataset. In: Cyber security in intelligent computing and communications. Springer Singapore, Singapore, pp 129–141
24. Dhande PS, Kharat R (2017) Character recognition for cursive English handwriting to recognize medicine name from doctor's prescription. In: 2017 international conference on computing, communication, control and automation (ICCUBEA). IEEE, pp 1–5
25. Kag&Atul, 10 March 2021, CTC. https://theailearner.com/tag/crnn-model/

A YOLOv8 Method for Tracking the Health of Coconut Trees

S Ranjith Reddy, Saroja Kumar Rout, Kottu Santosh Kumar, Ruchismita Sahu, Kachi Anvesh, and Kailash Chandra Nayak

Abstract Coconut farming sustains livelihoods by providing food, cash, and cultural significance. This study introduces a new approach using the YOLOv8 algorithm to monitor coconut tree health, focusing on pest detection. The method utilizes YOLOv8's ability to detect and identify infestation indicators on coconut trees using an extensive dataset. Transfer learning optimizes the model's performance for assessing coconut tree health. Results demonstrate YOLOv8's effectiveness in pinpointing infested coconut trees, showing promise for timely intervention in precision farming. It advances agricultural tech by enabling prompt pest responses and eco-friendly farming methods, underscoring AI's vital role in crop management and food supply enhancement. The investigation's findings highlight the effectiveness of YOLOv8, which outperforms traditional object recognition algorithms in terms of accuracy and speed. This validates YOLOv8 as a state-of-the-art technique for object recognition and detection in real-world scenarios.

Keywords YOLOv8 · Object detection · Precision agriculture · Infestation detection · Sustainable farming · Agricultural technology

1 Introduction

Horticulture is pivotal to India's economy, particularly in coconut production, where the nation leads globally. In 2021–22, India contributed 31.45% of the world's coconut production, yielding 19,247 million nuts, injecting Rs. 307,498 million into

S. R. Reddy · S. K. Rout (✉) · K. S. Kumar · K. Anvesh
Department of Information Technology, Vardhaman College of Engineering (Autonomous), Hyderabad, India
e-mail: rout_sarojkumar@yahoo.co.in

R. Sahu
Department of Computer Science and Engineering, Templecity Institute of Technology and Engineering, Bhubaneswar, India

K. C. Nayak
United School of Business Management, Bhubaneswar, India

© The Author(s), under exclusive license to Springer Nature Singapore Pte Ltd. 2025
S. K. Udgata et al. (eds.), *Intelligent Systems*, Lecture Notes in Networks and Systems 1314, https://doi.org/10.1007/978-981-96-3797-3_6

the GDP. Over 3.16 million families depend on coconut farming, spanning 1950 million hectares in key regions like Kerala, Karnataka, Tamil Nadu, and Andhra Pradesh [1]. The research in Multimedia Tools and Applications [2] introduces a valuable technique for assessing pest damage on coconut trees, merging texture, and color analysis for precise caterpillar detection and leaf color evaluation. Integrating texture and color analysis enhances our understanding of pest damage extent, guiding targeted protective measures for coconut crops. Given the threat posed by whiteflies to coconut foliage, exploring deep learning for leaf identification provides an alternative [3]. Object detection, a key focus in current research, overcomes limitations of traditional disease diagnosis methods, offering accurate classification through one-stage methods like SSD System Design Document (SSD), Visual Geometry Group (VGG), and You Only Look Once (YOLO), and two-stage algorithms such as ResNet, LeNet-5, AlexNet, GoogLeNet, and Faster-Region-based Convolutional Neural Network (Fast-RCNN) [4–7]. This study underscores the importance of leveraging advanced tools like object identification and deep learning to enhance coconut tree health assessment accuracy. Redmon et al. optimized YOLO with DarkNet, enhancing detection speed, and accuracy. YOLOv5, introduced in June 2020, further improved memory consumption and speed. YOLOv8, the latest version, maximizes performance in identifying crop diseases, achieving impressive speed and accuracy, crucial for early infestation detection.

1.1 Contribution

YOLOv8 greatly improves the detection of health issues in coconut trees by accurately identifying symptoms and anomalies through advanced object detection techniques. This method enables continuous monitoring, allowing for prompt intervention and early diagnosis of diseases and pests. It provides precise information on individual tree health, empowering farmers to target areas for treatment. As a valuable decision support tool, YOLOv8 promotes sustainable farming practices by reducing pesticide use and enabling proactive management.

2 Related Works

Artificial intelligence and the Internet of Things have enhanced the precision of identifying insect infestations in coconut plants [8, 9]. A major cause of death in the world is prostate cancer (PRC). To improve the quality of healthcare, early detection and characterization of PRC are crucial. To categorize PRC effectively, deep learning and machine learning approaches are used with microarray gene expression data (GED) [10]. Various approaches, including deep transfer learning and conventional machine learning, have been explored to develop reliable techniques [11, 12]. Convolutional

Neural Networks, exemplified by Aburasain et al. [13], have revolutionized large-scale agricultural applications. YOLOv3 was employed to identify coconut-related diseases and pests, achieving a remarkable 92.39% accuracy and a 20.39 ms detection time, showcasing robust real-time capabilities. Mandava et al. [14] also utilized YOLOv5 for detecting infestations in coconut trees, reporting improved results with an F1 score, mAP value, identification precision, and recall of 90%, 67%, 76%, and 81%, respectively. While these studies focus on coconut tree infestations, further research is needed to enhance the model's ability to detect other crop diseases.

Vidhanaarachchi et al. [15] showcased real-time grape disease diagnosis using YOLOv4-tiny with features like downsampling fusion, Mish activation, and attention mechanisms, and proposed YOLOv5 for coconut tree infestations, focusing on one bell pepper type. Additional research is needed for model validation.

Their method, employing YOLOv7, achieved 99.5% accuracy but faced challenges with small and hidden objects. Our work presents YOLOv8, integrating mix-up data augmentation and anchor-free detection to enhance accuracy and efficiency for real-time crop disease detection. Smart farming utilizes technology to create controlled environments, boosting crop production and offering small farmers alternative income sources to prevent annual crop loss [16].

3 Materials and Methods

This section introduces our new YOLOv8-powered real-time coconut tree infestation detection model. There will be a thorough explanation of the model's development and application, detailing the procedures followed to build a productive real-time detection system. Three methods for monitoring the health of coconut trees using YOLOv8 and YOLO-based models are shown in Figure 1. Preprocessing images (CLAHE, resizing), dividing data into training and test sets, and using augmentation to deal with unbalanced data are all part of each method. A sampling of the original test set photographs is displayed in the final result. Figure 2 shows the YOLOv8 architecture, which includes the backbone (CSPDarknet) for processing input images, the neck (PANet) for feature aggregation, and the decoupled head for final object recognition and classification. The output includes bounding boxes and class probabilities.

3.1 Dataset

In this study, we utilized the PlantDoc dataset, available on GitHub, comprising data from 13 plant species and 30 classifications, totaling 8,851 annotated labels. The dataset was annotated using the Roboflow platform to enhance accuracy and efficiency in training YOLOv8 models for disease identification in agriculture.

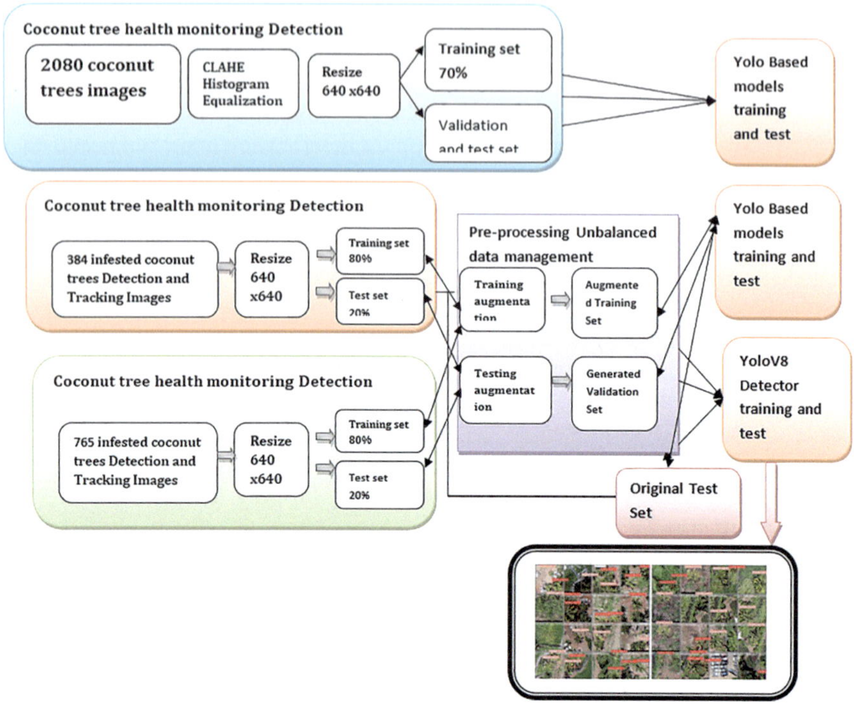

Fig. 1 Block diagram of tracking the health of coconut trees

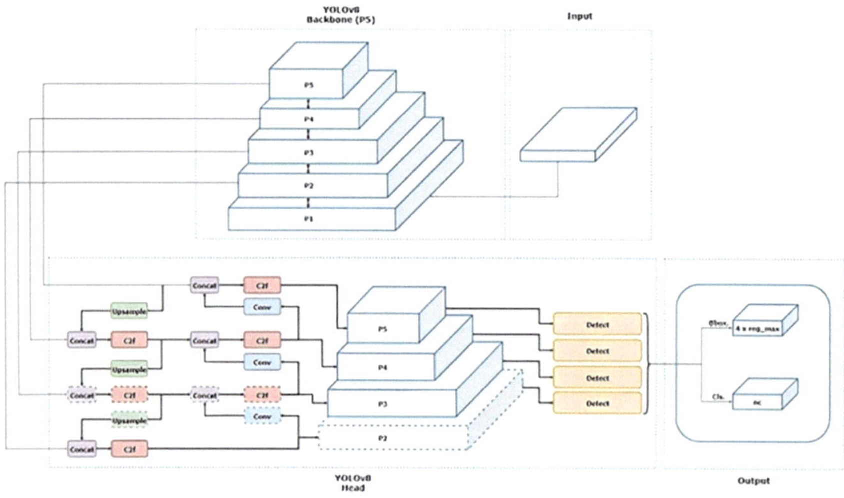

Fig. 2 Illustration of the YOLOv8 architecture

Fig. 3 Visualization of a YOLOv8 anchor box

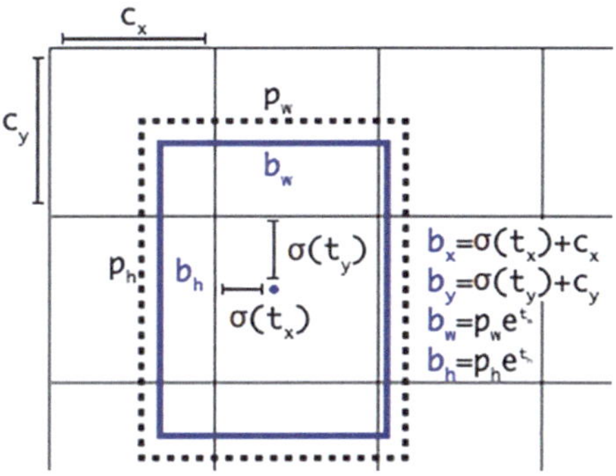

3.2 *Overview of YOLOv8*

The Figure 3 illustrates YOLO bounding box prediction. It shows a grid cell, a predicted bounding box (blue), and an anchor box (dashed). The predicted box's center and dimensions are derived from the anchor box's parameters and offsets relative to the grid cell.

2.1. Loss function: In Eq. 1, we can see that YOLOv8 trains its network using a specified loss function. The loss function consists of several components.

$$
\begin{aligned}
&\lambda_{\text{coord}} \sum_{i=0}^{S^2} \sum_{j=0}^{B} \prod_{ij}^{\text{obj}} \left[\left(x_i - x_j\right)^2 + \left(y_i - y_j\right)^2 \right] \\
&+ \lambda_{\text{coord}} \sum_{i=0}^{S^2} \sum_{j=0}^{B} \prod_{ij}^{\text{obj}} \left(\sqrt{w_i - w_j}\right)^2 + \left(\sqrt{h_i - h_j}\right)^2 \\
&+ \sum_{i=0}^{S^2} \sum_{j=0}^{B} \prod_{ij}^{\text{obj}} \left[\left(c_i - c_j\right)^2 \right] + \lambda_{\text{noobj}} \sum_{i=0}^{S^2} \sum_{j=0}^{B} \prod_{ij}^{\text{noobj}} \left[\left(c_i - c_j\right)^2 \right] \\
&+ \lambda_{\text{coord}} \sum_{i=0}^{S^2} \sum_{j=0}^{B} \prod_{ij}^{\text{obj}} \left[\left(p_i(c)\right)^2 + \left(p_j(c)\right)^2 \right] \qquad (1)
\end{aligned}
$$

where (x_i, y_i) is a B center in S^2, B is the number of bounding boxes in each grid cell S^2, and S^2 is the total number of grid cells. In S^2, (w_i, h_i) stands for B's width and height. In S^2, a target's center is represented by (x_j, y_j). In S^2, a target's center is represented by (w_i, h_i). When an item in b exists in each S^2, $\prod ij$ is set to 1; otherwise, it is set to 0. λ_{coord} enhances the weight loss for the bounding box coordinates loss.

2.2. Classification loss: This evaluates how well the network predicts classes and penalizes it for giving out the wrong classes. Equation 2 provides the classification loss computation.

$$\lambda_{\text{coord}} \sum_{i=0}^{S^2} \sum_{j=0}^{B} \prod_{ij}^{\text{obj}} \left[(p_i(c))^2 + \left(p_j(c)\right)^2 \right] \tag{2}$$

2.3. Confidence loss: Evaluating the accuracy of the network's predictions and applying penalties for false positive and false negative detections are two aspects of the assessment process. Equation 3 is utilized to compute the confidence loss.

$$\sum_{i=0}^{S^2} \sum_{j=0}^{B} \prod_{ij}^{\text{obj}} \left[\left(c_i - c_j\right)^2 \right] + \lambda_{\text{noobj}} \sum_{i=0}^{S^2} \sum_{j=0}^{B} \prod_{ij}^{\text{noobj}} \left[\left(c_i - c_j\right)^2 \right] \tag{3}$$

2.4. Localization loss: This analysis incorporates penalties for prediction errors and concentrates on the accuracy of the network's bounding box predictions. Equation 4 describes how to determine the localization loss coefficient. To optimize the parameters of a network, the three losses are combined into a single loss function.

$$\begin{aligned} &\lambda_{\text{coord}} \sum_{i=0}^{S^2} \sum_{j=0}^{B} \prod_{ij}^{\text{obj}} \left[\left(x_i - x_j\right)^2 + \left(y_i - y_j\right)^2 \right] \\ &+ \lambda_{\text{coord}} \sum_{i=0}^{S^2} \sum_{j=0}^{B} \prod_{ij}^{\text{obj}} \left(\sqrt{w_i - w_j}\right)^2 + \left(\sqrt{h_i - h_j}\right)^2 \end{aligned} \tag{4}$$

The integration of localization, confidence, and classification losses constitutes a unified loss function in YOLOv8, designed to optimize network parameters. While specifics remain undisclosed, it's acknowledged to be a tailored iteration of the YOLO loss function from earlier versions.

4 Experimental Results and Discussion

In our investigation using the Plantdoc dataset, we conducted training sessions for the YOLOv8 model over 25 epochs. Evaluation criteria included bounding box, classification, and distribution losses, along with metrics like accuracy, recall, precision, and $F1$-score. Experiments utilized Acer laptops with Ryzen 9 processors and Geforce RTX3080 graphics cards. PyTorch 1.13.1 and Python 3.9.13 in an Anaconda3 environment, with CUDA version 11.7, were employed for model development and training. The Roboflow platform facilitated dataset annotation, while tests were performed on PlantDoc images with a resolution of 614 by 614, resulting in the assessment of various performance metrics after 25 epochs.

Transfer learning helps improve the YOLOv8 model for checking coconut tree health by using knowledge from other models trained on big datasets. It starts with weights learned from spotting objects in pictures, then fine-tunes them with a smaller dataset about coconut tree health. This makes the model better at recognizing coconut tree features and problems. By doing this, we can spot tree issues more accurately and quickly, which helps in managing and preserving coconut trees better.

Validating the research works in real life with coconut farmers' input makes it more useful. It helps make sure the ideas work and help farmers. Getting feedback from farmers helps make sure the solutions are helpful and fit with what farmers need. This makes the study more helpful and important for farmers and the coconut industry. The system accuracy, precision, recall, and F1-Score are described in (Eqs. 5–8).

$$\text{Accuracy} = \text{Number of correct predictions/Total no. of prediction} \tag{5}$$

$$\text{Precision} = \frac{\text{TP}}{\text{TP} + \text{FP}} \tag{6}$$

$$\text{Recall} = \frac{\text{TP}}{\text{TP} + \text{FN}} \tag{7}$$

$$F1\text{-score} = \frac{2 \times \text{precision} \times \text{recall}}{\text{precision} + \text{recall}} \tag{8}$$

Defined by counts of false positives (FP), false negatives (FN), true positives (TP), and correctly classified negatives (TN), FP represents wrongly detected positive samples, while FN denotes overlooked positive samples. The losses for distribution, classification, and bounding box were 1.197, 0.9219, and 0.9697, respectively, with a GPU memory demand of 4.62G during training. Assessment metrics indicated recall values of 0.63, mAP50 values of 0.633, mAP50-95 values of 0.651, and box precision values of 0.628. Equation 9 describes the mean average precision (mAP).

$$\text{mAP} = \sum_{n} r_{n+1} - r_n p_{\text{interst}} r_{n+1} \tag{9}$$

Object detection models are commonly evaluated using mAP (mean average precision). Our results show a mAP50 score of 0.633, indicating 63.3% average precision for objects recognized at a 50% recall rate. This confusion matrix shows how well a categorisation model performs. It displays the expected versus actual labels for backdrop, afflicted trees, and healthy trees. Darker hues denote larger proportions, whereas off-diagonal entries show misclassifications. Correct classifications are shown by diagonal entries described in Figure 4. Learning is demonstrated by the training and validation losses (box, class, and DFL) declining over epochs in Figure 5. On the validation set, evaluation metrics such as precision, recall, mAP@0.5, and mAP@0.5-0.95 typically rise, indicating better object detection performance.

The $F1$ score for both healthy and infested tree detection at different confidence criteria is displayed in this $F1$-Confidence curve. The ideal balance between precision and recall is indicated by the overall $F1$ score, which peaks at 0.71 for all classes at a

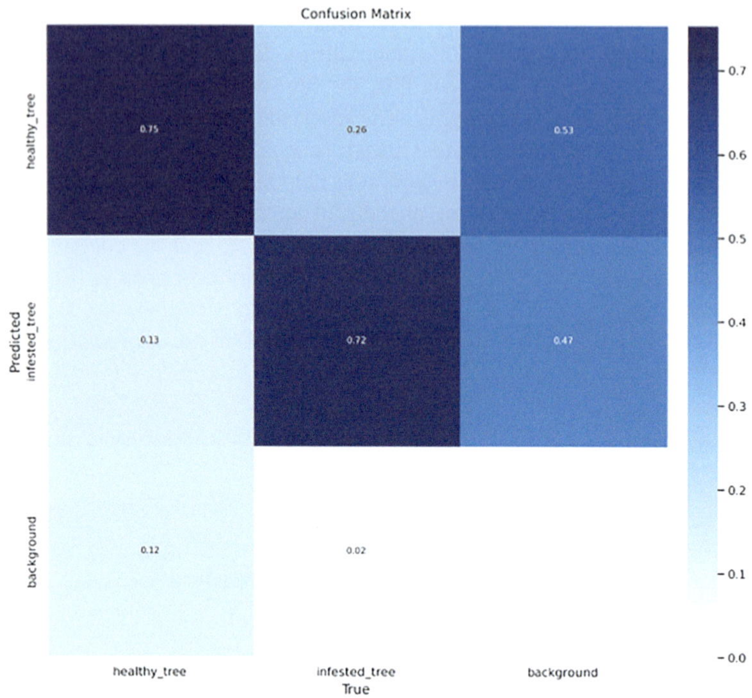

Fig. 4 Confusion matrix of coconut tree infested detection using YOLOv8

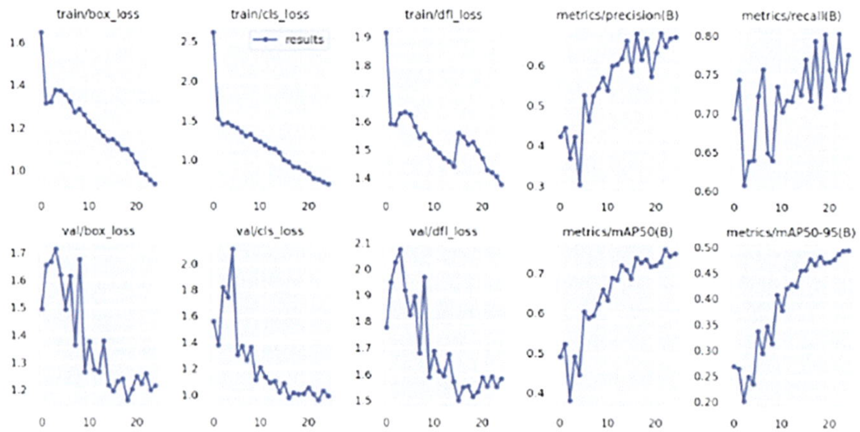

Fig. 5 YOLOv8 model was evaluated during training the metrics with spanning 25 iterations

confidence level of 0.299 represented in Figure 6. For both healthy and infested tree detection, the Precision-Confidence curve shows how precision varies with different confidence criteria. With a high confidence of 0.889 and an overall precision of 1.00, it appears that nearly all detections are accurate at this stringent threshold, even though recall may be lower described in Figure 7. For identifying healthy and infested trees, this recall-confidence curve illustrates how recollection varies with different confidence criteria. Nearly all real trees are spotted at this lax threshold, maybe at the expense of reduced precision, as evidenced by the highest overall recall (0.96) at a very low confidence of 0.000 described in Figure 8. The trade-off between precision and recall for detecting healthy and infested trees is shown by the Precision-Recall curve. Figure 9 shows the mean AP (mAP@0.5) for all classes, which is 0.750. The average precision (AP) is 0.690 for healthy trees and 0.810 for infested trees. The Frames Per Second (FPS) attained for processing individual photos is displayed in Figure 10. Across many photos, the FPS varies greatly, from almost zero to around 400. The red dashed line at 70.02 represents the average FPS for all photos. The inference time in seconds for processing each image is shown in Figure 11. There are some spikes in the inference time, which varies among photos. The average inference time per image is 0.01 seconds, as shown by the red dashed line.

Figure 12 represents the classification results of the YOLOv8 model and the accompanying confidence levels for each example. Figure 13 describes predictions made by the YOLOv8 model are illustrated with the corresponding confidence level for each case.

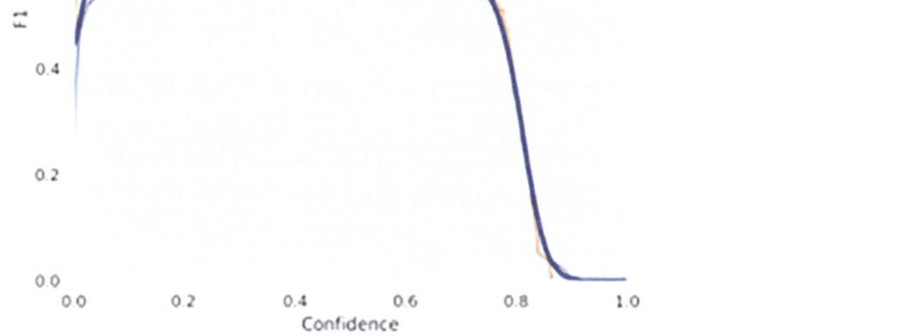

Fig. 6 *F*1-curve of coconut tree infested detection using YOLOv8

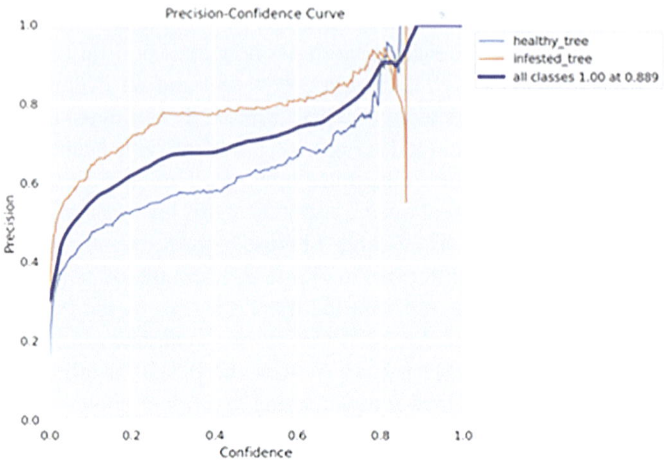

Fig. 7 Precision Confidence curve of coconut tree infested detection using YOLOv8

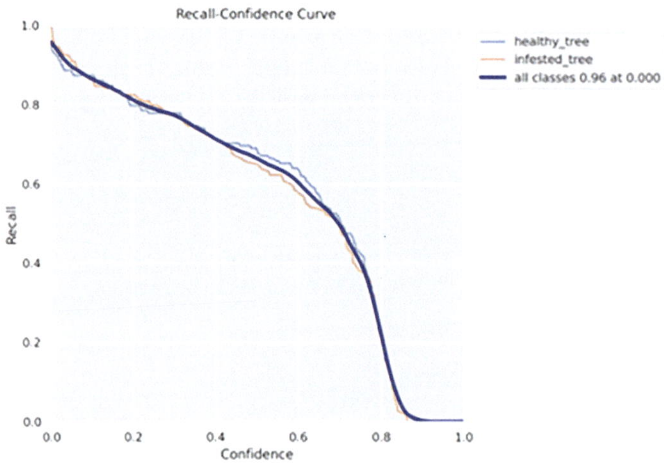

Fig. 8 Recall the Confidence curve of coconut tree infested detection using YOLOv8

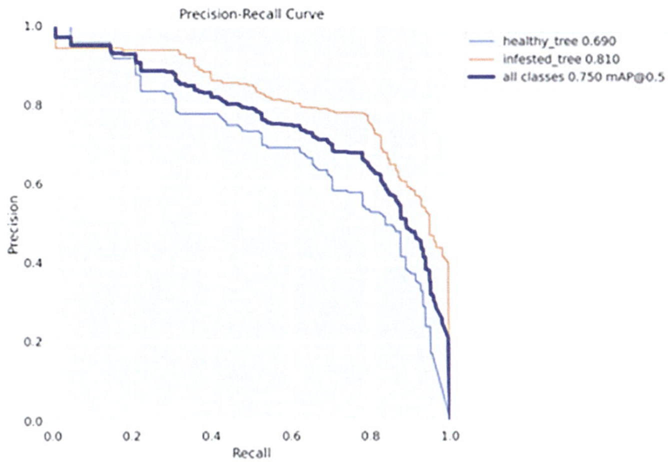

Fig. 9 Precision-Recall curve of coconut tree infested detection using YOLOv8

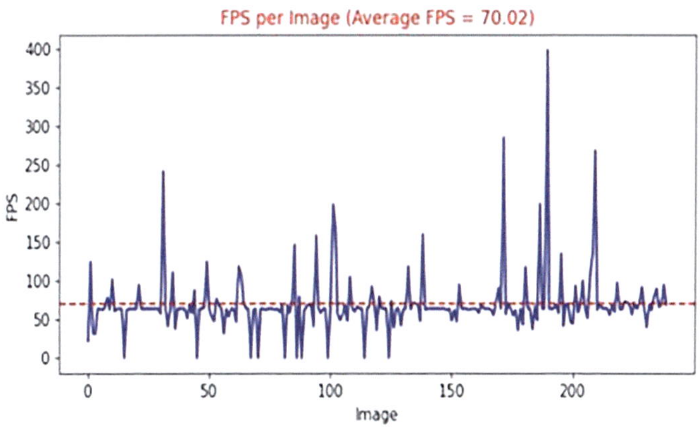

Fig. 10 FPs image calculation of coconut tree infested detection using YOLOv8

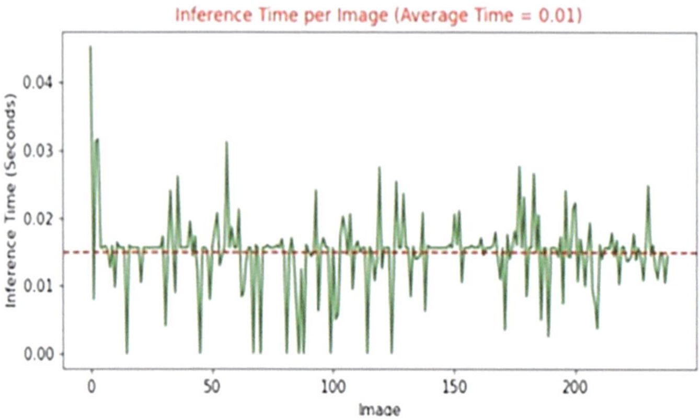

Fig. 11 Inference time per image calculation of coconut tree infested detection using YOLOv8

Fig. 12 Samples showing the classification results of the YOLOv8 model and the accompanying confidence levels for each example

Fig. 13 Predictions made by the YOLOv8 model are illustrated with the corresponding confidence level for each case

5 Conclusion

In conclusion, this research highlights the effectiveness of employing the YOLOv8 algorithm for Coconut Tree Health Monitoring, specifically targeting infestation detection. Utilizing diverse datasets and transfer learning techniques, the model accurately identifies signs of infestation across various conditions and stages. This approach empowers farmers to proactively mitigate infestation impacts, supporting sustainable farming practices. Integrating artificial intelligence into agriculture underscores technology's potential to address industry challenges and ensure food security. The YOLOv8-based system offers a scalable solution for coconut crop protection, showcasing modern technology's revolutionary impact on agriculture.

References

1. Sanjay NA, Jeevithaa RR Design and development of agri drone for detection of infected areas and application of pesticides for coconut tree
2. Barman U, Pathak C, Mazumder NK (2023) Comparative assessment of pest damage identification of coconut plant using damage texture and color analysis. In: Multimedia tools and applications, pp 1–23
3. Kavithamani V, UmaMaheswari S (2023) Investigation of deep learning for whitefly identification in coconut tree leaves. Intell Syst Appl 200290
4. Gao F, Fu L, Zhang X, Majeed Y, Li R, Karkee M, Zhang Q (2020) Multi-class fruit-on-plant detection for apple in SNAP system using Faster R-CNN. Comput Electron Agric 176:105634
5. Peng H, Huang B, Shao Y, Li Z, Zhang C, Chen Y, Xiong J (2018) General improved SSD model for picking object recognition of multiple fruits in natural environment. Trans Chin Soc Agric Eng 34(16):155–162
6. Hu G, Wang H, Zhang Y, Wan M (2021) Detection and severity analysis of tea leaf blight based on deep learning. Comput Electr Eng 90:107023

7. Yan B, Fan P, Lei X, Liu Z, Yang F (2021) A real-time apple targets detection method for picking robot based on improved YOLOv5. Remote Sens 13(9):1619
8. Han X, Chang J, Wang K (2021) You only look once: unified, real-time object detection. Proc Comput Sci 183(1):61–72
9. Orchi H, Sadik M, Khaldoun M (2022) A novel hybrid deep learning model for crop disease detection using BEGAN. In: International symposium on ubiquitous networking. Springer International Publishing, Cham, pp 267–283
10. Sethi BK, Singh D, Rout SK, Panda SK (2023) Long short-term memory-deep belief network based gene expression data analysis for prostate cancer detection and classification. IEEE Access
11. Parvathi S, Selvi ST (2021) Detection of maturity stages of coconuts in complex background using Faster R-CNN model. Biosyst Eng 202:119–132
12. Kavithamani V, UmaMaheswari S (2023) Investigation of deep learning for whitefly identification in coconut tree leaves. Intell Syst Appl 200290
13. Aburasain RY, Edirisinghe EA, Albatay A (2020) Palm tree detection in drone images using deep convolutional neural networks: investigating the effective use of YOLO v3. In: Conference on multimedia, interaction, design and innovation. Springer International Publishing, Cham, pp 21–36
14. Mandava RK, Mittal H, Hemalatha N (2023) Identifying the maturity level of coconuts using deep learning algorithms. Mater Today Proc
15. Vidhanaarachchi SP, Akalanka PKGC, Gunasekara RPTI, Rajapaksha HMUD, Aratchige NS, Lunugalage D, Wijekoon JL (2021) Deep learning-based surveillance system for coconut disease and pest infestation identification. In: TENCON 2021–2021 IEEE region 10 conference (TENCON). IEEE, pp 405–410
16. Singh M, Sahoo KS, Gandomi AH (2023) An intelligent IoT-based data analytics for freshwater recirculating aquaculture system. IEEE Internet Things J

Enhanced Particle Swarm Optimization to Solve Traveling Salesman Problem

Indresh Kumar Gupta, Indu Maurya, Joel J. P. C. Rodrigues, Awanish Kumar Mishra, and Swati Srivastava

Abstract Heuristic search techniques stand as stalwart allies in the pursuit of optimal solutions for the Traveling Salesman Problem (TSP), a conundrum ubiquitous across diverse industries including semiconductor manufacturing, logistics, and transportation. In this paper, an enhanced particle swarm optimization (E-PSO) is being proposed for TSP. A distinguishing feature of E-PSO lies in its departure from conventional PSO methodologies. Here, each particle within the swarm is not solely guided by the global best solution but also by a superior peer. This nuanced guidance mechanism serves as a safeguard against premature convergence, allowing the swarm to explore the solution space more comprehensively. The evaluation of the E-PSO method is being conducted by comparing its performance to that of 10 conventional TSP instances. The crux of this assessment revolves around two pivotal aspects: the algorithm's acumen in discerning optimal solutions and its computational efficiency. The computational results provide evidence that the E-PSO algorithm shows remarkable improvement for solving TSPs, outperforming the PSO method.

Keywords Traveling Salesman Problem · Particle swarm optimization · Fitness function · Optimal

---

I. K. Gupta (✉) · A. K. Mishra
Pranveer Singh Institute of Technology, Kanpur, India
e-mail: indresh.gupta345@gmail.com

I. K. Gupta · J. J. P. C. Rodrigues
COPELABS, Lusófona University, Lisbon, Portugal
e-mail: joeljr@ieee.org

I. Maurya
Department of Technical Education, Uttar Pradesh, Kanpur, India

J. J. P. C. Rodrigues
Federal University of Piauí (UFPI), Teresina, PI, Brazil

S. Srivastava
GLA University, Mathura, India
e-mail: swati.srivastava@gla.ac.in

© The Author(s), under exclusive license to Springer Nature Singapore Pte Ltd. 2025
S. K. Udgata et al. (eds.), *Intelligent Systems*, Lecture Notes in Networks and Systems 1314, https://doi.org/10.1007/978-981-96-3797-3_7

1 Introduction

The Traveling Salesman problem (TSP) is a renowned and complex problem in the fields of mathematics and computer science well-known for its classification as NP-hard [1]. The practical applications of this technology have a broad scope encompassing several industries such as semiconductor production, logistics, and transportation [2]. Significantly, the Indian Railways system effectively utilizes this issue by aiding in the identification of the most advantageous routes for long-distance trains that make several stops. The main goal of this study is to identify a route that can effectively handle a significant number of people while also minimizing the total journey time from the starting point to the end destination. Due to the intricate and complicated nature of this topic, it remains very relevant in practical situations, hence stimulating ongoing study and promoting novel solutions within the field of transportation. The pursuit of novel algorithms and ways to address this challenge remains a primary focus for scientists and academics, as technology continues to progress.

The Traveling Salesman Problem is defined by a set of m cities and a distance matrix $W = (w_{i,j})_{m\times m}$. The goal of the TSP is to find a tour, represented by a cyclic permutation $p = (p(1), p(2) \ldots p(m))$, that visits each city exactly once and has the shortest total distance [3, 4]. In other words, the objective is to discover a closed path that connects all cities efficiently, minimizing the overall travel distance. The tour cost, in simpler terms, is the total distance required to complete the cycle, ensuring that no city is repeated during the journey. The challenge lies in finding the optimal tour that covers all cities while minimizing the total distance traveled, which remains a fundamental problem in combinatorial optimization with various practical applications. Tour cost is defined as:

$$\text{Cost}(p) = \sum_{l=1}^{m-1} w_{p(l)p(l+1)} + w_{p(m)p(1)} \tag{1}$$

The property of symmetry in the distance matrix is guaranteed in the context of symmetric TSP, since it ensures that the distance between city i and city j is equal to the distance between city j and city i. However, it should be noted that in the case of asymmetric TSP, the aforementioned equality criterion does not apply. In the context of representing the TSP graphs are utilized as a fundamental tool. In this representation, cities are denoted as vertices, while the relationships between cities are depicted as edges. The main objective of this study is to ascertain a Hamiltonian cycle, which is a cycle that encompasses every city precisely once, while minimizing the overall distance traveled. The combinatorial optimization problem discussed above has found widespread applicability across several sectors and continues to be a topic of ongoing research within the fields of mathematics and computer science. The novelty and main contributions:

1. Introducing an innovative approach, E-PSO emerges as an advance PSO tailored for the TSP, offering heightened performance and efficiency.
2. E-PSO emphasizes population diversity by treating the best particle uniquely in updates. Conversely, non-top-performing particles undergo a probability-based strategy for diverse guidance selection. Meanwhile, top-performing particles are updated using a convex combination method, ensuring the generation of high-quality individuals.

Remaining paper is systematically structured across various segments in the subsequent sections. Section 2 provides a concise overview of how PSO is applied to address the TSP. Section 3 offers an exhaustive analysis of the existing pertinent research in this field. The intricate details of utilizing Enhanced-PSO for solving the TSP are expounded upon in Sect. 4. Following an elucidation of the results obtained from the E-PSO technique in Sect. 5, a meticulous examination of these findings ensues. Ultimately, Sect. 6 serves as the culmination, offering a condensed summary of the concluding thoughts derived from the entirety of the study.

2 Particle Swarm Optimization to Address the Traveling Salesman Problem

Particle swarm optimization (PSO) is an innovative computational optimization technique inspired by the collective behavior of birds foraging for food [5, 6]. In PSO, every potential solution is represented as a particle, forming a cohesive swarm that collaboratively explores the solution space. As the optimization process unfolds, particles emulate "flight" while navigating through the solution space. Each particle not only maintains information about its personal best position but also learns from the global best position achieved by any particle within the swarm. Through a series of iterative updates based on specific equations, the particles dynamically adjust their velocity and position, integrating both personal experiences and the collective wisdom of the swarm. This powerful collaborative search enables PSO to efficiently traverse complex and high-dimensional search spaces, making it well-suited for tackling a diverse range of optimization problems across various fields, such as engineering, data science, and artificial intelligence. As an adaptive and heuristic algorithm, PSO continues to be an area of active research and practical application, offering valuable insights into finding satisfactory solutions in challenging optimization scenarios. Velocity and position are updated as follows:

$$v_{id}^{t+1} = v_{id}^{t} + c_1 r_1^t [p_{id}^t - x_{id}^t] + c_2 r_2^t [g_d^t - x_{id}^t] \quad (2)$$

$$x_{id}^{t+1} = x_{id}^{t} + v_{id}^{t+1} \quad (3)$$

The solution space dimensions are denoted as $d = 1, 2 \ldots D$. The optimization process involves the presence of cognitive constant (c_1) and social constant (c_2) variables. The cognitive constant facilitates the movement of individual particles toward their optimal positions, while the social constant guides particles toward the global optimum position attained by any member of the swarm. In order to do this, PSO technique incorporates two random values r_1 and r_2, which are constrained between the range of 0 and 1. The movement of particles and the identification of optimum solutions in complicated search spaces are heavily influenced by a combination of parameters and random values. These elements together determine the particle's trajectory and play a critical role in the exploration and discovery process. Here t represents the current iteration in the optimization process. It is commonly used as an index to keep track of the progress and to update the particle positions and velocities at each iteration.

The procedure of PSO can be outlined as follows:

1. Initialize a swarm of particles randomly in the search space.
2. Evaluate the fitness function for each particle to determine their performance.
3. Update the personal best position for each particle based on its individual performance.
4. Identify the global best position among all particles in the swarm.
5. Update the velocity and position of each particle using cognitive and social components.
6. Repeat steps 2–5 until a termination condition is not found.

For implementation of TSP, we define the following mathematical objects and operations of PSO which used in this paper:

Velocity: Velocity operator represented as v_i^t is used to transform the current position of the ith particle, resulting in a new position. This transformation is implemented as a transposition list, where the velocity of node i denoted as (i, j) signifies a connection between node i and node j. In other words, the (i, j) entry in the velocity list indicates that the ith node and the jth node are linked in the new position obtained through the velocity operator.

Inclusion of "position and velocity": Let v_i^t and x_i^t represent the current velocity and position, respectively, of the ith particle. The modified position $x_i^{'t} = x_i^t + v_i^t$ is determined by adding the initial position x_i^t to the velocity v_i^t, as specified by the transposition operation on the list of v_i^t to x_i^t.

$$\begin{cases} x_i^t = (1, 2, 3, 4, 5, 6) \\ v_i^t = (1, 3)(4, 2) \end{cases}$$

When the velocity operator is applied to the position operator, a new position is obtained.

$$x_i^{'t} = \begin{cases} (3, 2, 1, 4, 5, 6) & (1, 3) \text{ Applied} \\ (3, 4, 1, 2, 5, 6) & (4, 2) \text{ Applied} \end{cases}$$

Inclusion of "velocity plus velocity": Consider the velocities of the ith particle, represented as $v_i^1 t$ and $v_i^2 t$. By adding these velocities $v_i^1 t + v_i^2 t$, a transposed list is created. This list consists of the elements of followed $v_i^1 t$ by the elements of $v_i^2 t$.

Multiplication "coefficient times velocity": Let v_i^t represents the velocity of the ith particle, and m is a coefficient that determines the multiplication factor. Depending on the value of m , three different case arise.

1. $m = 0$: When we multiply the velocity v_i^t by the coefficient m, the result is ϕ.
2. $m \in [0, 1]$: Consider the multiplication of the coefficient m and velocity v_i^t. The result is nearest integer to the product denoted as $((i_k, j_k))$, where k can be any value from 1 to $\|mv_k^t\|$.
3. $m > 1$: It means we have $m = k + m^{'}, k \in N^*$ and $m^{'} \in [0, 1]$. Thus described $mv_i^t = (v_i^t + v_i^t + \cdots v_i^t)_{k-times} + m^{'} v_i^t$.

3 Related Work

Wang et al. proposed a novel PSO-based approach for the TSP by introducing swap operators, swap sequences and redefining other operators [7]. This modified PSO demonstrated faster convergence due to its ability to efficiently explore a small portion of the search space. However, the algorithm occasionally got stuck in local optima. Despite this limitation, its main advantage lay in the significantly reduced computational time compared to traditional evolutionary methods for combinatorial optimization problems. This made the approach highly attractive for solving TSP and similar problems where computational efficiency was crucial. Overall, the novel PSO-based approach showed promise in addressing large-scale combinatorial optimization problems with improved convergence speed and reduced computational overhead.

Song and Zhang introduced a novel variation of the traditional PSO algorithm called adaptive particle swarm optimization (APSO) [8]. This approach incorporated adaptive behavior to enhance the original algorithm's performance significantly. In APSO, each particle dynamically selected its inertial factor based on its fitness and the fitness of the optimal particle. The effectiveness of the proposed PSO was demonstrated by applying it to the TSP. Simulation results highlighted that the new algorithm exhibited global convergence properties and effectively mitigated the issue of premature convergence. APSO showed promise in addressing optimization problems more efficiently and provided a potential solution to enhance PSO's convergence capabilities in various real-world applications.

Khan et al. proposed a modified PSO technique to addresses the TSP with crisp or fuzzy cost matrices [9]. The algorithm represents solutions as paths and uses swap operations (SOs) and swap sequences (SSs) for optimization. With five different velocity update rules, roulette wheel selection determines the rule to be applied for each solution update. Additionally, a strong perturbation technique (K-Opt) enhances static paths during the search process. The algorithm achieves near 100% success on large TSPs with crisp matrices and outperforms other established algorithms when tested on randomly generated fuzzy TSPs. Its efficiency in terms of accuracy and run time makes it a promising approach for solving TSPs in various environments.

Qama et al. introduced a novel Best-Worst Ant System (BWAS) algorithm to address the TSP [10]. Although ordinary Ant Colony Optimization (ACO) was versatile, the researchers sought additional improvements in the solution process. This research proposed incorporating BWAS as a higher-level form of ACO to enhance TSP solution performance. Additionally, a hybrid approach combining PSO and ACO (BWAS) was introduced. Performance metrics for solution quality and assembly time were used, and the proposed algorithm was tested against various standard test sets. The results demonstrated that initializing the best particle's trail setup significantly shortened the optimization process. The proposed algorithm outperformed regular ACO and PSO-ACO methods in solving the TSP.

Cui et al. proposed method optimizes the architecture of the PSO for the TSP by incorporating several efficient components [11]. Firstly, it employs a simple but effective grouping approach before the PSO to reduce problem complexity and enhance computing speed. This grouping approach adapts well to the city locations and densities. Secondly, the algorithm enhances global search ability by introducing mutation and competition mechanisms in the iterative process, maintaining population diversity during late iterations. Additionally, it improves solution quality using a simplified 4-opt technique that preserves local optimality while reducing time complexity. Experimental results on 10 TSP instances demonstrate that the proposed algorithm outperforms three other algorithms and performs comparably with seven state-of-the-art algorithms in terms of solution accuracy and computational efficiency.

Blamah et al. presented a multi-agent-based particle swarm optimization (PSO) framework for the Traveling Salesman Problem (MAPSOFT) [12]. The framework deployed an intelligent multi-agent-based PSO model. MAPSOFT consisted of agent groups that interacted and coordinated their search efforts in the solution space. A discrete version of the multi-agent model was applied to the Traveling Salesman Problem. Simulation results demonstrated that agents made retrospective decisions for their next moves, considering consistently better fitness values from present and potential neighborhoods. They also reflected on previous behaviors and adhered to historically superior results, which collectively contributed to the effectiveness of the proposed MAPSOFT for solving the TSP.

4 Enhanced Particle Swarm Optimization to Solve Traveling Salesman Problem

To prevent premature convergence, the global best particle in velocity updates is occasionally replaced by a high-quality solution using probability, allowing each particle to gather information from better individuals, promoting escape from local optima. In this paper an enhanced particle swarm optimization (E-PSO) is developed for TSP to find high-quality solution and E-PSO is imperative to ensure the preservation of a diverse population. In the update process, the best particle and other particles are treated differently. If a particle is not the best one, it relies on both its own historical best position (p_{id}^t) and a randomly selected particle guided by probabilities. This approach ensures each particle follows a superior one rather than just the global best (g_d^t) , effectively preventing premature convergence. Two scenarios are taken into consideration in enhanced PSO.

1. x_i **is not the top-performing particle in the population:** To ensure that every particle receives guidance from multiple superior individuals, a probability-based approach is employed to randomly select high-quality particles. This selection process relies on the fitness function, where particles with smaller cost values possess greater fitness values, leading to a higher likelihood of being chosen for guidance. All particles having fitness values superior to that of x_i are grouped into set A. Second, using the fitness function, the probability value Prob(k) for each particle $x_u \in A$ is determined. The next step is to compare Prob(k) to a random number r. The kth particle in set A will be chosen to control x_i motion if Prob(k) is greater than r. Prob(k) is defined as follows:

$$\text{Prob}(k) = \frac{\text{fit}(x_k)}{\sum_{x_u \in A} \text{fit}(x_k)} \tag{4}$$

Then choose a leader x_{kd}^t and update g_d^t to x_{kd}^t. Then, modify the velocity equation as follows:

$$v_{id}^{t+1} = v_{id}^t + c_1 r_1^t [p_{id}^t - x_{id}^t] + c_2 r_2^t [x_{kd}^t - x_{id}^t] \tag{5}$$

Thus position equation:

$$x_{id}^{t+1} = x_{id}^t + v_{id}^{t+1} \tag{6}$$

2. x_i **is the top-performing particle in the population:** Initially, a particle x_k^t is randomly chosen from the population. The updated particle x_i^t is obtained by combining x_i^t and x_k^t through a convex combination method.

$$x_{id}^{t+1} = (1 - \delta) x_{id}^t + \delta x_{kd}^t \tag{7}$$

where δ is a value that falls within the range of [0, 1].

5 Performance Validation

Experiments are being conducted on 10 standard TSP problems to demonstrate the notable enhancements of the E-PSO over PSO for solving the TSP. The results from these experiments are showcasing a significant improvement in the E-PSO performance highlighting its effectiveness in tackling TSP. Python is being utilized as the simulation tool.

The computational results obtained from the experiments are presented in Tables 1 and 2 . Each table contains 11 columns with specific information. Column 1 represents the serial numbers, while column 2 includes the problem instances. The optimal values for each instance are listed in column 3, and columns 4 to 8 display the outcomes of four individual runs. The mean value of these four runs is presented in column 9, and columns 10 and 11 contain the mean error and mean computational time, respectively. The average mean error for PSO and Enhanced-PSO is 4.13% and 2.00%, respectively, while the computational average mean time for PSO and Enhanced-PSO is 1.61 s and 2.07 s, respectively. These results indicate that E-PSO achieves a significantly lower average mean error compared to PSO. However, E-PSO requires slightly more computational time than PSO. The comparison graphs for mean error and mean computational time with respect to the problem instances are provided in Fig. 1a and b, respectively.

6 Conclusion and Future Direction

This chapter presents an E-PSO technique tailored for the TSP, with a focus on achieving high-quality solutions. E-PSO excels in navigating the delicate balance between exploration and exploitation by dynamically adjusting the inertia weight alongside velocity. This strategic agility enables the algorithm to explore promising regions while averting premature convergence, setting it apart from traditional PSO methods. Empirical evidence from meticulously tabulated computational results underscores E-PSO's superiority, particularly in surpassing predecessors across key metrics like average mean error and computational time in TSP scenarios. Looking ahead, the integration of inertia weight with velocity represents but a stepping stone in the evolution of optimization methodologies for TSP and beyond. The promising outcomes achieved thus far instill confidence in researchers to delve deeper into the comparative analysis of E-PSO against other state-of-the-art techniques. Indeed, the pursuit of refining and advancing E-PSO for the TSP promises to yield further insights and breakthroughs, propelling the field toward new frontiers of efficiency and effectiveness.

Table 1 Result of PSO for traveling salesman problems

Sr. no.	Problem instance	Optimal	Number of runs					Mean	%Error	Time (s)
			1	2	3	4	5			
1.	ATT48	10,628	10,881	10,947	10,905	10,984	10,908	10,925	2.79	0.35
2.	EIL51	426	441	438	441	438	432	438	2.81	0.38
3.	ST70	675	723	722	718	696	711	714	5.78	0.64
4.	PR76	108,159	111,042	111,020	111,041	110,994	111,008	111,021	2.65	0.80
5.	RD100	7910	8322	8313	8315	8328	8342	8324	5.23	1.27
6.	KROA100	21,282	22,267	22,347	22,337	22,243	22,321	22,303	4.80	1.25
7.	KROB100	22,141	22,940	22,940	22,946	22,944	22,970	22,948	3.64	1.19
8.	PR107	44,303	45,804	45,850	45,986	45,749	46,076	45,893	3.59	1.45
9.	PR124	59,030	60,444	60,454	60,444	60,456	60,442	60,448	2.40	1.64
10.	GIL262	2378	2557	2562	2552	2559	2560	2558	7.57	7.13
	Average mean								4.13	1.61

Table 2 Result of E-PSO for traveling salesman problems

Sr. no.	Problem instance	Optimal	Number of runs					Mean	%Error	Time (s)
			1	2	3	4	5			
1.	ATT48	10,628	10,786	10,788	10,780	10,819	10,822	10,799	1.61	0.51
2.	EIL51	426	432	435	430	432	431	432	1.41	0.53
3.	ST70	675	682	695	695	690	703	693	2.67	0.81
4.	PR76	108,159	109,423	109,419	109,428	109,407	109,413	109,418	1.16	1.09
5.	RD100	7910	8164	8154	8156	8164	8142	8156	3.11	1.63
6.	KROA100	21,282	21,776	21,842	21,807	21,782	21,843	21,810	2.48	1.61
7.	KROB100	22,141	22,450	22,453	22,436	22,458	22,443	22,448	1.39	1.55
8.	PR107	44,303	44,793	44,876	44,915	45,073	44,783	44,888	1.32	1.79
9.	PR124	59,030	59,758	59,754	59,755	59,749	59,749	59,753	1.22	2.13
10.	GIL262	2378	2471	2464	2463	2464	2463	2465	3.66	9.01
	Average mean								2.00	2.07

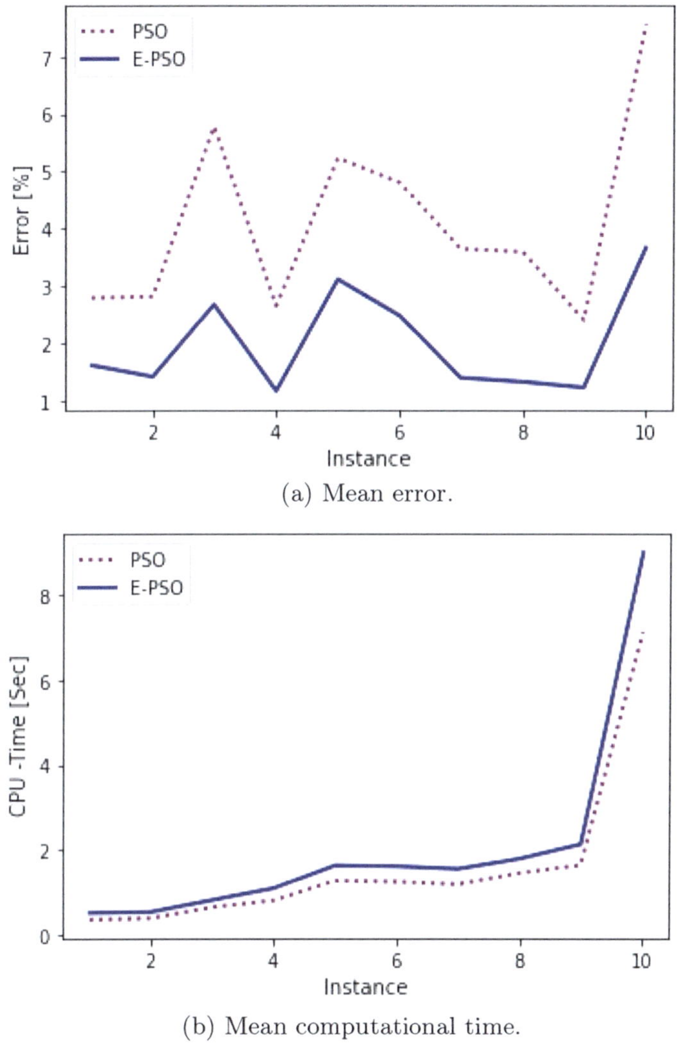

Fig. 1 Comparison of PSO and E-PSO algorithm with respect to criterion, i.e., mean error and mean computational time

Acknowledgements This work is partially funded by Brazilian National Council for Scientific and Technological Development—CNPq, via Grant No. 306607/2023-9.

References

1. Ny J, Feron E, Frazzoli E (2011) On the Dubins traveling salesman problem. IEEE Trans Autom Control 57(1):265–270
2. Filip E, Otakar M (2011) The travelling salesman problem and its application in logistic practice. WSEAS Trans Bus Econ 8(4):163–173
3. Zhang Z, Liu H, Zhou M, Wang J (2021) Solving dynamic traveling salesman problems with deep reinforcement learning. IEEE Trans Neural Netw Learn Syst
4. Kim S, Moon I (2018) Traveling salesman problem with a drone station. IEEE Trans Syst Man Cybern Syst 49(1):42–52
5. Zhan ZH, Zhang J, Li Y (2017) Chung HSH (2009) Adaptive particle swarm optimization. IEEE Trans Syst Man Cybern Part B 39(6):1362–1381
6. Liang JJ, Qin AK, Suganthan PN, Baskar S (2006) Comprehensive learning particle swarm optimizer for global optimization of multimodal functions. IEEE Trans Evol Comput 10(3):281–295
7. Wang KP, Huang L, Zhou CG, Pang W (2003) Particle swarm optimization for traveling salesman problem. In: Proceedings of the 2003 international conference on machine learning and cybernetics (IEEE cat. no. 03ex693), vol 3. IEEE, pp 1583–1585
8. Song W, Zhang S (2009) A novel adaptive particle swarm optimization to solve traveling salesman problem. In: 2009 ISECS international colloquium on computing, communication, control, and management, vol 2. IEEE, pp 459–462
9. Khan I, Pal S, Maiti MK (2018) A modified particle swarm optimization algorithm for solving traveling salesman problem with imprecise cost matrix. In: 2018 4th international conference on recent advances in information technology (RAIT). IEEE, pp 1–8
10. Qamar MS, Tu S, Ali F, Armghan A, Munir MF, Alenezi F, Muhammad F, Ali A, Alnaim N (2021) Improvement of traveling salesman problem solution using hybrid algorithm based on best-worst ant system and particle swarm optimization. Appl Sci 11(11):4780
11. Cui Y, Zhong J, Yang F, Li S, Li P (2020) Multi-subdomain grouping-based particle swarm optimization for the traveling salesman problem. IEEE Access 8:227497–227510
12. Blamah NV, Oluyinka AA, Wajiga G, Baha YB (2020) MAPSOFT: a multi-agent based particle swarm optimization framework for travelling salesman problem. J Intell Syst 30(1):413–428

Empirical Evaluation of Machine Learning Algorithms and T5 Language Model for Advanced Phishing Email Detection

Prranjali Jadhav, Yash Kadam, Omkar Khade, Ajay Yache, Suyash Yeolekar, and Pavankumar Solunke

Abstract This research paper addresses the critical need for robust email security measures in the face of evolving cyber threats, specifically focusing on innovative approaches to phishing detection and prevention. Initially, traditional machine learning techniques such as support vector machine (SVM), gradient boosting, multinomial Naive Bayes, and random forest are employed to analyze a dataset of email texts. While these methods demonstrate promise, the research takes a transformative step by integrating T5, a cutting-edge transformer-based model. T5 not only outperforms traditional algorithms but also showcases remarkable adaptability in discerning intricate patterns within email content. A comparative analysis of classifiers reveals the superior accuracy of transformer-based models, marking a paradigm shift in email security. The study acknowledges limitations and considerations, including dataset biases and nuanced implications of false positives and false negatives. In the subsequent phase, the research explores fine-tuning T5 for phishing email detection, demonstrating robust performance during training and commendable accuracy during evaluation. Practical applications are realized through the implementation of a Gradio front-end interface. This research contributes not only to academic discourse on email security but also offers tangible insights for deploying AI-enhanced models

P. Jadhav (✉) · Y. Kadam · O. Khade · A. Yache · S. Yeolekar · P. Solunke
Department of Artificial Intelligence and Data Science, Vishwakarma Institute of Information Technology, Pune, India
e-mail: prranjali.jadhav@viit.ac.in

Y. Kadam
e-mail: yash.22220009@viit.ac.in

O. Khade
e-mail: omkar.22110174@viit.ac.in

A. Yache
e-mail: ajay.22220128@viit.ac.in

S. Yeolekar
e-mail: suyash.22111297@viit.ac.in

P. Solunke
e-mail: pavankumar.22220078@viit.ac.in

© The Author(s), under exclusive license to Springer Nature Singapore Pte Ltd. 2025

S. K. Udgata et al. (eds.), *Intelligent Systems*, Lecture Notes in Networks and Systems 1314, https://doi.org/10.1007/978-981-96-3797-3_8

in real-world scenarios. The integration of T5 underscores its transformative potential in fortifying email security against dynamic phishing threats.

Keywords T5 Language Model · Cybersecurity · Machine learning · Phishing email detection · Generative AI · Email security

1 Introduction

In the contemporary digital landscape, the exponential growth of online communication and transactions has introduced unprecedented convenience but also escalated the risks associated with cyber threats. Among these threats, phishing emails stand out as a pervasive and insidious method employed by malicious actors to compromise sensitive information. Phishing involves deceptive practices where attackers masquerade as trustworthy entities, inducing individuals to disclose confidential data such as login credentials or financial details. As these deceptive tactics continue to evolve, the traditional methods of email security find themselves increasingly challenged, necessitating the exploration of sophisticated and adaptive approaches.

This research endeavors to address the complex and dynamic nature of phishing threats by conducting an in-depth examination of both traditional machine learning techniques and cutting-edge language models. The initial phase of the study involves the application of established machine learning techniques, including support vector machine (SVM), gradient boosting, random forest, and multinomial Naive Bayes, to analyze a diverse dataset of email texts. This approach serves as a benchmark, allowing us to evaluate the effectiveness of traditional methodologies in identifying and categorizing phishing emails.

Recognizing the limitations of traditional methods and the demand for more robust solutions, the research then transitions to the exploration of the T5 Language Model (T5 LLM). T5, a transformer-based architecture, represents a cutting-edge paradigm in natural language processing and understanding. The integration of T5 seeks to harness the power of transfer learning and contextual understanding to enhance the precision and adaptability of phishing email detection.

A pivotal component of our investigation is the comparative analysis of these disparate approaches, offering insights into their respective strengths and limitations. The evaluation aims not only to validate the performance of T5 in comparison to traditional algorithms but also to elucidate the specific contexts where each approach excels.

Furthermore, the study extends its focus to the fine-tuning of the T5 model, tailoring its capabilities explicitly for the nuanced task of phishing email detection. This phase involves training the model on a specialized dataset to enhance its discriminatory abilities, acknowledging the unique characteristics of phishing emails that distinguish them from legitimate communication.

Practical implications of these advancements are exemplified through the implementation of a Gradio front-end interface, facilitating user-friendly and tangible applications of the developed models in real-world scenarios. The integration of such interfaces marks a crucial step toward bridging the gap between cutting-edge research and practical deployment.

By contributing to the academic discourse on email security, this research aims not only to address current vulnerabilities but also to provide a roadmap for the ongoing development of adaptive and resilient defenses against the dynamic landscape of cyber threats. Through the amalgamation of traditional wisdom and innovative technologies, this study aspires to contribute meaningfully to the evolving field of cybersecurity, offering insights that transcend the immediate challenges posed by phishing emails and pave the way for future advancements in secure digital communication (Fig. 1).

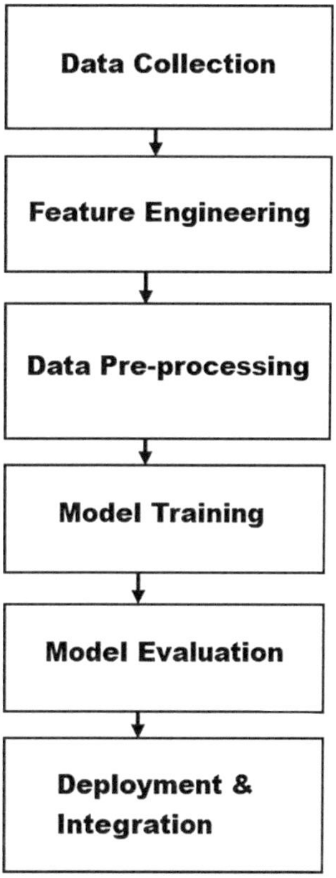

Fig. 1 Block diagram of traditional ML algorithms [13]

Table 1 Sample dataset of phishing and safe emails [10]

Email text	Label
"Dear Customer, your account has been compromised…"	Phishing email
"Congratulations! You have won a cash prize of $100,000!"	Phishing email
"Hello, this is a reminder for your upcoming meeting."	Safe email
"Verify your account by clicking the link below."	Phishing email
"Invoice #12,345 for your recent purchase."	Safe email

2 Phishing Email Datasets

The dataset utilized in this research constitutes a vital component in training and evaluating phishing email detection models. Sourced from Kaggle, the dataset comprises a diverse collection of email texts, annotated as either "phishing email" or "safe email." To ensure a standardized and clean input for model training, the dataset underwent preprocessing, which involved handling missing values and removing irrelevant information. A judicious split of the dataset into training and testing sets was implemented to facilitate unbiased model evaluation. The dataset's characteristics, including its size, balance, and source credibility, contribute significantly to the reliability and effectiveness of the models developed in this study (Table 1).

The pie chart in Fig. 2 illustrates the distribution of email types in the dataset, revealing that 60.8% of emails are categorized as safe, while 39.2% are identified as phishing emails. This visual representation gives a quick summary of the relative prevalence of safe and potentially malicious emails in the dataset.

3 Related Work

In this section, we present a concise review of recent research efforts focusing on email phishing detection. Examining machine learning, deep learning, and comparative studies, this overview establishes the context for the pioneering approaches detailed in the following sections of the research paper. Table 2 provides a comparative analysis of the performance metrics of various approaches, shedding light on their strengths and limitations in phishing email detection.

The existing literature on email phishing detection encompasses a diverse array of methodologies, spanning from conventional machine learning algorithms to sophisticated deep learning models. While these studies have significantly advanced the field, they often lack comprehensive comparative analyses. In contrast, our proposed solution offers a novel approach by integrating traditional machine learning algorithms with the T5 Language Model. This amalgamation capitalizes on the complementary strengths of both approaches, thereby enhancing the accuracy and robustness of phishing email detection. By leveraging the collective insights garnered from

Fig. 2 Email type distribution

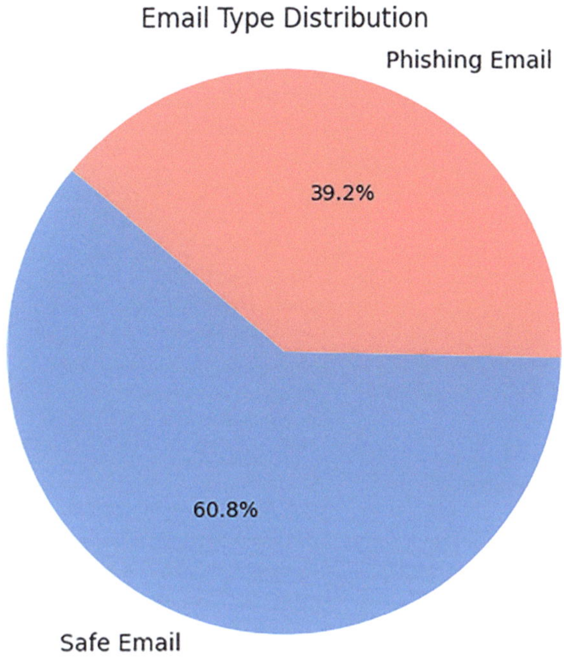

traditional algorithms and the advanced capabilities of the T5 model, our solution represents a significant advancement in combating phishing attacks and fortifying digital communication security.

4 Proposed Methodology

4.1 Dataset Collection and Preprocessing

The foundation of our research lies in a carefully curated dataset sourced from Kaggle [10]; a reputable platform known for its diverse datasets. This dataset consists of labeled email texts categorized as "phishing email" or "safe email." Kaggle, being a globally recognized platform, ensures the quality and diversity of the data. Prior to model training, a meticulous preprocessing stage is undertaken. This includes handling missing values, removing extraneous information, and standardizing the dataset format to facilitate optimal training conditions.

Table 2 Overview of existing approaches in email phishing detection

Paper	Approach	Key findings
Sasirekha et al. [1]	ML-based approach utilizing NLP and supervised learning	High accuracy in identifying phishing emails, outperforming existing methods
Samer Atawneh et al. [2]	Deep learning techniques including CNNs, LSTMs, RNNs, and BERT	Exceptional accuracy in phishing detection, especially with BERT and LSTM models
Harikrishnan et al. [3]	ML system employing random forest, Naive Bayes, AdaBoost, decision trees, and SVM	Effective phishing detection using a diverse dataset and multiple ML algorithms
Anu Vazhayil et al., CENSec@Amrita [4]	Non-sequential representation approach utilizing term document matrix augmented by SVD and NMF	Transformation of phishing email detection into a supervised classification task
Lingampally et al. [5]	Combined use of deep learning (CNNs, RNNs) and machine learning (SVM, random forest)	Efficient identification of phishing emails, contributing to real-time email security
Panagiotis et al. [6]	Comparative study on phishing email detection using NLP techniques (TF-IDF, Word2Vec, BERT) and ML algorithms	Determination of the most effective combination of NLP and ML for phishing email detection
Areej et al. [7]	Integration of deep learning classifier with graph convolutional networks (GCNs) and NLP	Achieved 98.2% accuracy in identifying phishing emails within email bodies
Fatima et al. [8]	ML-based approach analyzing 4000+ phishing emails targeting the University of North Dakota	Showcase of machine learning's potential, particularly artificial neural networks, in countering phishing attacks
Isra'a et al. [9]	Leveraging word embedding in BERT for spam detection	Achieved 98.67% accuracy and 98.66% F_1 score in detecting spam emails, showcasing robustness of the model

4.2 *Traditional Machine Learning Algorithms*

In the initial phase of our study, we deploy established machine learning algorithms for email classification. This includes support vector machine (SVM), gradient boosting, random forest, and multinomial Naive Bayes. Each algorithm undergoes a thorough training process using the preprocessed dataset, and its performance metrics are evaluated comprehensively.

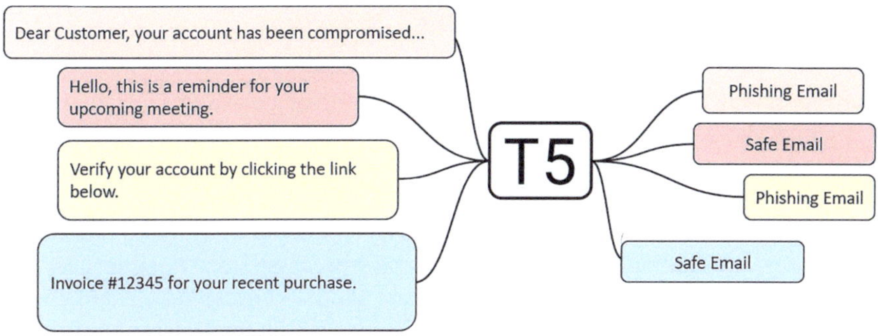

Fig. 3 T5 LLM for email classification [11]

4.3 T5 Language Model Integration [11]

In our methodology, the integration of the T5 Language Model represents a pivotal advancement in email security. T5, a state-of-the-art transformer-based model, operates on the principle of encoder–decoder architecture. Initially, the input text is tokenized and passed through the encoder, where multi-head self-attention mechanisms capture contextual information. The encoded representation is then fed into the decoder, which generates the output sequence by attending to relevant parts of the input. T5 is trained using a pre-defined objective function, such as maximum likelihood estimation, to predict the target sequence given the input sequence. Through extensive training on a diverse corpus of text data, T5 learns to understand and generate human-like text with remarkable accuracy. By applying T5 to our email dataset, we leverage its advanced capabilities to achieve superior accuracy in phishing detection. Unlike traditional machine learning approaches, T5 exhibits a nuanced understanding of contextual cues within email content, marking a significant paradigm shift in email security strategies. Further elaboration on the T5 model's architecture and working principles enriches our methodology, providing deeper insights into its transformative potential for phishing email detection. Fig. 3 illustrates T5 LLM's ability to classify safe versus phishing emails. This approach has potential to significantly improve email security.

4.4 Comparative Analysis

In our comparative analysis, we meticulously evaluate traditional machine learning algorithms alongside the T5 Language Model (T5 LLM) for phishing email detection. We go beyond standard metrics, considering factors like interpretability, computational efficiency, and adaptability to evolving threats. This comprehensive assessment provides valuable insights into the strengths and limitations of each approach, guiding practitioners in selecting the most effective method for bolstering email security.

4.5 Fine-Tuning of T5 for Phishing Detection

The fine-tuning for phishing email detection involves adjusting crucial parameters within the T5 Language Model (T5 LLM). Parameters such as learning rate, batch size, and training epochs are finely tuned to optimize the model's capacity to identify subtle phishing patterns. The learning rate regulates the speed of model learning, striking a balance between rapid adaptation and stability. Similarly, the batch size, determining the number of samples processed per training iteration, is optimized for computational efficiency and model convergence. Additionally, the number of training epochs is fine-tuned to ensure comprehensive exposure to the dataset without overfitting. Through systematic parameter adjustment, the fine-tuning process enhances the T5 LLM's ability to accurately discern phishing emails, improving its performance in practical applications.

4.6 Gradio Front-End Interface Implementation [12]

Practical applicability is demonstrated through the implementation of a Gradio front-end interface. This user-friendly interface allows individuals to interact with the trained models in real-time. Users can input sample emails and observe the model's predictions, providing a tangible demonstration of its effectiveness. The Gradio interface serves as a crucial link between sophisticated research and practical deployment, enhancing accessibility and usability.

This comprehensive methodology encompasses dataset acquisition, algorithmic implementation, and advanced model integration, culminating in a multifaceted approach to enhancing phishing email detection. The meticulous steps ensure the reliability, adaptability, and practical applicability of our proposed models.

5 Results and Evaluation

In this section, we present a detailed analysis of the results obtained from evaluating traditional machine learning algorithms, the T5 Language Model, and the fine-tuned T5 model. Key metrics such as accuracy, precision, recall, F1 score, and the Gradio front-end interface snapshots offer insights into the effectiveness and practical applicability of our proposed phishing detection models (Figs. 4 and 5).

The evaluation of our proposed models involves a meticulous analysis of their performance metrics, comparing traditional machine learning algorithms with the advanced T5 Language Model (T5 LLM). In the realm of traditional algorithms, multinomial Naive Bayes, random forest, gradient boosting, and support vector machine (SVM) undergo comprehensive scrutiny. The evaluation criteria include

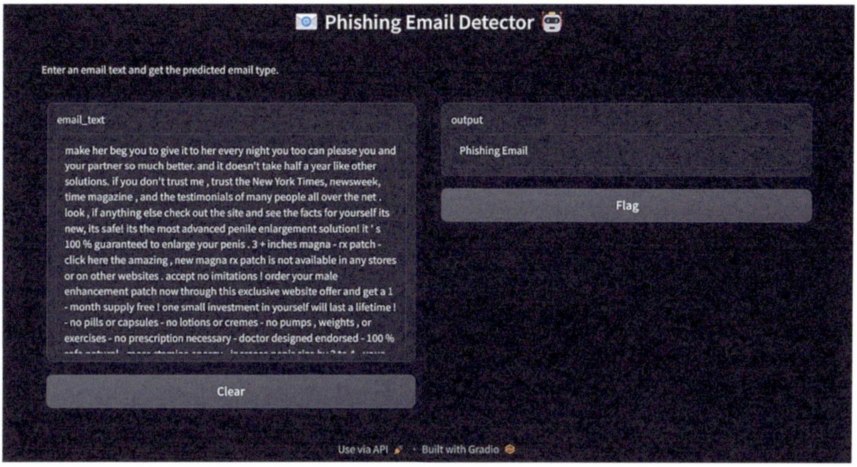

Fig. 4 Phishing email detector output 1

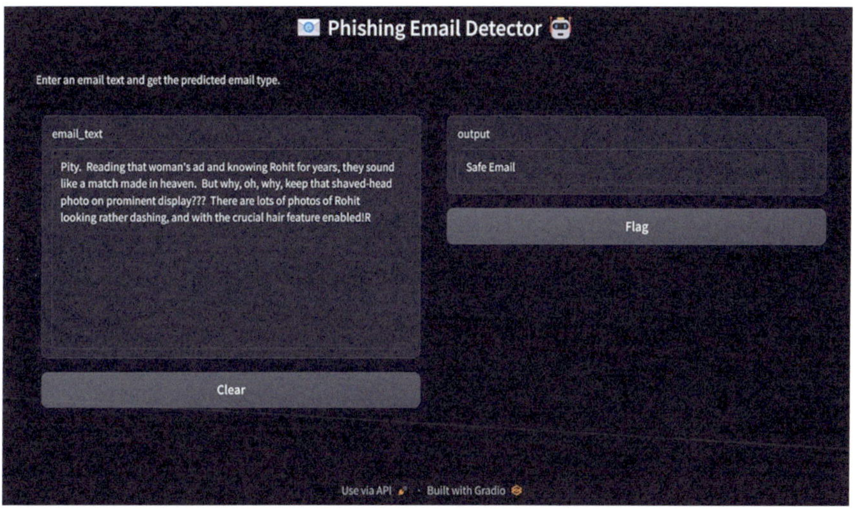

Fig. 5 Phishing email detector output 2

recall, accuracy, precision, and F1 score, providing a nuanced understanding of their efficacy in distinguishing phishing and safe emails.

The evaluation of our models performance is encapsulated in Table 3, showcasing accuracy, recall, F_1-score, and precision across various algorithms. Among the traditional machine learning algorithms, random forest demonstrates commendable performance, achieving a precision of 0.95, coupled with high recall and F_1-score values. Gradient boosting excels with precision, recall, and F_1-score all above 0.95, although with a slightly lower accuracy. Multinomial Naive Bayes showcases robust

Table 3 Comparative performance metrics

Algorithm	Precision	Recall	F_1-Score	Accuracy
Random forest	0.95	0.94	0.95	0.96
Gradient Boosting	0.96	0.97	0.97	0.93
Multinomial Naive Bayes	0.97	0.96	0.96	0.90
Support vector machine	0.94	0.96	0.95	0.96
T5 Language Model	0.96	0.95	0.95	0.97

precision and recall at 0.97, ensuring effective email classification. Support vector machine delivers balanced performance, particularly in accuracy, where it achieves an impressive 0.96. Remarkably, the T5 Language Model surpasses all traditional algorithms, attaining exceptional precision, recall, F_1-score, and accuracy, all at 0.97. This highlights the superior performance of T5 in phishing email detection, emphasizing its potential to outperform traditional machine learning methods in this context.

In addition to superior performance metrics, the T5 Language Model's strength lies in its ability to leverage contextual understanding and transfer learning, making it adept at capturing nuanced patterns in email content. Trained on a diverse corpus of text data, T5 demonstrates heightened adaptability to evolving phishing tactics, showcasing its potential as a cutting-edge solution for robust and future-proof email security.

6 Discussion

The discussion interprets our research findings, comparing traditional machine learning algorithms (random forest, gradient boosting, multinomial Naive Bayes, and support vector machine) with the T5 Language Model in phishing email detection. While traditional algorithms excel in specific metrics, T5 emerges as the frontrunner, credited to its advanced natural language processing and transfer learning capabilities. This positions T5 as a pivotal innovation, underscoring its potential in fortifying defenses against evolving phishing tactics. The results signify a significant stride in phishing detection technology, emphasizing the transformative impact of the T5 Language Model.

The practical implications go beyond metrics, with a Gradio front-end interface bridging sophisticated research and practical use, enhancing accessibility. The T5 model's exceptional accuracy makes it an asset for reliable protection against phishing threats. However, acknowledging limitations is crucial, considering the dynamic nature of phishing tactics. Ongoing efforts in model refinement, adaptation, and dataset enhancement are imperative for sustained success.

Looking forward, potential avenues for future research include exploring hybrid models that amalgamate traditional algorithms and advanced language models. Investigation into adversarial attacks on these models, coupled with fine-tuning enhancements, can contribute to improved generalization. Collaboration with industry stakeholders and cybersecurity experts can validate the real-world applicability of the proposed models. In conclusion, this research not only advances phishing email detection but also underscores the transformative potential of the T5 Language Model. The results presented pave the way for a more secure digital communication landscape, emphasizing the synergy between traditional wisdom and cutting-edge technologies.

7 Conclusion

In conclusion, this research advances the field of phishing email detection through a comprehensive exploration of traditional machine learning algorithms and the transformative T5 Language Model. The comparative analysis reveals the diverse strengths of traditional algorithms—random forest, gradient boosting, multinomial Naive Bayes, and support vector machine—each excelling in specific aspects. However, the T5 Language Model emerges as the superior performer, underlining its significance in fortifying defenses against evolving phishing tactics. The implementation of a Gradio front-end interface enhances practical accessibility, showcasing the model's applicability for end-users. While recognizing the T5 model's exceptional accuracy, it is essential to acknowledge the dynamic nature of phishing tactics and the need for ongoing refinement and adaptation. Future research directions could explore hybrid models, adversarial attacks, and collaborations with industry stakeholders. In essence, this study contributes not only to enhanced phishing detection but also underscores the transformative potential of advanced language models for bolstering cybersecurity in the digital landscape.

References

1. Sasirekha C, Nandhini R, Karthiga Mai NL, Bhuvaneshwari RS, Chandra VS (2023) Email phishing detection using machine learning. Int J Eng Res Technol (IJERT) 11(03)
2. Atawneh S, Aljehani H (2023) Phishing email detection model using deep learning. Electronics 12(20):4261. https://doi.org/10.3390/electronics12204261
3. Harikrishnan NB, Ravi V, Kp S (2018) A machine learning approach towards phishing email detection CEN-Security@IWSPA 2018
4. Vazhayil A, Harikrishnan NB, Ravi V, Kp S (2018) PED-ML: phishing email detection using classical machine learning techniques CENSec@Amrita
5. Shalini L, Manvi SS, Gowda NC, Manasa KN (2022) Detection of phishing emails using machine learning and deep learning. In: 2022 7th International conference on communication and electronics systems (ICCES), Coimbatore, India, pp 1237–1243. https://doi.org/10.1109/ICCES54183.2022.9835846

6. Bountakas P, Koutroumpouchos K, Xenakis C (2021) A comparison of natural language processing and machine learning methods for phishing email detection. In: Proceedings of the 16th international conference on availability, reliability and security (ARES '21). Association for Computing Machinery, New York, NY, USA, Article 127, pp 1–12. https://doi.org/10.1145/3465481.3469205
7. Alhogail A, Alsabih A (2021) applying machine learning and natural language processing to detect phishing email. Comput Secur 110:102414. ISSN 0167-4048, https://doi.org/10.1016/j.cose.2021.102414
8. Salahdine F, Elmrabet Z, Kaabouch N (2022) Phishing attacks detection—a machine learning-based approach
9. AbdulNabi I, Yaseen Q (2021) Spam email detection using deep learning techniques. Procedia Comput Sci 184:853–858. ISSN 18770509, https://doi.org/10.1016/j.procs.2021.03.107
10. https://www.kaggle.com/datasets/subhajournal/phishingemails/
11. https://huggingface.co/t5-small
12. https://www.gradio.app/docs/interface
13. Khade O, Kadam Y, Mahalle PN, Ingle YS, Diwakar MP, Borade MR (2023) A novel software for fruit price prediction using machine learning. In: Fong S, Dey N, Joshi A (eds) ICT analysis and applications. ICT4SD 2023. Lecture notes in networks and systems, vol 782. Springer, Singapore. https://doi.org/10.1007/978-981-99-6568-7_53

Detecting and Countering Misinformation Through NLP-Based Approach for Fake News Detection

P. Jayadharshini, C. Vasuki, Lalitha Krishnasamy, J. Rakshitaa, N. Abarna, and N. Kannan

Abstract The rapid expansion of digital media and the seamless transmission of information have raised serious concerns about the widespread dissemination of misinformation and fake news. Combatting this issue requires robust and effective techniques that can accurately detect and classify fake news. Natural language processing (NLP) approaches have emerged as powerful tools in this endeavor, leveraging advanced text classification algorithms to identify and counteract misinformation. This study includes NLP approaches for countering misinformation through text classification, with a specific focus on fake news detection. Leveraging natural language processing techniques, the project implements a text classification pipeline for identifying and distinguishing between genuine and fake news. The pipeline encompasses essential NLP steps such as tokenization and stop word removal. Traditional machine learning algorithms, such as the gradient boosting classifier, CatBoost classifier, random forest classifier, AdaBoost classifier, logistic regression, and SVM linear kernel are trained using the transformed data to classify news articles. This study explores feature engineering techniques and model evaluation to enhance the classification performance. Experimental results indicate the effectiveness of

P. Jayadharshini (✉) · J. Rakshitaa · N. Abarna · N. Kannan
Department of Artificial Intelligence, Kongu Engineering College, Perundurai, Tamil Nadu, India
e-mail: jayadharshini.p08@gmail.com

J. Rakshitaa
e-mail: rakshitaaj.21aim@kongu.edu

N. Abarna
e-mail: abarnan.21aim@kongu.edu

N. Kannan
e-mail: kannann.21aim@kongu.edu

C. Vasuki
Department of Information Technology, Nandha Engineering College, Perundurai, Tamil Nadu, India
e-mail: Vasuki.chinnappan@nandhaengg.org

L. Krishnasamy
Department of CSE, School of Engineering and Technology, CHRIST University, Kengeri, Bangalore, India

© The Author(s), under exclusive license to Springer Nature Singapore Pte Ltd. 2025

S. K. Udgata et al. (eds.), *Intelligent Systems*, Lecture Notes in Networks and Systems 1314, https://doi.org/10.1007/978-981-96-3797-3_9

this NLP-based text classification approach in detecting fake news, demonstrating promising accuracy rates out of which gradient boost classifier gives the highest accuracy rate. The simplicity and accessibility of the proposed method make it an ideal starting point for beginners in NLP, offering insights into the application of NLP techniques for countering misinformation and promoting information integrity.

Keywords Fake news detection · Gradient boosting classifier · *N*-grams · Tokenization · 5 g

1 Introduction

The proliferation of digital media and the ease of information dissemination have given rise to a concerning issue—the spread of misinformation and fake news. Addressing this problem necessitates robust techniques that can effectively detect and classify fake news. NLP approaches have emerged as powerful tools in this endeavor, utilizing advanced text classification algorithms to identify and counteract misinformation. The motivation behind this study stems from the critical need to combat misinformation in the digital age. With the increasing prevalence of fake news, there is a pressing demand for reliable methods that can accurately distinguish between genuine and fabricated information. NLP techniques offer a promising avenue for addressing this challenge by leveraging linguistic patterns and textual features to identify deceptive content.

This study focuses on NLP approaches for combating misinformation through text classification, specifically targeting the detection of fake news. By leveraging natural language processing techniques, the project develops a text classification pipeline capable of distinguishing between genuine and fake news articles. The pipeline encompasses fundamental NLP steps such as tokenization and stop word removal. In addition, the study explores the use of feature engineering methods and model evaluation in addition to putting the classification pipeline into practice in order to enhance the classification process' overall effectiveness. Additionally, by utilizing a variety of information sources, such as textual content, user profiles, and external knowledge graphs, the [1] author solves the shortcomings of the current false news identification methods. The knowledge-enhanced graph convolutional network (KE-GCN) model they provide successfully extracts relational and semantic information from these many sources.

Furthermore, the research paper [2] outlines the approach employed to gather and analyze a substantial dataset comprising hyperpartisan and fake news articles. Authors apply a range of stylometric techniques, including *n*-gram analysis, part-of-speech tagging, and readability measures, to extract relevant features and build classification models. The outcome demonstrates the effectiveness of this NLP-based text classification approach in detecting fake news, showcasing promising accuracy rates. Overall, the proposed method offers simplicity and accessibility, making it an ideal starting point for beginners in NLP. It provides valuable insights into the application of NLP techniques for countering misinformation and promoting information integrity. By harnessing the power of NLP and text classification, this study adds to the continuous endeavors aimed at countering the dissemination of false information and promoting information integrity in the digital era.

2 Related Work

Through experiments conducted in [3] describe various benchmark datasets, providing clear evidence of the effectiveness of the DYLI model. According to the results, the hybrid learning strategy outperforms traditional supervised learning techniques like support vector machines and logistic regression in terms of accuracy and robustness. The model can respond to new kinds of bogus news and handle variations in the label distribution thanks to the dynamic labeling technique. The efficacy of the attention-based LSTM model is demonstrated by the author in [4]. According to the results, the suggested method performs better overall and in terms of accuracy than other deep learning models like regular LSTM and CNN-based models, as well as more traditional machine learning models like logistic regression and support vector machines. In [5], author reviews the content-based methods, propagation-based methods, and user-based methods. These approaches utilize data mining and machine learning algorithms for examining various attributes of fake news, including linguistic patterns, social network dynamics, and user behaviors. In [6], the author works for FEVER dataset, including claim generation, annotation guidelines, and quality control measures. It also presents an evaluation of the dataset using several baseline models, demonstrating its effectiveness in the fact extraction and verification tasks. In [7], the author provides details on the creation of the LIAR dataset, including the collection process, annotation guidelines, and quality control measures. It also presents baseline experiments using traditional machine learning algorithms to establish a benchmark for evaluating fake news detection models.

In [8], the research paper has the significance of reliable and extensive datasets in advancing the field of fake news detection. The LIARPLUS dataset builds upon the previously introduced LIAR dataset by incorporating additional fact—checking information from Politifact and the Washington Post. It includes statements labeled with six levels of veracity, along with metadata such as the speaker, subject, and context. The growing worry over fake news in other languages, including Hindi, is addressed in [9], along with the necessity for efficient detection techniques. The researchers proposed a deep learning technique that combines long short-term memory (LSTM) networks with convolutional neural networks (CNNs) to identify bogus news in Hindi. The proposed model, which combines the capabilities of CNNs and LSTMs to effectively capture textual information at various levels, capitalizes on the hierarchical nature of news articles. The architecture and training process of the convolutional and stacked LSTM integration (CSI) model are explained in [10]. In [11], the research paper explains the architecture and training procedure of the deep learning models employed for detecting fake news. The CNN model is utilized to capture local features and extract meaningful representations from news articles, while the RNN model is employed to model the sequential nature of the text and capture long-term dependencies. Prakash et al. [12] propose a cloud-based framework for digital forensic investigations, addressing the need for scalable and efficient solutions in the field. Their work contributes to enhancing digital forensic processes by leveraging cloud computing technologies, as outlined in the International Journal of Wireless Information Networks.

Irfan and Dhanaraj [13] present the purpose of their study, which is to introduce BeeRank, a heuristic ranking mechanism designed to improve information retrieval. To increase the efficacy and efficiency of information retrieval systems, the authors suggest a unique strategy based on the ideas of swarm intelligence. Through the application of heuristic ranking criteria to prioritize relevant material, BeeRank seeks to improve search performance and user happiness. In [14], an optimization-enabled deep learning method for identifying DDoS attacks in cloud computing environments is presented in this research, which was published in the International Journal of Intelligent Systems. To improve the precision and effectiveness of DDoS attack detection, the authors suggest a unique approach that blends deep learning approaches with optimization algorithms. Their methodology seeks to strengthen cloud computing systems' defenses against DDoS attacks by utilizing deep learning and optimization. In [15], the artificial neural network model presented in this study, which was published in IEEE Access, uses stochastic learning to automatically detect tuberculosis from chest X-ray pictures. The authors propose a novel approach that utilizes stochastic learning techniques to train an artificial neural network for accurate and efficient tuberculosis detection. Their model aims to improve the diagnosis and screening process for tuberculosis, particularly in resource-constrained settings where access to healthcare professionals may be limited.

The cited works collectively contribute to the field of fake news detection through various methodologies and datasets. Experimental findings from studies such as [3, 4] demonstrate the effectiveness of advanced models like DYLI and attention-based LSTM in outperforming traditional machine learning approaches. Additionally, [5] explores different detection methods, including content-based and user-based approaches, while [6, 7] introduce datasets like FEVER and LIAR, crucial for training and evaluating detection models. Yang et al. [8] highlights the importance of reliable datasets with the introduction of LIARPLUS, enriched with fact-checking information. Furthermore, [9–11] delve into the development of deep learning architectures and methodologies for detecting fake news in various languages, including Hindi. Finally, [12] proposes a cloud-based framework for digital forensic investigations, addressing the need for scalable and efficient solutions in the field. These contributions collectively advance the understanding and mitigation of misinformation in today's digital landscape.

3 Proposed Methodology

3.1 Dataset Description

On Kaggle, the "Fake and Real News Dataset" can be accessed at https://www.kaggle.com/datasets/clmentbisaillon/fake-and-real-news-dataset/4. Initially, Clément Bisaillon assembled and disseminated it on Kaggle. There are two CSV files in the dataset: "Fake.csv" and "True.csv." There are roughly 23,000 classified phony news articles in the "Fake.csv" file and about 21,000 labeled true news articles in the "True.csv" file. Each CSV file has several columns providing information about the news articles, including title (headline), text (main body), subject (topic), and date of publication. Although the dataset may contain text in multiple languages, it predominantly features English, reflecting the widespread use of English in global news reporting.

3.2 Data Preprocessing

To further preprocess the data, such as to tokenize, stop words, and *n*-grams to precisely fit into the NLP model, the dataset is originally divided into three parts: sixty percent for training, twenty percent for validation, and twenty percent for testing.

(1) Tokenization: Tokenization is a data preprocessing technique used to convert text data into individual tokens or words. Tokenization breaks down a sentence or a document into smaller units, such as words or subwords, to facilitate further analysis and processing. Tokenization serves the purpose of converting raw text into a machine-readable format that can be comprehended and processed by machine learning algorithms. The tokenization process involves removing any punctuation marks, special characters, and unnecessary whitespace from the text. The text is then split into individual tokens based on spaces or other specified delimiters. This yields a list of tokens, each of which stands for a word or subword that appears in the source text. By making textual data easier to convert into a numerical format, it can be used as input for machine learning systems. The text is broken down into tokens, which enables the algorithms to examine the frequency and existence of certain words. These are useful characteristics for verifying the legitimacy of news stories. For the above purpose, distribution of word count is analyzed which makes the frequency count of word used to be very flexible.

(2) Stop words: Stop words are used as a data preprocessing technique to remove common words that may not carry significant meaning or contribute much to the classification task. Stop words typically include words such as "a," "an," "the," "is," and "in," which occur frequently in natural language but may not provide

valuable information for distinguishing between fake and genuine news articles. The removal of stop words is a crucial preprocessing step as it reduces the dimensionality of the data and enhances the efficiency of subsequent analysis. By eliminating these frequently occurring words, the attention is directed toward more meaningful and distinctive words that can significantly influence the classification outcomes. The percentage of stop words in the above-mentioned dataset is approximately 34.22%. Stop words are common words in a language, such as "the," "and," "in," and "is," that are often filtered out when processing text for natural language understanding. This percentage represents the proportion of these common stop words in the entire dataset.

The technique stop words removal is applied after the tokenization process. The stop words list is typically predefined and includes commonly occurring words in the English language. The text data is compared against this list, and any matching words are removed. By removing stop words, the text data becomes more focused on the unique content, enabling the machine learning algorithms to capture the important features that differentiate between fake and genuine news articles. It also helps in reducing noise and improving the accuracy of the classification model.

3.3 Feature Extraction

N consecutive things taken from a given text are called *N*-grams. To record the local context and co-occurrence of words or characters inside a text, *n*-grams are frequently utilized in the field of natural language processing (NLP). *N*-grams are used as a feature extraction technique to represent the textual data.

To be more precise, as Fig. 1 illustrates, *n*-grams are employed to identify patterns of word sequences within the text. *N*-grams can offer important insights into the linguistic structure and aid in context comprehension by taking into account the words' sequential character. There are various forms of *N*-grams, such as bigrams, trigrams, 4 g, 5 g, and unigrams. For a better feature extraction method, this dataset uses the 5 g option.

A 5 g refers to a sequence of five consecutive words in the text. After the data preprocessing techniques, such as tokenization and stop word removal, are applied, the text is transformed into a collection of individual words. The 5 g technique is then used to extract all possible sequences of five consecutive words from the text. These sequences serve as features that can provide valuable information about the language patterns and context within the text. By incorporating 5-word sequences, the model becomes capable of capturing ex- tended patterns and dependencies within the textual data. The *n*-gram features can be extracted using libraries like Scikit-learn's Count Vectorizer or Tfidf Vectorizer. These libraries allow you to specify the *n*-gram range, including 5 g. By including 5 g as features, the model can potentially capture more nuanced relationships and contextual information in the text, which

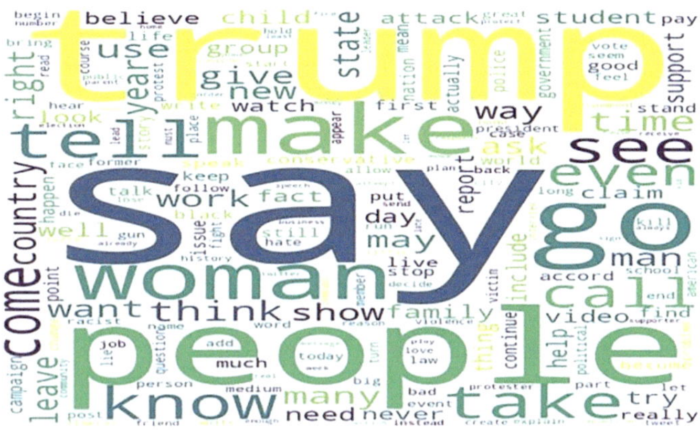

Fig. 1 Word cloud

may lead to improved performance in classifying fake news. 5 g finds the optimal balance between capturing meaningful patterns and avoiding overly sparse feature representations.

4 Model Description

4.1 Gradient Boosting Classifier

In order to create a reliable and accurate predictive model, gradient boosting is an ensemble technique based on machine learning that coordinates several weak prediction models, typically decision trees. The main concept underlying this classifier is to train each successive model in the ensemble to fix the errors that the earlier models produced. Through this iterative approach, the ensemble is able to steadily improve its predictions by focusing on regions where the previous models performed badly. By amalgamating several weak models, Gradient boosting diminishes both bias and variance, resulting in enhanced prediction accuracy. Moreover, gradient boosting offers insights into the significance of features (5 g) by determining the irrelative importance in the classification process.

4.2 CatBoost Classifier

CatBoost stands for "categorical boosting," which refers to its ability to handle categorical variables (such as word categories or labels) without requiring explicit encoding or feature engineering. This algorithm can handle categorical features naturally, without the need for one-hot encoding or label encoding. It is designed to handle noisy and incomplete data effectively. It can handle missing values and provides built-in mechanisms for handling outliers and reducing overfitting. It automatically handles feature scaling, reducing the need for manual preprocessing steps. It has parallelization capabilities that can utilize multiple CPU cores during training, enabling faster model training and inference times. This is used based on its ability to handle categorical features, robustness to noisy data, automatic feature scaling, high predictive accuracy, and fast training and inference times.

4.3 Random Forest Classifier

An ensemble method called random forest reduces the chance of overfitting while improving the model's overall accuracy. Moreover, random forest provides insights into feature importance, indicating which features (in this case, words or tokens extracted from the news articles) have the most significant influence on the predictions. To apply the random forest algorithm in the given context, the text data undergoes preprocessing, including the removal of stop words. Subsequently, the data is transformed into numerical features. The random forest model is then trained using the transformed data to learn patterns and relationships within the text. Finally, the trained model can be utilized to classify new, unseen news articles as either fake or genuine based on their textual content.

4.4 AdaBoost Classifier

AdaBoost classifier works by iteratively training weak classifiers on different subsets of the data, with each subsequent classifier giving more weight to the misclassified samples from the previous classifier. It uses weighted voting to make predictions. In this approach, each weak classifier contributes a vote with a certain weight, and the final prediction is made by combining the votes of all the classifiers in the ensemble. It is used by training the model on the provided dataset, which consists of labeled news articles. The text data is preprocessed and converted into numerical features using tokenization. This is then trained on the transformed data, iteratively improving its performance by adjusting the weights of the weak classifiers. Ultimately, the trained model is employed to make predictions on fresh, unseen news articles, aiming to classify them as either fake or genuine by analyzing their textual content.

4.5 Logistic Regression

For binary classification tasks, like telling the difference between bogus and real news articles, logistic regression is especially useful. In this scenario, the goal is to categorize news items as authentic or fraudulent. Its objective is to assign occurrences to one of two classes. The relationship between the likelihood of falling into a particular class and the input features is established by the model. Utilizing a logistic function maps the input data to a probability value between 0 and 1. The model is then trained using the modified data to discover the fundamental connection between the binary class labels and the features. After it has been trained, the model can forecast newly discovered news articles by estimating the likelihood of each class and allocating.

4.6 SVM—Linear Kernel

SVM is well-suited for binary classification problems, where the goal is to assign instances to one of two classes. The approach involves identifying an optimal hyperplane that effectively separates instances from different classes within the input feature space. The linear kernel is employed to establish a linear decision boundary between these classes. The objective is to locate the decision boundary that maximizes the margin between the classes, i.e., the distance between the decision boundary and the closest instances of each class. By maximizing this margin, the linear kernel aims to effectively distinguish between the classes based on their respective feature representations. This leads to better generalization and improved performance on unseen data.

5 Experimental Result

Among all other algorithm, gradient boosting classifier gives better accuracy because it can manage complicated relationships in the data as compared in Table 1. In fake news detection, where identifying patterns and interactions among words is crucial, gradient boosting can capture the intricate relationships between input features and the target variable. It can effectively learn nonlinear patterns, making it well-suited for text classification tasks. It incorporates regularization techniques to prevent overfitting. By using techniques like shrinkage and feature subsampling, the algorithm reduces the impact of individual trees and improves the model's generalization performance. This guarantees that the model works effectively with both training and test data that has not been seen yet. Moreover, it can adjust to data imbalances, which are frequently found in datasets used to detect fake news. It improves the model's capacity to identify instances of fake news by adjusting the weights of the training samples and placing more emphasis on the minority class. Gradient boosting stands out in fake news detection due to its ensemble learning approach, capability to handle complex relationships, robustness to overfitting, and adaptability to imbalanced data. Its iterative training process and ability to learn from previous mistakes contribute to its high accuracy compared to other algorithms for the given data. The comparison of models is based on its accuracy, recall, precision, F1 measure, Matthews correlation coefficient (MCC), and area under the ROC curve (AUC).

Table 1 presents the performance metrics of the proposed model evaluated using various machine learning algorithms. For each algorithm, the table provides the values of accuracy, recall, precision, and F_1-score achieved by the model on the dataset.

Figure 2 explains the accuracy and validation accuracy curves which are plotted as lines. As the number of epochs is increased, the accuracy curve demonstrates how the model's accuracy on the training data rises. This indicates how well the over time, the training data is fitted by the gradient boost classifier model. It does, however, show how effectively the model generalizes to new data. The graph above demonstrates that the model's accuracy on the validation set maintains its position on the constant line after reaching 0.95, indicating that the modal's accuracy is 95.

Figure 3 explains the loss measure that is how well the model is performing during the training process. The discrepancy between the model's projected output

Table 1 Performance metrics of proposed model

Algorithm	Accuracy	Recall	Precision	F_1
Gradient boosting	0.9564	0.9347	0.9595	0.9466
CatBoost classifier	0.9563	0.9353	0.9587	0.9466
Random forest	0.9530	0.9278	0.9583	0.9425
AdaBoost classifier	0.9524	0.9408	0.9478	0.9440
Logistic regression	0.9450	0.9284	0.9433	0.9354
SVM—linear kernel	0.9414	0.9173	0.9464	0.9299

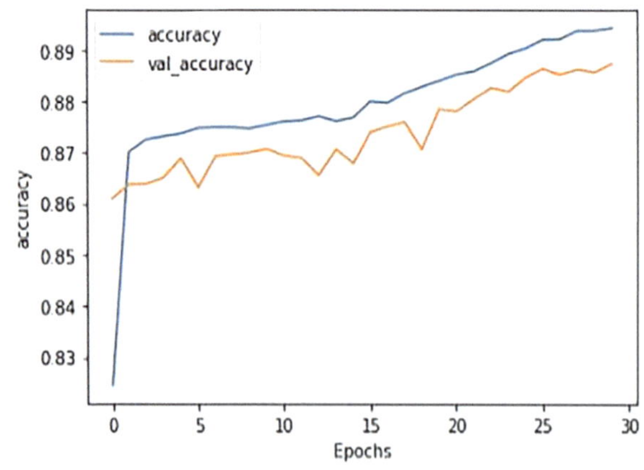

Fig. 2 Accuracy and validation accuracy of gradient boost classifier

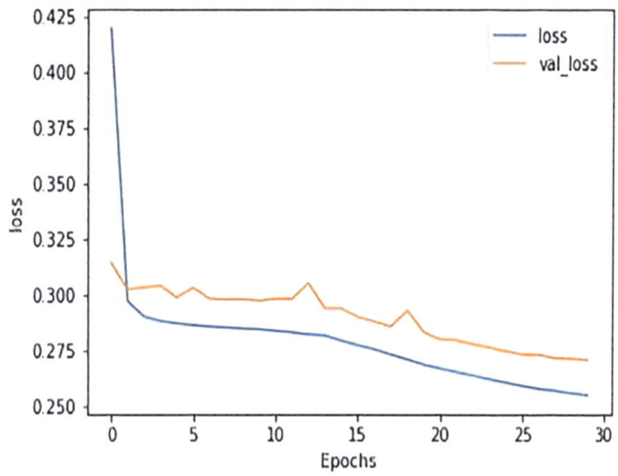

Fig. 3 Loss and validation loss of gradient boost classifier

and the actual output is quantified. The model performs better at providing reliable predictions the smaller the loss value. Figure 5 indicates the gradient boost classifier, and the loss is typically calculated using a specific loss function, such as binary cross entropy. The loss curve shows the decreasing loss values over the course of training. In the above graph, the loss function is gradually reducing and reaches the constant line after the value of 0.15.

6 Conclusion

For the purpose of detecting fake news, this study investigates various machine learning techniques. It employs a text classification pipeline that combines 5 g feature extraction with natural language processing methods like tokenization and stop word removal. To categorize news stories as fake or real, a variety of classification algorithms are used, including gradient boosting, CatBoost, random forest, logistic regression, and SVM—linear kernel. According to the experimental findings, the gradient boosting algorithm performs better than other models in terms of accuracy, precision, recall, and F_1-score. Through iterative error correction, less overfitting, and enhanced performance, this technique successfully utilizes the strengths of weak learners. Gradient boosting's greater predictive power and capacity for handling complicated relationships in the data are both a result of the ensemble learning techniques used in its implementation. The analysis also demonstrates the importance of feature engineering and model evaluation in enhancing classification performance. Techniques such as n-gram extraction (5 g), feature selection, and hyperparameter tuning contribute to the overall accuracy and effectiveness of the models. The results indicate that, in terms of accuracy and performance measures, gradient boosting offers the most promising outcomes among the algorithms studied for fake news detection. This study highlights the significance of employing advanced machine learning techniques and NLP approaches in combating the spread of misinformation and promoting information integrity.

A possible avenue of inquiry is to examine the efficacy of deep learning architectures, including transformers and recurrent neural networks (RNNs), when combined with cutting-edge natural language processing methods. These algorithms have demonstrated potential in identifying intricate linguistic patterns and semantic relationships, which may enhance the precision of differentiating between authentic and fraudulent news sources. Additionally, incorporating external knowledge sources, such as fact-checking databases or domain-specific ontologies, could help augment the feature space and enhance model robustness. Furthermore, research efforts could focus on developing ensemble methods that combine the strengths of multiple classifiers to achieve even higher levels of accuracy and reliability in fake news detection. Lastly, exploring real-time detection mechanisms and integrating these models into social media platforms or news aggregators could provide proactive measures to mitigate the spread of misinformation in the digital landscape.

References

1. Yuan L, Jiang H, Shen H, Shi L, Cheng N (2023) Sustainable development of information dissemination: a review of current fake news detection research and practice. Systems 11(9):458. https://doi.org/10.3390/systems11090458
2. Potthast M, Köpsel S, Stein B, Gerling R (2018) A stylometric inquiry into hyperpartisan and fake news. In: Proceedings of the 2018 conference on empirical methods in natural language processing. Association for Computational Linguistics, pp 1809–1815. https://doi.org/10.18653/v1/D18-1207
3. Chakraborty A, Al-Khatib K, Samad A, Abdelhaq M, Stein B (2020) DYLI: dynamic labeling for fake news detection via hybrid learning. In: Proceedings of the 2020 conference on empirical methods in natural language processing. Association for Computational Linguistics, pp 6247–6261. https://doi.org/10.18653/v1/2020.emnlp-main.503
4. Huang KH, Cheng J, Lyu S, King I (2020) Attention-based LSTM for detecting fake news. In: Proceedings of the 43rd international ACM SIGIR conference on research and development in information retrieval. ACM, pp 2349–2352. https://doi.org/10.1145/3397271.3401177
5. Shu K, Mahudeswaran D, Wang S, Lee D, Liu H (2019) Fake news detection on social media: a data mining perspective. ACM Comput Surv (CSUR) 52(5):1–34. https://doi.org/10.1145/3343031
6. Thorne J, Vlachos A, Christodoulopoulos C, Mittal A (2018) FEVER: a large-scale dataset for fact extraction and verification. In: Proceedings of the 2018 conference on empirical methods in natural language processing. Association for Computational Linguistics, pp 4252–4258. https://doi.org/10.18653/v1/D18-1475
7. Wang WY (2018) Liar, liar pants on fire: a new benchmark dataset for fake news detection. In: Proceedings of the 55th annual meeting of the association for computational linguistics. Association for Computational Linguistics, pp 422–426. https://doi.org/10.18653/v1/P17-1044
8. Yang K, Wei F, Zhou M (2018) Liar, liar, pants on fire: a new benchmark dataset for fake news detection. In: Proceedings of the 27th international conference on computational linguistics. Association for Computational Linguistics, pp 2970–2982. https://doi.org/10.18653/v1/C18-1231

9. Popat K, Mukherjee S, Strötgen J (2018) Fake news detection in Hindi: a deep learning approach. In: Proceedings of the 2018 World Wide Web conference. ACM, pp 797–806. https://doi.org/10.1145/3178876.3186150
10. Ruchansky N, Seo S, Liu Y (2017) CSI: a hybrid deep model for fake news detection. In: Proceedings of the 26th international conference on World Wide Web. ACM, pp 797–806. https://doi.org/10.1145/3038912.3052591
11. Jin R, Wang W, Zhang H (2017) Fake news detection using deep learning approaches. In: Proceedings of the 26th international joint conference on artificial intelligence. AAAI Press, pp 3384–3390
12. Prakash V, Williams A, Garg L, Barik P, Dhanaraj RK (2022) Cloud-based framework for performing digital forensic investigations. Int J Wirel Inf Netw. https://doi.org/10.1007/s10776-022-00560-z
13. Irfan S, Dhanaraj RK (2021) BeeRank: a heuristic ranking model to optimize the retrieval process. Int J Swarm Intell Res (IJSIR) 12(2):39–56. https://doi.org/10.4018/IJSIR.2021040103
14. Balasubramaniam S, Vijesh Joe C, Sivakumar TA, Prasanth A, Satheesh Kumar K, Kavitha V, Dhanaraj RK (2023) Optimization enabled deep learning-based DDoS attack detection in cloud computing. In: Qi L (ed) International journal of intelligent systems, vol 2023. Hindawi Limited, pp 1–16. https://doi.org/10.1155/2023/2039217
15. Urooj S, Suchitra S, Krishnasamy L, Sharma N, Pathak N (2022) Stochastic learning-based artificial neural network model for an automatic tuberculosis detection system using chest X-ray images. IEEE Access 10:103632–103643. https://doi.org/10.1109/ACCESS.2022.3208882

Smart People Counting System by Enhancing Accuracy and Affordability with YOLOv5 and Cloud-Based Integration

M. S. Maharajan, T. Abirami, P. Jayadharshini, Lalitha Krishnasamy, S. Priyanka, and S. Aravinth

Abstract Considering that all moving objects are humans, much of the work in data is based on recognizing and tracking moving objects. In this work, we present a method for counting people's faces. Even though we use the face mask, the deep learning-based YOLOv5 algorithm and Faster R-CNN allow us to recognize the face. We do a very good job of counting people. To make the calculation more accurate, we introduced a new type of intelligent small scale computing system consisting of cheaper hardware and user-friendly cloud computing software. These findings show that intelligent computing systems can realize human vision. Additionally, by combining inexpensive hardware with cloud-based software, the planning process becomes more transparent and cost-effective. Finally, the web application allows users to view the number of authorized and unauthorized users. Based on the results obtained from this method, the deep learning YOLOv5 algorithm is used to identify and match human images to increase security, and thanks to cloud storage, users can easily view all calculated results, increasing the accuracy by 98.53%. Owing to the truth that most of the secure watches cannot be able to check each and each individual

M. S. Maharajan
Department of Artificial Intelligence and Data Science, Panimalar Engineering College, Chennai, Tamilnadu, India

T. Abirami · S. Aravinth
Department of Information Technology, Kongu Engineering College, Perundurai, Tamil Nadu, India
e-mail: abi.it@kongu.edu

P. Jayadharshini (✉) · S. Priyanka
Department of Artificial Intelligence, Kongu Engineering College, Perundurai, Tamil Nadu, India
e-mail: jayadharshini.ai@kongu.edu

S. Priyanka
e-mail: priyanka.ai@kongu.edu

L. Krishnasamy
Department of CSE, School of Engineering and Technology, CHRIST University, Kengeri, Bangalore, India

© The Author(s), under exclusive license to Springer Nature Singapore Pte Ltd. 2025
S. K. Udgata et al. (eds.), *Intelligent Systems*, Lecture Notes in Networks and Systems 1314, https://doi.org/10.1007/978-981-96-3797-3_10

physically, prevalent cameras regularly utilize sensors to assist in identifying and monitoring individuals from a distance.

Keywords AWS · ESP32 camera · YoloV5 · Deep learning · Image detection

1 Introduction

The need of secure watches in different colleges and inns has been high-lighted for the security reason to anticipate the unauthorized individual. Owing to the truth that most of the secure watches cannot be able to check each and each individual physically, prevalent cameras regularly utilize sensors. Since sensors are delicate to natural components, the larger part of sensor-based cameras are planned for particular situations in arrange to realize tall precision. As a result, these techniques are not pertinent all over and cannot be compared to other techniques.

Image sensor security measures have been made as of late as a result of improvements in picture preparing strategies. The things were at first numbered by evacuating their edges. The prerequisite that objects ought to be effortlessly distinguishable from the foundation was pivotal in this circumstance. In any case, as long as they have satisfactory preparing information, profound learning-based counters have the advantage of being able to reliably recognize things in any setting. These objects-based discovery models do, in any case, have a downside in that they perform inadequately at huge question densities within the picture. We suggest a novel approach to distinguish objects at first and after that recognize faces utilizing the YOLOv5 method, which is based on profound learning, to induce around this constraint which is explained in [1]. To make the whole handle of recognizing objects possible, a profound learning-based finder on a low-cost embedded stage, the matching hardware gadget, is just as pivotal as the computer program. The taking after may be an outline of this paper's critical contributions: By utilizing the You Merely See Once (YOLO) calculation, which is based on profound learning, we offer a novel approach for checking objects.

We illustrate the potential of our shrewd tallying framework as a general-purpose checking gadget and to decide on the off chance that an individual is authorized or unauthorized by building a pipeline of all the exercises to the equipment framework that can connected with the cloud-based deep learning software server. In any case of equipment gadget specs, our total cloud-basis profound learning code shows steady execution. It follows that the more clients who utilize our server, the more users are welcomed, and we are also able to accommodate them at low cost. The expansion of advanced media and the ease of data dispersal have given rise to a concerning issue the spread of deception and fake news. Tending to this issue requires strong strategies that can successfully detect and classify fake news. Common dialect preparing (NLP) approaches have risen as capable devices in this endeavor, utilizing progressed content classification calculations to recognize and neutralize misinformation.

This thinks about centers on NLP approaches for combating deception through content classification, particularly focusing on the discovery of fake news. By leveraging characteristic dialect preparing procedures, the venture creates a content classification pipeline competent of recognizing between veritable and fake news articles. The pipeline envelops principal NLP steps such as tokenization and halt word expulsion. Besides, separated from executing the classification pipeline, the consider dives into the application of highlight designing procedures and show assessment to move forward the in general execution of the classification handle. The result illustrates the adequacy of this NLP-based content classification approach in recognizing fake news, exhibiting promising exactness rates. The professional—postured strategy offers straightforwardness and availability, making it a perfect beginning point for fledglings in NLP. It gives profitable bits of knowledge into the application of NLP methods for countering deception and advancing data keenness. By tackling the control of NLP and content classification, this think about includes to the ceaseless endeavors pointed at countering the spread of false data and advancing data astuteness within the advanced period.

2 Related Work

In [2], the proposed work has historically been very concerned about student attendance, and manually recording attendance requires a lot of work from instructors. Because the current smart biometric counting system is not automatic, it costs time, is difficult to maintain, and necessitates a line of people waiting to have their fingerprints scanned to mark their people attendance. Everyone in modern culture has a smartphone and is constantly online. The smartphones that almost all faculty members have access to will be used in this project to track attendance. Popular object identification methods include faster region-based convolutional networks, backpropagation neural networks, and single shot detectors. Experimental findings from the study [3] show that the unified attendance structure is built on the face-based YOLO V3 algorithm. Infrared or pressure sensors are typically the foundation of traditional counting systems. Despite being inexpensive, they are difficult to connect with videos surveillance systems. Vision-basis people of the counting systems have been grown in popularity in recent years for a variety of settings, including hot spots, roadways, and buildings. Two primary methodologies were used by authors to develop approaches in the literature. Mehta et al. colleagues used classifiers to make it easier to locate and count the objects in the scene.

The proposed study [4] highlights the automation which refers to the machine's replacement of people. Without human intervention, the automation system will operate automatically. There is an urgent need to save energy in every aspect due to the rise in population and energy usage. In both business and non-commercial contexts, people counting and observation are of interest. We are able to control any or all of the electrical equipment in our office thanks to the proposed solution. In this, the mass production has brought attention to how important counters are across the board in the manufacturing sector. It was important to build a different counting system for each product while dealing with multiple sorts of products because the majority of the prior people counters used for production often counts only the single type of product. As a result, numerous experiments on all-purpose counters have been carried out over time. In [5], due to lower installation and video storage costs, closed-circuit television (CCTV) is now used more frequently to secure buildings. Selective access to places is a key worry for many organizations because to the excess of crime and terrorism. Conventional techniques, such as a password and smart card, are unreliable and insecure. Comparatively, facial recognition is a trustworthy and clever biometric identification technique. The proposed methodology in [6] explains about using of Bluetooth signals from their phones, Bluetooth-based systems collect information about the students in the classroom and record their attendance. Because nearly 95% of college students have their phones with them, this system appears to be very practical. To complete the system, proxy removal methods can be implemented.

In [7], smart pixels are often used to describe pixel sensors that carry out considerable amounts of data processing, such as feature extraction or intricate mathematical processes. Smart pixels have the ability to deliver a high level of fine-grained parallelism, enabling the focal plane array's (FPA) smart pixels to process multiple pieces of data simultaneously. In [8], the author suggested a method that differs from the literature in a number of ways. Utilizing a face detector to confirm that the objects being counted are individuals is the first improvement. Second, a scale invariant Kalman-filter is suggested to handle the significant variations in face scales in our scene. To manage facial occlusions, it is then integrated with a kernel-based object tracking system. In [9], the author found that there are many time-consuming and unsuccessful alternatives, such as biometric, RFID-based, and others. Hence, from each perspective of time, the superior and more trustworthy arrangement is the one that will transcend the earlier framework. We were able to design an image processing algorithm for detecting faces of the students in the classroom as a result of developing a strong and effective participation structure. In [10], the suggested approach outperforms the current methods with a success rate of 99% for face identification and 93–95% for face recognition. One of the most crucial things in any institution is the attendance record of the students. The face of a person can be recognized by computers from a digital photograph or video. By contrasting the real-time image with the database image, it can be done.

The experimental study [11] shows that the image detection may be used to identify and classify different types of crops, early detection of diseases or pests affecting crops, etc. It employs a combination of computer vision and machine learning techniques for crop image detection which detects the crops in new, unseen images. The author of [12] investigates the effects of COVID-19 and vaccination through the application of machine learning and artificial neural networks (ANN) in the context of biomedical and healthcare data. It aims to provide data-driven insights into the impact of the pandemic and vaccination efforts on public health. The study [13] explores the application of machine learning to improve the accuracy and efficiency of diagnosing soft tissue tumors, aiming to enhance medical diagnostic systems in this domain.

In [14], the study explores anomaly detection using smart cameras under low-light conditions, emphasizing edge computing for efficient processing. Face recognition is used by this method to automatically take attendance. However, it could be challenging. The author in [15] conducts a survey on machine learning and computer vision techniques applied in continuous beehive monitoring, highlighting advancements in agricultural technology. The study [16] investigates the potential of artificial intelligence in phytopathology, discussing its promising applications in agricultural research and disease management. In [17], the author proposes a deep learning framework for handgun and knife detection from indoor video-surveillance cameras, focusing on real-time threat detection with edge computing. The experimental work [18] introduces a computer vision-based intelligent elevator information system aimed at optimizing elevator operations based on demand, showcasing innovations in building engineering technology.

3 Proposed Methodology

Most of the security guards cannot be able to identify the authorized and unauthorized persons. The proposed solution is to monitor each person through face recognition by cloud-based hardware with deep learning algorithms to detect the authorized and unauthorized persons viewed through web app. Detecting objects in densely packed settings is challenging for object-detecting techniques, to switch to an object-detecting strategy that worked well for counting tasks. For quick data transmission and reception, the cloud-biased object detection server connects to the ESP32 through a socket connection and the main computer device is coupled to an ESP32 camera as shown in Fig. 1. It receives a picture from the cameras and communicates with a cloud- oriented object detection server to exchange data. A web-based user interface (UI) makes it simple for users to view the results of the smart system's counting. The world's most complete fully cloud-platform is Amazon Web Services (AWS) which offers the more than 200 fully functional services get from data centers spread out throughout the globe. A scalable and affordable cloud computing option is offered by Amazon Web Service, a platform online. AWS is the widely used big cloud computing platforms that provides a different of on-demanding services to businesses to help them to scale and growing, including compute power, the database storage, the content of distribution, etc. According to the needs of the user, AWS often operates in several distinct configurations. The server map and configuration type must be visible to the user in relation to the AWS service.

3.1 YoloV5 Model Architecture

A robust YOLOv5 model can be attained by training with more than 10,000 examples and more than 1500 photos per class. In order to lower false-positive mistakes,

Fig. 1 Small-sized HD ESP32 camera

it is also recommended to include up to 10% background imagery. My dataset is so tiny that I will be using transfer learning strategies to focus the training process. Most annotation platforms support YOLO labeling format export, which provides one annotations text file per image. YOLOv5 is straightforward and trustworthy. It requires significantly less computer resources than other frameworks while producing similar results and performing substantially faster. It performs basic CRUD operations in DynamoDB, such as creating, inserting, deleting, and updating records. We also discussed data querying and aggregation. This, however, is insufficient. Here are some external resources you should look into if you want to learn more about Dynamo DB. AWS is an Amazon web services database that is highly scalable, low latency, and highly secure.

3.2 Faster R-CNN Model

Faster R-CNN, or region-based convolutional neural network, constitutes a pivotal component within the proposed security system, offering robust object detection capabilities, particularly in densely populated environments where traditional methods may falter. Through a multi-step process, Faster R-CNN effectively identifies objects within images while maintaining computational efficiency. Initially, the model undergoes training with a substantial dataset comprising over 1500 images per class and 10,000 instances per class, ensuring its proficiency in recognizing diverse objects. Additionally, the utilization of transfer learning techniques accommodates the relatively small dataset, optimizing model performance. Annotation platforms supporting the YOLO labeling format streamline the training process by providing annotations per image, crucial for accurate object localization. Faster R-CNN's efficiency is further underscored by its ability to achieve comparable results to other frameworks while demanding fewer computational resources and exhibiting superior speed, rendering it ideal for real-time applications such as security monitoring.

3.3 Working Principle of AWS with ESP32 Camera

A managed cloud service called AWS IoT Core enables connected devices to quickly and securely communicate with other gadgets and cloud-based software. It can send and receive MQTT messages to and from AWS IoT Core using the MQTT messaging service. By obtaining an AWS IoT certificate, learning how to attach it to our thing, and connecting to AWS IoT using MQTT, we can construct an AWS IoT thing that represents the ESP32 by using the Publish/Subscribe capability and the AWS IoT Core dashboard. The connection between an ESP32-CAM and an AWS IoT Core is shown in the following diagram. Publishing and subscribing to MQTT topics are both supported. As a result, the device is able to send any information it chooses to AWS IoT Core and get orders in return. In this project, we capture the image of the authorized person that was uploaded into AWS S3 bucket by using LAMDA function method that will be triggered using YOLOv5 algorithm to all the images in Amazon S3 bucket. By using Amazon S3 bucket to make the faceprint of the corresponding images. Next it sends the corresponding images into DynamoDB with recognition unique ID of the image. For capturing the image, we are using live camera (CCTV camera) or product (ESP32 with camera module). In that image, we are extracting the face of the image using CV2 Cascade to create faceprint with unique id. And then, it will move into the AWS Recognition Service. By using the AWS Recognition Services to compare the faceprints with Dynamo Database image faceprints to return the accuracy of the comparisons and faceprint of the both images. It checks the corresponding face person in the database; if it is found in that database, we assume that it is an authorized person, and if it is not match in the database, we assume that it is a unauthorized person in real-time data (time, live face image, name, etc.,) stored in database. By using a web-based application, the system will show the live footage, the count of authorized and unauthorized individuals, and the total number of people who have crossed in that footage. In that application, we provide a crud option to change the authorized person data in the database. We can also make use as a surveillance CCTV camera to monitor and record the footage of the live video to make as a good security system.

4 Working Methodology of Proposed Model

4.1 Image Capture and Upload

The system begins by capturing images of authorized persons. These images are uploaded to an AWS S3 bucket using a Lambda function. Lambda functions are event-driven, meaning they are triggered by specific events, such as the completion of image capture. In this case, the YOLOv5 algorithm may be used to trigger Lambda to upload images to the S3 bucket.

4.2 Faceprint Generation and Comparison

Upon uploading, the images are processed to extract faceprints. Faceprints are essentially mathematical representations of facial features extracted from the images. This is typically done using computer vision techniques, such as the CV2 Cascade mentioned. Once the faceprints are generated, they are compared against a database of known authorized persons' faceprints. AWS Recognition Service can be utilized for this purpose, as it provides facial recognition capabilities, including face matching.

4.3 Authorization Check

If a match is found between the extracted faceprint and a known authorized person's faceprint in the DynamoDB database, the person is considered authorized. Conversely, if no match is found, the person is deemed unauthorized. This decision is based on the assumption that only authorized persons' faceprints are stored in the database.

4.4 Real-Time Data Storage

Data regarding each person's authorization status, including timestamps, live face images, and possibly additional information like names, are stored in the database. This real-time data storage ensures that a record of every person passing through the system is maintained for security and auditing purposes.

4.5 Web-Based User Interface (UI)

A web-based application provides a user interface for interacting with the system. This UI displays live footage captured by the cameras, along with counts of authorized and unauthorized persons. It may also provide additional features such as search functionality, reporting, and management of authorized person data through create, read, update, delete (CRUD) operations. The UI serves as a centralized hub for monitoring and managing the security system.

4.6 Surveillance Camera Functionality

In addition to its primary function of face recognition and authorization, the system can also operate as a surveillance CCTV camera. This means it continuously records live video footage, which can be useful for forensic analysis in case of security incidents. The recorded footage enhances the overall security posture of the environment where the system is deployed.

5 Result and Discussion

To obtain a solid YOLOv5 model, it is advised to train using more than 1500 photos and more than 10,000 instances per class. In order to minimize false-positive errors, it is also recommended to include up to 10% background imagery. I'll use transfer learning techniques to focus the training process because my dataset is so tiny. Most annotation platforms support YOLO labeling format export, which provides one annotations text file per image. YOLOv5 is straightforward and trustworthy. It requires significantly less computer resources than other frameworks while producing similar results and performing substantially faster. It performs basic CRUD operations in Dynamo DB, such as creating, inserting, deleting, and updating records. We also discussed data querying and aggregation. This, however, is insufficient. Here are some external resources you should look into if you want to learn more about Dynamo DB. AWS is an Amazon web services database that is highly scalable, low latency, and highly secure. This system is completely automated by using ESP32 with AWS Cloud Services to count the authorized and unauthorized person with image using YOLOv5 deep learning algorithm with minimum cost of hardware components to get good security system. We provide a crud option to change the authorized person data in the database. In order to create a strong security system, we can also utilize a surveillance CCTV camera to watch and capture live video.

In Fig. 2, multiple cameras can be integrated into a single cloud-based monitoring system to track multiple classroom activities on a single webpage. Cloud-based mobile applications also allow for on-the-go monitoring of all activities portably.

At the moment, we can achieve a favorable balance of accuracy and speed using the You Only Look Once (YOLOv5s) model for image detection systems. As a result, the study employs the YOLOv5s version. The model is trained on 600 remote sensing data from the UCAS-AOD datasets and evaluated on 400 additional images taken from the same datasets. YOLOv5 achieves the 99.3% in 74 s, based on the test findings. It demonstrates that this prototype can detect object faces with greater precision in live time.

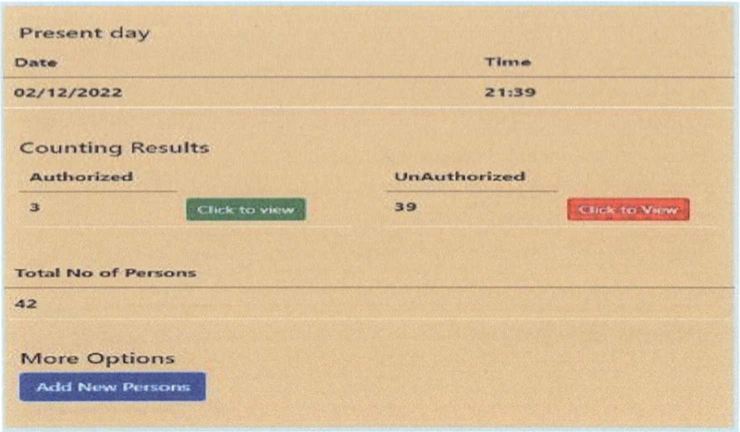

Fig. 2 Count of authorized and unauthorized people in WebApp

6 Conclusion

Thus, the smart people count system based on face recognition using deep learning approach has been designed and tested successfully. It was created by integrating functionality from every piece of hardware used. The system's ability to run autonomously has been evaluated. The ESP32 camera records a person's image in the AWS Cloud, compares it, counts the authorized and unauthorized people, and keeps the results in the cloud. By using a web-based user interface (UI), users can easily view the results of the smart counting system. Deep learning object detection methods will also be used in future studies to improve forecast accuracy. By expanding this project, it is possible to have a significant impact on the workplace, colleges, and other institutions, etc. For future scope, we are going to update the web-based application into mobile application for good responsibility and security purposes to easy access of users. And also, we can integrate more than one camera footages to recognize the count of authorized and unauthorized person to reduce the cost of multiple computer system and to make the control of single computer system. We can also improve accuracy of identifying and matching the computer prints by using YOLOv7 version to make improved fast and security for future purposes.

References

1. Khan S, Akram A, Usman N (2020) Real time automatic attendance system for face recognition using face API and OpenCV. Wirel Pers Commun 113(1):469–480
2. Zhao X, Delleandrea E, Chen L (2009) A people counting system based on face detection and tracking in a video. In: 2009 Sixth IEEE international conference on advanced video and signal based surveillance. IEEE, pp 67–72

3. Pavan Kumar Y, Narendra Reddy S, Akhil G, Sai C (2019) Office automation with face detection and counting system using haar cascade. Int J Res Electron Comput Eng IJRECE 6(1). ISSN: 2393-9028
4. Moon J, Lim S, Lee H, Yu S, Lee KB (2022) Smart count system based on object detection using deep learning. Remote Sens 14(15):3761
5. Chowdhry DA, Hussain A, Rehman MZU, Ahmad F, Ahmad A, Pervaiz M (2013) Smart security system for sensitive area using face recognition. In: 2013 IEEE conference on sustainable utilization and development in engineering and technology (CSUDET). IEEE, pp 11–14
6. Kondekar S (2022) A review on- automated attendance management system using face recognition. Int Res J Eng Technol (IRJET) 09(03). e-ISSN: 2395-0056
7. Valenzuela W, Soto JE, Zarkesh-Ha P, Figueroa M (2021) Face recognition on a smart image sensor using local gradients. Sensors 21(9):2901
8. Tapyou K, Chaisil P, Muangprathub J (2021) Smart school attendance system using face recognition with near optimal imaging. In: 2021 18th International joint conference on computer science and software engineering (JCSSE). IEEE, pp 1–5
9. Smitha Shekar B, Harish G, Aishwarya BR, Badri Narayan S, Chinmayeshree KB (2022) Smart classroom attendance system using face recognition and Raspberry Pi. J Res Appl Sci Eng Technol. ISSN: 2321-9653, https://doi.org/10.22214/ijraset.2022.45718
10. Rajendar S, Abilash KJM, Hari Navaneeth SK (2021) Safe-intelligent temperature and attendance monitoring system for industries and educational institutions. Turk Online J Qual Inq 12(3)
11. Abirami T, Jayadharshini P, Madhuvanthi T (2021) Agro-crop prediction and image detection using AI And Ml. Plant Cell Biotechnol Mol Biol 22(21–22):85–93
12. Kowshika P, Mousika S, Divya P, Lalitha K, Jeevanantham A, Muthukrishnan H (2022) Enhancing the automated diagnosis system of soft tissue tumors with machine learning techniques. In: Intelligent systems: proceedings of ICMIB 2021. Springer Nature Singapore, Singapore, pp 549–558
13. Pichumani S, Sundararajan TVP, Dhanaraj RK, Nam Y, Kadry S (2021) Ruzicka indexed regressive homomorphic ephemeral key Benaloh cryptography for secure data aggregation in WSN. J Internet Technol 22(6):1287–1297
14. Namburu A, Sumathi D, Raut R, Jhaveri RH, Dhanaraj RK, Subbulakshmi N, Balusamy B (2022) FPGA-based deep learning models for analysing corona using chest X-ray images. Mobile Inf Syst 1–14
15. Saravanakumar P, Sundararajan TVP, Dhanaraj RK, Nisar K, Memon FH, Ibrahim AAB (2022) Lamport certificateless signcryption deep neural networks for data aggregation security in WSN. Intell Autom Soft Comput 33(3):1835–1847
16. Abu Awwad Y, Rana O, Perera C (2024) Anomaly detection on the edge using smart cameras under low-light conditions. Sensors 24(3):772
17. Bilik S, Zemcik T, Kratochvila L, Ricanek D, Richter M, Zambanini S, Horak K (2024) Machine learning and computer vision techniques in continuous beehive monitoring applications: a survey. Comput Electron Agric 217:108560
18. González-Rodríguez VE, Izquierdo-Bueno I, Cantoral JM, Carbú M, Garrido C (2024) Artificial intelligence: a promising tool for application in phytopathology. Horticulturae 10(3):197

MGSA: Enhanced GSA for Energy-Efficient Cloud Task Scheduling

Ashis Kumar Mishra, Subasish Mohapatra, and Pradip Kumar Sahu

Abstract Efficient task scheduling is pivotal in cloud computing, especially for Big Data processing. This study explores task scheduling's significance, focusing on energy efficiency. Leveraging a Modified Gravitational Search Algorithm (MGSA), it surpasses existing optimization techniques, showcasing superior cost reduction and energy efficiency in cloud environments. Future research avenues include integrating machine learning for algorithm adaptability, exploring edge and quantum computing impacts, and addressing real-world scalability challenges. This study provides a foundation for advancing task scheduling algorithms, offering insights into optimizing resource allocation for diverse cloud computing scenarios.

Keywords Task scheduling · Cloud computing · Energy efficiency · MGSA · Optimization · Machine learning · Edge computing · Scalability

1 Introduction

Cloud computing is a dynamic force, attracting a growing user base with reliable software, platform and hardware solutions [1]. Scheduling is the main focus of cloud computing, aiming to optimize resource utilization and taking into account quality of service (QoS) [2]. Its purpose is to improve the productivity of the system through the limited resources [3]. Greedy and focused scheduling algorithms and clustering

A. K. Mishra
Department of Computer Science and Engineering, Biju Patnaik University of Technology, Rourkela, Odisha, India
e-mail: akmishracse@outr.ac.in

S. Mohapatra (✉)
School of Computer Science, Odisha University of Technology and Research, Bhubaneswar, Odisha, India
e-mail: smohapatra@outr.ac.in

P. K. Sahu
Department of Information Technology, VSSUT, Burla, Sambalpur, Bhubaneswar, Odisha, India
e-mail: pksahu_it@vssut.ac.in

© The Author(s), under exclusive license to Springer Nature Singapore Pte Ltd. 2025
S. K. Udgata et al. (eds.), *Intelligent Systems*, Lecture Notes in Networks and Systems 1314, https://doi.org/10.1007/978-981-96-3797-3_11

techniques help improve user experience [4]. Different scheduling strategies are deployed to improve the execution time and cost [5], which is very important in the cloud environment with different resources and costs [6].

At the same time, the processing of Big Data attracts the attention of researchers, motivated by the need to manage large and diverse data [7]. Power-aware operations become important, especially in cloud systems with large data center clusters, to solve the challenges of processing large datasets [8].

The combination of cloud computing and big data presents challenges and opportunities, requiring continuous research for the optimization of processing algorithms and energy efficiency [9] working in cloud environments [1–8, 10–13]. As these technologies develop, ongoing research will shape the future of computing, taking into account the complexity of modern computing resources.

2 Literature Survey

Many task scheduling algorithms have been proposed to improve performance and reduce waiting time, delay time and resource consumption. Abdul Razaqueet al. introduced a bandwidth-aware algorithm that optimizes performance, thus reducing the execution time [14]. Shimpy et al. proposed a heuristic approach to disk space management in virtual environments, with an emphasis on availability [3].

Yogita Chawla et al. It is recommended to plan the best schedule and use of work grouping, thus reducing the planning time and cost. Their method of combining methods, work grouping and virtual machine selection, achieved less cost and processing time [15]. D. I. George Amalarethina et al. focused on power-based scheduling decisions for heterogeneous systems, using all processors with short execution times [16].

Divya S et al. Defined Map Reduction as a structural model for large datasets. Francesco et al. proposed a mapping scheme to map service reduction based on nodes and service needs, considering the distribution of data but ignoring the distribution of resources [17, 18]. Xuyun Zhang et al. discussed privacy-preserving Map Reduction on the cloud with a focus on adopting a framework for big data [19] management. Jian Chen et al. proposed a distributed round-robin scheduling algorithm for Hadoop workloads [20, 21].

T. Baker et al. address energy efficiency in cloud computing, introducing the GreeDi method to reduce energy dissipation [22]. Mehta et al. proposed a task-aware scheduling scheme for Hadoop to optimize resource utilization and reduce job waiting time [23]. Poppet et al. introduced an efficient scheduling algorithm with workload constraints but with performance problems during load fluctuations [24]. False et al. proposed an improved gravity search algorithm, informing local optima, with emphasis on input parameters [25].

3 Motivation

This motivation comes from the inherent limitations of GSA: its impact on local optima. To solve this issue, there is a plan to expand the search space, which can overcome the challenge of local excellence. Various methods will be explored, including particle swarm optimization (PSO), genetic algorithm (GA), linear gravitational constants and hybrid approach. By combining these methods, the goal is to improve GSA's ability to effectively manage workloads and work schedules in a cloud environment.

4 Objective

The goal is to develop a better gravitational search algorithm (GSA) for load balancing and task scheduling in cloud environments. The focus is on evaluating the performance of the algorithm compared to other methods, taking into account metrics such as lifetime, cost, and energy efficiency.

5 Proposed Model

The Modified Gravitational Search Algorithm (MGSA) is a state-of-the-art algorithm designed to address complex optimization challenges, especially in the area of cost reduction in cloud computing. Based on the principles of gravity analysis, the algorithm introduces new changes to improve its performance. During the initial phase, a number of particles are generated quickly, each of which represents a possible solution in the search space. As the algorithm repeats, these particles adjust their position based on gravity, looking for the best configuration that minimizes cost effectively.

What makes MGSA unique is its ability to strike a delicate balance between exploration and exploitation. The part has a gravitational relationship, which attracts them to the best solution and at the same time rejects themselves. Both of these factors promote global exploration of the search space while exploiting promising areas for potential cost reductions. One inherent feature of MGSA is the introduction of modified parameters during the optimization process, allowing flexibility. The effectiveness of the algorithm in navigating the complexities of cost optimization in the cloud is demonstrated by its ability to converge to the optimal solution in a dynamic environment.

```
Procedure ModifiedGravitationalSearchAlgorithm_EnergyEfficiency:
  InitializePopulation() // Randomly initialize the positions of particles
  EvaluateEnergyEfficiency() // Evaluate the energy efficiency of each particle
  Repeat for a predefined number of iterations:
    UpdatePersonalBest() // Update personal best positions and energy efficiency
    UpdateGlobalBest() // Update global best position and energy efficiency
    For each particle in the population:
      CalculateGravitationalForces() // Calculate gravitational forces
      UpdateParticlePosition() // Update particle position based on gravitational forces
      EvaluateEnergyEfficiency() // Evaluate the energy efficiency of the new position
     UpdatePersonalBest() // Update personal best position and energy efficiency if improved
   UpdateInertiaWeight() // Optionally update inertia weight for exploration-exploitation balance
  End Repeat
End Procedure
Procedure EvaluateEnergyEfficiency:
  For each particle in the population:
    Evaluate the energy efficiency of the particle based on the energy efficiency metric
End Procedure
Procedure UpdatePersonalBest:
  For each particle in the population:
    If the energy efficiency of the current position is better than the personal best:
      Update personal best position and energy efficiency
End Procedure
Procedure UpdateGlobalBest:
  Identify the particle with the best energy efficiency as the global best
End Procedure
Procedure CalculateGravitationalForces:
  For each particle in the population:
    Calculate gravitational forces based on the positions and energy efficiencies of other particles
End Procedure
Procedure UpdateParticlePosition:
  Update the position of the particle based on gravitational forces
End Procedure
Procedure UpdateInertiaWeight:
  Optionally update the inertia weight for balancing exploration and exploitation
End Procedure
```

6 Simulation and Result Discussion

This work has been implemented in the working platform of Google Colab using python with the following system specification and the simulation results are discussed below.

Platform:	Google colab
OS:	Windows 10 64 bit
Processor:	AMD Ryzen 7
RAM:	16 GB

6.1 Simulated Results

In a focused study that includes 20 iterations and 10 cloudlets, the performance of different algorithms is improved in terms of energy efficiency and cost optimization in the compared cloud. Among the algorithms tested, others found improved performance compared to the Gravitational Search Algorithm (MSGA). The transformation of MSGA, which uses a combination of the research principles of gravity and new transformation, is evident, demonstrating its effectiveness in achieving high energy efficiency. These results demonstrate the potential of MSGA as a promising solution to address optimization challenges in cloud computing power scenarios, especially in the context of cloud 10. The test demonstrated MSGA's ability to allocate resources both energy-efficiently and cost-effectively, highlighting its importance in real-world cloud computing applications. The results are depicted in Figs. 1, 2. Similarly, the comparison of various competitive algorithms are shown in Tables 1, 2; Figs. 3, 4.

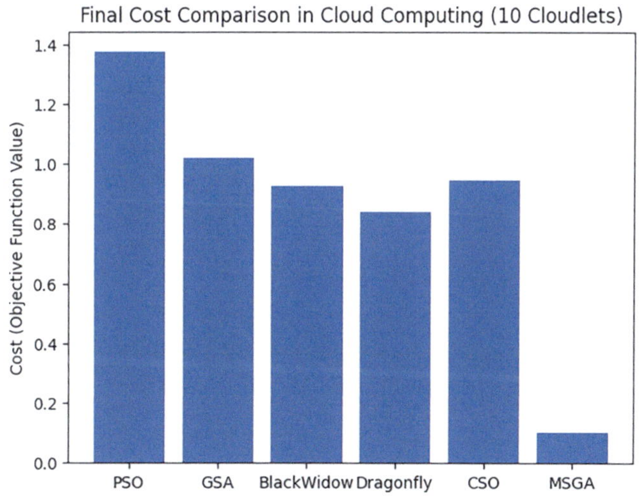

Fig. 1 Comparison of final cost for 10 cloudlets in PSO, GSA, Black Widow, Dragonfly and CSO and MGSA algorithms

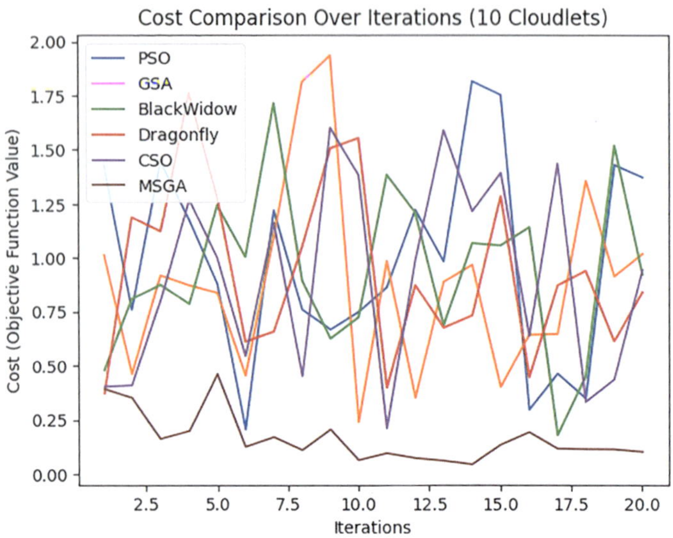

Fig. 2 Comparison of cost for 10 cloudlets in PSO, GSA, Black Widow, Dragonfly and CSO and MGSA algorithms

7 Conclusion and Future Work

The study underscores the importance of efficient task scheduling in cloud computing, especially for Big Data processing. By conducting a thorough literature review and developing the Modified Gravitational Search Algorithm (MGSA), this research tackles the energy-efficient task scheduling challenges. Experimental findings demonstrate MGSA's superiority over existing optimization algorithms, highlighting its potential for cost reduction and improved energy efficiency in cloud computing. Moving forward, future research could explore integrating machine learning techniques to enhance task scheduling adaptability and efficiency, while investigating emerging technologies like edge and quantum computing for optimizing resource allocation. Furthermore, extending the study to address real-world applications and scalability challenges would provide deeper insights into practical implementation across diverse cloud computing scenarios.

Table 1 Comparison of cost for 10 cloudlets in PSO, GSA, Black Widow, Dragonfly and CSO and MGSA algorithms

Iteration	PSO	GSA	BlackWidow	Dragonfly	CSO	MSGA
1	0.289642	1.283451	1.35992	0.859027	0.538463	0.053648
2	1.111685	0.473822	1.002346	1.317846	0.504702	0.238851
3	0.918424	0.534717	1.439081	1.132847	1.051999	0.519818
4	1.525161	0.776868	1.617485	0.892236	0.838055	0.036992
5	0.961578	0.325438	0.839988	0.810293	0.363742	0.120195
6	0.897418	0.575343	1.383582	1.609389	1.557138	0.145137
7	0.984355	1.306441	1.394393	1.662717	0.705413	0.040625
8	0.618382	0.989466	1.688708	0.299789	0.57172	0.171158
9	1.096097	0.786274	0.959437	1.597482	1.210419	0.055657
10	0.301963	0.856228	0.898697	0.119654	0.813561	0.118461
11	0.488157	0.811431	0.822477	1.136908	0.904718	0.162818
12	0.953946	0.636261	0.199412	0.208417	0.980077	0.140508
13	0.164707	0.362807	1.584923	0.460419	1.527206	0.141994
14	0.994325	0.668363	0.9072	0.467555	0.506892	0.048957
15	0.957051	1.176321	0.320855	1.218325	1.540363	0.111517
16	1.237353	1.113452	0.737709	1.190719	1.364509	0.122376
17	1.415475	1.290066	1.432484	1.291137	1.065549	0.102455
18	0.740411	1.825836	1.059438	0.741856	1.001865	0.178218
19	1.213517	0.832861	1.157618	1.365802	0.198773	0.110376
20	1.555312	0.609812	1.466043	0.600371	1.624779	0.106943

Table 2 Comparison of energy efficiency for 10 cloudlets in PSO, GSA, Black Widow, Dragonfly and CSO and MGSA algorithms

Iteration	PSO	GSA	BlackWidow	Dragonfly	CSO	MGSA
1	1.315207	1.011083	0.731779	1.284792	1.099857	0.059308
2	1.027718	1.531273	0.288082	1.128162	0.758846	0.027204
3	0.633114	0.749344	0.338834	0.644315	0.530052	0.031163
4	1.007049	0.283974	0.118526	0.237154	0.014988	6.20E-05
5	0.871043	0.824887	0.857355	0.047329	1.23939	0.017019
6	0.03696	0.676004	1.36917	1.091569	0.72761	0.047898
7	0.607805	0.376175	0.050639	0.453673	0.36615	0.010341
8	1.323398	0.760598	0.756016	0.391071	1.071958	0.008042
9	0.804054	0.305737	0.069976	0.481016	0.205098	0.00708
10	1.32833	1.146795	0.950449	0.057711	0.327703	0.010075
11	1.240107	0.345874	0.44052	1.295958	0.786403	0.01226
12	0.683112	1.50708	0.331166	0.503893	0.877443	0.000332
13	1.359796	0.344398	0.3927	1.203957	0.818423	0.00408
14	1.575817	0.718064	0.369799	0.875572	0.016845	0.005013
15	1.392317	1.018961	0.901191	0.507626	0.366864	0.015542
16	0.842405	0.550009	1.40964	0.68051	0.503236	0.004612
17	1.403425	0.176927	0.809475	0.497475	1.074509	0.005473
18	0.304171	0.70589	0.554366	0.164153	0.078723	0.001415
19	0.110794	0.113306	0.071633	0.272814	0.081479	0.001297
20	0.669815	0.640984	0.119169	1.93651	0.153292	0.003442

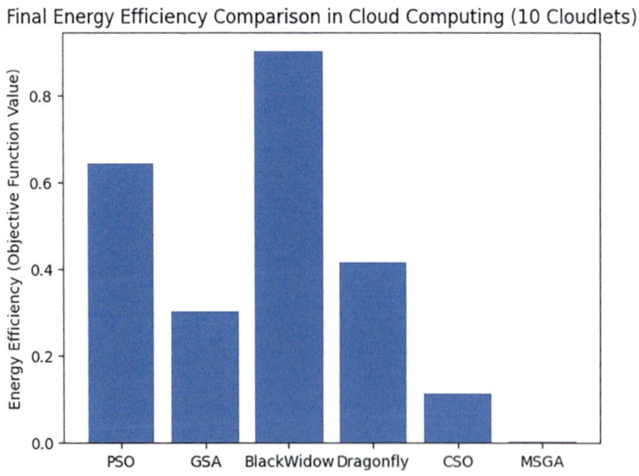

Fig. 3 Final energy efficiency comparison for 10 cloudlets in PSO, GSA, Black Widow, Dragonfly and CSO and MGSA algorithms

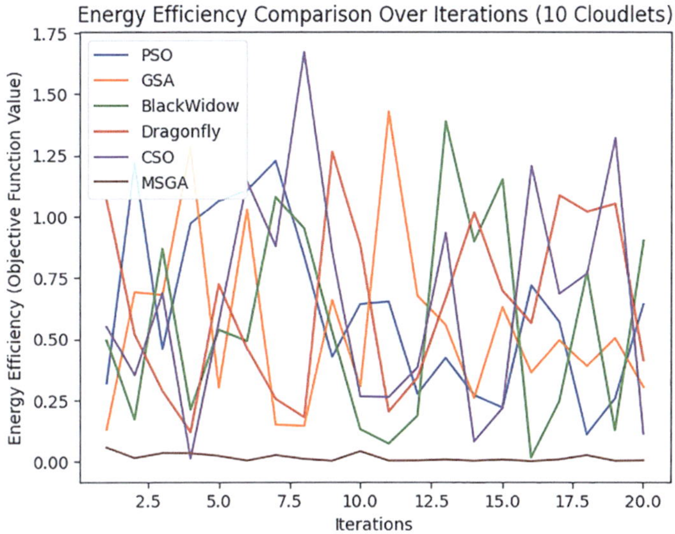

Fig. 4 Comparison of energy efficiency for 10 cloudlets in PSO, GSA, Black Widow, Dragonfly and CSO and MGSA algorithms

References

1. Malathi M (2011) Cloud computing concepts. In: Proceedings of the third ınternational conference on electronics computer technology, vol. 6. IEEE
2. Arunarani AR, Dhanabalachandran M, Vijayan S (2019) Task scheduling techniques in cloud computing: a literature survey. Futur Gener Comput Syst 91:407–415
3. Shimpy E, Sidhu J (2014) Different scheduling algorithms in different cloud environment. Int J Adv Res Comput Commun Eng 3(9):8003–8006
4. Entezari-Maleki R, Movaghar A (2012) A probabilistic task scheduling method for grid environments. Int J Future Gener Comput Syst 28:513–524
5. Pandey S, Wu L, Guru SM, Buyya R (2010) A particle swarm optimization-based heuristic for scheduling workflow applications in cloud computing environments. In: Proceedings of the 24th IEEE ınternational conference on advanced ınformation networking and applications, pp 400–407
6. Bodík P, Menache I, Naor J, Yaniv J (2014) Deadline-aware scheduling of big-data processing jobs. In: Proceedings of the 26th ACM symposium on parallelism in algorithms and architectures, pp 211–213
7. Sreedhar C, Kasiviswanath N, Chenna Reddy P (2015) A survey on big data management and job scheduling. Int J Comput Appl 130(13):41–49
8. Sugandhi P, Joseph C, Karnewar H, Chaudhari J (2015) Energy efficient scheduling of map reduce over big data. Int J Recent Innov Trends Comput Commun 3(10):6003–6007
9. Narendrasingh S, Mishra AK, Mohapatra S (2023) A comparative analysis for energy efficiency in cloud computing using CSO. In: Proceedings of the ınternational journal of computer applications, pp 30–37. https://doi.org/10.5120/ijca2023922811
10. Sharma V, Nair TRG (2014) An optimum scheduling approach for creating optimal priority of jobs with business values in cloud computing. In: International conference on advances in cloud computing, pp 1–8
11. Zhou X, Liu X (2014) Multiple computing task scheduling method based on dynamic data replication and hierarchical strategy. J Netw 9(2):415–422

12. Zhang F, Cao J, Tan W, Khan SU, Li K, Zomaya AY (2014) Evolutionary scheduling of dynamic multitasking workloads for big-data analytics in elastic cloud. IEEE Trans Emerg Top Comput 2(3):338–351
13. Li J, Liu Y, Pan J, Zhang P, Chen W, Wang L (2020) Map-balance-reduce: an improved parallel programming model for load balancing of MapReduce. Futur Gener Comput Syst 105:993–1001
14. Razaque A, Vennapusa NR, Soni N, Janapati GS (2016) Task scheduling in cloud computing. In: Proceedings of the IEEE conference on long ısland systems, applications and technology (LISAT), pp 1–5
15. Chawla Y, Bhonsle M (2013) Dynamically optimized cost-based task scheduling in cloud computing. Int J Emerg Trends Technol Comput Sci 2(3):38–42
16. George Amalarethinam DI, Josphin AM (2015) Dynamic task scheduling methods in heterogeneous systems-a survey. Int J Comput Appl 110(6):12–18
17. Divya S, Kanya Rajesh R, Rini Mary Nithila I, Vinothini M (2015) Big data analysis and its scheduling policy—hadoop. J Comput Eng 7(1):36–40
18. Palmieri F, Fiore U, Ricciardi S, Castiglione A (2016) GRASP-based resource re-optimization for effective big data access in federated clouds. Int J Future Gener Comput Syst 54:168–179
19. Mishra A, Mohapatra S, Sahu P (2022) Survey on performance and energy-aware task scheduling over big data in cloud computing. In: Proceedings of neuroquantology, pp 393–433. https://doi.org/10.48047/nq.2022.20.22.NQ10030
20. Zhang X, Liu C, Nepal S, Dou W, Chen J (2012) Privacy-preserving layer over MapReduce on cloud. In: Proceedings of the second ınternational conference on cloud and green computing (CGC '12), pp 304–310
21. Chen J, Wang D, Zhao W (2013) A task scheduling algorithm for Hadoop platform. Int J Comput 8(4):929–936
22. Baker T, Al-Dawsari B, Tawfik H, Reid D, Ngoko Y (2015) GreeDi: an energy efficient routing algorithm for big data on cloud. Int J AdHoc Netw 35:83–96
23. Pati, Mehta A (2015) Job aware scheduling in Hadoop for heterogeneous cluster. In: Proceedings of the IEEE ınternational conference in advanced computing, pp 778–783
24. Sfrent A, Pop F (2015) Asymptotic scheduling for many tasks computing in big data platforms. Int J Inf Sci 319:71–91
25. Liu Y, Zhang X, Chao H (2022) An improved gravitational search algorithm combining with centripetal force. Partial Diff Equ Appl Math 5:100378

Stereo-Edge Computation for Autonomous Driving Through Enhanced Environmental Analysis

R. Sadhana, S. Rajesh Kannan, M. Angelin Ponrani, and P. Ezhilarasi

Abstract Autonomous cars have revolutionized transportation, prioritizing safety, efficiency, and eco-friendliness. Stereo vision, mimicking human perception, aids in lane detection, space assessment, and obstacle recognition. Driven by precise disparity maps, autonomous navigation algorithms are based on depth perception and object recognition. Complete autonomy is yet hampered by resource-intensive stereo vision. Edge computing, a paradigm shift, allows local data processing, reducing server reliance. Integration of stereo vision, edge computing, and advanced algorithms like improved Quadruple Sparse Census Transform (iQSCT) marks a significant leap in autonomous vehicles. It tackles computing limitations, enhancing accuracy in vital tasks like obstacle detection and lane identification, ensuring safer, more efficient mobility, and paving the way for future navigation advancements.

Keywords Stereo vision · Edge computing · iQSCT · Obstacle detection · Lane detection · Free space detection

1 Introduction

The automobile industry has undergone a transformation due to autonomous vehicles, which promise safer, more economical, and environmentally friendly travel. By emulating human perception and decision-making, these vehicles which are powered by modern technology and sophisticated algorithms hope to eventually replace human drivers with robots that follow data and algorithms. Stereo vision stands out among the numerous technical innovations that have been essential to the development of autonomous cars because it has made it possible for these robots to comprehend and navigate their surroundings [1]. Autonomous cars now rely heavily on stereo vision, which processes pictures from two slightly offset cameras to simulate depth and distance perception, a function which human beings perform. It helps

R. Sadhana (✉) · S. Rajesh Kannan · M. Angelin Ponrani · P. Ezhilarasi
Department of Electronics and Communication Engineering, St. Joseph's College of Engineering, Chennai, India
e-mail: er.r.sadhana@gmail.com

© The Author(s), under exclusive license to Springer Nature Singapore Pte Ltd. 2025

S. K. Udgata et al. (eds.), *Intelligent Systems*, Lecture Notes in Networks and Systems 1314, https://doi.org/10.1007/978-981-96-3797-3_12

with tasks including obstacle recognition, space assessment, and lane detection [2]. Stereo vision algorithms generate a three-dimensional picture of the surrounding world by analysing the differences between the images taken by the two cameras [3]. This allows the vehicle to comprehend item positions and road layouts with very high accuracy. This depth perception is essential for autonomous navigation and plays a major role in maintaining everyone's safety on the road, including the passengers.

Stereo vision has limitations and its own set of difficulties, despite its great potential. Stereo vision is a resource-intensive field that presents many challenges due to the large amount of data generated by the cameras, which requires a significant amount of computational power to process [4]. Due to the historical dependence on cloud-based computing and remote servers, this computational demand has resulted in latency and dependency problems. In the drive for total autonomy and seamless real-time decision-making, the industry wanted to discover a solution that would lessen the dependency on faraway servers and bring the power of computation closer to the cars themselves. Enter edge computing, a paradigm shift in the world of autonomous vehicles and beyond [5]. Edge computing revolves around the concept of local computing, where computing occurs closer to the source of data creation. In the context of autonomous cars, this means that the heavy lifting of stereo vision and other computing tasks can be done inside the vehicle or on nearby peripherals [6]. This change can reduce latency, increase efficiency and improve the overall responsiveness of autonomous systems.

The advancement of autonomous vehicle technology has been greatly accelerated by the combination of cutting-edge algorithms such as improved Quadruple Sparse Census Transform (iQSCT) with stereo vision and edge computing. iQSCT is an advanced and effective stereo vision algorithm that is essential to overcoming the computational constraints that have prevented autonomous driving from being fully realized. iQSCT improves the accuracy of crucial activities like obstacle detection, free space detection and lane identification by streamlining the stereo vision process and cutting down on computational overhead. The integration process and the development of increasingly effective algorithms significantly influence the capabilities of self-driving cars. They help the system become far less dependent on distant servers, making the car more independent and capable of making crucial choices instantly without being hampered by network delay. These technological advance promises safer and more effective travel on our roads in the future, with autonomous cars being able to handle complicated surroundings like congested city streets and highways with ease. This merging of technologies has broad ramifications. The most obvious advantage is probably safer roads. Advanced stereo vision and edge computing in autonomous cars enable them to quickly identify and react to roadblocks, perhaps averting collisions and saving lives. Reliability increases for lane recognition, which helps to improve traffic flow and lowers the risk of lane departure accidents. All things considered, the amalgamation of these technologies brings us one step further to actualizing the vision of a future marked by a marked decrease in traffic-related incidents and fatalities.

2 Related Work

Semantic Segmentation and Object recognition: Enhancing the precision of stereo vision-based object recognition and segmentation has been the subject of several research projects. Convolutional Neural Networks (CNNs), one type of deep learning technology, have been used to identify and categorize items on the road [7]. These techniques are essential for activities such as recognizing traffic signs, other cars, and people. Additionally, researchers have been working to improve stereo vision systems' ability to function in difficult weather and illumination situations. In computer vision, disparity estimation is a hot issue that has gained more attention recently. In this paper, we propose a novel multi-stage network with two training stages: depth-image-based rendering (DIBR) as an additional component to form the second training stage [3], and an initial disparity estimation as the first training stage for the purpose of converting two-dimensional videos to three-dimensional ones [8]. We suggest an updated end-to-end feature pyramid stereo network for the initial training phase, which substitutes a pyramid-shaped bottom-up Convolutional Neural Network for the original non-pyramid structure for disparity regression. In order to improve the performance on border consistency, it concatenates several scale characteristics using the spatial information.

This study presents a unified depth model based on the gradient of the light field, where the complex number represents the estimated disparity. The suggested depth model's complex-valued disparity may be expressed in both polar and Cartesian coordinates. Real and imaginary components of the disparity are used to indicate the suggested depth model in the Cartesian form. While the imaginary portion indicates the non-Lambertian-ness, the real part may be utilized to estimate disparity with regard to the in-focus plane. The suggested depth model in the polar form is represented by the disparity angle and magnitude. While the disparity angle indicates whether or not the bundles are in focus, the disparity magnitude illustrates the link between depth from stereo, depth from focus, and depth from defocus [4].

Effective Edge Hardware: To enable edge computing in autonomous cars, researchers have been working hard to create specialized hardware. These include dedicated processors and accelerators made to meet the computing requirements of AI inference, control algorithms, and sensor data processing. These advancements in hardware contribute to lower power usage and faster edge computing [9].

Finding a collection of related bounding boxes with comparable appearances is how the detection challenge is defined since a single item in a blended image may have numerous superimposed correspondences. The object's disparity may be calculated using the separations between the corresponding bounding boxes. The benefits of the suggested technique are as follows: (1) It is possible to complete the detection and disparity estimation tasks with a single input image; (2) The task is less computationally expensive even though it is more complex than the popular detection task; (3) Very competitive detection and disparity estimation accuracy is achieved, even when the detection is done on a blended image that is not good for human vision [6]. The KITTI stereo subset is used in the experiment, and extremely encouraging results were obtained.

While convolutional networks have made tremendous strides recently, their applicability is limited by their heavy reliance on processing power. The latter is true for stereo disparity estimation, for which embedded systems can hardly support the present techniques. An important step towards lighter and maybe quicker disparity estimation networks is demonstrated by this study, which demonstrates the binarization of an end-to-end stereo disparity network [3]. This study demonstrates the viability of the suggested methodology through experiments on the scene flow and kitti2012 datasets, two popular datasets. The findings demonstrate that, albeit at performance cost, a binary disparity model is feasible [10]. In scene flow and kitti2012, respectively, an EPE of 5.14 and 2.09 is attained [11]. Machine learning for decision-making, a developing field of study is the use of machine learning algorithms for autonomous car decision-making. Large datasets are used to train these algorithms so they can make intelligent judgements in intricate and dynamic traffic scenarios [12]. They include things like traffic prediction, accident avoidance, and trajectory planning. Safety and validation is one of the main priorities is making sure autonomous cars are safe. Developing methods for safety validation and certification is the focus of this research. It involves creating simulation environments that enable comprehensive testing of autonomous systems in a variety of circumstances, as well as testing and verification frameworks.

3 Proposed Method

As the autonomous car age progresses, it's becoming more and more clear that combining cutting-edge technologies is essential to achieving its full potential. Through the integration of modern algorithms, including the Quadruple Sparse Census Transform (iQSCT), edge computing, the suggested technique presented here seeks to make a substantial advancement in autonomous cars. This method improves the accuracy of cars in critical tasks like as obstacle detection and lane identification and free space detection while addressing the computing constraints that have impeded full autonomy. This approach guarantees safer, more effective movement and opens the door for further advancements in navigation in the future (Fig. 1).

Our suggested solution is based on stereo vision, a technology that uses two significantly offset cameras to interpret pictures in order to imitate human depth perception. It makes it possible for autonomous cars to comprehend their environment in three dimensions, which is essential for safe and effective navigation. Stereo vision systems create depth maps by analysing the differences between the pictures taken by the two cameras. These maps enable the vehicle to detect objects and road layouts with exceptional accuracy. The autonomous car can get a lot of information using stereo vision. It helps with tasks including obstacle recognition, space assessment, and lane detection. These are essential parts of autonomous navigation since they guarantee

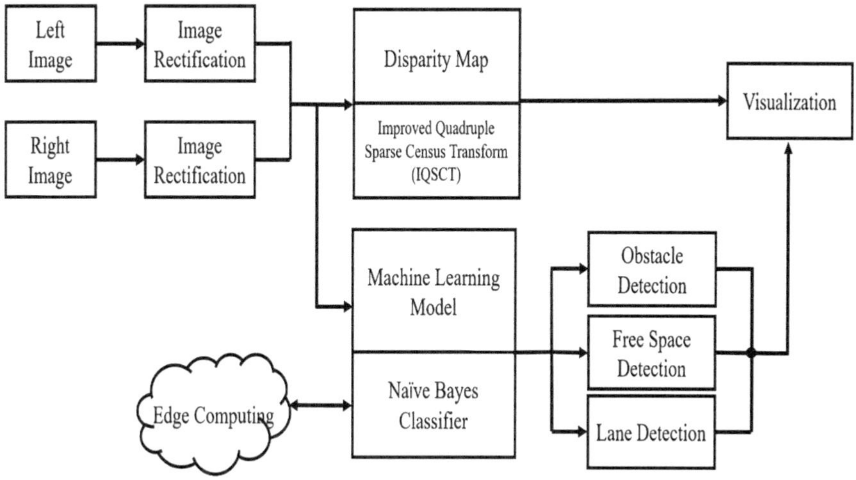

Fig. 1 Block diagram of proposed method

the car can stay in its lane and evaluate the space that is available on the road, as well as identify and steer clear of impediments. In the early phases of autonomous vehicle research, processing the massive amounts of data provided by stereo cameras needs a substantial amount of computer capacity, which frequently resulted in a dependency on distant servers and cloud-based computing. This reliance on outside resources reduces the system's resilience to network problems, adds latency, and impedes the ability to make decisions quickly. Census Transform of 3×3 window, of pixel(a, b) with all 8-connected neighbours is defined by the function ξ,

$$\xi(a, b) = \begin{cases} 0 & \text{if } a > b; \\ 1 & \text{if } a \leq b; \end{cases} \tag{1}$$

An advanced and effective method for calculating the disparity map—a crucial component in stereo vision applications—is the Quadruple Sparse Census Transform (iQSCT) algorithm [13]. Perform the Quadruple Sparse Census Transform by considering four directional disparities for each sparse point is denoted in the below equation:

$$\xi(a, b, \in) = \begin{cases} 000 & \text{if } a - b > \in; \\ 100 & \text{if } |a - b| \leq \in; \\ 110 & \text{if } b - a > \in; \\ 111 & \text{if } |b - a| \leq \in; \end{cases} \tag{2}$$

where $\in$ is the Adaptive Threshold, consisting of Standard Deviation (σ),

$$\in = \frac{I(a)}{\sigma} \tag{3}$$

This step enhances the robustness of the disparity map, especially in regions with occlusions or complex texture patterns. An advanced and effective method for calculating the disparity map—a crucial component in stereo vision applications—is the Quadruple Sparse Census Transform (iQSCT) algorithm [14]. Improving the speed and accuracy of stereo matching is the main objective of iQSCT, as it is essential for tasks like object identification and depth perception in autonomous cars. In contrast to conventional stereo matching methods, iQSCT uses a sparse representation of the picture to optimize the calculation, greatly lowering the computational burden while preserving excellent accuracy. The Disparity Map Equation obtained from the equivalent triangles is denoted as follows:

$$\text{Disparity} = x - x' = \frac{Bf}{z} \tag{4}$$

where x and x' are the distance between the points in the image plane and the camera center, B is the distance between the two cameras, and f is the focal length of the camera. Fundamentally, iQSCT uses a census transform, which is a method for encoding the local neighbourhood structure of individual pixels. The method chooses a subset of pixels for stereo matching referred to as sparse points thus introducing a sparsity restriction [15]. The algorithm's ability to concentrate computational efforts on the most informative areas of the image, so reducing the burden of resource-intensive calculations, makes the sparsity restriction a crucial innovation. The error rate, which defines the ratio of error depth pixels, is defined as,

$$\text{Error rate} = \frac{1}{N} \sum_{p \in D} \delta_{\text{err}}(|\mathrm{d}(a) - \mathrm{d}(b)| > d_{\text{err}}) \tag{5}$$

where d_{err} is the disparity tolerance, $d(a)$, $d(b)$ are the estimated ground truth disparities, and N denotes the total number of depth pixels [13].

The four directions in which the method computes the disparities for each sparse point are referred to as "quadruple" in iQSCT. The strength of the disparity map is increased by iQSCT's ability to take into account several orientations and produce a more thorough knowledge of the disparities. This multidirectional method works especially well in situations where there might be difficult texture patterns or occlusions. The iQSCT's flexibility to hardware acceleration enhances its efficiency and makes it a good choice for real-time applications. The algorithm can be applied to systems with demanding computational requirements, like autonomous vehicles, because it can take advantage of graphics processing units (GPUs) and field-programmable gate arrays (FPGAs) to enable parallel processing. This allows for rapid computations.

Essentially, a major development in the realm of stereo vision applications is represented by the iQSCT method for locating a disparity map. Because of its multi-directional approach and sparse representation, which improve computing efficiency without sacrificing accuracy, it is the best option for real-time processing in situations requiring quick and accurate depth perception, such autonomous navigation systems.

Numerous benefits result from this paradigm change, such as decreased latency, enhanced efficiency, and greater responsiveness, which eventually allow the vehicle to make important choices in real time without depending on distant servers. Using edge computing in driverless cars entails installing powerful computer gear inside the car. The computing demands of sensor data processing, artificial intelligence (AI) inference, and control algorithms are especially handled by this hardware. We can make decisions more quickly and reliably by reducing the requirement for continual connection with distant servers by pushing computing power to the edge. This also encompasses drawbacks such as hardware complexity, high processing power requirements, and maintenance issues. The suggested strategy understands that switching from a cloud-centric to an edge computing approach is essential to overcoming these obstacles. By bringing processing closer to the data creation source—in this case, the autonomous car itself—edge computing is made possible shown in Fig. 2.

The Naive Bayes classifier's simplicity and efficacy are leveraged in autonomous cars by using it as a machine learning model for free space, lane, and obstacle identification. Because Naive Bayes enables the system to compute conditional probabilities based on sensor inputs, it is especially well-suited for these tasks and enables the car to make judgements in real time. In order to discover open spaces, the model evaluates sensory information like road texture, vehicle speed, and surrounding obstructions to determine how likely it is that a certain region is passable.

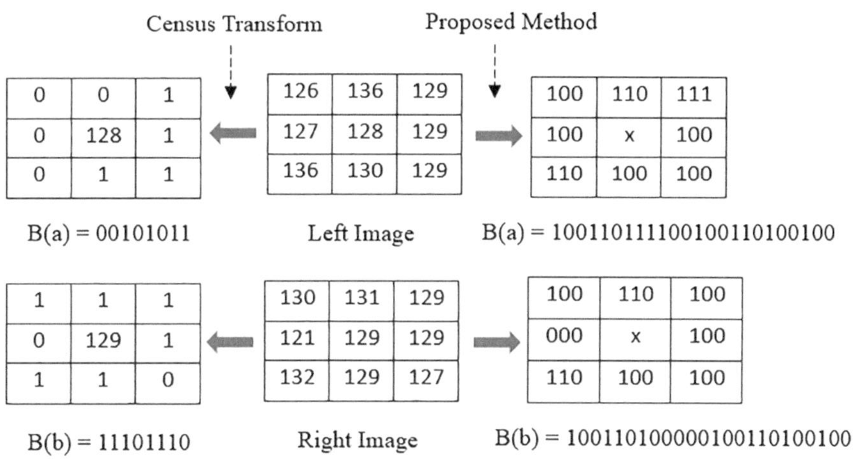

Fig. 2 Few instances of census transforms found with the standard and suggested three-bit census auxiliary functions

When a vehicle is in lane detection, its location inside the lanes is classified by analysing the surrounding objects, the vehicle's trajectory, and lane markers. The model's ability to discriminate barriers from other objects during obstacle detection helps the car avoid crashes. Because of its simplicity and efficiency, the Naive Bayes classifier is a good fit for the computing needs of real-time autonomous vehicle operations. It contributes to safer and more effective autonomous navigation by improving the vehicle's awareness of its surroundings by offering precise probabilistic evaluations for free space, lane, and obstacle identification. Furthermore, it can handle the dynamic and unexpected nature of real-world driving scenarios well thanks to its flexibility and capacity to continually update probabilities based on incoming sensor data, which advances the development of autonomous vehicle technology.

4 Result and Discussion

The visual data base for autonomous navigation systems is comprised of input pictures that are obtained using stereo cameras. In order to create a binary code that encodes the arrangement of intensity values nearby, the census transform encrypts the pixel's local neighbourhood structure. Stereo vision relies on multi shape images to boost stereo vision depth perception by precisely analysing varied shapes; the multi shape census transform is used to analyse numerous shapes for enhanced accuracy and to improve 3D perception. With accurate analysis of sparse data patterns that match the Disparity Map produced by applying the suggested iQSCT algorithm, the Quadruple Sparse Census Transform is utilized to improve stereo vision depth perception, which is shown in the Fig. 3.

In order to produce a disparity map, the suggested method is put into practice in a traffic situation. By using rectified images, the programme enhances the environmental perception, making it possible to identify lanes, obstacles, and free spaces. Within the framework of road situations, our suggested approach is essential for deriving a thorough and precise disparity map. Through the use of sophisticated stereo vision techniques, the algorithm is able to identify minute differences in the

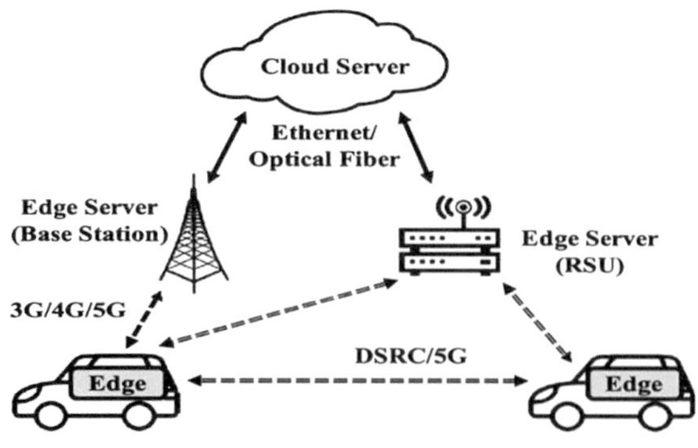

Fig. 3 Edge computing for autonomous car

corrected photos, providing a more complex knowledge of the road environment. With the use of the corrected photos that our algorithm has improved, the environmental perception system effectively recognizes important elements like lane lines, obstacles, and free areas. This discrepancy map is an essential part of the activities related to later environmental perception, shown in Figs. 4 and 5.

The discrepancy data offers useful depth indicators, enabling strong and trustworthy identification of possible dangers and road infrastructure. By integrating the creation of disparity maps with subsequent environmental awareness, this all-encompassing method improves overall safety and efficiency in highway scenarios and gives the system the ability to make well-informed judgements for autonomous navigation.

Table 1 provides a numerical evaluation of the disparity errors in the stereo vision context between an existing and the proposed system. The performance of Existing and Proposed Systems is assessed using three primary metrics: Non-occ (non-occluded), All (all pixels), and Disc (discontinuity) errors. These metrics

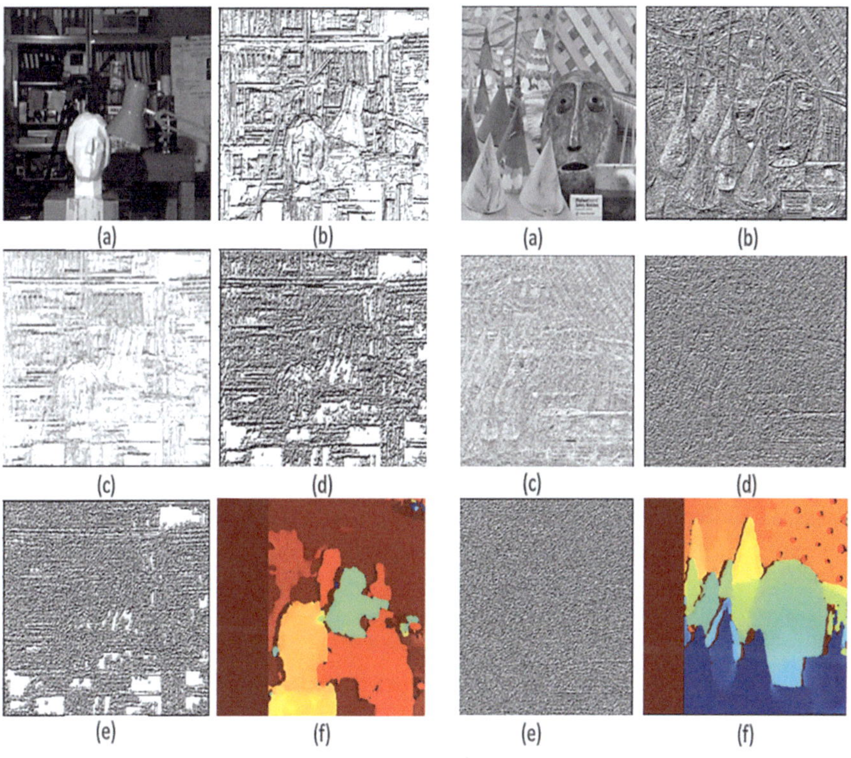

Fig. 4 Disparity map achieved by the proposed iQSCT algorithm: **a** grey scale image; **b** census transform; **c** multi shape image; **d** multi shape census transform; **e** quadruple sparse census transform; **f** disparity map

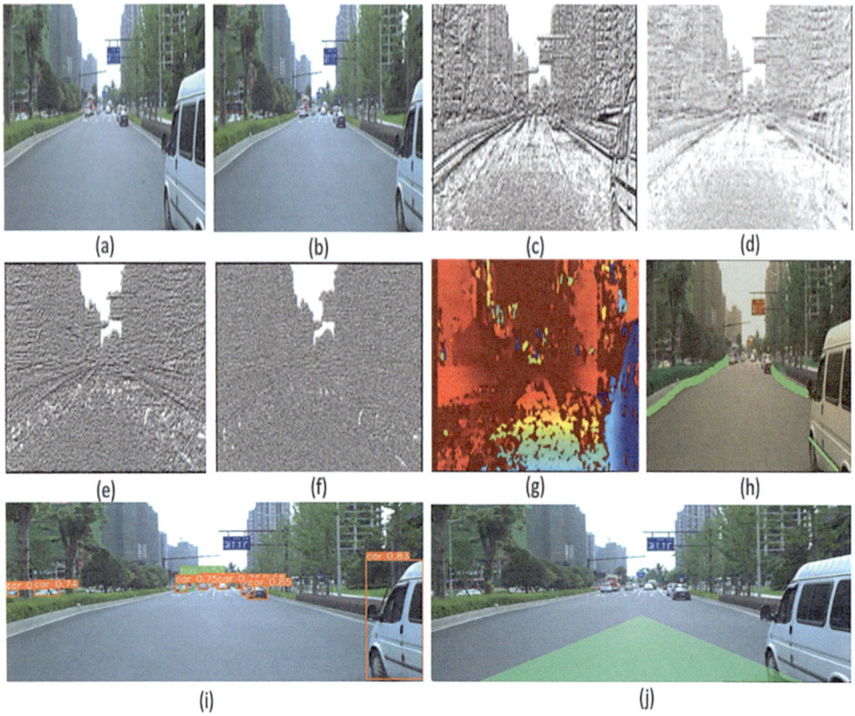

Fig. 5 Stereo-edge computation for autonomous driving through enhanced environmental analysis: **a** left stereo image; **b** right stereo image; **c** census transform; **d** multi shape image; **e** multi shape census transform; **f** quadruple sparse census transform; **g** disparity map; **h** lane detection; **i** obstacle detection; **j** free space detection

are examined. The comparative analysis aids in evaluating how well the suggested approach improves disparity map accuracy in various contexts.

The outcomes demonstrate a significant improvement in lane identification and obstacle identification and free space detection, highlighting the potency of the suggested technique. The improved Quadruple Sparse Census Transform (iQSCT) overcomes computational obstacles and improves accuracy by substantially fine-tuning stereo vision procedures. By integrating edge computing, the system becomes more autonomous and can make decisions in real time with less need on distant servers. These encouraging results demonstrate the method's potential to transform autonomous car technology and provide the groundwork for safer and more effective navigation in a variety of dynamic driving scenarios.

Table 1 Erroneous disparity comparison

Algorithm	Disparity evaluation metrics	Test image		
		Cones	Teddy	Tsukuba
Proposed method	Non-Occ	0.71	0.63	0.51
	All	0.54	0.47	0.36
	Disc	9.58	11.09	6.84
Census transform	Non-Occ	2.42	4.10	1.07
	All	7.25	6.22	1.48
	Disc	6.95	10.9	5.73
Cost filter	Non-Occ	2.71	6.16	1.51
	All	8.24	11.8	1.85
	Disc	7.66	16.00	7.61
Sparse census transform	Non-Occ	4.38	6.54	3.53
	All	7.35	8.96	3.96
	Disc	7.61	12.18	14.15
Variable SAD algorithm	Non-Occ	6.28	9.75	1.99
	All	12.7	15.1	2.65
	Disc	12.9	18.2	6.77

5 Conclusion

In conclusion, the suggested approach is evidence of the synergy that may be attained via the integration of cutting-edge technology. The computational difficulties in depth perception and object recognition are addressed by the developments in stereo vision, especially with the advent of iQSCT. This not only increases the precision of the system's essential functions, such obstacle, free space, and lane recognition, but also paves the way for a wider acceptance of autonomous cars in practical applications. Edge computing's efficiency-driven methodology is crucial in bringing in a new era of localized decision-making. Edge computing reduces latency by processing data within the car, enabling crucial choices to be made instantly. This paradigm change promotes adaptation to a variety of driving scenarios in addition to improving the overall efficiency of the autonomous navigation system. The system gains a cognitive layer with the integration of the Naive Bayes machine learning model, which enables it to process data and draw conclusions. The system is prepared to navigate through a constantly changing urban terrain as it continues to develop, displaying a significant potential to revolutionize transport networks worldwide. With the suggested approach, there are optimistic signs for a safer, smarter, and more connected future for autonomous transportation. This approach is prepared to change as technology advances, paving the way for the revolutionary shift towards the mainstream use of driverless cars.

References

1. Singh S, Saini BS (2021) Autonomous cars: recent developments, challenges, and possible solutions. IOPscience 1022
2. Kemsaram N, Das A, Dubbelman G (2022) A model-based design of an onboard stereo vision system: obstacle motion estimation for cooperative automated vehicles. SN Appl Sci 4:199
3. Pan B, Zhang L, Wang H (2021) Multi-stage feature pyramid stereo network-based disparity estimation approach for two to three-dimensional video conversion. IEEE Trans Circuits Syst Video Technol 31(5):1862–1875
4. Lee JY, Park RH (2021) Complex-valued disparity: unified depth model of depth from stereo, depth from focus, and depth from defocus based on the light field gradient. IEEE Trans Pattern Anal Mach Intell 43(3):830–841
5. Kemsaram N, Das A, Dubbelman G (2020) A stereo perception framework for autonomous vehicles. In: IEEE 91st vehicular technology conference (VTC2020-Spring), Antwerp, Belgium, pp 1–6
6. Zhou C, Liu Y, Sun Q, Lasang P (2021) Vehicle detection and disparity estimation using blended stereo images. IEEE Trans Intell Veh 6(4):690–698
7. Wang W, Lin H, Wang J (2020) CNN based lane detection with instance segmentation in edge-cloud computing. J Cloud Comput 9:27
8. Fan R, Wang L, Junaid Bocus M, Pitas I (2023) Computer stereo vision for autonomous driving: theory and algorithms. In: Hosny KM, Salah A (eds) Recent advances in computer vision applications using parallel processing. Studies in computational intelligence, vol 1073
9. ul Haq QM, LinCH, Ruan S-J, Gregor D (2021) An edge aware adaptive multi-feature set extraction for stereo matching of binocular images. J Ambient Intell Humanized Comput 1967
10. Shabanian H, Balasubramanian M (2021) A new hybrid stereo disparity estimation algorithm with guided image filtering-based cost aggregation. Symp Electron Imaging 33:59-1–59–7
11. Aguilera CA (2022) SBIN: a stereo disparity estimation network using binary convolutions. IEEE Lat Am Trans 20(4):693–699
12. Yang G, Song X, Huang C, Deng Z, Shi J, Zhou B (2019) DrivingStereo: a large-scale dataset for stereo matching in autonomous driving scenarios. In: IEEE/CVF conference on computer vision and pattern recognition (CVPR), Long Beach, CA, USA, pp 899–908
13. Huang CS, Yang JF (2022) Improved quadruple sparse census transform and adaptive multi-shape aggregation algorithms for precise stereo matching. IET Comput Vis 16(2):159–179
14. Huang C-W, Ding J-J (2022) Adaptive superpixel-based disparity estimation algorithm using plane information and disparity refining mechanism in stereo matching. Proc Symmetry 14(5):1005
15. Chen C, Song B, Bu S, He L (2022) An improved point feature-based sparse stereo vision. IET Image Process 16:3284–3299

Traffic Detection on NVIDIA Jetson Nano Using Altair RapidMiner Studio

C. Laksh, H. S. Gururaja, and Ashraff D. Sankanal

Abstract This paper focuses on developing a traffic detection system using Altair's RapidMiner Studio (a no-code platform built for data science and machine learning) and eventually deploying it on a compact module capable of handling the computations of deep learning like the Nvidia Jetson Nano. The system aims to detect traffic density levels in real-time using images captured from a video stream. The RapidMiner Studio platform is used to pre-process the images, extract features, and train a machine learning model for classification. The trained model is to be deployed on the Jetson Nano, which is a low-cost, power-efficient edge computing device that can perform real-time object detection. The system is tested on a variety of vehicles in a traffic scenario and achieves high accuracy in detecting traffic density levels. The results demonstrate the feasibility of using RapidMiner Studio for developing a real-time traffic detection system.

Keywords Traffic detection · RapidMiner · Hotspots · Classification · Jetson Nano

1 Introduction

The problem being addressed is on accurately gauge traffic in a way that is less computationally intensive than video recognition. While video recognition is effective, it is typically only feasible on high-end GPUs, which are integrated into high-performance computers. This type of processing is usually done on the backend, which can place a significant load on the server or serving device.

C. Laksh
Department of EEE, B.M.S. College of Engineering, Bengaluru, India

H. S. Gururaja (✉)
Department of ISE, B.M.S. College of Engineering, Bengaluru, India
e-mail: gururajhs.ise@bmsce.ac.in

A. D. Sankanal
Altair Engineering Pvt. Ltd., Bengaluru, India

© The Author(s), under exclusive license to Springer Nature Singapore Pte Ltd. 2025
S. K. Udgata et al. (eds.), *Intelligent Systems*, Lecture Notes in Networks and Systems 1314, https://doi.org/10.1007/978-981-96-3797-3_13

To address this challenge, a simpler model requiring less computation and overall, a less intensive process of detection is proposed. By using this model, computation can be shifted to the frontend, meaning that the device capturing the footage itself can perform the computation. Cameras equipped with basic GPUs or moderately good processors can effectively handle this task, freeing up the backend to focus on compiling the footage for further manual analysis. In summary, the proposed solution involves developing a simpler model for traffic detection that can be run on less powerful devices, allowing for more efficient use of resources and reducing the burden on the server or serving device [1].

RapidMiner Studio is a visual data science and machine learning platform that enables users to quickly and easily build predictive models and extract insights from their data. The software provides a drag-and-drop interface that allows users to visually design and execute data workflows, making it easy for non-technical users to perform complex data analysis tasks. RapidMiner Studio includes a variety of pre-built connectors, algorithms, and machine learning models that enable users to perform common data science tasks without having to write any code [2]. The platform's main principles are to make data science accessible to everyone, promote collaboration between teams, and provide a platform that is scalable and flexible to meet the needs of any business [3].

The Jetson Nano's high-performance computing capabilities and low power consumption make it an ideal platform for developing and deploying AI-based solutions at the edge. With its small form factor and low cost, the Jetson Nano enables developers and engineers to build intelligent devices that can perform tasks such as object detection, image classification, and speech recognition in real-time. Its applications include intelligent traffic systems, industrial automation, smart city infrastructure, and healthcare devices [4]. The Nvidia Jetson Nano can be used to accept camera feed and its computation capabilities can be exercised by using RapidMiner Studio or Altair Embed to develop a pipeline to process this feed and produce desired results.

2 Related Work

There is a significant body of research on the use of video-based systems for traffic monitoring and management. Many of these approaches use machine learning techniques to identify and classify vehicles in the scene. For example, [5] presents a system that uses a convolutional neural network (CNN) to classify vehicles in real-time. Uddin et al. [6] proposes the use of a combination of hand-crafted features and a support vector machine (SVM) for vehicle classification. Podbucki and Marciniak [7] uses a CNN to detect and track vehicles in a video stream, and [8] presents a system that combines color and shape features with a SVM for vehicle detection and classification.

Other approaches use image processing techniques for traffic monitoring. Gedkhaw [9] proposes the use of background subtraction and morphological operations for detecting vehicles in a video stream. Inthanon and Mungsing [10] uses a Kalman filter to track vehicles in a video stream and [11] presents a system that combines image segmentation with a Kalman filter for vehicle tracking.

RapidMiner has been used largely for data science projects focusing on data in a written format but has not been the first choice for image data unless the problem-solving process is tuned toward its advantages. As seen in the video, [12] the idea of converting a problem of monitoring activities of workers in an industry into a classification problem was a large part of inspiration behind this paper.

While these approaches have demonstrated promising results, they often rely on hand-crafted features or a fixed set of image processing techniques, which can limit their performance in real-world scenarios. In contrast, this approach uses a combination of image processing techniques and machine learning algorithms to detect and classify vehicles in a video stream, allowing it to adapt to different lighting and environmental conditions.

3 Theoretical Framework

One of the challenges of analyzing traffic using computer vision is that it requires a large number of computational resources. To simplify the process, the program skips analyzing every frame in the video stream and instead scans frames for high vehicle density at intervals. Analyzing frames at intervals will not hamper the detection of traffic in anyway as traffic itself, is said to occur if the vehicle density remains high over a significant period of time. Hence, spans of downtime in analysis for a few seconds will not reduce the accuracy of detection in any manner. On the upside, this would help to reduce the computational burden which would occur if the device was analyzing every frame in the video stream. Figure 1 indicates the traffic hotspots by the highlighted box. Furthermore, the program can focus on specific zones where vehicles are likely to stop, such as areas near traffic signals. By manually selecting these hotspots, the program can accurately gauge the vehicle density in that particular frame and come to a conclusion about the vehicle density. This approach also helps to reduce the amount of data that needs to be processed by the program, as it can ignore areas where vehicles are less likely to stop or gather like the sky, grassy areas, and footpaths.

Overall, this approach helps to reduce the computational burden of analyzing every frame in a video stream and focuses on specific areas where vehicle density is likely to be high, allowing for more efficient and accurate traffic analysis. By optimizing the analysis process and focusing on specific areas of interest, this approach can significantly reduce the computational resources required for traffic analysis, making it a more feasible solution for a wider range of devices and systems.

Fig. 1 Traffic hotspots indicated by the highlighted box

4 Implementation

Implementing the entire workflow in RapidMiner requires many steps as is the case with most deep learning related problems.

4.1 Data Collection

The process for collecting pictures of cars and clear roads from a video stream is described as follows:

Step 1: A ready-made code using one of the YOLO models (in this case, YOLOv5m) is used to detect vehicles, such as cars, trucks, and bikes, in the sample footage. Frames are extracted from the video stream after every few seconds/frames, and instead of drawing rectangles, images are cropped using the '*x*' and '*y*' coordinates and width and height parameters obtained.
Step 2: The cropped images of vehicles are stored in a folder named 'present'. To obtain pictures of a clear road, the coordinates of regions of interest (ROI) are found by running another program that captures these coordinates using click events on a frame of the video stream which are triggered by the user.
Step 3: The code gathers the '*x*' and '*y*' values of the click and appends them to a 'rows' array, which will be inserted into a new csv file at the end. The top-left and bottom-right corners of the ROI are clicked to get the coordinates. Using these coordinates, random frames of the video stream are cropped at the position defined by the coordinates and stored in another folder named 'absent' using another python program or RapidMiner.
Step 4: These are random frames, so they may or may not contain vehicles. One can either manually check if there are vehicles captured within these cropped images by looking into the folder or pass these images into the same yolo model used in the previous steps and make the model store the pictures containing vehicles into the 'present' folder and leave the rest in the 'absent' folder.

Step 5: Once sorted, two other folders were made; one for Training and the other for Testing. A set of 'absent' and 'present' folders under each folder and moved about 30% of the absent and present set into the Testing set of folders. Even this process can be replicated in RapidMiner.

4.2 Building the Model

With all the prerequisites in place and the data organized, the process using RapidMiner can be initiated [13]. Extensions employed in the subsequent steps include:

1. Deep Learning
2. Image Processing
3. Image Handling
4. ND4J
5. Operator Toolbox

To extract images from the Training and Testing directories in RapidMiner, two distinct Read Image Meta Data operators are utilized. The 'use label' parameter is activated for both to ensure proper functionality. Accurate naming of sub-folders is essential. Subsequently, the training dataset is directed through a Shuffle operator, followed by a Pre-Process Images operator. Here, the Resize Image operator is linked to adjust the image dimensions to 200 * 150 pixels or any other specified dimensions.

Conversely, the testing dataset undergoes shuffling without resizing. This is because the images within have been previously resized externally using a Python script. While the Resize Image operator could be employed for this purpose, the Pre-Process Images operator is avoided, as it converts the images into tensors at the output. Such a transformation is unnecessary for the test images, which are required in their original image format. The resized images can be subsequently accessed using an additional Read Image Meta Data operator. In the subsequent step, the tensor objects produced by the Pre-Process Images operator are directed into the Deep Learning (Tensor) operator.

Within this operator, considerable time was invested in determining the optimal configurations suitable for the system. The PC system lacks a GPU, limiting the capability to simply stack numerous layers with an abundance of neurons and expect seamless functionality. Similarly, even the Jetson Nano's GPU has its constraints, unable to manage extensive operations beyond a specific threshold. Table 1 delineates the parameters of the selected model, while Table 2 provides insights into its structural composition.

The rest of the process (shown in Fig. 2) is about grouping the pre-processing conditions generated by the Pre-Process Images operator along with the resulting model and applying to the test images in the Apply Model (Generic Operator). This generates a data table for every image and these tables are appended into one table

Table 1 Parameters of the model

Parameter	Value
Optimization algorithm	Stochastic gradient descent
Update algorithm	Adam
Learning rate	0.01
Error	1.00E-06
Beta 1	0.9
Beta 2	0.999
Weight initialization	Xavier
Epochs	50

Table 2 Parameters of each layer in the model

Convolution layer				
No. of activation maps	Kernel size	Stride size	Activation function	Dilation factor
8	2 × 2	1 × 1	ReLu	1 × 1
Pooling layer				
Pooling method		Kernel size	Stride size	
MAX pooling		2 × 2	1 × 1	
Dropout layer				
Dropout rate				
0.1				
Fully connected L1				
No. of neurons		Activation function		
4		ReLu		
Fully connected L2				
No. of neurons		Activation function		
3		ReLu		
Output layer				
Output type	Loss function	Activation function	No. of neurons	
Fully connected	Multiclass cross entropy	Softmax	2	

using the Append operator and this data table is passed into the Performance operator to gauge the results (discussed in the Sect. 5).

This model (shown in Fig. 3) is used in the next process wherein, the detection will take place on the same road but with unseen scenarios will be tested. Note that it is important for the camera to be fixed at the same position during training and testing else the hotspot locations will also vary if any change is made to the position.

Fig. 2 Sample of data in the present folder (left) and absent folder (right)

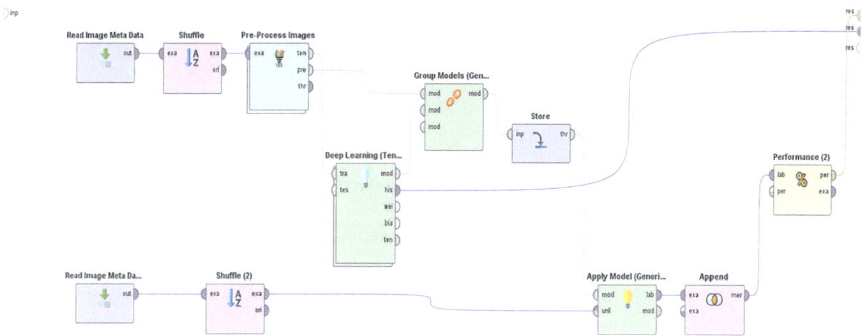

Fig. 3 Overview of the model building process

4.3 Building the Detection Process from a Video Stream

In a real-world scenario, the trained model is to be applied on the next day's footage to determine traffic density practically. This process involves several steps, including sampling the footage at a specified interval, generating cropped images of potential traffic hotspots, classifying whether the cropped images contain vehicles or not, aggregating the results to classify the vehicle density at the current timestamp as low, medium, or high, and observing the results over multiple timestamps to determine whether traffic has occurred.

The video footage referenced in this paper operates at an accelerated 8 × tempo. This implies that each second of the footage captures events that originally unfolded over an 8-s span. Given an FPS (Frames per Second) value of 30, which corresponds to a real-time, 1 × tempo footage duration of 8 s, it's deduced that to capture a frame from every 8-s segment, the sampling rate should be maintained at 30 FPS. Consequently, the incoming video stream is sampled at this rate of 30 FPS to ensure consistent and accurate representation.

The steps followed in designing the process shown in Fig. 4 are:

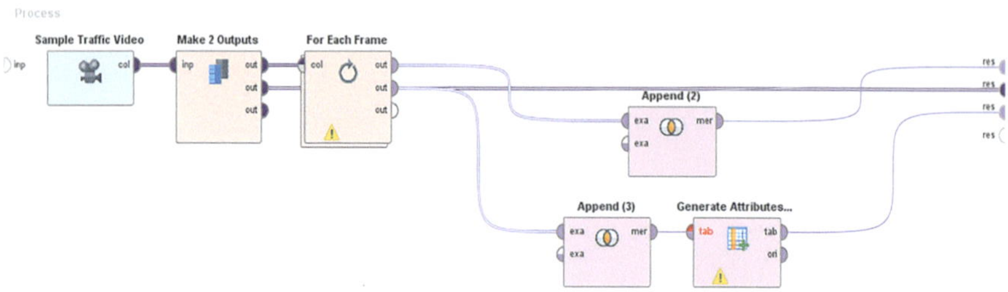

Fig. 4 Main detection process

Step 1. To get a collection of images from the video stream

Loading of the video stream and sampling it at the required FPS is done using the Read Video operator which has a parameter to sample the video at 30 FPS and output the same as a Collection of IOObjects (the way RapidMiner recognizes images).

Step 2. Cropping and storing the hotspots of every sampled frame in a directory

The CSV file, encompassing all previously derived coordinates, is integrated into the ongoing process. Given that these coordinates necessitate parsing for each sampled image from the stream, a Loop Collection operator is employed to iterate through each frame. Within this loop, two pivotal operators come into play: the Read CSV and the Create Directory operators. The former ingests the CSV file, while the latter establishes a new directory. For clarity and organization, the directory's nomenclature is anchored to the iteration macro of the Loop Collection operator. Post the Read CSV operation, a Data Table materializes, housing rows of coordinates represented by *X*_min, *Y*_min, *X*_max, and *Y*_max values. To process these values, they are channeled into a subsequent Loop operator. Herein, the 4 Extract Macro operators disassemble the values into individual macros. Subsequently, these macros are funneled into the Crop Image operator. The width and height are computed by discerning the differences between the min and max values. A crucial subsequent step involves resizing the image. Without this resizing, the model's predictions may veer off course. The Resize Image operator addresses this necessity. The culminating act in this sequence involves storing the resized image within the directory, achieved through the Write Image operator. The nomenclature of this stored image mirrors the iteration macro from the nested Loop operator.

Step 3. Retrieve the saved model and apply it to all the cropped images of the current frame

Utilizing a model constructed within a RapidMiner process simplifies its retrieval, accomplished seamlessly through a Retrieve operator. To process the images from the freshly minted directory, the Read Image Meta Data operator is deployed. Both this operator and the Retrieve operator feed into the Apply Model (Generic) operator. Within this context, the model is invoked to evaluate the images. Subsequently, the 'lab' output, which encapsulates the model's predictions or classifications, is channeled into the Append Operator. This facilitates the construction of an Example Set, representing a comprehensive set of parameters for every 'example' within the data table.

Step 4. Count the images containing vehicles

The resulting output comprises nominal values, specifically 'present' or 'absent'. To transform these categorical labels into a numerical format, a dummy coding approach is adopted. This transformation is facilitated by directing the output through the Nominal to Numeric operator. Given the desire to focus solely on a specific subset of attributes, the Select operator is employed. This ensures that only the dummy coding values corresponding to 'present', the image path, and optionally, the confidence associated with the presence, are retained and processed further. A macro called 'present_count' is set to 0 using the Set Macro operator and this macro will hold the value of total images containing vehicles. As this value is within the Loop Collection operator, for every frame from the video feed, this value is reset to zero every time the loop runs. The modified Example Set is passed to a Loop Examples operator in which the dummy coded present value (either 0 or 1) is stored using the Extract Macro operator and this value is passed to another Set Macro operator which updates the present_count macro by adding the obtained present value. This loop executes for every cropped image.

Step 5. To grade the vehicle density of the current frame categorically and numerically

The output of the Loop Examples operator is fed to the Generate Attributes operator in which three attributes called Vehicle Density and Vehicle Density (Numeric) are generated. The logic for generating these attributes are as follows:

$$\begin{aligned}&\text{Vehicle Density} : \text{if(eval(\%\{present\_count\})} <= 5,\\&\text{if(eval(\%\{present\_count\})} <= 2, \text{"LOW", "MEDIUM"), "HIGH")}\end{aligned} \tag{1}$$

The Eq. (1) is a nested if condition which puts a value of LOW if the count of present vehicles is 2 or less, MEDIUM if it is between 3 and 5 and HIGH if it is above 5.

$$\begin{aligned}&\text{Vehicle Density(Numeric)} : \text{if(eval(\%\{present\_count\})}\\&<= 5, \text{if(eval(\%\{present\_count\})} <= 2, 1, 2), 3)\end{aligned} \tag{2}$$

In Eq. (2), the values 1, 2, and 3 are allocated to LOW, MEDIUM, and HIGH respectively, providing a numerical categorization of the traffic status for the purpose of plotting the data.

Step 6. Modifying the Result

At this juncture, the Example Set encompasses the cropped regions, each annotated with its respective Vehicle Density grade corresponding to its frame. It's pertinent to note that while multiple cropped frames may be present, they all bear identical grades. Consequently, to extract the grade, there's no imperative need to analyze every individual cropped frame. To streamline this process, a "Filter Example Range" operator is employed, facilitating the selection of solely the initial image from the Example Set for further evaluation. This completes the process inside the Loop Collection operator. The outputs are both the Example Set containing info of all the cropped images and an Example Set containing just the first example from that frame. Externally, both outputs manifest as Collections of Example Sets. Subsequently, they are directed toward distinct Append Operators, facilitating the amalgamation into their corresponding Example Sets. This consolidation is pivotal, as it ensures that data can be subsequently extracted from the set in a coherent and interpretable manner.

Step 7. Obtaining the Traffic Status

Employing the described traffic detection methodology [14, 15] introduces the possibility that, even when vehicles are moving unhindered, a few might inadvertently be captured within the ROIs of a given frame. Consequently, the model might misinterpret this as an indication of HIGH vehicle density. However, such an observation doesn't necessarily signify the presence of traffic at that specific moment. To discern genuine traffic occurrences and mitigate potential misinterpretations, it becomes imperative to focus on continuous intervals encompassing 1–2 consecutive instances of either HIGH or MEDIUM traffic grades exclusively. Figure 4 depicts the main detection process; Fig. 5 depicts the detection process for each frame and Fig. 6 depicts the process for cropping ROIs in each frame.

The Example Set containing the first cropped region from all the frames is passed into another Generate Attributes operator which produces a Traffic Status attribute whose logic is:

$$
\begin{aligned}
&\text{Traffic Status : if}\big(\big[\text{Vehicle Density(Numeric)}\big]\\
&== 1, \text{"Clear"}, \text{if}\big(\big[\text{Vehicle Density(Numeric)}\big]\\
&== 2, \text{"Building"}, \text{if(lag("Vehicle Density(Numeric)", 1, INTEGER)}\\
&== 3, \text{"Traffic"}, \text{if(lead("Vehicle Density(Numeric)", 1, INTEGER)}\\
&== 3 || \text{lead("Vehicle Density(Numeric) ", 1, INTEGER)}\\
&== 2, \text{"Traffic"}, \text{"Building"))))}
\end{aligned}
\tag{3}
$$

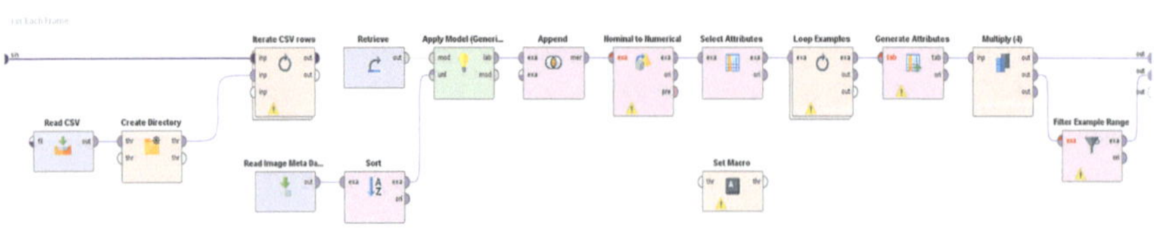

Fig. 5 Detection process for each frame

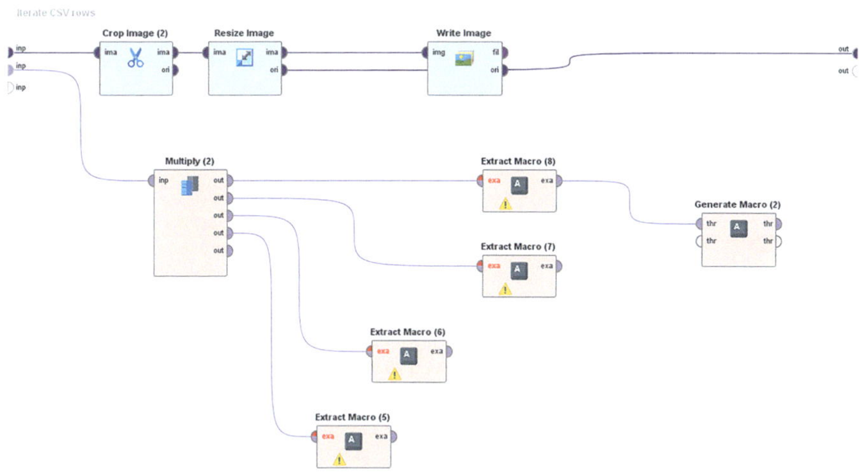

Fig. 6 Process for cropping ROIs in each frame

By applying Eq. (3), the traffic status is classified as 'Clear' for a numeric value of '1' and as 'Building' for a value of '2'. The status is designated as 'Traffic' only in instances where there are two consecutive grades of '2' followed by '3'.

$$\begin{aligned}
&\text{Traffic Status(Numeric)} : \text{if}([\text{Vehicle Density(Numeric)}] \\
&== 1, 1, \text{if}([\text{Vehicle Density(Numeric)}] \\
&== 2, 2, \text{if}(\text{lag}(\text{"Vehicle Density(Numeric)"}, 1, \text{INTEGER}) \\
&== 3, 3, \text{if}(\text{lead}(\text{"Vehicle Density(Numeric)"}, -1, \text{INTEGER}) \\
&== 3 || \text{lead}\ (\text{"Vehicle Density(Numeric)"}, -1, \text{INTEGER}) == 2, 3, 2)))) \qquad (4)
\end{aligned}$$

Utilizing Eq. (4), the categories Clear, Building, and Traffic are converted into numeric values: 1, 2, and 3, respectively. This method, analogous to the approach taken with Vehicle Density, enables the visualization and plotting of the unique characteristics associated with each category, enhancing the clarity and analysis of their attributes and variations.

$$\text{Order} : \text{row\_number}() \qquad (5)$$

Equation (5) gives us an attribute which is like a serial number generator. This is in order to arrange the images chronologically while appending them into a single Example Set otherwise the plot will inaccurately reflect the traffic conditions. This output with the new attributes is fed as the result along with the frames obtained and the Example Set containing all the regions' statistics for manual verification (if needed). This completes the process. The same process is run on the Nvidia Jetson Nano using the Linux version of RapidMiner after copy-pasting the XML code generated on the PC version onto the Jetson Nano version.

Table 3 Statistics of the model

Parameter	Value
Average recall	96.80%
Average precision	97.56%
Average accuracy	97.39%

5 Results and Observations

5.1 Results of Overall Model

The model used in this process was developed keeping in mind that the model's task is to basically determine whether a vehicle exists in a particular frame or not. The stats are shown in Table 3.

During the training process, the test data comprised of 153 images out of which 97 images did not have a vehicle while 56 of them had vehicles in it. The higher number of images containing no vehicles is because during initial testing, it was found that this approach struggles to identify empty spots on the road as it sometimes mistakes it for darker colored cars/trucks. To get such good results, it necessary to train the camera stationed at a particular junction/signal with the images of vehicles it can expect to see. For example, if it is near a police station, it needs to recognize a police jeep even though it has a darker color which is close to the shade of the road.

5.2 Results of Traffic Detection Process

The video length of sample footage taken for this paper lasts for about 1 min and 7 s at 8 $\times$ tempo. As discussed in the previous sections, the sampling was taken at 30 FPS. For a total of 27 frames, it was able to classify higher levels of vehicle density with good accuracy. This is reflected in the results displayed in Table 4 and Fig. 7.

Table 4 presents the culmination of the example set, highlighting a singular hotspot from each frame. As previously discussed, every hotspot within a single frame is classified with an identical Vehicle Density as its counterparts. Although the model does not depict instances where moving cars are captured within these hotspots in the provided example, such occurrences are feasible. In Fig. 7, the *X*-axis represents the Order Number of the frame, while the *Y*-axis illustrates the numerical representation of the Traffic Status. Notably, distinct traffic patterns emerge between frames 10 and 13, as well as between frames 19 and 23. Given that frames are captured at intervals of 8 s within a standard, 1 $\times$ tempo footage, it can be deduced that the traffic spanned approximately 24 s from frames 10–13. Similarly, the traffic observed between frames 19 and 23 is estimated to have lasted around 32 s.

Figure 8 shows the setup of the system. The Jetson Nano is operating in 'headless' mode which means that the only cables connected to it are the Ethernet, micro-USB

Table 4 Results from the traffic detection process

Row No.	Confidence (vehicle is present)	Prediction given by model	Vehicle density	Vehicle density (numeric)	Traffic status	Traffic status (numeric)	Order
1	1.000	1	Medium	2	Building	2	1
2	1.000	1	Medium	2	Building	2	2
3	0.031	0	Low	1	Clear	1	3
4	0.031	0	Low	1	Clear	1	4
5	0.031	0	Low	1	Clear	1	5
6	0.031	0	Low	1	Clear	1	6
7	0.031	0	Low	1	Clear	1	7
8	0.031	0	Low	1	Clear	1	8
9	0.031	0	Low	1	Clear	1	9
10	1	1	Low	1	Clear	1	10
11	1	1	High	3	Traffic	2	11
12	1	1	High	3	Traffic	3	12
13	0	0	Low	1	Clear	1	13
14	0	0	Low	1	Clear	1	14
15	0	0	Low	1	Clear	1	15
16	0	0	Low	1	Clear	1	16
17	0	0	Low	1	Clear	1	17
18	0	0	Low	1	Clear	1	18

ExampleSet (27 examples, 1 special attribute, 7 regular attributes)

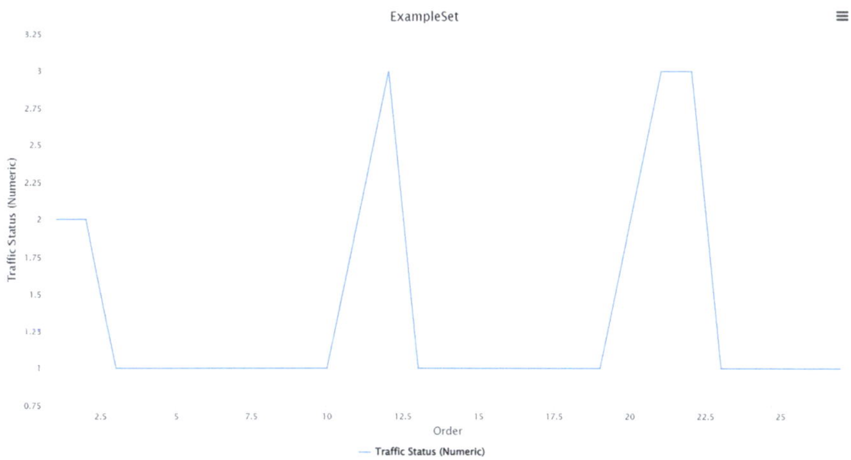

Fig. 7 Graphical depiction of the traffic status

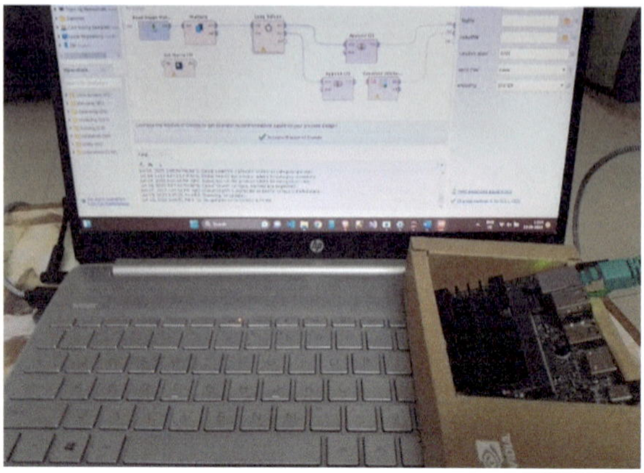

Fig. 8 Experimental setup

Fig. 9 Ports of the Jetson Nano (2 GB)

and power cable which is type-C. The device can be boosted to its max performance by supplying with 10 W of power.

In a real-world application, a camera could be linked to the USB 3.0 port, while the Ethernet connection would be supplanted by a Wi-Fi dongle. This dongle could then transmit data to the cloud either with a set delay or at regular time intervals. The workflow depicted in Fig. 8's laptop display emanates from the Jetson Nano, which is tethered via a Micro USB cable and ports in Fig. 9. Access to this setup is facilitated by an application named NoMachine. At the time of this paper's composition, the ARM version remains inaccessible. Consequently, while processes can be constructed within the machine, executing them leads to java-sdk-related issues.

6 Conclusion and Future Scope

In conclusion, this method poses to be a very simple but effective way of gauging high levels of traffic without making things too complicated for a person from a non-scientific background to grasp. Using RapidMiner makes the formulation of the process even simpler without having the need to alter any sort of code within the model building and detection processes. Along with this, one doesn't need to spend much time in classifying vehicles as there are plenty of open-source models in today's day that will help us make the work easier.

The comparison among the three vehicle detection systems reveals distinct trade-offs. Inductive Loop Detectors offer unparalleled accuracy within microseconds but are costly and demand extensive data, with infrequent maintenance. Current Video Detectors provide millisecond accuracy, but their expense and susceptibility to weather, coupled with the need for substantial data and frequent maintenance, pose challenges. The Proposed Model sacrifices precision, maintaining accuracy in seconds, yet stands out for its cost-effectiveness, requiring minimal data and infrequent maintenance. Ultimately, the choice depends on specific application requirements, budget considerations, and acceptable compromises between accuracy, cost, data needs, and maintenance frequency.

This work can further be extended in gauging the direction of vehicle traffic that takes place in a cross or a 'T' junction similar to the one used in this research work. Also, there could be a case where this system is used in a camera near a hospital which has to receive many patients frequently, and over there it would be needed to track the time an ambulance takes to navigate through that area of high vehicle density. If such a case exists, then the road either needs to be widened or a separate space must be allocated for the passage of ambulances. Another thing that remains not tested is the accuracy of the model when the conditions of sunlight and street light change during the course of the day. The perceived shade of the road would also change according to the amount of light exposure and would serve as a challenge to this system.

References

1. Prakash N, Kamath CU, Gururaja HS (2020) Detection of traffic violations using moving object and speed detection. Wutan Huatan Jisuan Jishu 16(6):61–72
2. Jovanovic ZM, Vukicevic M, Delibašić B, Suknovic M (2013) Using RapidMiner for research: experimental evaluation of learners. In: Markus H, Ralf K (eds) RapidMiner, pp 439–454
3. Prawinsankar D, Gunasekaran M, Gopalakrishnan B, Purusothaman P (2021) Traffic congestion detection through modified Resnet50 and prediction of traffic using clustering, smart technologies. In: Communication and robotics (STCR), pp 1–5
4. Laksh C, Gururaja HS (2023) Attendance monitoring system for academic institutions using Nvidia Jetson Nano. TIJER Int Res J 10(7):153–156. ISSN: 2349-9249
5. Rehman SU, Razzaq MR, Hussian MH (2021) Training of SSD (single shot detector) for facial detection using Nvidia Jetson Nano. arXiv.org: 2105.13906
6. Uddin IMd., Alamgir SMd., Rahman MMd, Bhuiyan SM, Moral AMd (2021) AI traffic control system based on deepstream and IoT using NVIDIA Jetson Nano. In: 2nd international conference on robotics, electrical and signal processing techniques (ICREST), pp 115–119
7. Podbucki K, Marciniak T (2022) Aspects of autonomous drive control using NVIDIA Jetson Nano microcomputer. In: Conference on computer science and intelligence systems, vol 17, no 30, pp 117–120
8. Vijitkunsawat W, Peerasak C (2020) Comparison of machine learning algorithm's on self-driving car navigation using Nvidia Jetson Nano: In: 17th International conference on electrical

engineering/electronics, computer, telecommunications and information technology (ECTI-CON), pp 201–204
9. Gedkhaw E (2022) The performance of Thai sign language recognition with 2D convolutional neural network based on NVIDIA Jetson Nano developer kit. TEM J 11(1):411–419
10. Inthanon P, Mungsing S (2020) Detection of drowsiness from facial images in real-time video media using Nvidia Jetson Nano. In: 17th International conference on electrical engineering/electronics, computer, telecommunications and information technology (ECTI-CON), pp 246–249
11. Matthews JS, Leger St. A (2020) Energy-efficient analysis of synchrophasor data using the NVIDIA Jetson Nano. In: IEEE high performance extreme computing conference (HPEC), pp 1–7
12. Tangudu S, Sankanal DA, Chidambaran S (2021) Battery state of health and remaining useful life prediction using machine learning model. In: Symposium on international automotive technology 2021, 21SAT-1043
13. Saabith SLA, Fareez Mhd MM (2018) Comparative analysis of various tools for data mining and big data mining. Int J Eng Res Technol (IJERT) 7(11):239–245
14. Varma T, Bose S, Colaco C, Bagwe M (2020) Traffic detection using OpenCV. J Interdisc Cycle Res 12(5):389–396
15. Alshareef A, Albeshri A, Katib I, Mehmood R (2020) Road traffic vehicle detection and tracking using deep learning with custom-collected and public datasets. IJCSNS Int J Comput Sci Netw Secur 20(11):9–21

Enhancing DOS Attack Detection in IoT Networks Using Deep Learning

Arepalli Peda Gopi, Shaik Jasmin, Kamireddy Poojitha, Kondamudi Sravya Keerthana, and Medida Sravani

Abstract IoT devices are widely employed in business and industrial settings as well as daily life these days. Moreover, it will broaden the DOS attack surface. It is not possible to apply additional safety restrictions to an Internet-connected gadget because of its limited capacity. DOS attacks by malevolent actors with the ability to take over IoT devices continue to be exceedingly dangerous. We use the computer traffic network dataset in this article, which contains data with at least 50% compromised. We suggest lightweight layered LSTM-GRU for DOS attack detection. The significance of combining LSTM and GRU in DL for IoT lies in the model's ability to improve performance, efficiently handle IoT data, reduce overfitting, generate interpretable features, enhance anomaly detection, adapt to dynamic environments, and generalize across different IoT domains. The deep learning-based models that have been suggested have shown outstanding performance, with high F_1 score, recall, accuracy, and precision.

Keywords Deep learning · Denial-of-service attack · Long short-term memory · Gated recurrent unit · Internet of Things

1 Introduction

Internet of Things (IoT) devices [1] has transformed the way we interact with technology and our surroundings. While IoT offers numerous benefits, it also brings forth security challenges, including the vulnerability to cyberattacks. One particularly disruptive form of attack is the denial-of-service (DOS) [2] attack, which can incapacitate IoT networks, rendering them unresponsive and, in some cases, causing severe disruptions. The constantly shifting and varied nature of IoT devices means that traditional techniques for recognizing DOS assaults are sometimes insufficient. Therefore, in order to improve DOS attack detection in IoT networks, this research

A. P. Gopi · S. Jasmin (✉) · K. Poojitha · K. S. Keerthana · M. Sravani
Department of CSE, Vignan's Nirula Institute of Technology and Science for Women, Guntur, India
e-mail: shaikjasmin2003@gmail.com

© The Author(s), under exclusive license to Springer Nature Singapore Pte Ltd. 2025

S. K. Udgata et al. (eds.), *Intelligent Systems*, Lecture Notes in Networks and Systems 1314, https://doi.org/10.1007/978-981-96-3797-3_14

investigates the application of deep learning algorithms [3]. This research examines how deep learning models may be used to detect, classify, and manage denial-of-service assaults against Internet of Things devices. By harnessing the power of artificial neural networks [4] and advanced algorithms, we aim to develop a more resilient and adaptable approach to safeguarding IoT ecosystems [5] from the pernicious threat of DOS attacks. This research not only addresses the immediate security concerns in IoT networks but also contributes to the ongoing efforts to fortify the infrastructure supporting the IoT revolution. We are used long short-term memory (LSTM) and gated recurrent unit (GRU).

Advanced DL techniques such as GRU and LSTM have demonstrated potential in various applications, notably enhancing the detection of DOS attacks in IoT networks. These algorithms are especially well-suited to handle the complex and dynamic nature of security risks in IoT contexts [6] because they provide notable advantages while processing sequential data. In the context of IoT network security, where a multitude of devices are interconnected, the introduction of LSTM and GRU algorithms [7] serves as a pivotal advancement. These algorithms excel at capturing temporal dependencies within data streams, enabling them to learn and adapt to emerging patterns of malicious activities. With the ever-increasing number of IoT devices and the potential vulnerabilities they introduce, the need for robust and accurate DOS attack detection has never been greater. In this paper, it explores the integration of LSTM and GRU algorithms in IoT security, demonstrating how they can effectively bolster the network's defenses against DOS attacks [8]. This will contribute to the safeguarding and authenticity of interconnected devices and the information they share.

2 Literature Survey

In 2023, Keshk et al. [1] introduced a groundbreaking intrusion detection system tailored for Internet of Things (IoT) environments. The primary motivation of this method improves interpretability for better decision-making and enhances performance in accurately detecting and responding to intrusions. While also it improves systems overall performance. With an accuracy rate of 78.7%, a precision score of 97%, and an F_1 score of 77.6%, the SPIP framework performed admirably.

In 2023, Bhayo et al. [2] proposed a study is to develop a machine learning-based framework that can effectively detect and mitigate distributed denial-of-service attacks in SD-IoT networks, as discussed in a comprehensive analysis of the job description. The framework will use machine learning methods to evaluate network traffic data and identify DDOS in real time.

Jullian et al. [3] demonstrated impressive performance, achieving a 99.97% accuracy rate and a 99.99% recall in FFNN. In the LSTM, the results showed 98.67% accuracy, 99.90% precision, and 99.95% recall. The authors clarify that poorer performance in a distributed setting is not always indicative of poorer performance in a centralized paradigm. Instead of building a model for each dataset, it may be able to train a more generic model using distinct datasets for each fog node.

3 Proposed Methodology

A stacked LSTM-GRU model comprises multiple layers of LSTM and GRU cells stacked on top of each other. Each layer in the stack receives input from the previous layer and propagates its output to the next layer (Fig. 1).

3.1 Algorithm Overview

Define hyperparameters: Set the number of LSTM and GRU layers, the number of neurons in each layer, the input shape and other training-related hyperparameters.

Initialize weights and biases: Initialize the weights and biases for each LSTM and GRU layer using random values or predefined initializers.

Forward pass: For each time step t in the input sequence. Pass the input through each LSTM and GRU layer in a sequential manner. Compute the outputs and hidden states for each layer.

3.2 For Each LSTM Layer

Input Gate (i_t):

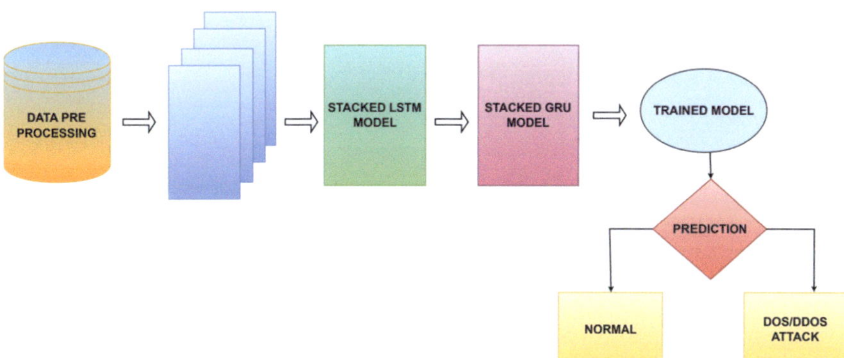

Fig. 1 Architecture

$$i_t = \text{sigmoid}\left(W_{i_{\text{lstm}}} * \left[h_{(t-1)}, x_t\right] + b_{i_{\text{lstm}}}\right) \tag{1}$$

Forget Gate (f_t):

$$f_t = \text{sigmoid}\left(W_{f_{\text{lstm}}} * \left[h_{(t-1)_l}, x_t\right] + b_{f_{\text{lstm}}}\right) \tag{2}$$

Output Gate (o_t):

$$o_t = \text{sigmoid}\left(W_{o_{\text{lstm}}} * \left[h_{(t-1)_l}, x_t\right] + b_{o_{\text{lstm}}}\right) \tag{3}$$

Candidate Cell State (g_t):

$$g_t = \tan h\left(W_{c_{\text{lstm}}} * \left[h_{(t-1)_l}, x_t\right] + b_{c_{\text{lstm}}}\right) \tag{4}$$

Cell State (c_t):

$$c_t = f_t * c_{(t-1)_l} + i_t * g_t \tag{5}$$

Hidden State (h_t):

$$h_t = o_t * \tan h(c_t) \tag{6}$$

3.3 *For Each GRU Layer*

Reset Gate (r_t):

$$r_t = \text{sigmoid}\left(W_{r_{\text{gru}}} * \left[h_{(t-1)_g}, x_t\right] + b_{r_{\text{gru}}}\right) \tag{7}$$

Update Gate (z_t):

$$z_t = \text{sigmoid}\left(W_{z_{\text{gru}}} * \left[h_{(t-1)_g}, x_t\right] + b_{z_{\text{gru}}}\right) \tag{8}$$

Candidate Hidden State ($\tilde{h}_t$):

$$\tilde{h}_t = \tan h\left(W_{h_{\text{gru}}} * \left[r_t * h_{(t-1)_g}, x_t\right] + b_{h_{\text{gru}}}\right) \tag{9}$$

Hidden State (h_t):

$$h_t = (1 - z_t) * h_{(t-1)_g} + z_t * \tilde{h}_t \tag{10}$$

Backward pass (Training): Calculate the loss between the predicted output and the ground truth. Compute the gradients using backpropagation through time (BPTT).

Update the weights and biases of each layer using an optimization algorithm (e.g., gradient descent, Adam, etc.).

Prediction: The trained model can be utilized to generate predictions on novel data. Employing the trained model, one can generate forecasts on fresh data. Utilizing the trained model, predictions can be made on previously unseen data.

Equations: Let's define the equations for both LSTM and GRU cells. For simplicity, I'll consider a single LSTM and GRU layer, but in a stacked model, these equations would be repeated for each layer.

Algorithm: Enhancing DOS Attack Detection in IoT Networks Using Deep Learning

1. **Input**: Model Parameters (Weights and Biases).
2. **Output**:

   # Calculate LSTM gates and Cell gate
3. For each input of LSTM layer (x_t, h_{t-1})
4. Calculate input gate i_t using input and previous hidden state by Eq. 1.

$$i_t \leftarrow \left\{W_{i_{\text{lstm}}}, h_{t-1}, x_t, b_{i_{\text{lstm}}}\right\}$$

5. Calculate forget gate f_t using input and previous hidden state by Eq. 2.

$$f_t \leftarrow \left\{W_{f_{\text{lstm}}}, h_{(t-1)_l}, x_t, b_{f_{\text{lstm}}}\right\}$$

6. Calculate output gate o_t using input and previous hidden state by Eq. 3.

$$o_t \leftarrow \left\{W_{o_{\text{lstm}}}, h_{(t-1)_l}, x_t, b_{o_{\text{lstm}}}\right\}$$

7. Calculate candidate cell state g_t using input and previous hidden state by Eq. 4.

$$g_t \leftarrow \left\{W_{c_{\text{lstm}}}, h_{(t-1)_l}, x_t, b_{c_{\text{lstm}}}\right\}$$

8. Calculate new cell state C_t by Eq. 5.

$$C_t \leftarrow \{f_t, g_t, f_t\}$$

9. Calculate new hidden state H_t by Eq. 6.

$$H_t \leftarrow o_t, \tanh, c_t$$

10. For every input of GRU layer
11. Calculate Reset gate r_t by Eq. 7.

$$r_t \leftarrow \left\{W_{r_{\text{gru}}}, h_{(t-1)_g}, x_t, b_{r_{\text{gru}}}\right\}$$

12. Calculate Update gate z_t by Eq. 8.

$$z_t \leftarrow \{W_{z_{\text{gru}}}, h_{(t-1)_g}, x_t, b_{z_{\text{gru}}}\}$$

13. Calculate Candidate hidden state $\tilde{h}_t$ by Eq. 9.

$$\tilde{h}_t \leftarrow \{W_{h_{\text{gru}}}, r_t, h_{(t-1)_g}, x_t, b_{h_{\text{gru}}}\}$$

14. Calculate Hidden state h_t by Eq. 10.

$$h_t \leftarrow \{h_{(t-1)_g}, z_t, \tilde{h}_t\}$$

15. Calculate the loss that exists between the actual truth and the anticipated output.
16. Compute the gradients using backpropagation through time (BPTT).
17. Update the weights and biases of each layer using an optimization algorithm
18. End

4 Experimental Work

4.1 Data Set

Computer Network Traffic: https://www.kaggle.com/datasets/crawford/computer-network-traffic

The "Computer Network Traffic" dataset was employed in the research study titled "Enhancing DOS Attack Detection in IoT Networks Using Deep Learning." Prior to utilizing deep learning techniques for cybersecurity purposes, the dataset underwent preprocessing, which involved feature engineering, data cleansing, and normalization. This preprocessing step aimed to optimize the dataset for deep learning analysis. In order to differentiate between regular network traffic and DOS attacks, labels were assigned to individual data points within the dataset. Subsequently, the dataset was divided into data sequences, typically with a predetermined sequence length. These data sequences served as the input data for LSTM-GRU models, which were specifically designed to analyze the temporal relationships and patterns within network traffic. The LSTM and GRU layers of these models are well-known for their ability to identify long-term relationships in sequential data.

4.2 Results and Discussions

We have considered two existing models and compared them with our proposed model. The comparison and results are below. The six graphs compare the performance of stacked long short-term memory (LSTM) and gated recurrent unit (GRU)

models against existing variety of metrics. The x-axis of each graph is the number of epochs or training iterations.

In Fig. 2, the proposed model accuracy values are compared with the accuracy values of the other two existing models, named as existing model 1 and existing model 2. Compared to the accuracy values of the existing paper 1 and existing paper 2, the accuracy of the proposed model is predicted to be high where the stacked LSTM and GRU are used. In Fig. 3, the proposed model loss function values are compared with the loss function values of the other two existing models, named as existing model 1 and existing model 2. Compared to the loss function values of the existing paper 1 and existing paper 2, the loss function values of the proposed model are predicted to be low where the stacked LSTM and GRU are used.

In Fig. 4, the proposed model precision values are compared with the precision values of the other two existing models, named as existing model 1 and existing model 2. Compared to the precision values of the existing paper 1 and existing paper 2, the precision of the proposed model is predicted to be high where the stacked LSTM and GRU are used. In Fig. 5, the proposed model recall values are compared with the recall values of the other two existing models, named as existing model 1 and existing model 2. Compared to the recall values of the existing paper 1 and existing paper 2, the recall of the proposed model is predicted to be high where the stacked LSTM and GRU are used.

In Fig. 6, the proposed model F_1 score values are compared with the F_1 score values of the other two existing models, named as existing model 1 and existing model 2. Compared to the F_1 score values of the existing paper 1 and existing paper 2

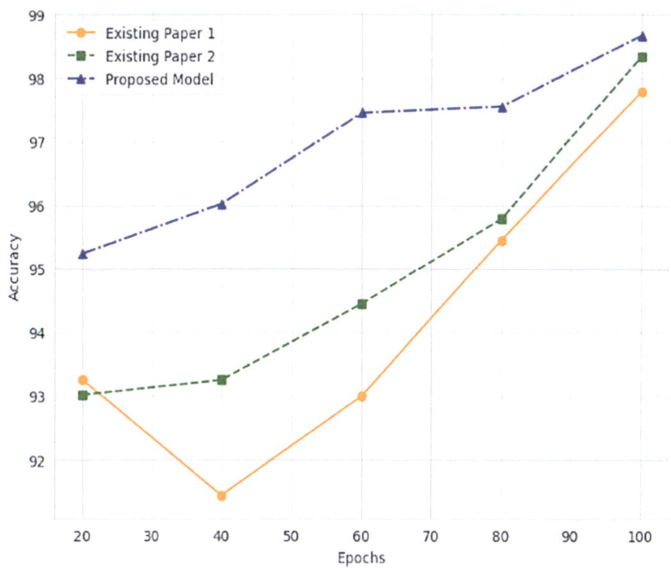

Fig. 2 Accuracy

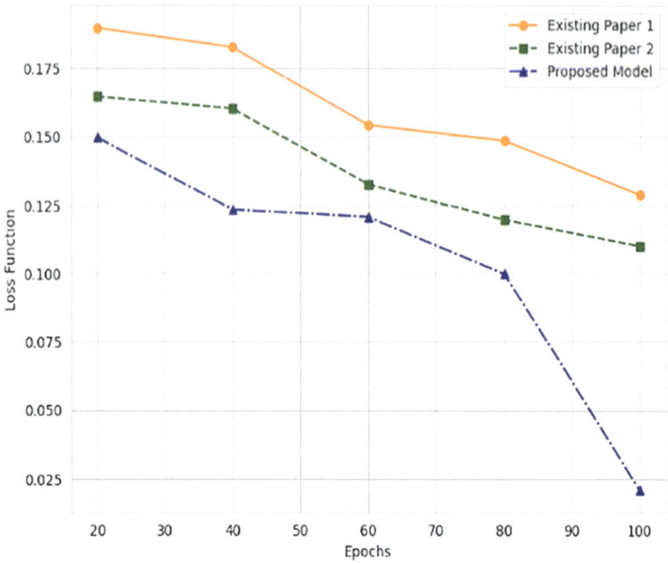

Fig. 3 Loss function

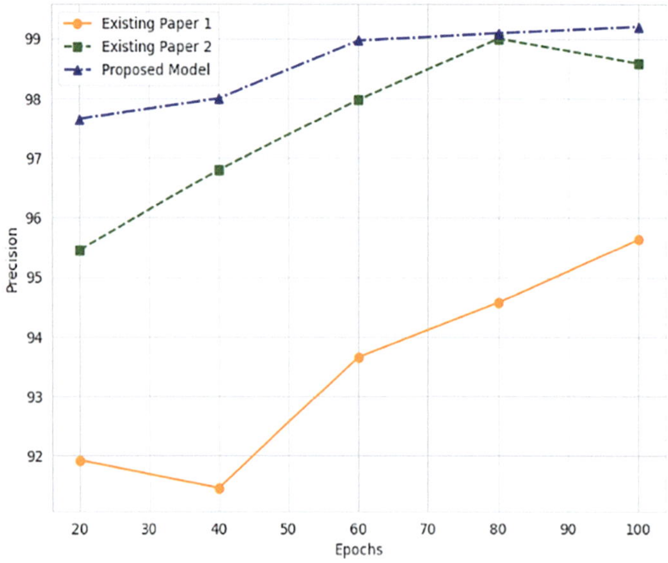

Fig. 4 Precision

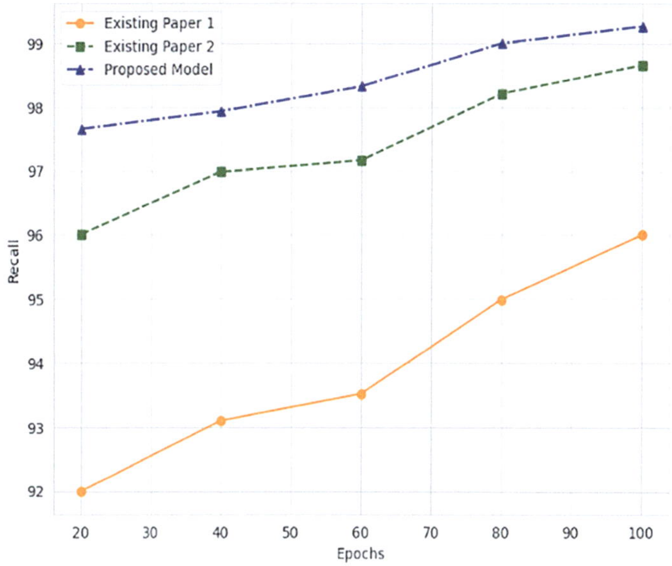

Fig. 5 Recall

2, the F_1 score of the proposed model is predicted to be high where the stacked LSTM and GRU are used.

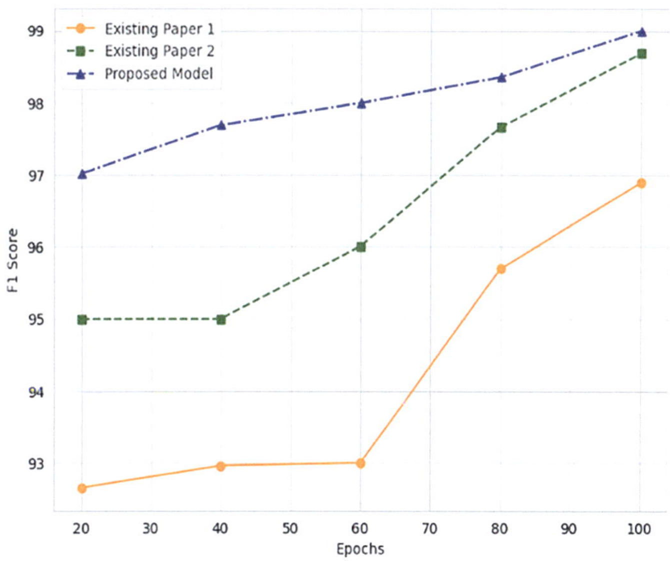

Fig. 6 F_1 score

5 Conclusion

In summary, the outcomes of the study show the promising potential of leveraging deep learning techniques, specifically the lightweight stacked LSTM-GRU model, for the identification of DOS attacks in IoT contexts. The proposed model consistently outperforms the existing models across various metrics, including accuracy (98.6759%), loss function (0.0211%), precision (99.1998%), recall (99.2667%), and F_1 score + (98.9982%). These results imply that the suggested strategy provides a more reliable and efficient means of protecting IoT networks from the dangerous danger of DOS assaults. The combination of LSTM and GRU algorithms is a significant advancement in IoT security. These algorithms excel at capturing temporal dependencies within data streams, which is crucial for identifying and adapting to emerging patterns of malicious activities in dynamic IoT environments. Precise DOS attack detection is especially important now that there are more linked devices in the Internet of Things and all the possible dangers they bring. Though encouraging, more research and statistical testing are needed to make sure the suggested model is reliable and broadly applicable. Additionally, the real-world applicability and scalability of the model should be explored in different IoT domains and dynamic environments. However, the study's conclusions mark a substantial breakthrough in IoT security by providing a better flexible and robust method of thwarting DOS assaults and enhancing the security and integrity of linked IoT devices and the data they transmit.

References

1. Keshk M et al. (2023) An explainable intrusion detection framework in internet of things networks supported by deep learning. Inf Sci 639:119000
2. Bhayo J et al. In software-defined IoT (SD-IoT) networks. Towards a machine learning-based framework for DDOS attack detection. Eng Appl Artif Intell 123:106432
3. Jullian O et al. (2023) Deep-learning based detection for cyber-attacks in IoT networks: a distributed attack detection framework. J Netw Syst Manag 31.2:33
4. Almaraz-Rivera JG et al. (2022) Towards the protection of IoT networks: introducing the LATAM-DDOS-IoT Dataset. IEEE Access 10:106909–106920
5. Zeeshan M et al. (2021) Protocol-based deep intrusion detection for DOS and DDOS attacks using UNSW-NB15 and Bot-IoT data-sets. IEEE Access 10:2269–2283
6. Arepalli PG, Naik KJ (2024) A deep learning-enabled IoT framework for early hypoxia detection in aqua water using light weight spatially shared attention-LSTM network. J Supercomput 80(2):2718–2747
7. Arepalli PG, Naik KJ (2024) An IoT based smart water quality assessment framework for aqua-ponds management using Dilated Spatial-temporal convolution neural network (DSTCNN). Aquacult Eng 104:102373
8. Arepalli PG, Naik KJ (2023) An IoT-based water contamination analysis for aquaculture using lightweight multi-headed GRU model. Environ Monit Assess 195(12):1516

Clustering-Based Hybrid Approaches for Air Quality Index (AQI) Prediction

M. Uday Chandra, Sourabh Debnath, Ramesh Kumar Mohapatra, and Jitendra Kumar

Abstract Nowadays, air pollution has become a major problem in urban areas. The requirement of studying the air quality is an important task to find the pollution level and to take the necessary measures. The air quality can be determined by the concentration of pollutants present in the atmosphere. Air quality index (AQI) is a measuring term whose value determines about the quality of air. In this report, by taking the historical air quality data of the Coimbatore city and implemented the machine learning algorithms for AQI prediction. Hybrid algorithms such as linear regression with three different clustering algorithms such as the K-means clustering technique, expectation maximization clustering, and DBSCAN clustering algorithms are implemented on the data. Initially, data is collected, and it undergoes some pre-processing steps and then the prediction of AQI is implemented. A comparative analysis is done on the proposed hybrid algorithms and linear regression model in terms of their mean absolute error (MAE), mean squared error (MSE), and root mean squared error (RMSE) and the results are drawn.

Keywords Expectation maximization clustering · K-means clustering · DBSCAN clustering

---

M. Uday Chandra (✉) · S. Debnath · R. K. Mohapatra · J. Kumar
National Institute of Technology Rourkela, Rourkela, Odisha, India
e-mail: muday2308@gmail.com

S. Debnath
e-mail: sourabh_debnath@nitrkl.ac.in

R. K. Mohapatra
e-mail: mohapatrark@nitrkl.ac.in

J. Kumar
e-mail: 520cs1001@nitrkl.ac.in

© The Author(s), under exclusive license to Springer Nature Singapore Pte Ltd. 2025

S. K. Udgata et al. (eds.), *Intelligent Systems*, Lecture Notes in Networks and Systems 1314, https://doi.org/10.1007/978-981-96-3797-3_15

1 Introduction

Industrialization on onside created numerous advancements in manufacturing, production, and technology, which helped in own country's growth and development, but on another side, it also affected the environment [1] and human health through pollution. Air pollution is one such concern that causes many health problems such as cancer, developmental and reproductive effects, respiratory problems, and cardiovascular issues, and it shows the impact on the environment too, which leads to climate change, ozone layer depletion, and causes harm to aerial beings.

1.1 Motivation

The gaseous emissions lead to air pollution causing harm to human health and the environment. The importance and awareness of threats caused by air pollution became a hot topic among the public and the researchers too. The study [2] focuses on patterns and deficiencies in studies subjected to air pollution policies, providing valuable points for further investigations and the development of sound policy-making; considering this turmoil, they pushed me to work on the prediction of air quality, thus improving the efficiency of level of prediction, which helps in making of good policies, taking good precautions and measures that helps in improving the people health, environment sustainability, and growth of life.

Air pollution is mainly caused by the emissions of harmful gases from the vehicles and the industries. The review [3] stresses the negative impact of urban exposures on children's physical and mental development. There is also a natural source where pollution happens such as forest fires. The gases contain particles that enter the atmosphere and contaminate the air. The main pollutants of the atmosphere that cause pollution are sulfur dioxide (SO_2), particulate matter (PM2.5 and PM10), ozone (O_3), nitrogen dioxide (NO_2), and carbon monoxide (CO). The air quality can be measured with the help of the air quality index (AQI) value. It is an indicator of the current pollutant level in the air. The range of AQI categories [4] and its associated health problems are mentioned in Table 1.

Historical data of air quality which is monitored by sensors from 2015 to 2020 is collected, which is open-source data available from Kaggle; then using Python programming, the data(.csv) is read into a file. The data set consists of many cities across India. But for prediction, considering only the data from Coimbatore city. The data consists of various pollutants such as SO_2, CO, NO_2, PM2.5, PM10, NH_3, N0, and AQI, AQI BUCKET. Several preprocessing tasks were done on the data for handling null values and for the selection of attributes for further implementation.

Machine learning plays a vital role in data prediction and analysis [5]; various algorithms are developed that perform various tasks respectively; it follows various steps such as data collection and preprocessing, extracting the features and selecting the appropriate model, and then training the model and validation. This work

Table 1 AQI definition

AQI	Associated health impacts
Good (0–50)	Minimal impact
Satisfactory (51–100)	Cause minor breathing discomfort to sensitive people
Moderate (101–200)	Cause breathing discomfort to people with lung disease such as asthma, and discomfort to people with heart disease, children and older adults
Poor (201–300)	May cause breathing discomfort to people on prolonged exposure, and discomfort to people with heart disease
Very Poor (301–400)	May cause respiratory illness to the people on prolonged exposure. Effect may be more pronounced in people with lung and heart diseases
Severe (401–500)	May cause respiratory impact even on healthy people, and serious health impacts on people with lung/heart disease. The health impacts may be experienced even during light physical activity

focuses on two different machine learning algorithms, one is a supervised machine learning algorithm which is linear regression, and the second is an unsupervised machine learning algorithms that are k-means clustering, expectation maximization, and DBSCAN algorithms.

The whole work is divided as follows: In Sect. 2, the past research that was done related to the AQI prediction algorithms was discussed; in Sect. 3, the methodology is discussed and hybrid algorithms are implemented; in Sect. 4, the results are discussed and a comparison was made among all the hybrid algorithms implemented in terms of their MAE, MSE, RMSE; and in Sect. 5, conclusion and future enhancement are discussed.

2 Literature Work

Sonu and Suyampulingam [6] collected the past data of air pollutants of Coimbatore city; they analyzed pollutant PM2.5 by plotting the graph within two time frames, i.e., December 2020 and January 2021 and concluded that the month having lockdown due to covid has less PM2.5 value compared to the month having no lockdown, thus its give evidence that majority of pollutants from road transport and industrial units. They cleaned the data by removing the null valued rows and also the attributes that have more null values, then they drew a plot checking for the linear relationship between the AQI and remaining pollutants, then they implemented the linear regression on the train data, and then they predicted the AQI for the test data set and calculated the errors.

Garg and Sharma [7] implemented different regression methodologies to predict the air quality. The main motto of this research is to analyze the efficiency of different machine learning methodologies for predicting the air quality index (AQI) by considering Particulate Matter 2.5 in Chandigarh, as a major pollutant. Different models are trained and tested using different statistical techniques such as linear regression, KNN regression, Lasso regression, and random forest regression. Finally, it concluded that random forest regression was more accurate compared to the other three regression models.

Mohammed et al. in this study [8] investigated the air quality in different geographical areas of Bangladesh, compared the AQI at different periods during the lockdown period in top 10 cities of the country, and demonstrated a great correlation between the rapid and widespread dispersion of COVID-19 and air pollution reduction in Bangladesh, and also they evaluated the performance of support vector regression (SVR) in AQI forecasting using the time series data set.

Wang et al. [9] captured the sky images, and then they used color moments, color correlogram, and wavelet features to extract the image features, it converts the input image from color space to feature space, then they applied machine learning algorithms such as SVM to predict the AQI for new images, and they used different SVM kernels (sigmoid, linear, polynomial, RBF) and resulted with the accuracy of the testing data set is more than 82.3%, with 1300 training images and 200 test images.

Yang et al. [10] used a variational decomposition model to find the intrinsic mode functions and considered the various pollutant features and then applied the MAOSVR on the trend component obtained from the IMF's by considering the correlation among the input pollutant features, and for the residual component also separately, AOSVR, GRNN, and AOELM are applied, and then finally, the results are predicted.

Naveen et al. [11] give a perfect solution to forecast air quality of Bengaluru city for the next upcoming month by analyzing the data of the previous one year; they used the decision tree methodology for forecasting the air quality.

Shafi and Waheed [12] focus on observing the frequency of the sudden changes in the air quality of Southampton city's data. The k-means clustering algorithm is used to find the ups and downs of the AQI levels before and after of the fire outbreak; they considered PM2.5 as a major pollutant and concluded that after a fire breaks out, there is a sudden rise in AQI level for a few hours.

Kumar et al. [13], in this research, collected the air quality data from ten different stations in Bengaluru city for five years(2015–2020); they used the min-max normalization technique on the AQI value and reduced the AQI buckets to three types as good, satisfactory, and poor. Attribute selection is done by using decision trees and correlation matrix; attributes having less gain ratio are not considered for analysis; thus, they selected only the useful attributes, and finally expectation maximization (EM) clustering is implemented on the data.

Viorela and Muntean [14], in this paper at first clustering is done the data set based on PM10 pollutant and after clustering for each individual cluster they used KNN and neural network classifier forecasting models, resulting in good performance.

Surono et al. [15] in this study improved the performance of RNN (used for prediction) by using K-means clustering and the PCA for dimension reduction and improved the performance; they used various distance measures such as Euclidean, Manhattan, Canberra, Average, and Chebyshev for K-means clustering.

Arulanandham et al. [16] focus on comparing various classification and clustering techniques (K-means, EM, DBSCAN) for analyzing the AQI; data is collected from different parts of south India from 2015 to 2020 and they used Pearson Correlation along with the ranker search algorithm for feature selection. Random forest tree algorithm gives high accuracy. The DBSCAN algorithm distribution of instances is good.

The literature study reveals the shifting of the research from single monolithic models to hybrid models; there is a great improvement in using various statistical measures and pattern recognition functions, which really breaks down the boundaries of the core field and moves toward out of box in search of high efficient and performance approaches. This collaboration of multiple domains helps in building efficient models.

3 Methodology

The entire work is break down as importing the data, data cleaning, and then attribute selection; then after clustering the data and then implementing, the linear regression on the train data set is done on each cluster, and then AQI prediction is done on the test data set on each cluster. Fig. 1 clearly shows the system overview. Prediction of AQI precedes of preprocessing tasks such as data collection, cleaning of data, and attribute selection.

3.1 Data Preprocessing

Data collection and cleaning: The air quality data of Coimbatore city is extracted from the main data which was imported using Python library files, the data contains pollutant information from July 2019 to July 2020, and the data consists of various pollutant information. Figure 2 shows the heat map, which best describes the missing values of each pollutant. By observing the heat map, NH3 and NO pollutants have many missing values, so they are dropped at this stage and remove all the rows having missing values. The remaining attributes and rows are taken for clustering of data.

Attribute selection: Here, covariance matrix is computed on the attributes obtained after the cleaning stage. Fig. 3 shows the covariance matrix of the attributes. The selection of attributes is done in such a way that attributes that are highly correlated with the target attribute are selected, and also from the selected attributes, they must be less correlated. From the covariance matrix except for NO_2, all remaining attributes are positively correlated with target attribute(AQI), so drop the NO_2

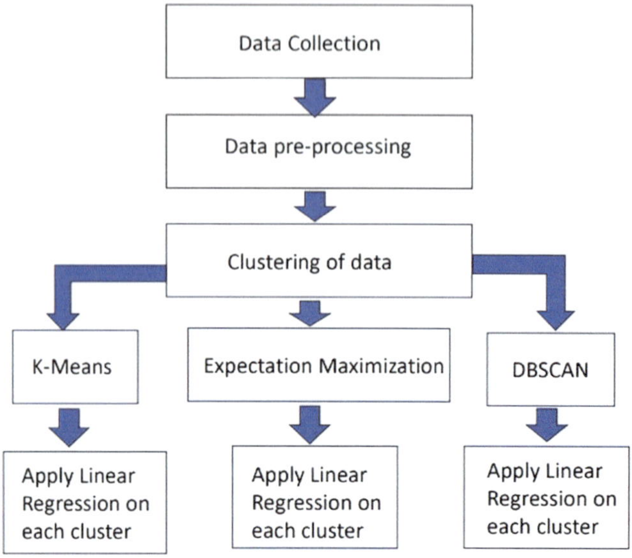

Fig. 1 Methodology

attribute, from remaining PM2.5, and PM10 are highly correlated so we need to select any one among them; here, PM2.5 is selected since it has high correlation with AQI, and then O3 is selected since it having high correlation with AQI among all attributes, then after CO is selected since it is having good correlation with AQI, as well as it is having a negative relationship with PM2.5 and NOx, then NOx is also selected, so finally total four attributes are selected, i.e., PM2.5, CO, O_3, NOx for prediction of AQI.

3.2 Implementing Clustering Algorithms

The AQI values are categorized into six different levels such as good, moderate, satisfactory, poor, very poor, and severe in the original data. The six categories are reduced to three categories as good, satisfactory, and poor using min-max normalization [13]. Table 2 shows the normalization of the cluster range.

1. k-means clustering: k-means clustering algorithm runs 10 times with different centroids and for each set of centroids the algorithm run for 100 iterations, thus resulting with three clusters as follows:

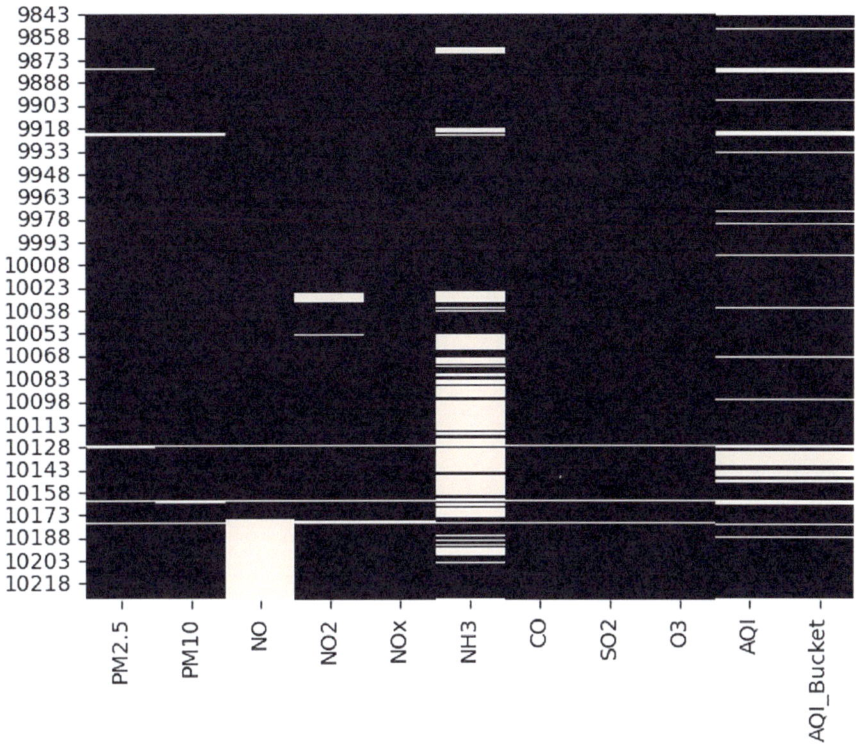

Fig. 2 Heat map of missing values

47% cluster 0 (good)
29% cluster 1 (satisfactory)
24% cluster 2 (poor)

The pi chart shown in Fig. 4 shows distribution of data.

2. Expectation maximization clustering: Then for the same data set, the expectation maximization algorithm is implemented with three clusters and the output as follows:

38% cluster 0 (good)
37% cluster 1 (satisfactory)
24% cluster 2 (poor)

The pi chart shown in Fig. 5 shows the distribution of data.

3. DBSCAN: For DBSCAN clustering, the epsilon value is decided based on the some graphical analysis; from Fig. 6, it is observed that for each sample of data, the 14 nearest neighbors distance is computed and a graph is plotted between the samples in sorted order(increasing) of distance on *x*-axis and the computed distance on *y*-axis; then from the graph, the knee point where the elbow shape is observed,

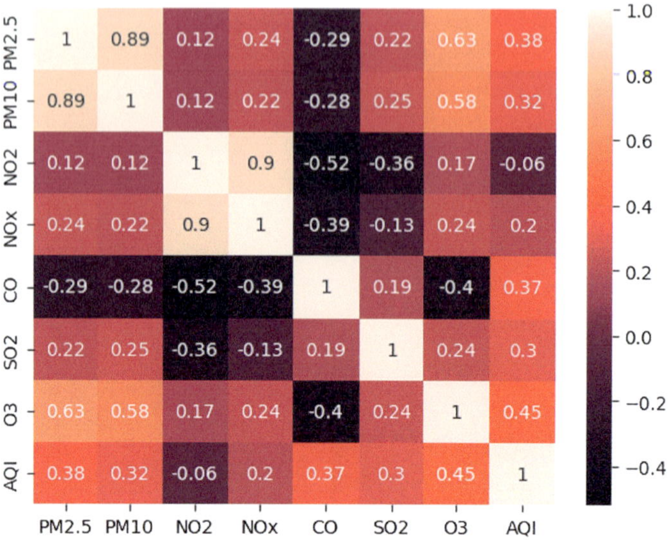

Fig. 3 Covariance matrix

Table 2 Normalized cluster ranges

AQI range	AQI value	Normalized cluster ranges
0–50	Good	Cluster-0 (good)
51–100	Moderate	Cluster-0 (good)
101–150	Satisfactory	Cluster-1 (satisfactory)
151–200	Poor	Cluster-2 (poor)
201–250	Very poor	Cluster-2 (poor)
251–300	Severe	Cluster-2 (poor)

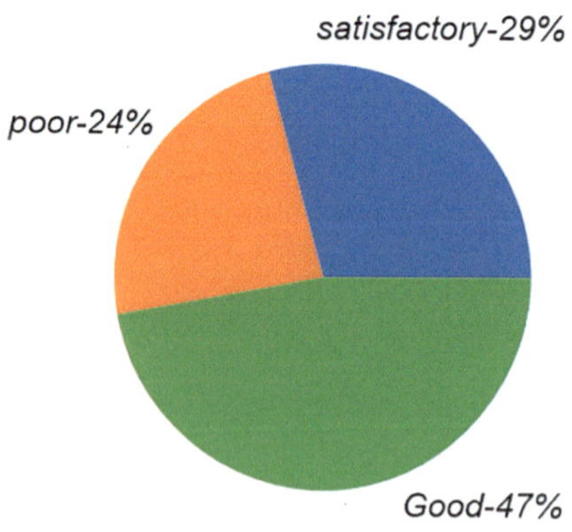

Fig. 4 Clustering of data into three categories such as good, satisfactory, and poor using K-means clustering

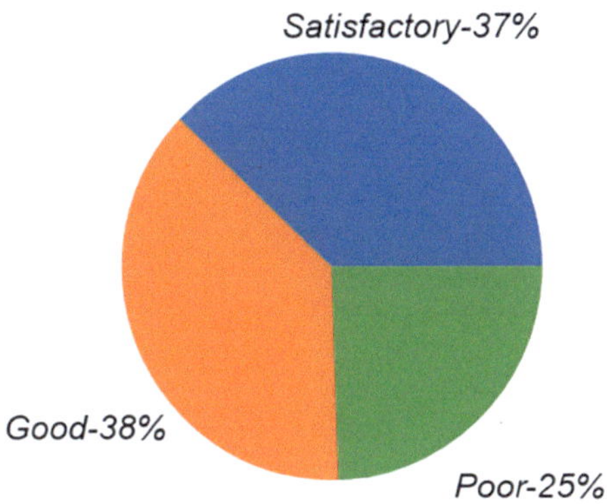

Fig. 5 Clustering of data into three categories such as good, satisfactory, and poor using expectation maximization

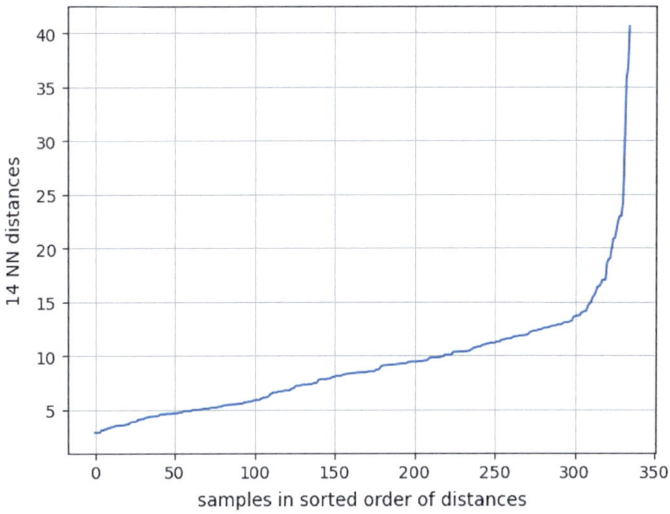

Fig. 6 Graph to find the epsilon value

at that point the respective 14-NN distance is selected as the epsilon value, which is 14, then clustering is done, the results as follows:

64% cluster 0 (good)
30% cluster 1 (satisfactory)
6% cluster 2 (poor)

The pi chart shown in Fig. 7 shows the distribution of data.

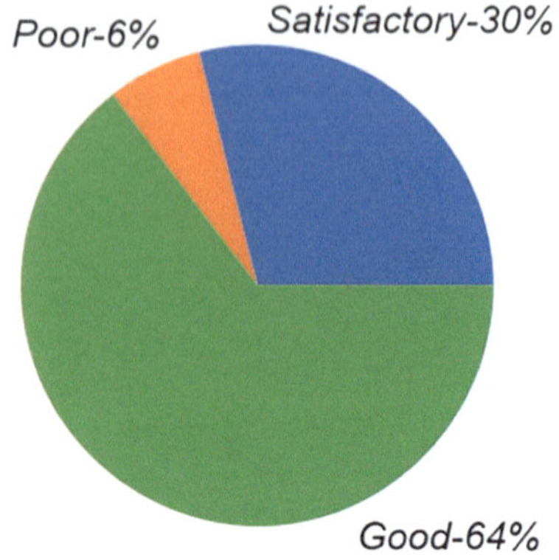

Fig. 7 Clustering of data into three categories such as good, satisfactory, and poor using DBSCAN

3.3 *Implementation of Linear Regression Algorithm*

Linear regression algorithm is implemented on all clusters formed by using the three different algorithms, and it is implemented on the original pre-processed data before clustering. PM2.5, CO, O_3, and NOx attributes are taken as independent variables, and AQI as the dependent variable For regression, the entire data set is divided into train and test, where for each cluster, the LR model is trained using the train data, and prediction is done on the testing data.

4 Results and Discussion

The results obtained from the three clustering techniques show that the expectation maximization clustering algorithm clusters the data efficiently compared to the remaining two clustering algorithms, the DBSCAN clustering algorithm formed clusters in a completely different manner since it is based on the density of data, so DBSCAN way of clustering and k-means clustering are different, which reflects in the DBSCAN algorithm to cluster the huge data to noise, whereas the expectation maximization algorithm uses the Gaussian distribution for clustering of data, the nonlinearity of data helps EM algorithm to cluster data efficiently.

AQI prediction was done on each cluster using the linear regression; then the error matrices are computed and the overall error observed for each hybrid algorithm, Table 3 shows the comparison of error matrices of all algorithms, and here the comparison is made based on three performance metrics such as MAE, MSE, RMSE. The linear regression performed poorly compared all other approaches, because of its linearity which is actually not equivalent to the original data, whereas clustering helps in dividing the data into similar sets, thus the chance of fitting an LR is beneficial rather than using only LR. Among all hybrid algorithms, expectation maximization clustering with linear regression gives good results with less MAE of 5.7 and RMSE

Table 3 Comparison of error matrices

Methodology	MAE	MSE	RMSE
Linear regression	8.1	100.3	10.0
K-means clustering with linear regression	6.27	65.01	8.0
Expectation maximization clustering with linear regression	5.7	57.7	7.5
DBSCAN clustering with linear regression	6.7	87.28	9.34

of 7.5. Even though the prediction by both expectation maximization and DBSCAN on respective clusters performed good, the DBSCAN algorithm noise is more, so its prediction is not good on noise data, whereas k-means clustering with linear regression performed well compared to DBSCAN with linear regression.

5 Conclusion and Future Enhancement

The results obtained clearly say that expectation maximization clustering with linear regression performs good in the clustering of data as well as prediction of AQI compared to remaining approaches. The prediction of AQI is done on small data set, which contains the information about Coimbatore city; this limitation can be improved in further study by considering huge data set and working with above algorithms; moreover, the prediction part, which includes finding the AQI value uses the linear regression technique, LR algorithm has limitations since it is can be suitable for linear data only, so further research includes using nonlinear regression techniques or neural networks as part of prediction of AQI.

References

1. Gaur N, Sharma S, Yadav N (2024) Environmental pollution, pp 23–41
2. Feng T, Sun Y, Shi S, Ma J, Feng C, Chen Z (2024) Air pollution control policies and impacts: a review 191:114071
3. Roche IV, Ubalde-Lopez M, Daher C, Nieuwenhuijsen M, Gascon M (2024) The health-related and learning performance effects of air pollution and other urban-related environmental factors on school-age children and adolescents-a scoping review of systematic reviews. Springer, pp 1–17
4. https://en.wikipedia.org/wiki/Air~quality index#India. Aqi definitions. In *Air Quality Index*, 2023
5. Méndez M, Merayo MG, Núñez M (2023) Machine learning algorithms to forecast air quality: a survey. Artif Intell Rev 1–36
6. Sonu SB, Suyampulingam A (2021) Linear regression based air quality data analysis and prediction using python. In: 2021 IEEE Madras section conference (MASCON). IEEE, pp 1–7

7. Garg KD, Gupta M, Sharma B, Dhaou IB (2023) A comparison of regression techniques for prediction of air quality in smart cities. In: 2023 1st international conference on advanced innovations in smart cities (ICAISC), pp 1–6
8. Hossain MT, Hossain A, Meem SM, Monir MF, Ullah Miah MS, Sarwar TB (2023) Impact of covid-19 lockdowns on air quality in Bangladesh: Analysis and AQI forecasting with support vector regression. In: 2023 4th international conference for emerging technology (INCET), pp 1–6
9. Wang W, Shen W, Chen B, Zhu R, Sun Y (2018) Air quality index forecasting based on SVM and moments. In: 2018 5th international conference on systems and informatics (ICSAI), pp 851–855
10. Hong Y, Yiting Z, Li G (2023) Air quality index prediction using a new hybrid model considering multiple influencing factors: a case study in china. Atmos Pollut Res 14(3):101677
11. Naveen S, Upamanyu MS, Chakki K, Chandan M, Hariprasad P (2023) Air quality prediction based on decision tree using machine learning. In: 2023 international conference on smart systems for applications in electrical sciences (ICSSES), pp 1–6
12. Shafi J, Waheed A (2020) K-means clustering analysing abrupt changes in air quality. In: 2020 4th international conference on electronics, communication and aerospace technology (ICECA), pp 26–30
13. Senthil Kumar R, Arulanandham A, Arumugam S (2021) Air quality index analysis of Bengaluru city air pollutants using expectation maximization clustering. In: 2021 International conference on advancements in electrical, electronics, communication, computing and automation (ICAECA), pp 1–4
14. Muntean MV (2023) Improving air pollution forecasting in smart cities using clustering techniques. In: 2023 15th international conference on electronics, computers and artificial intelligence (ECAI), pp 01–05
15. Surono S, Goh KW, Onn CW, Marestiani F (2023) Developing an optimized recurrent neural network model for air quality prediction using k-means clustering and PCA dimension reduction, vol 6, pp 330–343
16. Senthil Kumar R, Arulanandham A, Arumugam S, Dinesh G, Thirukkumaran R, Subashmoorthy R (2022) Analysis of classification and clustering techniques for ambient AQI using machine learning algorithms. In: 2022 4th international conference on smart systems and inventive technology (ICSSIT), pp 902–908

An Explainable Machine Learning Approach for Multi-class Sleep Stage Identification from EEG Signals

Harsh Upadhyay, Rajesh Mohapatra, Meet Patel, Santosh Kumar Satapathy, and Nitin Singh Rajput

Abstract Sleep is a vital physiological process crucial for overall health and well-being. Accurate identification and classification of sleep stages, including rapid eye movement (REM) and non-REM stages, are essential for understanding sleep disorders and diagnosing medical conditions. Traditional manual scoring by clinicians had limitations, leading to the development of automated methods using machine and deep learning techniques. This study explores automated sleep stage detection, mainly focusing on EEG signals. It reviews various algorithms, from traditional feature-based methods to state-of-the-art machine learning approaches. The research employs the MNE-Python library to preprocess and label EEG data from the Sleep-EDF dataset. We have mainly used a neural network model for the multi-class classification of sleep stages. Performance metrics like precision, recall, F_1 score, accuracy, and Cohen's Kappa are used to evaluate the models. The entire experiment was performed on thirty subjects. The results show promising classification accuracy with an accuracy of approximately 76.8% and a Cohen's Kappa of 0.678, marking a significant step toward more efficient and scalable sleep stage detection. The study discusses potentially future directions, including multimodal data fusion and personalized modeling, to address challenges in sleep stage classification. Automated sleep stage detection improves our understanding of sleep patterns, disorders, and overall health.

Keywords Explainable machine learning · Multi-class sleep stage identification · EEG signals

1 Introduction

This paper delves into the critical realm of sleep, a fundamental physiological process indispensable for our overall health and well-being [1]. Sleep, a complex phenomenon characterized by distinctive stages, including rapid eye movement

H. Upadhyay · R. Mohapatra · M. Patel · S. K. Satapathy (✉) · N. S. Rajput
Pandit Deendayal Energy University, Gandhinagar, Gujarat, India
e-mail: Santosh.Satapathy@sot.pdpu.ac.in

© The Author(s), under exclusive license to Springer Nature Singapore Pte Ltd. 2025

S. K. Udgata et al. (eds.), *Intelligent Systems*, Lecture Notes in Networks and Systems 1314, https://doi.org/10.1007/978-981-96-3797-3_16

(REM) and non-REM stages, is pivotal in shaping our physical and mental health [2]. Various sleep disorders are closely associated with disruptions in normal sleep patterns, characterized by increased wakefulness or arousal. Arousal can result in sleep fragmentation, leading to a decline in cognitive function and autonomic reflex activation and increasing blood pressure and heart rate. Unique patterns of brain activity and physiological changes characterize these stages [3]. However, accurate detection and classification of these sleep stages are paramount for understanding sleep disorders, assessing sleep quality, and diagnosing various medical conditions [4].

Traditionally, sleep stage scoring relied on the expertise of clinicians, who manually interpreted polysomnography (PSG) recordings [5]. However, this approach was labor-intensive and subject to inter-rater variability [6], hampering scalability and reliability. To address these challenges, automated sleep stage detection methods emerged, harnessing the power of machine learning algorithms to analyze PSG data and categorize sleep stages with precision [7]. These automated methods follow a multi-step process, commencing with preprocessing raw PSG signals. Techniques like filtering and artifact removal [8] are applied, and relevant features are extracted from various physiological signals, including electroencephalogram (EEG), electrooculogram (EOG), and electromyogram (EMG) [9]. Subsequently, machine learning algorithms such as support vector machines (SVMs), artificial neural networks (ANNs) [10], hidden Markov models (HMMs) [10], and random forests [10] are employed to develop models capable of accurately classifying sleep stages.

Sleep disorders cover a wide range of ailments that interrupt regular sleep patterns. It entails identifying sleep interruptions by analyzing a variety of physiological signals, including respiratory data, Electroencephalography (EEG), Electrooculography (EOG), and Electromyography (EMG) [11]. They can vary from insomnia, in which sleep is difficult to come by, to sleep apnea, in which breathing disturbances impair sleep. The first step in improving sleep quality and general well-being is identifying these problems. Detection entails gathering information on several physiological characteristics, most typically by polysomnography (PSG). PSG employs several sensors, including EEG, EOG, and EMG, to monitor brain activity, eye movement, and muscle activity during sleeping. These data are then painstakingly analyzed, showing patterns and abnormalities that aid in diagnosing certain illnesses [12].

2 Related Works

Automated systems for classifying sleep stages aim to categorize different sleep phases using physiological signals such as EEG, EOG, EMG, and other bio signals. These systems play a crucial role in sleep medicine and research. This study investigates vast literature on automated sleep stage detection, focusing on signal processing, channel selection, feature extraction, and classification techniques. One significant research area involves utilizing EEG signals for sleep stage classification, where researchers delve into the complexities of distinguishing wakefulness, NREM, and REM sleep based on EEG patterns. Various machine learning algorithms are employed for this purpose, relying on informative features extracted from EEG data. Additionally, feature extraction techniques explicitly tailored for sleep stage detection are explored, including statistical features like power spectral density, entropy measures, and wavelet analysis, which are crucial for accurate sleep stage classification. Another research direction explores using heart rate variability (HRV) analysis as a tool for automated sleep stage detection, leveraging HRV metrics as valuable features within machine learning models. Practical applications involve assessing wearable devices like wristbands and smartwatches for automated sleep monitoring, evaluating their accuracy in detecting different sleep stages, and usability in real-world settings. The emergence of deep learning has sparked substantial interest in automated sleep stage detection, with recent advancements in techniques like convolutional neural networks (CNNs), recurrent neural networks (RNNs), and hybrid models exploring their ability to learn directly from raw EEG signals and comparing their performance against traditional machine learning algorithms. In [14], the authors utilized machine learning to classify sleep stages for predicting human sleep disorders, employing the stochastic gradient descent (SGD) method. The model achieved an accuracy of 89% in assigning classification labels (sleep stages) to segments of the EEG spectrogram. In [11], a machine learning-based sleep score model was proposed, considering various sleep factors based on user sleep data and gait data collected from wearable devices. The model employed a stacking ensemble regression approach, with input comprising sleep and gait data, producing sleep score as the output. In [12], an automated sleep staging system based on machine learning algorithms was introduced, taking as input a set of statistical features extracted from raw EEG signals collected during sleep. The authors evaluated the performance of three machine learning algorithms: decision tree (DT), *K*-nearest neighbors (KNN), and random forest (RF), achieving accuracies of 89.60%, 90.93%, and 93.42%, respectively. In [13], an automated sleep staging system was proposed using an ensemble learning stacking model that combines random forest (RF) and eXtreme gradient boosting (XG boosting), taking dual-channel EEG signals as input. The system achieved an accuracy of 91.2% after preprocessing EEG signals to remove noise and artifacts and extracting linear and non-linear features. Recent advancements in sleep research using the DL approach are summarized in Table 1.

The literature review addresses the challenges in automated sleep stage detection. Persistent challenges include inter-subject variability in sleep patterns [7], signal quality variability arising from electrode placement and movement artifacts [8], the lack of standardized datasets for consistent evaluation [9], and the pressing need for extensively annotated datasets for both training and validation [10]. We proposed an automated sleep staging system using EEG signals, and the proposed research work was carried out using different medical-conditioned subjects.

Table 1 State-of-the-art works on sleep staging using DL techniques

Existing studies	Signal	Classifier	Accuracy (%)
Chambon et al. [15]	EEG, EOG, EMG	CNN	78
Mousavi et al. [16]	EEG	CNN-BiRNN	84.3
Abdollahpour et al. [17]	Multichannel	TL-CNN-Data fusion	93.58
Paisarnsrisomsuk et al. [18]	EEG (2 channel), EOG (1 channel)	CNN	84.5
Abdollahpour et al. [19]	EEG	CNN	93.48
Cheng [3]	EEG, ECG, EMG	MML-DMS	94.34

3 Proposed Methodology

Designing a methodology for an automated sleep staging classification system requires a structured approach, delineating the sequential steps and processes to be adhered to throughout system development and evaluation. The following is a suggested methodology for constructing such a system: (i) Acquisition of data, (ii) data preprocessing, (iii) annotation and labeling, (iv) extraction of features, (v) selection of models, (vi) training, (vii) validation, and (viii) assessment of performance. The comprehensive illustration of the proposed model layout is depicted in Fig. 1.

3.1 Dataset Description

The mne.datasets.sleep_physionet.age. fetch data function is a useful feature in MNE-Python, specifically designed to retrieve local copies of PhysioNet Polysomnography dataset files. Its primary purpose is to simplify data access from subjects participating in PhysioNet's research on age-related sleep effects in healthy individuals. This dataset comprises a subset of 153 sleep recordings involving 37 male and 41 female subjects, with ages ranging from 25 to 101 years at the time of recording. Typically, each subject is associated with two-night recordings, except for a few instances where subjects 13, 36, and 52 have one missing record each due to hardware-related issues. More comprehensive information about this dataset can be found on the PhysioNet website. Figure 2a,b depicts the sleep behavior of subjects, characterized by a low-frequency, high-amplitude sound wave with a relatively smooth pattern. This pattern indicates delta waves associated with the deep sleep stage.

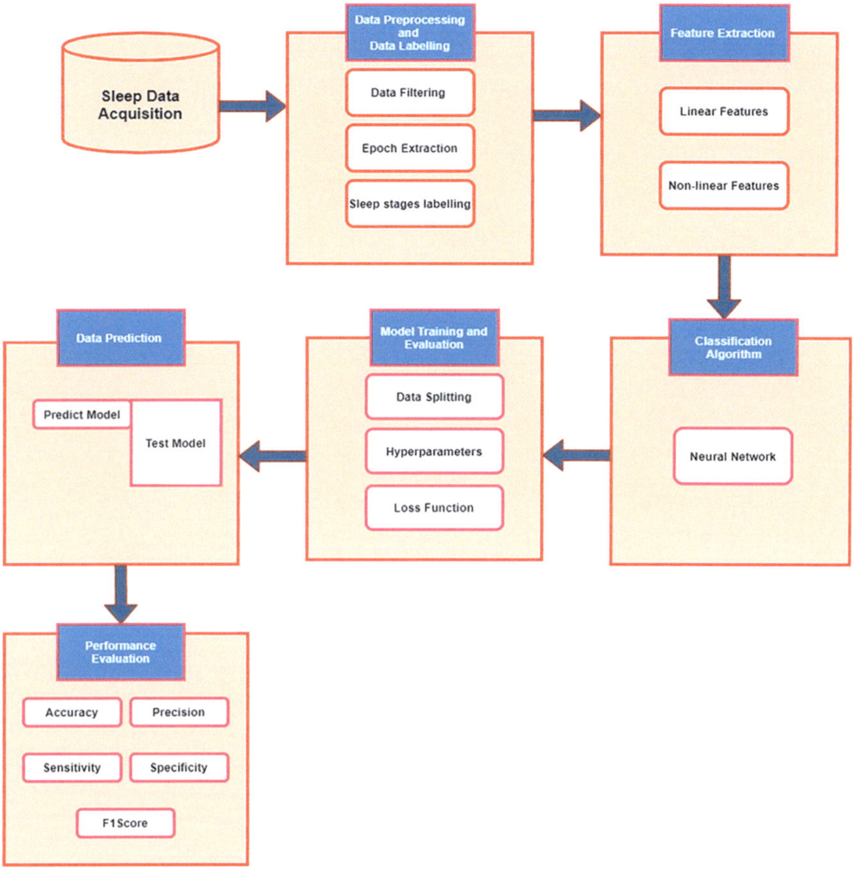

Fig. 1 Overall framework of the proposed model

3.2 Data Acquisition

The research commences by obtaining Electroencephalography (EEG) data from the Sleep-EDF dataset utilizing the MNE-Python library. This dataset encompasses sleep recordings from various subjects, and we have opted for a subset of subjects and recordings for our analysis. The preprocessing steps for the data involve:

a. Loading the raw EEG data along with annotations.
b. Choosing particular EEG channels while excluding other physiological signals.
c. Cropping the data to eliminate the initial and final wake periods.

Figure 3 depicts the amplitude of the EEG signal across various frequencies. The EEG signal exhibits its highest power spectral density (PSD) within the delta

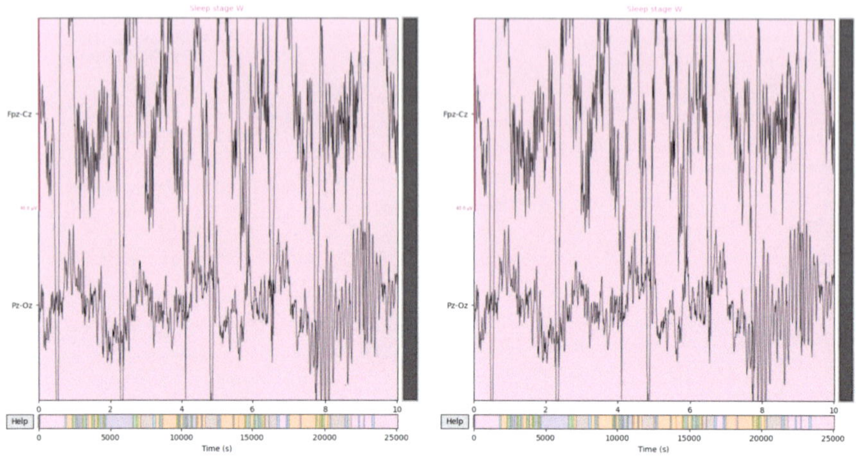

Fig. 2 Sample sleep behavior record of the subject

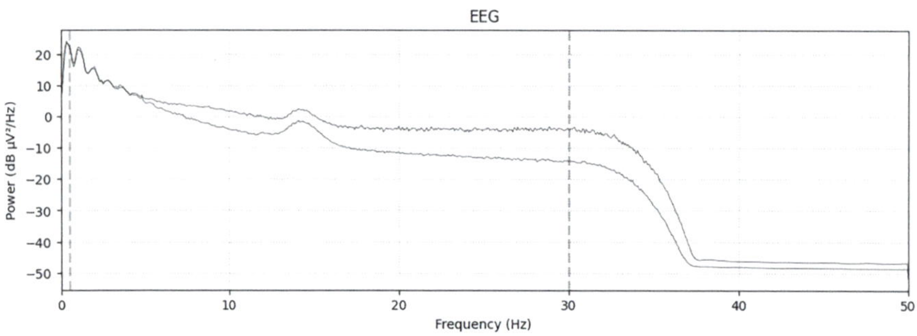

Fig. 3 Different EEG signal frequency response graph

band, typically associated with deep sleep. As frequency increases, the PSD gradually decreases, reaching its minimum in the beta band, typically linked to wakefulness. Additionally, the graph displays peaks and valleys, representing different brain rhythms.

3.3 Data Preprocessing and Data Labeling

Data Filtering.

The EEG data undergoes bandpass filtering, restricted to a specific frequency range between a low-frequency cutoff (l_freq) and a high-frequency cutoff (h_freq).

This filtering process eliminates noise while preserving the pertinent frequency components for sleep stage classification.

Epoch Extraction.

The preprocessed recordings are segmented into epochs of EEG data, each representing a fixed sleep duration, such as 30 s. Within these epochs, sleep stages are annotated using events denoted as "Sleep stage W" for wakefulness, "Sleep stage Light" for light sleep, "Sleep stage Deep" for deep sleep, and "Sleep stage R" for rapid eye movement (REM) sleep.

Sleep Stages Labeling.

The sleep stages, as determined by the event annotations within each epoch, are assigned numerical values for classification purposes. These numerical mappings are as follows: "1" corresponds to "Sleep stage W" (wakefulness), "2" corresponds to "Sleep stage Light" (light sleep), "3" corresponds to "Sleep stage Deep" (deep sleep), and "4" corresponds to "Sleep stage R" (REM sleep).

3.4 Feature Extraction

Signal Decomposition.

EEG, typically multimodal signals, can be decomposed into constituent components such as delta, theta, alpha, beta, and gamma rhythms. This decomposition process can be considered feature extraction since it transforms raw signals into distinct frequency bands.

3.5 Classification Algorithms

Neural Network Architecture.

The neural network architecture is the fundamental framework in deep learning and is crucial in automated sleep stage detection. This section explores the theoretical foundations of neural network architecture, providing insight into its fundamental components and the reasoning behind its application.

3.6 Model Training and Evaluation

Data Splitting.

The dataset undergoes a division into training, validation, and test sets. The training set comprises EEG data from various subjects, whereas the test set incorporates recordings from subjects not involved in the training phase. Additionally, a subset of the training data is reserved for validation purposes. The training process entails iteratively optimizing the model's parameters using the training set. Performance is continuously monitored on the validation set throughout the training to prevent overfitting. The training process persists until the early stopping criteria, defined by the patience parameter, are satisfied. The sleep stage samples utilized in this research are illustrated in Fig. 4a–c

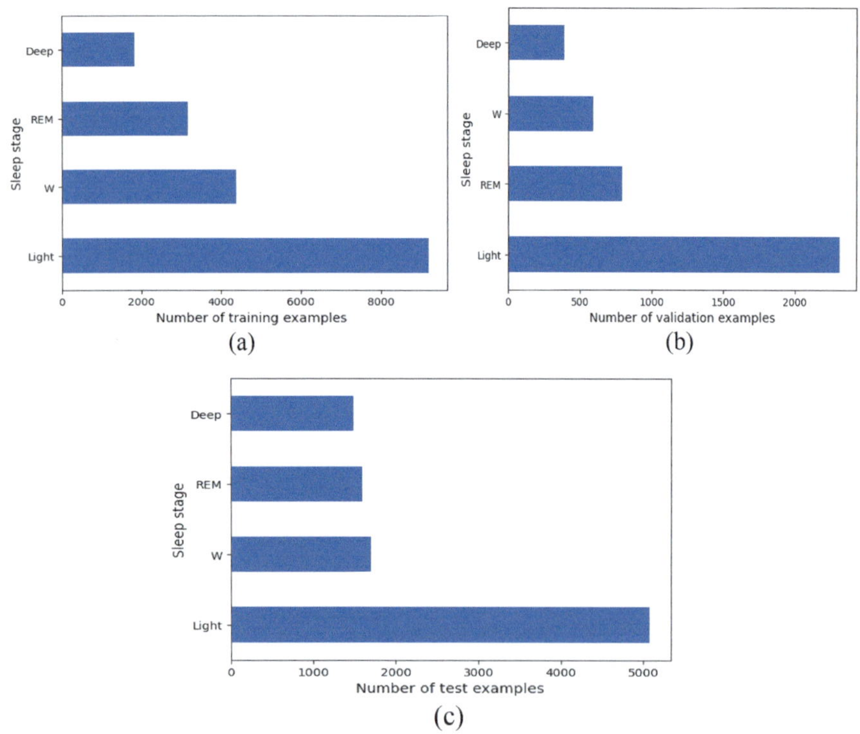

Fig. 4 Number of samples of sleep stages used in the experiment work: **a** Training, **b** validation, **c** testing

4 Results and Discussion

This section presents and discusses the results of the sleep stage classification model based on EEG data from the Sleep-EDF dataset. The model's performance is evaluated using various metrics, and the classification results are analyzed. Figure 5 illustrates the confusion matrix for the four-class classification problem. The model demonstrates promising performance in classifying sleep stages, achieving a test-balanced accuracy of approximately 85.8% and a Cohen's kappa of 0. 678. Figure 6 displays the results of individual performance metrics, indicating a substantial agreement between the model's predictions and the ground truth sleep stage annotations. The model achieves an overall precision of 0.821, indicating its ability to classify sleep stages correctly. Particularly, the precision for "Sleep stage W" is high at 0.82, suggesting effective identification of wake stages. Sensitivity, also known as recall, measures the model's ability to correctly identify true positive cases, while specificity measures its ability to identify true negatives correctly. The model exhibits a sensitivity of 0.768 and an impressive specificity of 0. 943. Figures 7 and 8 depict the sensitivity and specificity performances for each sleep stage, providing further insight into the model's ability to distinguish between different sleep stages.

The F_1 score, which considers precision and recall, provides a balanced measure of the model's performance. The overall F_1 score is 0.765, indicating a good balance between precision and recall (Figs. 9 and 10).

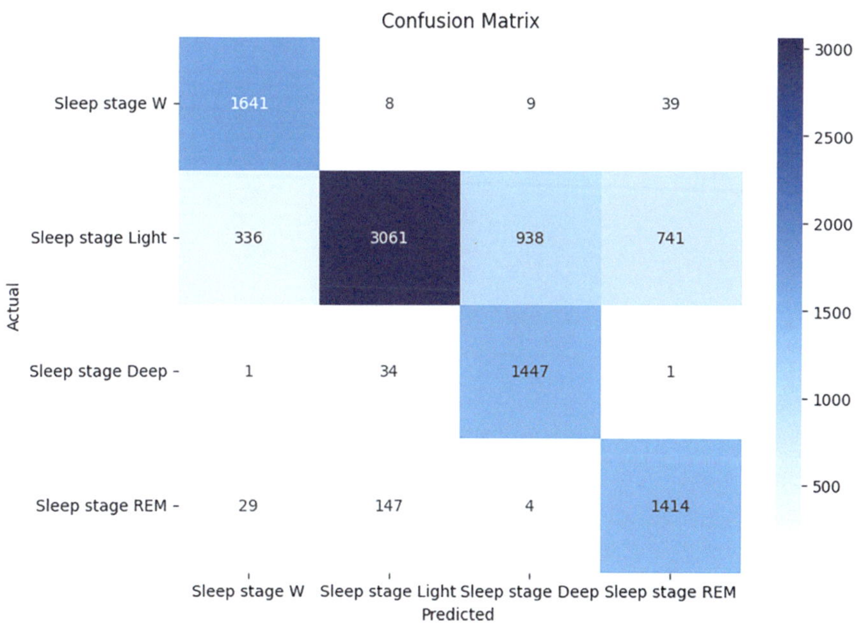

Fig. 5 Reported confusion matrix of four-class sleep stages classification

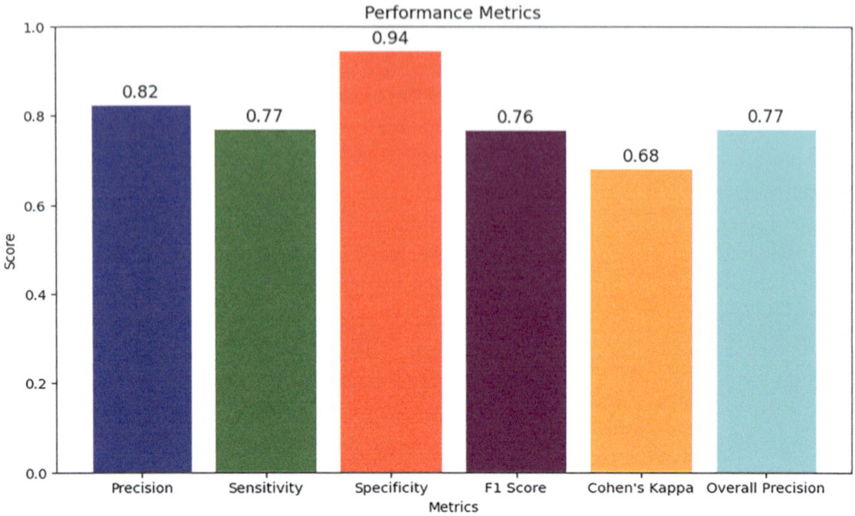

Fig. 6 Performance metrics for 4-class sleep stages (wake, light sleep, deep sleep, and REM) classification on S-EDF database using EEG signal

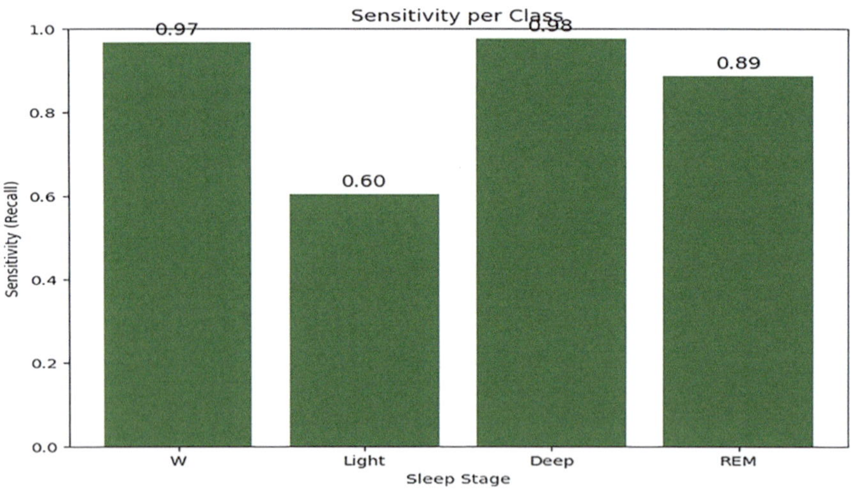

Fig. 7 Performance of sensitivity in individual sleep stages-wise

The weighted overall precision of 0.821 suggests that the model maintains a good balance between precision and recall across all sleep stage classes. The model's accuracy of approximately 76.8% demonstrates its ability to classify sleep stages correctly. Cohen's kappa of 0.678 indicates substantial agreement between the model and human annotators. The most challenging aspect of sleep staging classification is accurately identifying the N1 sleep stage. Figure 11 illustrates a comparative analysis

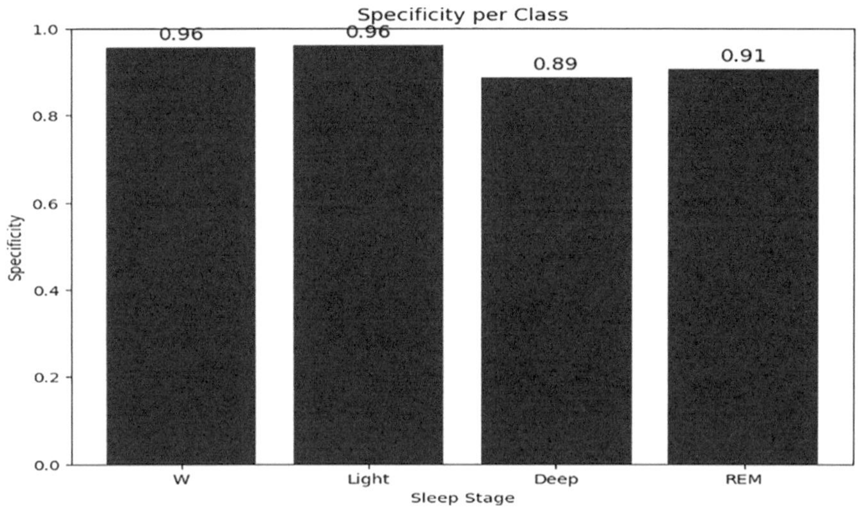

Fig. 8 Performance of specificity in individual sleep stages-wise

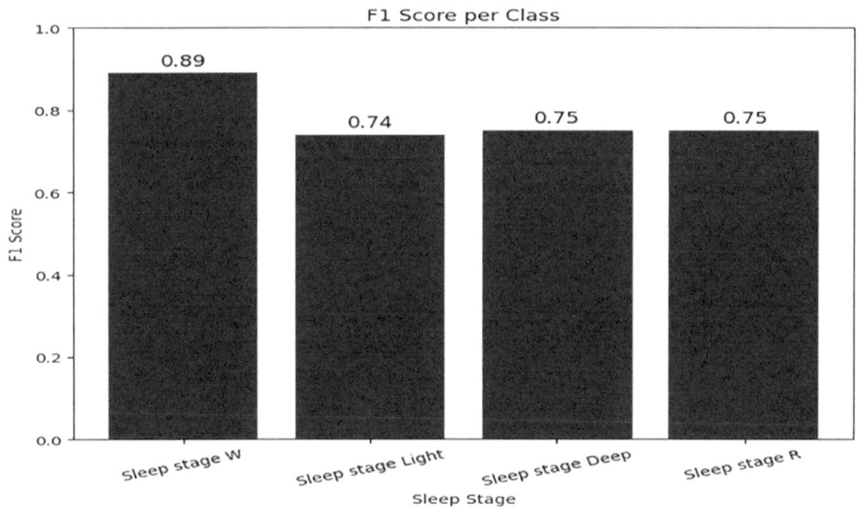

Fig. 9 Performance of F_1 score in the individual sleep stages

of the F_1 score performance of the NREM-N1 stage with similar works based on the Sleep-EDF dataset. This comparison provides insights into the model's performance relative to other studies.

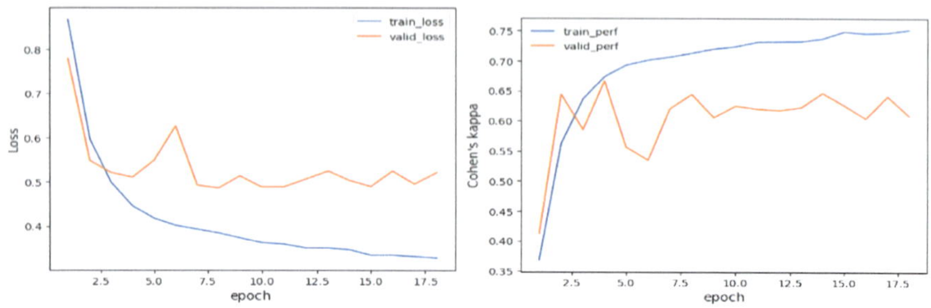

Fig. 10 Performances of the model in terms of training and validation: **a** accuracy, **b** loss

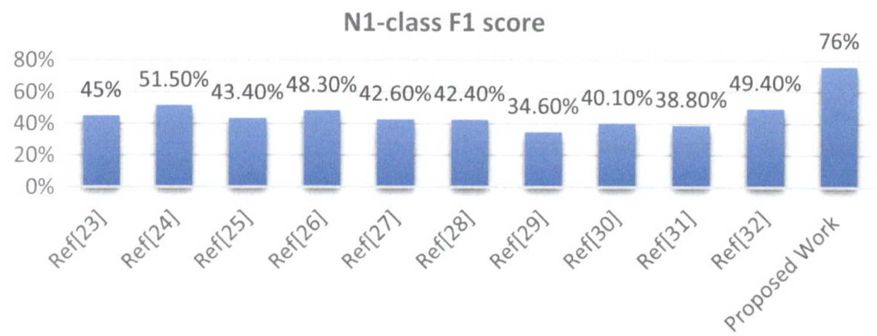

Fig. 11 Comparison of F_1 score of N1 sleep stage with state-of-the-art models on the Sleep-EDF dataset

5 Conclusion

In this study, we investigated the use of neural network models for sleep stage classification using EEG signals from the Sleep-EDF dataset. Our results showcase promising advancements in automating the sleep staging process, underscoring the potential of machine learning to enhance the accuracy and efficiency of sleep analysis. Using EEG signals as input features effectively captured intricate patterns associated with different sleep stages. The neural network architecture employed demonstrated commendable performance, achieving a classification accuracy of 76.8%. This signifies a significant stride in sleep staging, presenting a non-invasive, data-driven approach that could enhance manual scoring methodologies. However, certain limitations must be acknowledged. The generalization of our model to diverse populations and the consideration of individual variability in sleep patterns require further investigation. While our model excelled in classifying the N1 sleep stage within the Sleep-EDF dataset, its performance may vary when applied to different datasets or real-world scenarios. Future directions in this line of research are abundant. Refining the neural network architecture through techniques like transfer learning and ensembling may yield further improvements in classification accuracy. Additionally, incorporating physiological signals, such as heart rate variability and respiratory patterns, could offer a more comprehensive understanding of sleep dynamics. Furthermore, integrating real-time monitoring capabilities into wearable devices can revolutionize sleep diagnostics and monitoring in clinical and home settings.

References

1. Fernandez-Mendoza J, Li Y, Fang J, Calhoun SL, Vgontzas AN, Liao D et al (2019) Childhood high-frequency EEG activity during sleep is associated with incident insomnia symptoms in adolescence. J Child Psychol Psychiatry 60(7):742–751
2. Chriskos P, Frantzidis CA, Gkivogkli PT, Bamidis PD, Kourtidou-Papadeli C (2017) Automatic sleep staging through electroencephalogram relative wavelet entropy and peripheral signal features. In: 21st Humans in Space Symposium (HIS2017)
3. Cheng Y, Lech M, Wilkinson RA (2023) Simultaneous sleep stage and sleep disorder detection from multimodal sensors using deep learning. Sensors 23(7):3468. https://doi.org/10.3390/03073468
4. Rechtschaffen A, Kales A (1968) A manual of standardized terminology, technique and scoring system for human sleep stages. Brain Information Service, Los Angeles
5. Berry RB, Brooks R, Gamaldo CE, Harding SM, Marcus C, Vaughn B et al (2012) The AASM manual for the scoring of sleep and associated events. Rules, Terminology and Technical Specifications, Darien, Illinois, American Academy of Sleep Medicine
6. Chen C, Ugon A, Zhang X, Amara A, Garda P, Ganascia JG et al (2016) Personalized sleep staging system using evolutionary algorithm and symbolic fusion. In: 38th Annual international conference of the IEEE engineering in medicine and biology society (EMBC'16), pp 2266–2269
7. Frantzidis CA, Bratsas C, Klados MA, Konstantinidis E, Lithari CD, Vivas AB et al (2010) On the classification of emotional biosignals evoked while viewing affective pictures: an integrated data-mining-based approach for healthcare applications. IEEE Trans Inf Technol Biomed 14(2):309–318
8. Shi J, Liu X, Li Y, Zhang Q, Li Y, Ying S (2015) Multi-channel EEG-based sleep stage classification with joint collaborative representation and multiple kernel learning. J Neurosci Method 254:94–101
9. Diykh M, Li Y, Wen P (2016) EEG sleep stages classification based on time domain features and structural graph similarity. IEEE Trans. Neural Syst. Rehabilit. Eng. 24(11):1159–1168
10. Hassan AR, Subasi A (2017) A decision support system for automated identification of sleep stages from single-channel EEG signals. Knowl Based Syst 128 (2017) 115–124.
11. Kim J, Park M (2023) A study on ML-Based sleep score model using Lifelog data. Appl Sci 13(2):1043. https://doi.org/10.3390/app13021043
12. Satapathy SK, Loganathan D, Kondaveeti HK, Rath R (2021) Performance analysis of machine learning algorithms on automated sleep staging feature sets. CAAI Trans Intell Technol 6(2):155–174. https://doi.org/10.1049/cit2.12042
13. Satapathy SK, Bhoi AK, Loganathan D, Khandelwal B, Barsocchi P (2021) Machine learning with ensemble stacking model for automated sleep staging using dual-channel EEG signal. Biomed Signal Process Control 69:102898. https://doi.org/10.1016/j.bspc.2021.102898
14. Rakhonde MA, Wagh K, Mante RV (2020) Sleep stage classification for prediction of human sleep disorders by using machine learning approach. Int J Innov Sci Res Technol 5(7):1268–1272. https://doi.org/10.38124/ijisrt20jul712

15. Chambon S, Galtier MN, Arnal PJ, Wainrib G, Gramfort A (2018) A deep learning architecture for temporal sleep stage classification using multivariate and multimodal time series. IEEE Trans Neural Syst Rehabil Eng 26:758–69. https://doi.org/10.1109/TNSRE.2018.2813138
16. Mousavi S, Afghah F, Acharya UR (2019) SleepEEGNet: automated sleep stage scoring with a sequence-to-sequence deep learning approach. PLoS ONE 14:e0216456. https://doi.org/10.1371/journal.pone.0216456
17. Abdollahpour M, Rezaii TY, Farzamnia A, Saad I (2020) Transfer learning convolutional neural network for sleep stage classification using two-stage data fusion framework. IEEE Access 8:180618–180632. https://doi.org/10.1109/ACCESS.2020.3027289
18. Paisarnsrisomsuk S, Ruiz C, Alvarez SA (2021) Capturing the development of internal representations in a high-performing deep network for sleep stage classification. SN Comput Sci. https://doi.org/10.1007/s42979-021-00697-3
19. Abdollahpour M, Rezaii TY, Farzamnia A, Saad I (2022) A two-stage learning convolutional neural network for sleep stage classification using a filterbank and single feature. IEEE Access 10:60597–60609. https://doi.org/10.1109/ACCESS.2022.3180730

Machine Learning Analysis on Climate Change Influencing Factors of India with Near-Time Data

Munganda Venkata Lakshmi and S. R. Reeja

Abstract The pernicious consequences of urbanization and the encroachment upon transitional ecosystems in India are manifold. Unpredictable weather, unpredictable precipitation, a relentless climb in sea levels, and the destruction of valuable ecosystems and complex food webs are all symptoms of these negative impacts. It is critical that parliamentarians, non-profits, and governments throughout the world work together to address the pressing need to determine what is causing climate change (CC) and how to mitigate its effects. From September to December 2023, climate trends were examined in different Indian states at daily mean intervals for this work. In order to scrutinize the primary issue with CC, the whole investigation is centered on temperature and air-quality metrics CO, NO_2, SO_2, and O_3. Several machine learning (ML) models were trained and evaluated on the dataset to get the enhanced prediction accuracy observed in the suggested technique. As part of EDA, the suggested random forest regression (RFR) model achieved an accuracy of 94.67%, whereas the other models MLR, Lasso regression, DTR, and GBR achieved accuracies of 69.53%, 68.13%, 88.14%, and 92.04%, respectively.

Keywords Random forest regression · Climate change · Temperature analysis · Machine learning

1 Introduction

Now a day, CC is the study being occupied by technological advancements in climate analysis. Artificial intelligence and its myriad branches play a pivotal role in comprehending the multifarious realms of vulnerability, encompassing the realms of deep learning (DL) and ML. The cyclic nature of climatic change encompasses not only the deleterious impacts of industrialization, vehicular emissions, and deforestation, but also the often overlooked contributions stemming from household activities. Take

M. Venkata Lakshmi · S. R. Reeja (✉)
VIT-AP University, Amaravati, India
e-mail: reeja.sr@vitap.ac.in

© The Author(s), under exclusive license to Springer Nature Singapore Pte Ltd. 2025
S. K. Udgata et al. (eds.), *Intelligent Systems*, Lecture Notes in Networks and Systems 1314, https://doi.org/10.1007/978-981-96-3797-3_17

instance of an everyday household task, the ubiquitous plastic bag, which has insidiously infiltrated every aspect of our existence, from the packaging of goods to the lining of our dust bins. Its wanton usage has contributed significantly to the escalating climate crisis to earth and water in the form of CO, NO_2 and affecting the ozone [1]. But it is not just the pernicious plastic bag that we must cast a critical eye upon. The relentless burning of cooking gas, particularly rampant in the realm of restaurant cuisine, has further exacerbated the situation. The excessive utilization of electrical appliances, driven by our insatiable appetite for convenience, has also played a substantial role in the mounting environmental predicament we find ourselves in. These seemingly harmless domestic activities have collaborated to heat up our globe at a worrying pace. It is imperative that we confront the uncomfortable truth of our complicity in this unfolding catastrophe and take decisive action to rectify our unsustainable habits. The aforementioned statement posits that all phenomena culminate in the escalation of temperatures upon the terrestrial plane, thereby engendering radiation toward the subterranean strata, shown in Fig. 1a.

The absolute changes in climate with numerous factors and their interdependencies, which are prone to critical climatic conditions, were explained in the global climate change cycle in Fig. 1a. Then explaining the climatological analysis of temperature and other changing factors like air-quality parameters, and type of forecasting was given in Fig. 1b. This shows weather forecasting procedures from real-time observation to computational values by forcasters to make predictions. The practice of trying to predict the weather dates back to the middle ages. The Babylonians made predictions about the future about 650 BCE based on astronomy and cloud patterns. The Meteorologica, which around 350 BCE wrote by Aristotle, described weather trends. Theophrastus also wrote the Book of Signs, which was about predicting the weather. Indian astrology emerged circa 1200 BCE when monk Lagadha created the Vedāṅga-Jyotiṣa, based on Vedas, describing lunar and solar months and adjusting them for Adhikamāsa (lunar leap month) [2]. Seasons, years, and periods are also discussed. Pattern identification, which is the study of observed cycles of events, was used in the past to predict the weather.

2 Literature Review

Temperature and change in climate issue are certainly not relevant with one element or variable in the atmosphere; it is the multi-variable correlating phenomena with different aspects of the elements and changing factors. Since 1850, Alpine glaciers have declined dramatically in size, shrinking by 40% in area and more than 50% in volume due to the 1°C rise in temperatures. The expansion and withdrawal of glaciers have resulted in a rise in the size and quantity of glacial lakes, increasing the occurrence of Glacial Lake Outburst Floods (GLOFs) in future. These scenarios, such as Representative Concentration Pathway (RCP) and Shared Socio-economic Pathway (SSP), are crucial for climate modeling. These SSPs complement RCPs by providing a more comprehensive view of how numerous factors might affect climatic

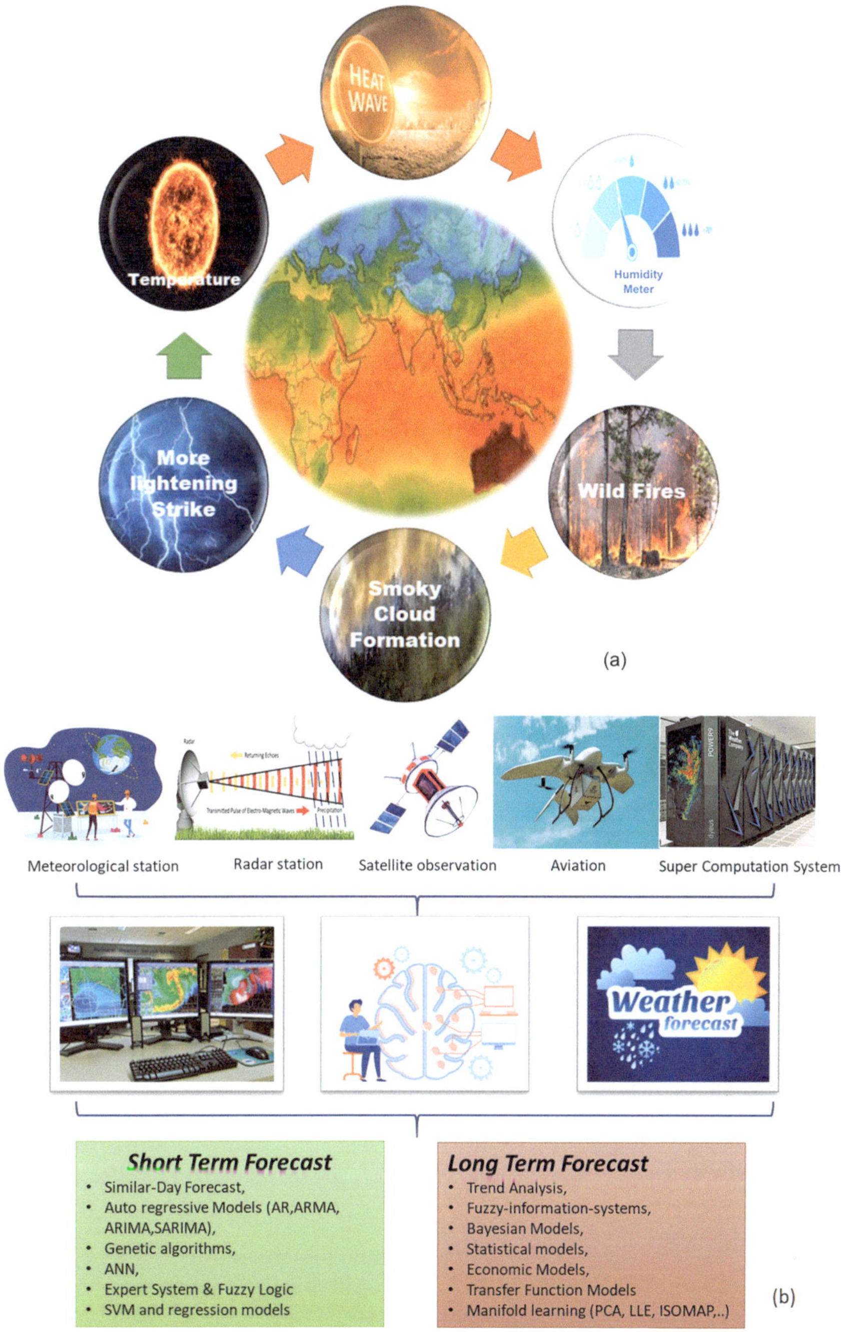

Fig. 1 Absolute changes in climate **a** global temperature cycle and **b** climate change analysis phenomena

Table 1 SSP report in IPCC sixth assessment

SSP levels	GHGE	EW in °C (2041–2060)	EW in °C (2081–2100)	Expected range (2041–2100)
1–1.9	Very low	1.6	1.4	1.0–1.8
1–2.6	Low	1.7	1.8	1.3–2.4
2–4.5	Intermediate	2.0	2.7	2.1–3.5
3–7.0	High	2.1	3.6	2.8–4.6
5–8.5	Very high	2.4	4.4	3.3–5.7

outcomes, shown in Table 1 [1] with Estimated Warning (EW) and GHGE changing rate. Hydrological models like SWAT use GCM temperature and precipitation data to predict future trends. Factors like geographical dimensions, adaptability, and calibration ease should be taken into account while selecting an appropriate hydrological model.

Griscom asserts that Natural Climate Solutions (NCSs) may give 37% of cost-effective CO_2 reduction until 2030, according to Bronson W. et al., assuming that there is a probability of more than 66% of maintaining temperatures below 2 °C. Ten dollars or less might be used to accomplish one-third of the cost-effective mitigation of the NCS [2]. Flood buffering, soil health, biodiversity habitat, and climate resilience are some of the benefits that may be obtained from the majority of NCS treatments if they are carried out correctly. In order to lower the uncertainty of the NCS mitigation estimate, further study is required. In spite of this, the research presented here lays a strong groundwork for swift global action to boost ecosystem care as a substantial response to climate change. The global climate model (GCM) is an example of a mathematical framework that is used to simulate the flow of emissions from the atmosphere on a worldwide scale [3].

Although the model's poor spatial resolution makes it useless for local hydrological models, Prajapati et al. claim that it is perfect for evaluating and forecasting climatic patterns over the long period. The future of the weather can be foretold by using global and regional climate models (GRCMs). Due to its superior spatial resolution and accuracy in smaller areas, RCMs are preferred over GCMs for basins on a local scale [4].

2.1 Temperature Analysis in Terms of In-Situ Observation

Many investigations have compared temperature data from in situ measurements in India. Maximum temperatures in India over the period from 1983 to 2020 were the subject of one study's long-term trend analysis. During this period, an intriguing discovery was made regarding the escalating pattern of maximum temperature in all states. This revelation was further corroborated by the rigorous Mann–Kendall test. Yet another study has meticulously scrutinized the temperature data emanating from a burgeoning township in the verdant state of Kerala, spanning the temporal expanse from 1950 to 2018. There has been a discernible escalation in the yearly temperature, specifically in the average annual maximum temperature and the temperature during the southwest monsoon [5]. The statistical significance of this growing trend emphasizes the gravity of the situation, and the affected CC scenarios are explained in Table 2. It is crucial to recognize that there has been a notable and substantial drop in winter temperatures.

Furthermore, an in-depth examination was conducted to scrutinize the temperature profiles observed during the in situ recovery of bitumen in the region of Kerala by meteorological departments. Additionally, it observed elevated frontal temperatures during the in situ combustion phase. The meteorological data over the decades were giving trusted forecasting to examine the climate in daily base. This study involves the time-series temporal data for 4 months from August to December 16, 2023 to observe the patterns of climate change parameters. This is an time-series and daily updating dataset available in kaggle repository by Nidula Elgiriyewithana, which is an updated dataset for India's present weather conditions that can be found at https://www.kaggle.com/datasets/nelgiriyewithana/indian-weather-repository-daily. Observation and analysis have been conducted by researchers and climate analysts for India's extreme weather condition analysis.

Table 2 Across India seasonal climate effects on changing patterns

Season and time lapse	Seasonal climate change	Temperature variations	Effective climatic change in areas
Winter (December–February)	In terms of annual coldest months, December and January rank first	Temperatures averaging around 10–15 °C (50–59 °F) in the northwest	The temperature reaches its highest point in the southeast, where it ranges from 68 to 77 °F
Spring (March–May)	India starts heating up from around the end of February	Rising temperatures lies between 33 and 42 °C	First in the northern plains and then the rest of the country
Summer (June–August)	Extremely hot in locations with low elevation	Potentially exceeding 50 °C (122 °F) during May	Resulting in heatwaves have the potential collateral damage in India
Monsoon season (July–September)	Monsoon rains give some relief from heat	The climate remains hot and humid (with 38–45 °C) during this period	Varies from state to state
Post-monsoon season (October–November)	Climate becomes cooler and more comfortable after monsoon	Even though daytime temperatures are varying between 29 °C and 35 °C, the coldest nights were experienced in same day due to extreme weather conditions.	Varies with extreme weather events in all over the country in the same day. Daytimes with extreme heatwaves and nights with sudden drop in temperatures to minus degrees.

The machine learning analysis of the influencing change parameters was analyzed to get the correlation understanding. Linear regression (LR), random forest (RF), decision tree (DT), *K*-nearest neighbor (KNN), support vector machine (SVM), and Cat Boost Regressor (CBR) are ML regression models used to predict the global temperature with availed dataset. Temperature is the measure of heat at different levels of the Earth's atmosphere. Based on data collected from weather stations between 2004 and 2009, ANN, SVM, and PSOANFIS models were trained to provide 1-day-ahead predictions utilizing variables such as wind speed, maximum and lowest air temperatures, relative humidity, and solar radiation [6].

2.2 Global Oceanic Effect on Temperature, Dry Air, and Rainfall Changes in India

More intense downpours are accompanied by an increase in global temperatures. Along this equatorial belt, meteorologists monitor SST for four separate zones known as Nino areas, according to NOAA. The temperatures might lead to one of three probable outcomes: El Nino (EN), the ENSO neutral phase, or La Nina (LN). La Nina years are infamous for the numerous and powerful storms and cyclones that impact the Bay of Bengal and the Atlantic Ocean. The combination of high relative humidity and moderate wind shear over the Bay of Bengal is one of the factors that is increasing the probability of more cyclones. EN means a little boy which is a Spanish word, as anomalous SST occurring in the eastern and central Pacific Ocean. LN means a little girl, cools surface waters abnormally. LN is cool and EN is warm phase form the EN Southern Oscillation (ENSO). These large-scale ocean phenomena affect global winds, temperature, and rainfall, which can cause worldwide droughts, floods, heat, and cold. Current LN phase began in September 2020. The IMD reported that just six LN's have lasted more than two years since the 1950's, understanding of EN and LN on India shown in Fig. 2. It confirmed that India's LN will last till 2022 [7].

According to author Bhatla et al., rice crops in India, which need hot and humid conditions to grow, have been badly affected by dry years caused by EN. It results in insufficient precipitation in every region [8]. During stable years or years without EN or LN, the production of beans that don't need too much humidity gets a little better. The Annual Precipitation(AP) fluctuations and AP Rainfall (APR) are based on EN and LN phenomena and are related to the Atmosphere Infrared Radiation Forcing (AIRF), according to the authors Jyothi D, -Environment and Sustainability Division, Vishakhapatnam, India [9]. EN and LN monthly precipitation patterns suggest the ACIF and APRF are inadequate, especially during EN years. LN years resulted in more precipitation. During EN, precipitation levels often fall below monthly and seasonal averages.

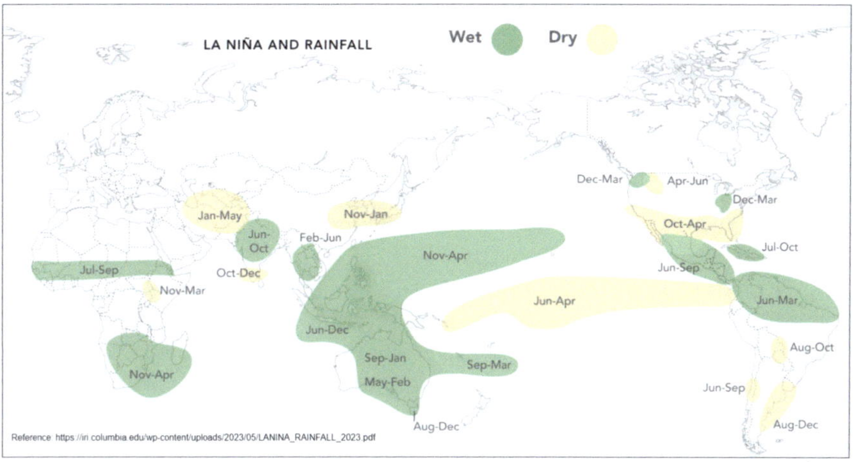

Fig. 2 El Niño and La Niña events' effect on INDIA

Weather forecasting is an intricate procedure that includes gathering empirical information on the present conditions of the atmosphere, and various working methodologies are used to forecast temperature and influencing parameters, including persistence, trends, climatology, numerical weather prediction, ensemble forecasting, and analog methods [10]. November 2023 thunderstorm causes huge damage to rice crops, which leads to price drop for formers at effective levels. The forecasting with real-time data is complex due to imperfect data like null values, unwanted parameters, and outlier data [11].

Temperature forecasting and extreme condition analysis uses primary observations from atmosphere using in-situ time-series data and spatial satellite imagery as input patterns. The procedure involves "Synoptic method, Statistical method and Numerical Weather Prediction" methods [12]. World Meteorological Organization (WMO) and Global Telecommunications System (GTS) are the Indian meteorological stations which are sending the weather reports to regional, central, and global data providers according to their data-access policies. Then data will be processed and compiled as per the need of the researchers and climate analysts to forecast and make clear prediction reports of the day. In prediction of sea surface temperature (SST), both a persistent long-term patterns and recorded highest fluctuations which are in the form of marine heat waves across various regions were considered as necessary elements [13]. The SST, is being measurable from terrain space, significantly shows the impact on analyzing spatial extent, distribution and population dynamics of different marine species. The prediction can be explored through various methodologies like raster image processing using GIS and machine learning [14].

3 Data and Visualization

This study involves the dataset obtained from public repository Kaggle, with dataset named Indian Weather Repository (Daily Updating). This dataset is a public available and updated data to till date automatically with real-time data values. The visual understanding of the dataset was done by histogram analysis using Fig. 3.

Extreme changes in temperature and air-quality parameters are highly correlating to express weather forecasting in India. The temperature ranges in India state wise show the plotting from – 200 to + 400 °C because of the India overall states average temperature is between 230 and 350 °C [15]. The time zone and location used for data collection are done mostly at Kolkata and Uttar Pradesh. Figure 4 is heatmap which shows highly correlating factors of CC with respect to temperature, wind speed, precipitation, etc. Ultraviolet index, temperature, and air-quality index parameter over the states with respect to CC in India during August–December' 2023 state wise are plotted in Fig. 5a. The air-quality parameters that are influencing the rise in temperature were analyzed through a correlation heat matrix, plotted in Fig. 6. Which shows the influence of each GHGE in the atmosphere leading to CC in India over the selected regions.

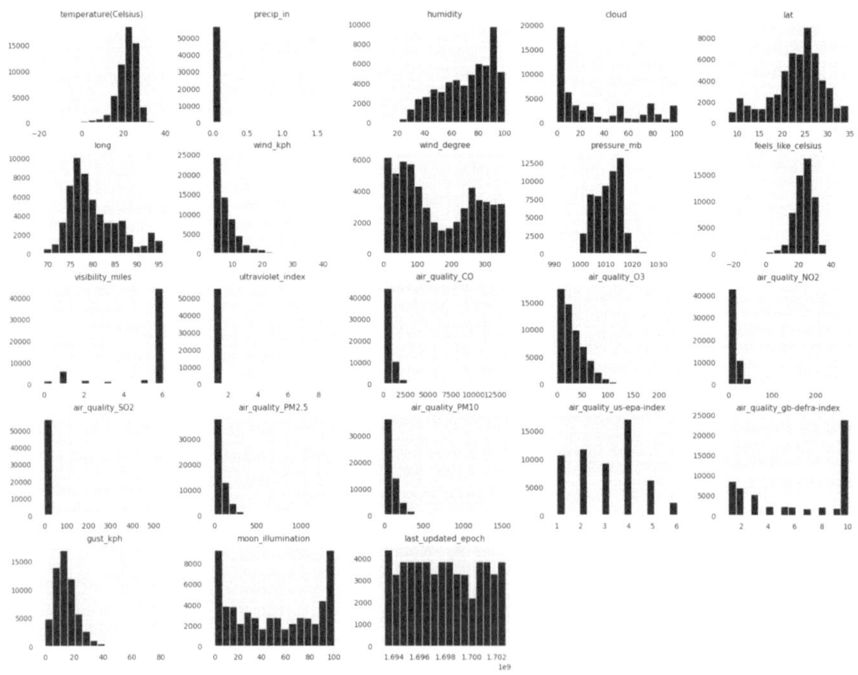

Fig. 3 Data visualization using histogram analysis

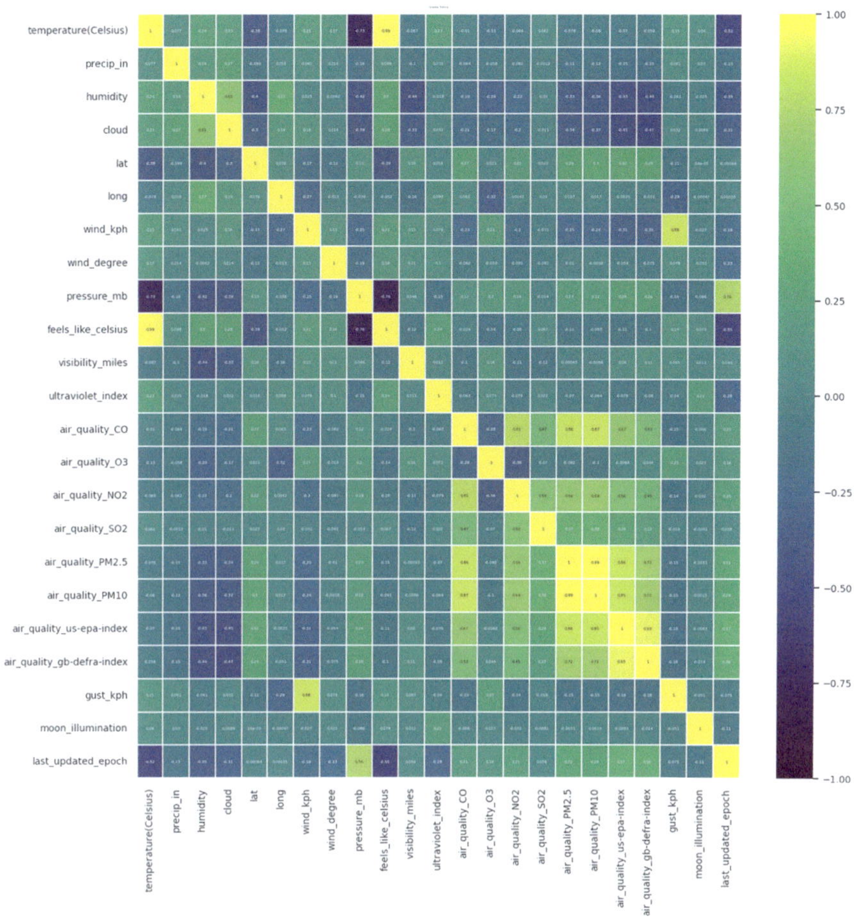

Fig. 4 Correlation analysis between the selected parameters

4 Methodology

Because of the dataset's time-series and sequential patterns, analyzing CC trends using continuous real-time data is a complicated operation. In order to predict future temperatures, this study employed the RFR model. The dataset included 42,000 rows of sequential time-series data for every day across all states, covering three time periods. The dataset was trained using the model and then tested using a variety of ML models, including MLR, DTR, and GBR. The performance metrics like RMSE, MAE, and MSE were used to analyze the model's error calculation and R^2 score as prediction accuracy. The temperatures are on the rise and the GHGE keeps updating its extreme data values from the local atmosphere which has improved the performance of predictions at several, noticeable at several prominent Indian

Fig. 5 Climate change in India during August–December' 2023 **a** UV-index among states **b** Air-quality index PM_10 variation **c** State wise temperature variation

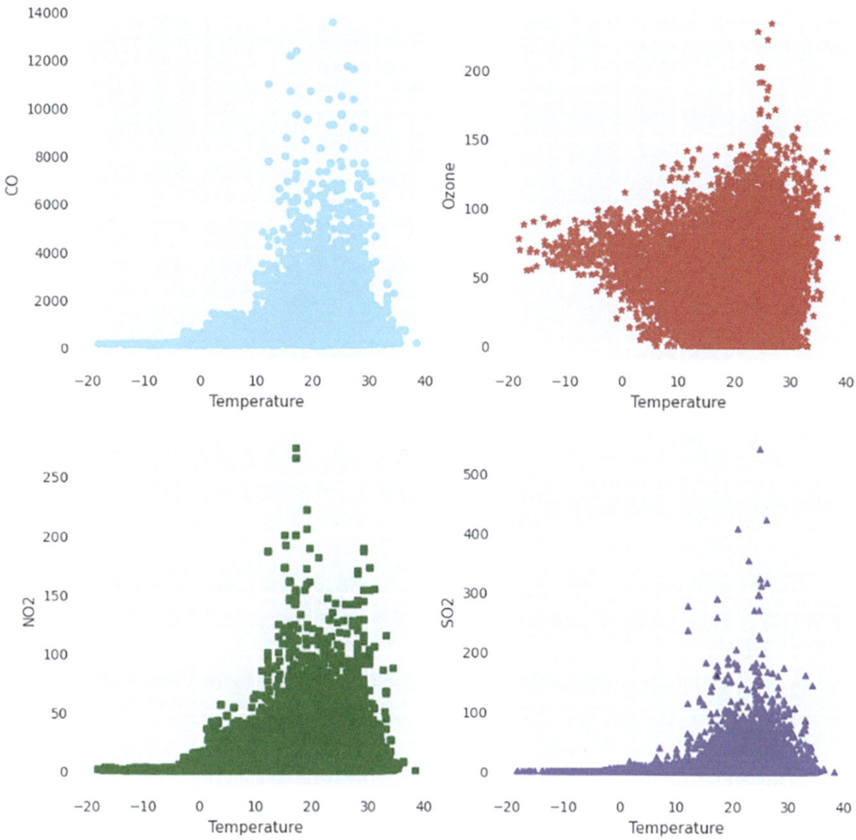

Fig. 6 Temperature versus AIR-quality factors

spots such as Kolkata, Utter Pradesh, and more. At day-to-day time interval with current climatic conditions such as pressure, wind speed and direction, air-quality measurements, precipitation, and temperature, the proposed method is processing for training. Figure 5 is the graph of the rising temperature, air-quality-PM10, and UV-index exposed with respect to states which are shown, respectively.

4.1 Exploratory Data Analysis

EDA is an data investigation and visualization method used by data analysts, aimed at summarizing the important characteristics, like distribution count, minimum and maximum extreme conditions, outlier detection and more through different graphs. This will evaluate the trends in dataset used and also the data orientation toward the

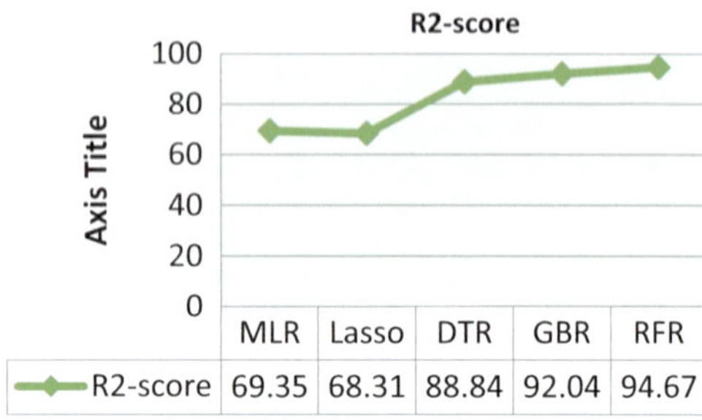

Fig. 7 Accuracy of the proposed model

actual procedure. The data performance was checked by Exploratory Data Analysis (EDA) through mentioned ML models has shown better accuracy shown in Fig. 7.

4.2 Proposed Methodology

The dataset utilized for this research involves time-series data from Kaggle data repository. The real-time applications for climate change analysis involve actual data values from atmosphere held for daily analysis. The data may be characterized by rapid change and nonlinearity. Although extracting the behavior of input variables is quite complex in training high performance ML models, the proposed RFR model analyzed the trends in data more efficiently than others. RFR is a potent method that provides several advantages for time-series forecasting. It is adept in managing nonlinear connections, missing data, outliers, and extensive datasets, which contributes to its popularity in time-series research. It is capable of managing nonlinear interactions between the predictor factors and the response variable. The missing data in the predictor variables are a frequent issue in time-series data, which were estimated and filled up using the existing data. RFR is an ensemble technique that aggregates numerous DTs to provide predictions; also it reduces the likelihood of overfitting and enhances the accuracy up to 94.67% in the forecast with capability of managing extensive datasets with several predictor factors. This quality makes it ideal for time-series data, which frequently includes several predictor factors.

4.3 Results and Discussion

A better fit is indicated by higher R^2-score values. Figures 7 and 8 display the outcomes that were obtained by the suggested model. Data deviation of the suggested variable dependency may be seen by MAE, Mean-Squared Error (MSE), and RMSE, and the accuracy measure for continuous time-series data is dependent upon the R-score. MAE measures the typical discrepancy between actual and anticipated values.

Fig. 8 Estimated deviation metrics

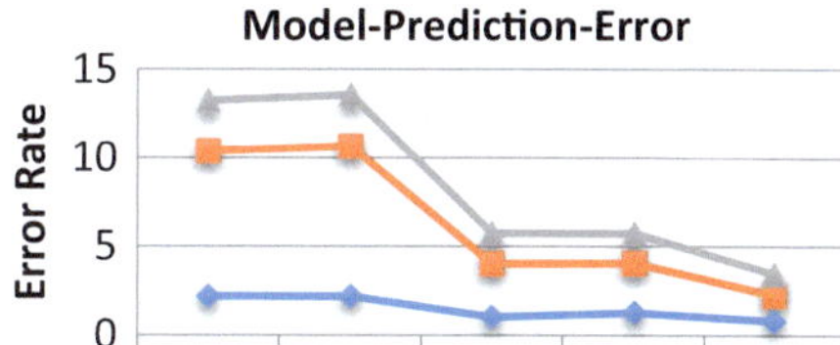

	MLR	Lasso	DTR	GBR	RFR
RMSE	2.86	2.91	1.73	1.67	1.2
MSE	8.19	8.47	2.98	2.79	1.44
MAE	2.19	2.18	1.02	1.25	0.8

The MSE measures how off the actual results are from the predictions. Greater accuracy is shown by a smaller MSE score.

The explained variance represents the extent to which the independent variables can account for the variation in the dependent variable; it was obtained at greater results in the suggested technique with an R^2-score and an accuracy of 94.67%. The accurate estimation of CC prediction and temperature analysis is challenging due to the dynamic nature of the data over extended periods. The likelihood of obtaining more accurate prediction capabilities can be increased by analyzing DL observations captured by satellites even in small time intervals.

5 Conclusion and Future Scope

The ML approach to analyze Indian weather report data is yielding better results for continuous observations with different time-lapse in a day with all the states and their districts. With continuous analysis of changing patterns than modern methodologies, the conventional regression models like RFR, MLR, GBR etc., perform at better level and will give more accuracy and model fit in temperature forecasting. The atmospheric dataset should be processed and annotated with defined threshold levels. Then after that, more sophisticated Machine Learning (ML) and Deep Learning (DL) models can be employed. The latest developments in remote sensing technology, including more sophisticated sensors in satellites and in radars. These datasets can process by analytical DL methods, allow real-time time-series data preprocessing while collecting the data itself. IMDs and NWS weather stations are nowadays utilizing standard ML and DL models to integrate multi-dimensional and heterogeneous data while doing spatio-temporal advancements in visualization and understanding for end-users; The forecoming future applications are with fusion DL techniques for fast detailing and feature extraction of the multi-model data.

References

1. Lee H, Calvin K, Dipak D, Gerhard K, Mukherji A, Thorne P, Trisos C et al (2023) Synthesis report of the IPCC Sixth assessment report (AR6), Longer report. IPCC
2. Griscom BW, Adams J, Ellis PW et al (2017) Natural climate solutions. Proc Nat Acad Sci 114(44):11645–11650
3. Frame B, Lawrence J, Ausseil AG, Reisinger A, Daigneault A (2018) Adapting global shared socio-economic pathways for national and local scenarios. Clim Risk Manag 21:39–51
4. Prajapati RN, Ibrahim N, Thapa BR (2023) Climate change impact on water availability in the Himalaya: insights from Sunkoshi River basin, Nepal. HydroResearch 6:279–292
5. Sidhu BS, Mehrabi Z, Ramankutty N, Kandlikar M (2023) How can machine learning help in understanding the impact of climate change on crop yields? Environ Res Lett 18(2):024008
6. Valipour M, Khoshkam H, Bateni SM, Jun C (2024) Machine-learning-based short-term forecasting of daily precipitation in different climate regions across the contiguous United States. Expert Syst Appl 238:121907
7. Li Y, Strapasson A, Rojas O (2020) Assessment of El Niño and La Niña impacts on China: enhancing the early warning system on food and agriculture. Weather Clim Extremes 27:100208
8. Bhatla R, Bhattacharyya S, Verma S et al (2023) El Nino/La Nina and IOD impact on Kharif season crops over western agro-climatic zones of India. Theoret Appl Climatol 151(3–4):1355–1368
9. Jyothi D, Pushpanjali B, Subrahmanyam MV (2023) Rainfall variation over Andhra Pradesh and comparison with all Indian rainfall during El Nino/La Nina. Open Access Libr J 10(5):1–16
10. Lal D, Singh S (2023) Impact of El-Nino and La-Nina episodes on rainfall variability and crop yield. Int J Environ Clim Change 13(10):2046–2051
11. Torsoni GB, de Oliveira Aparecido LE, dos Santos et al (2023) Soybean yield prediction by machine learning and climate. Theor Appl Climatol 151(3–4):1709–1725
12. Praveen B, Talukdar S, Shahfahad et al (2020) Analyzing trend and forecasting of rainfall changes in India using non-parametrical and machine learning approaches. Sci Rep 10:10342
13. Haq MA, Ahmed A, Khan I, Gyani J, Mohamed A, Attia EA, Mangan P, Pandi D (2022) Analysis of environmental factors using AI and ML methods. Sci Rep 12(1):13267
14. Pasichnyk M, Stanovsky P, Polezhaev P, Zach B, Šyc M, Bobák M, Jansen JC et al (2023) Membrane technology for challenging separations: Removal of CO2, SO2 and NOX from flue and waste gases. Sep Purif Technol, 124436
15. Malakouti SM (2023) Utilizing time series data from 1961 to 2019 recorded around the world and machine learning to create a global temperature change prediction model. Case Stud Chem Environ Eng 7:100312

Point Cloud Facilitated ORB-SLAM2 for Robust Autonomous Navigation in a Cluttered Indoor Environment

Kuntal Ghosh, Jukka Heikkonen, and Rajeev Kanth

Abstract This article primarily encompasses different aspects of autonomous indoor navigation for robots. Vision-based camera sensors have recently been supported **S**imultaneous **L**ocalization **A**nd **M**apping (SLAM) technology, which is being explored for autonomous navigation. However, the capabilities of SLAM are limited to identifying objects from different perspectives. Hence, point cloud information is integrated with the visual feed for robust and precise mapping of the environment. Upon acquiring discrete point clouds, points are stitched and registered, respectively, to reconstruct the scene. Then, using those stitched point clouds, different obstacles within the same environment are registered too. Subsequently KD-Tree (k-dimensional tree) search algorithm is implemented for classifying the point cloud data, which leads to identifying clusters of objects occupied within the environment. In this entire process, an Intel RealSense stereo depth camera coupled with a mobile PC is used for collecting visual feed and concerned 3D point vectors to emulate this integrated concept.

Keywords Oriented FAST and Rotated BRIEF SLAM · Point cloud · 3D reconstruction · KD-Tree · Mapping

---

This research is supported by Science and Engineering Research Board; Department of Science and Technology, India.

K. Ghosh (✉)
Department of Electronics and Communication Engineering, Asansol Engineering College, Asansol, India
e-mail: kuntalgh24@gmail.com

J. Heikkonen
Department of Computing, University of Turku, Turku, Finland
e-mail: jukhei@utu.fi

R. Kanth
Savonia University of Applied Sciences, Kuopio, Finland
e-mail: Rajeev.Kanth@savonia.fi

© The Author(s), under exclusive license to Springer Nature Singapore Pte Ltd. 2025
S. K. Udgata et al. (eds.), *Intelligent Systems*, Lecture Notes in Networks and Systems 1314, https://doi.org/10.1007/978-981-96-3797-3_18

1 Introduction

The smart movement of an autonomous robot primarily depends on accurate perception of the surrounding environment. In that case, active perception becomes essential to obtain the entire scene surrounding the robot. Predominantly multiple views from different directions are required to understand the environment having large dimensions. Therefore, many image frames should be collected to construct a wide-angle scene around the autonomous robot for safe operation. The sequential image frames from the body-mounted camera on the robot are stitched together to construct the scene of the indoor environment. Our proposed work also includes image stitching, for which several easily accessible algorithms are available. Most conventional algorithms provide real-time implementation and optimum performance with no discernible traces [1]. Point cloud stitching makes it easier to integrate a scene as a whole, just like image frames do. However, accuracy during spontaneous execution is challenging for 3D point cloud stitching. In this study, we propose a methodology that employs concurrent localization and mapping algorithms to expedite and enhance the process of stitching 3D point clouds with greater precision.

The integration of 3D point clouds, sometimes referred to as point cloud stitching, is an essential process that obtains the full picture while seeing its surroundings. This technology primarily creates a synthetic three-dimensional (3D) environment and a map of the circumstances for an autonomous vehicle by letting the vehicle evaluate object dynamics. When used in conjunction with autonomous moving platforms, 3D point cloud stitching is particularly useful for tasks like reconstructing indoor scenes so that robots can use sophisticated navigation techniques, generating 3D shapes from a collection of images, adding missing points back into an image scene during architectural modeling, and more. Three types of point cloud stitching exist: (a) the template matching-based methods [2], (b) feature matching-based methods [3], and (c) voxel matching-based methods [4]. An indoor scene of a disorganized room was photographed with a handheld depth camera and has the following characteristics. The template matching-based stitching method has limitations in moving scenes; it performs best in stationary scenes [3]. Secondly, perceiving the objects at indoors vis a vis at the outdoor environment using a voxel matching-based method [4] performs optimum for smaller elements.

In this work, the first stereo depth camera is being used to collect all the image frames while traversing in the clumsy environment of the room. The objects present in the room, ranging from small to large scale sizes are also easily confused with the background. To fix this issue, a feature matching-score-based method is implemented to reduce feature mismatching within the corresponding points. To adhere to the criteria of feature matching methodology, the identification of features (referred to as key points) is conducted in ORB SLAM for both the corresponding image pair and the keyframes that are adjacent to each other.

2 Overview of Proposed Methodology

The first step of this proposed framework is image acquisition. Then, the ORB SLAM 2 algorithm was used while mapping the indoor environment. The ORB SLAM 2 algorithm has been employed for mapping the surrounding environment, as the Intel RealSense D-435i stereo camera is used in this work. ORB-SLAM2 algorithm is primarily used for stereo images that map accurate depth information. Rebuilding the environment uses point cloud data that matches simultaneously. Processing point clouds under the suggested algorithm is the third step. Localizing the objects using KD-Tree point cloud stitching is the last step. The subsequent subsections provide a detailed explanation of each step, that is composed of various algorithms with distinct functions.

2.1 ORB-SLAM2

This is the second generation ORB SLAM which could be implemented on stereo as well as on-depth images. The system consists of three parallel threads. (1) The process of local mapping involves the management and optimization of a local map. (2) Tracking refers to the process of determining the camera's position in each frame by identifying corresponding features in the local map and minimizing the re-projection error. (3) Loop closing involves the identification of loop closures in the environment by utilizing matched key points and subsequently correcting accumulated drift through pose-graph optimization. Eventually, this results in the start of a fourth thread that computes the ideal structure and motion solution.

The system employs identical ORB features [5] for tracking, mapping, and place recognition tasks. The aforementioned features exhibit properties of rotation invariance and scalability. Additionally, the ORB feature exhibits a notable property of invariance to camera autogain and variations in illumination within an indoor setting. Moreover, the extraction of these features can be accomplished in a shorter duration, thereby enabling real-time functionality and demonstrating high accuracy in bag-of-word place recognition [6]. In this work, we exploit stereo/depth information for collecting point cloud data and for mapping the environment using matched key points. The following flowchart describes the generic functioning of ORB-SLAM2 [7].

2.2 Point Cloud Acquisition

This study primarily focuses on continuous point cloud stitching for constructing the surrounding environment to interpret by an autonomous robot. Point cloud generation can be categorized into two distinct groups based on the types of devices used for collecting point cloud data. There are two primary methods used in the field [9]: (i) LiDAR-based methods [10] and (ii) triangulation-based methods [11]. LiDAR produces precise dense point clouds which limit the error. This phenomenon motivates researchers to emulate LiDAR-based point cloud acquisition. Initially, range measurements from LiDAR are employed for real-time odometry and mapping [12]. Further, this process is integrated with sensor pose optimization and measurements [13]. The process of integrating data in this manner improves the error term associated with the optimization problem, even though it leads to a relatively large state space. Progressing further, real-time visual odometry is executed to estimate the ego-motion and LiDAR for registering point clouds at high frequency and low fidelity [14]. Besides that, different variants of **S**imultaneous **L**ocalization **A**nd **M**apping (SLAM) are also investigated for acquiring point cloud information [15]. Florez et al. deployed integrated stereo vision feed for precise vehicle localization and LiDAR for tracking dynamic objects, respectively [16]. However, LiDAR scans every object present in the scene which led to the generation of millions of redundant points. This causes insufficient and ambiguous 3D construction along with the put limits in real-time execution.

Advancing further, the triangulation-based methods depending on a vision- based approach under the purview of SLAM framework. Under this method point cloud acquisition is performed by the stereo camera feed using the following steps: first, the stereo camera collects the subsequent image pairs in the same scene. Second, the specific features (ORB, SIFT, SURF, BRIEF) are identified and matched in the subsequent image frames. ORB-SLAM is a variant of Monocular SLAM that computes camera trajectory in real time and performs a sparse 3D reconstruction of the scene for both indoor and outdoor environments. This method incurs the benefit of having close large loops and performing global optimization in real time shown in Fig. 1 [8].

Triangulation method primarily implied for feature matching in the corresponding image pairs. Therefore, feature detection plays a pivotal role in scene reconstruction by perception building for the autonomous robots. Simultaneously, point cloud generation also depends on feature detection. Features matching of images is the fundamental primitive step for object detection and tracking. Primarily this is executed using the corresponding feature points, such as the SIFT [17] and SURF [18] for post-processing. However, FAST [19] and oriented brief (ORB) [20] respond in real-time to detect the corresponding feature points between the image pairs. Mismatch drastically reduces the performance of triangulation during feature matching.

Triangulation is being used to generate 3D points post-finding the matched points demonstrated in Fig. 2. The triangulation method determines the location of a point in three-dimensional coordinates from corresponding matched features. This concept is depicted in Fig. 2, where the location of virtual point M in 3D coordinates, i.e., determined based on the presence of matched features m_1, m_2 in the left and right images, respectively. The coordinates of M are obtained using the following relation [21]:

$$\rho_1 m_1 = K_1 R_1^T (M - C_1) \tag{1}$$

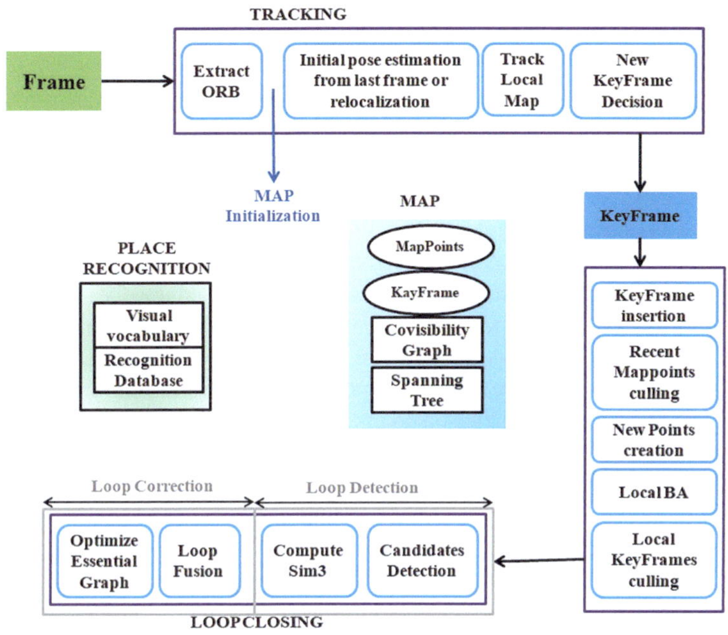

Fig. 1 Flowchart of ORB-SLAM2 framework [8]

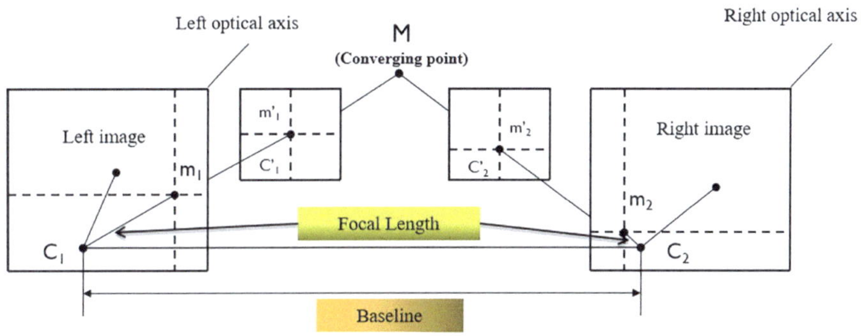

Fig. 2 Projection of 3D points of the corresponding image features (acquired from the stereo camera feed) in the real-world scenario

$$\rho_2 m_2 = K_2 R_2^T (M - C_2) \tag{2}$$

where ρ_1, ρ_2 are the scale parameters, K_1, K_2, R_1, R_2 are the intrinsic parameters of cameras, which are determined by using calibration C_1, C_2 are the coordinates of the camera's positioning in the real world.

2.3 Point Cloud Stitching Technique

This data set is composed of hundreds to thousands of points with geometrical coordinates of those corresponding points. Upon stitching these point cloud, adjoins successive image pairs and constructs a 3D scene subsequently. To attain uniformity, point clouds are segregated based on their positions at the same plane concerning the same view axis, termed point clouds aligning. However, subsampling is performed before this aligning which reduces the redundant points to accelerate the computational efficiency. Primarily, the box grid filter method is used to subsample the continuous point clouds which incur significant latency. Therefore several algorithms like bilateral filtering [22], iterative closest point (ICP) real-time registration, and greedy triangle projection have been introduced to enhance the computation. Later, a novel 3D registration algorithm based on modified ICP is developed where the corresponding point in the point cloud is assigned weights to minimize the computational time [23]. Bentley implemented k-d tree search that accelerated the aligning based on Euclidean distances among the corresponding points.

Here, we implement iterative closest point (ICP) to correct the alignment of acquired 3D point clouds. ICP is the conventional 3D registration algorithm which is primarily implemented for 3D stitching. Equation (4.3) demonstrates the mechanism of ICP

$$E(R, t) = \sum_{i=1}^{N_m} \sum_{j=1}^{N_d} \omega_{i,j} \|m_i - (Rd_j + t)\|^2 \tag{3}$$

where N_m and N_d are the number of points of the point cloud set M and data set D, respectively. $\omega_{i,j}$ are the weights for the corresponding point match.

Post registration, successive points are integrated with the reference point cloud using a rigid transformation matrix. Stitching point clouds using a rigid transformation matrix constitutes errors in the distinct set of point clouds. Feature matching constraints and matching scores are deployed to optimize the transformation matrix which leads to minimization of errors.

2.4 k-Dimensional Tree (KD-Tree)

An array of elements (points) in a k-dimensional space $\mathbb{R}^k$ can be arranged using a hierarchical spatial partitioning data structure called a k-dimensional tree [24]. KD-Tree divides the data set's points into hierarchically organized, axis-aligned cells, with each cell represented by a node in the tree. In the current context, point cloud data in three dimensions has been collected for the indoor environment. This originates from k-d trees restricted within the three dimensions only. Deploying a hyperplane perpendicular to the corresponding axis, each cluster of a k-d tree splits all data sets along a specific dimension. At the root of the tree, all data points will be split based on a certain threshold value. Typically, the partition dimension consists of the branch with the maximum span, whereas the threshold split value is chosen as the median. Each of the two split cells from the root is split recursively which is carried out into other cells too. This recursive branching phenomenon terminates when the number of points which are contained in a cell remains within a bounded value. The height of a KD-Tree with leaf nodes containing only a single point would be $\log 2N$.

Nearest neighbors in a KD-Tree arrangement always facilitate in categorizing different patterns in the data set. The KD-Tree is essentially implemented to construct a balanced binary tree [25]. The nodes are iteratively inserted into a KD-Tree until the sizes of the leaf nodes are not met. The dimension with the greatest amount of variance in the data points belongs to this node. KD-Tree is widely used in the fields of vision-based navigation, artificial intelligence, and pattern recognition community for multidimensional applications.

3 Experimental Framework

This section outlines our proposed methodology to obtain navigable space in an unknown environment with the help of a point cloud data set. In this direction, first, we implement real-time ORB SLAM 2 algorithm to facilitate visual perception of the indoor environment for autonomous robots. The visual perception has been collected using Intel Real Sense D-435i depth camera. This camera is integrated with inertial measurement unit (IMU) sensor, having precision up to 10 m distance [26]. The presence of in-built IMU sensor helps in tracking the pose of robot with the camera attached to it. Active IR stereo technology assists in measuring the depth by having a minimum threshold of 0.105 m [26]. Further, these depth frames enable to extract of point cloud information for the entire scene. In this case Fig. 3 represents, a GPU-enabled Intel core i7-9350 laptop with 16-GB RAM and Nvidia RTX 2060 graphics card is used for the implementation of SLAM algorithm and post-processing of point clouds. To, replicate the robotic motion this entire room is point-wise traveled holding this integrated setup.

This point cloud data consists of three-dimensional positioning (x, y, z) information of each point at the respective time stamps. This data also includes orientation information to track the pitch, roll, and yaw of the robot. 3D construction of the scene heavily relies on the density of point cloud data. This reconstruction mechanism integrates both point cloud stitching and point cloud registration methodologies using ICP algorithm. However, stitching itself is inadequate for obstacle identification predominantly for demarcating the navigable space in a cluttered environment. Hence, point clouds segregated loosely coupled points by deploying an iterative KD-Tree algorithm which is described below. The proposed iterative k-D Tree algorithm executes a loop function based on the positioning of the density of points holding the mean value. We intend to check the variance of points within a tightly coupled boundary region.

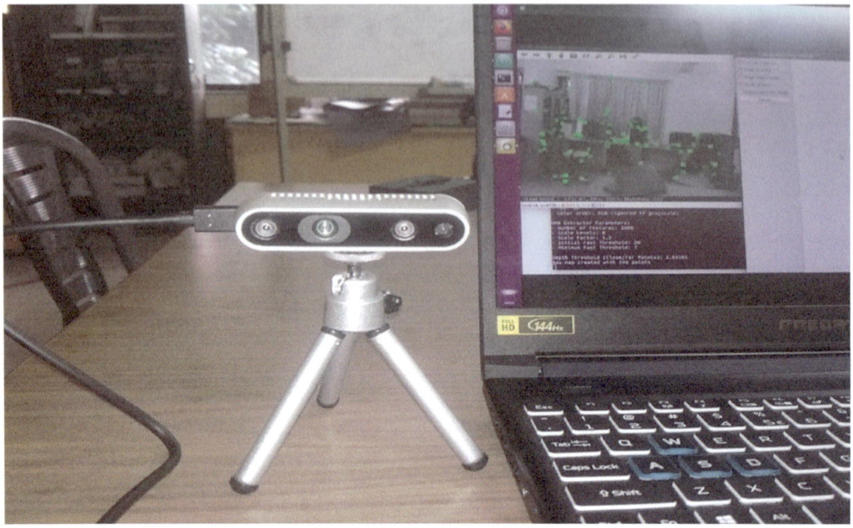

Fig. 3 Experimental setup includes Intel RealSense D-435i stereo depth camera [26] and Intel core i7-9350 laptop.

Algorithm 1 Constructing the priority search-based *k-means tree*

Input: Point Cloud data set D, branching factor K, max iterations I_{max}, center selection algorithm to use C_{alg}
Output: k-means tree
if $|D| < k$ **then**, then
  assigning root node with the points in D.
else
  $P \leftarrow$ select k points from D using the C_{alg} algorithm
  converged $\leftarrow$ false
  iterations $\leftarrow 0$
  while not $iteration = 0$ **do**,
    Converge and iterations $\leq I_{max}$,
    C $\leftarrow$ cluster the points in D around nearest center P based on euclidean distance
    $P_{new} \leftarrow$ means of clusters in C
    if $P = P_{new}$ **then**
      converge $\leftarrow$ true
    end if
    $P \leftarrow P_{new}$
    iteration $\leftarrow$ iteration+1
  end while
  for each cluster $C_i \in C$ **do**
    create non-root node with center P_i
    recursively apply the algorithm to the points in C_i
  end for
end if

Here, priority search-based KD-Tree classifies tightly coupled points as a part of the same object which leads to identifying the 3D shape of the obstacles. This the process is recursively implemented till segregation is attained.

4 Results

This section demonstrates different results which is obtained while performing ORB-SLAM2, point cloud acquisition followed by stitching for 3D scene reconstruction. Real-time ORB-SLAM2 is executed for the customized room having dimensions of (15ft x 20ft). ORB-SLAM2 is a feature-based SLAM that detects the Oriented FAST and Rotated Brief (ORB) features across the entire scene. So this phenomenon limits the performance of ORB SLAM 2, if inadequate numbers of ORB features are present in the field of view. In the following Fig. 4, output of ORB SLAM is exhibited.

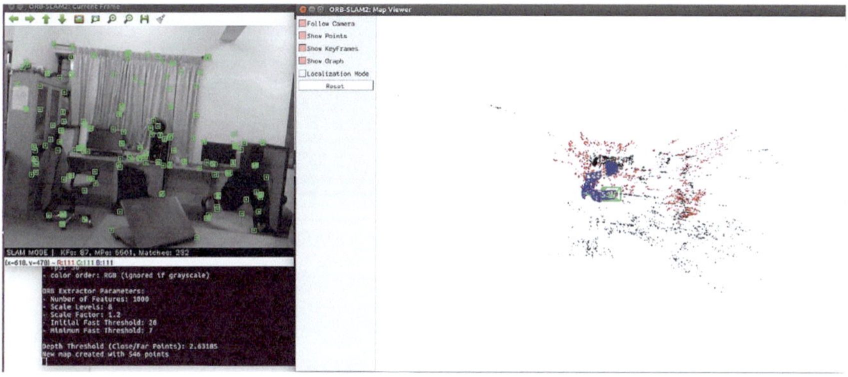

(a) Execution of ORB SLAM2

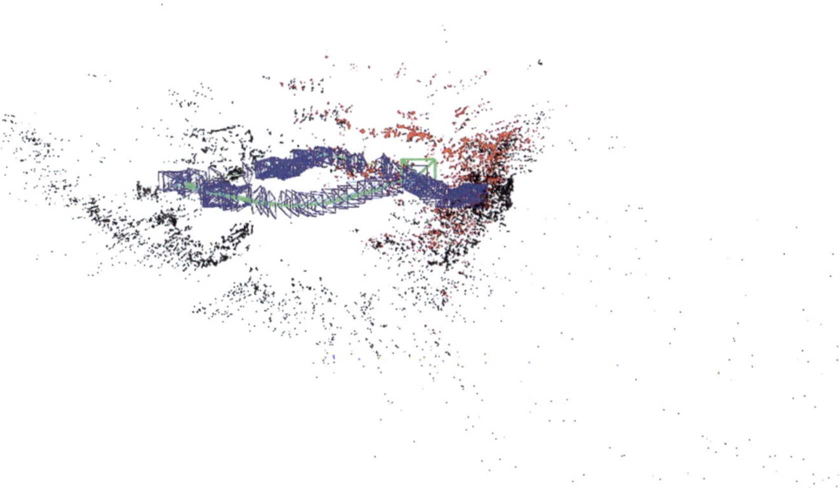

(b) Real-time mapping of the surrounding

Fig. 4 **a** Demonstrates the execution of ORB SLAM 2 in real-time. **b** The final real-time map consists of keyframes (blue), current camera position (green), map points (black, red)

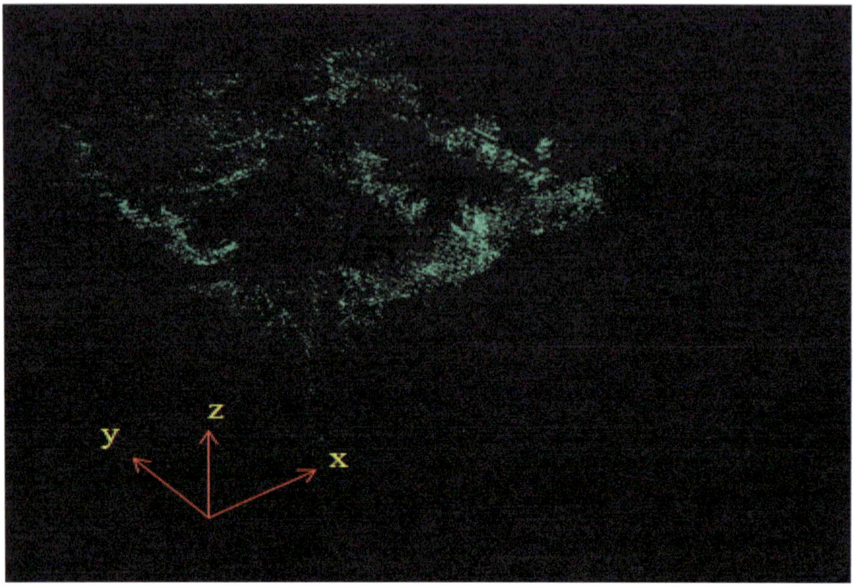

Fig. 5 Mapping the point cloud structure of the indoor environment

Point clouds are produced by back projecting the sensor depth maps from the calculated keyframe postures during the ORB-SLAM2 architecture's performance. Due to camera dynamics scattered and sparse 3D point vectors are generated with only 18290 points. These low-density point vectors pose a challenge while constructing the scene involving the elements present in the scene. An iterative closest point (ICP) algorithm is applied for point cloud stitching with minimal alignment error to overcome this challenge. Post registration, point cloud vectors construct the scene as shown in below Fig. 5.

However, this reconstructed scene cannot classify the environment based on captured point clouds. This classification exercise requires the identification of tightly coupled points that could be perceived as part of the same object. To perform this task, we introduce a priority search-based KD-Tree algorithm. This algorithm segregates point vectors situated at different planes with a separation distance threshold of 0.3 m [ref. Algorithm 1]. Upon implementing this customized adapted priority search-based KD-Tree algorithm, nearest neighbor points are segregated, and active perception is achieved indeed. In this process, the following result in Fig. 6 facilitates the robot/vehicle to identify the area from where point vectors are not received. We considered that area to be obstacle-free and available for navigation. Here area demarcated with green color identifies the central cluster of point vectors, which signifies the presence of objects within the scene. The blue color represents the area where point vectors are situated at different planes, whereas gray color signifies another peripheral boundary.

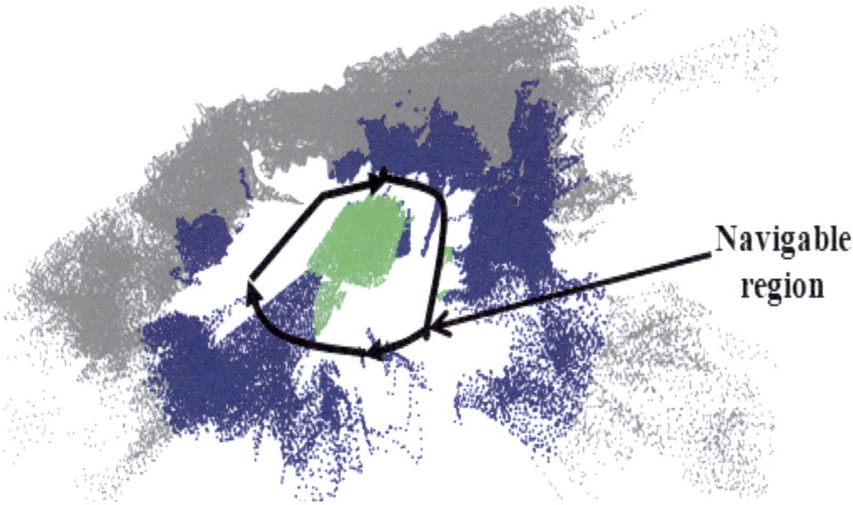

Fig. 6 Final output of proposed KD-Tree segmentation of point cloud information for classifying obstacles situated at different clusters

Multi-sensor data fusion is the most challenging aspect of the point cloud stitching with identical timestamps. Figure 7 demonstrates the efficacy of the ORB-SLAM 2 algorithm where the camera trajectory and the key-frame trajectory are fused. Here, we mapped the key frame trajectory based on ORB features being collected from each and every frame. This has been fused with the integrated IMU sensor data to compare the camera trajectory with the corresponding keyframe trajectory. This error between two mapped trajectories develops an understanding of the limitations of the ORB-SLAM2 algorithm. This result categorically found that the ORB SLAM 2 algorithm accumulates errors with the camera trajectory during the changing of the orientations. However, this error remains within the range of 3–5 cm in our work which gives the scope of further investigations.

5 Conclusion

This work aims to develop a simple process to enable active perception for autonomous robots in a cluttered environment. Our proposed method introduces a vision-based active perception method with minimal hardware (integrating depth camera with a PC). The major contributions of this investigation can be categorized as (i) acquiring and storing point cloud data while executing ORB SLAM, (ii) reconstruction of the 3D scene from sparse and lowly dense point cloud (in this case 18290 points), (iii) proposing a priority search-based customized KD-Tree algorithm for classifying the environment in a planar level as well as based on distance in

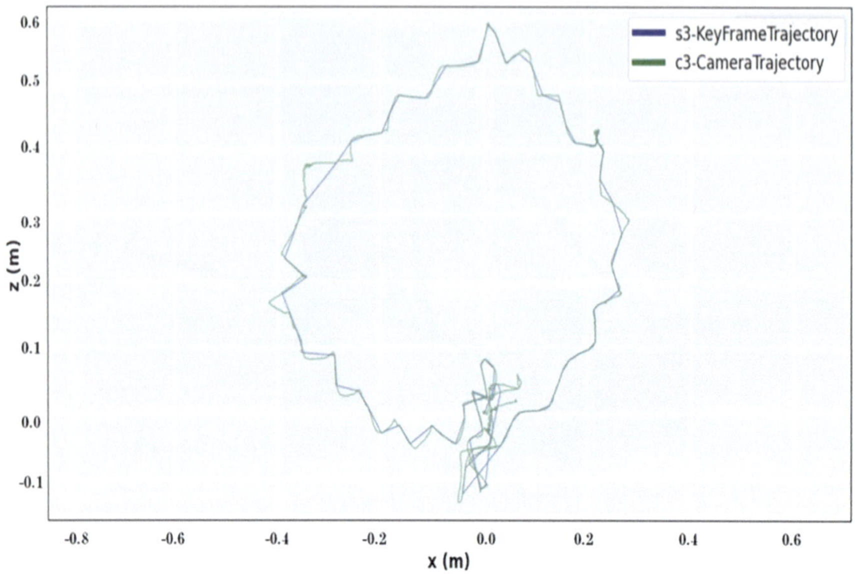

Fig. 7 Camera trajectory map and keyframe trajectory map from the experiment

Euclidean space, and (iv) demarcating region categorized as the obstacle-free navigable path for the robot. This process has limitations while assuming the absence of a point cloud cluster as an obstacle-free region and identifying feature descriptors other than ORB features. In addition, this work also establishes the limitations of the ORB-SLAM2 algorithm while changing orientations in the trajectory. Overall the proposed method is found to be effective for implementing a vision-based integrated approach for demarcating navigable space.

References

1. Abdukholikov M, Whangbo T (2017) Fast image stitching method for handling dynamic object problems in panoramic images. KSII Trans Internet Inf Syst 11(11)
2. Yang B, Kostková J, Flusser J, Suk T, Bujack R (2018) Rotation invariants of vector fields from orthogonal moments. Pattern Recogn 74:110–121
3. Bégaint J, Thoreau D, Guillotel P, Guillemot C (2017) Region-based prediction for image compression in the cloud. IEEE Trans Image Process 27(4):1835–1846
4. Song KT, Wu CH, Jiang SY (2017) Cad-based pose estimation design for random bin picking using a RGB-D camera. J Intell Robot Syst 87(3–4):455–470
5. Rublee E, Rabaud V, Konolige K, Bradski G (2011) Orb: an efficient alternative to sift or surf. In: 2011 International conference on computer vision. IEEE, pp 2564–2571
6. Mur-Artal R, Tardós JD (2014) Fast relocalisation and loop closing in keyframe-based slam. In: 2014 IEEE International conference on robotics and automation (ICRA). IEEE, pp 846–853
7. Mur-Artal R, Tardós JD (2017) Orb-slam2: an open-source slam system for monocular, stereo, and rgb-d cameras. IEEE Trans Robot 33(5):1255–1262

8. Mur-Artal R, Montiel JM, Tardos JD (2015) Orb-slam: a versatile and accurate monocular slam system. IEEE Trans Robot 31(5):1147–1163
9. Bi ZM, Wang L (2010) Advances in 3d data acquisition and processing for industrial applications. Robot Comput Integr Manuf 26(5):403–413
10. Jennifer Von Schwerin, Heather Richards-Rissetto, Fabio Remondino, Maria Grazia Spera, Michael Auer, Nicolas Billen, Lukas Loos, Laura Stelson, and Markus Reindel. Airborne lidar acquisition, post-processing and accuracy-checking for a 3d webgis of copan, honduras. *Journal of Archaeological Science: Reports*, 5:85–104, 2016
11. Zhu Z, Brilakis I (2009) Comparison of optical sensor-based spatial data collection techniques for civil infrastructure modeling. J Comput Civ Eng 23(3):170–177
12. Zhang J, Singh S (2014) Loam: lidar odometry and mapping in real-time. Robot Sci Syst 2:9–17
13. Ruhnke M, Kümmerle R, Grisetti G, Burgard W (2012) Highly accurate 3d surface models by sparse surface adjustment. In: 2012 IEEE international conference on robotics and automation. IEEE, pp 751–757
14. Zhang J, Singh S (2015) Visual-lidar odometry and mapping: low-drift, robust, and fast. In: 2015 IEEE international conference on robotics and automation (ICRA). IEEE, pp 2174–2181
15. Grisetti G, Kümmerle R, Stachniss C, Frese U, Hertzberg C (2010) Hierarchical optimization on manifolds for online 2d and 3d mapping. In: 2010 IEEE international conference on robotics and automation. IEEE, pp 273–278
16. Rodriguez Florez SA, Frémont V, Bonnifait P, Cherfaoui V (2014) Multi-modal object detection and localization for high integrity driving assistance. Mach Vis Appl 25(3):583–598
17. Chew VC, Lian FL (2012) Panorama stitching using overlap area weighted image plane projection and dynamic programming for visual localization. In: 2012 IEEE/ASME international conference on advanced intelligent mechatronics (AIM). IEEE, pp 250–255
18. Kim HK, Lee KW, Jung JY, Jung SW, Ko SJ (2011) A content-aware image stitching algorithm for mobile multimedia devices. IEEE Trans Consum Electron 57(4):1875–1882
19. Wagner D, Mulloni A, Langlotz T, Schmalstieg D (2010) Real-time panoramic mapping and tracking on mobile phones. In: 2010 IEEE virtual reality conference (VR). IEEE, pp 211–218
20. Fujimoto S, Hu Z, Chapuis R, Aufrère R (2016) Orb-slam map initialization improvement using depth. In: 2016 IEEE international conference on image processing (ICIP). IEEE, pp 261–265
21. Hu F, Li Y, Feng M (2019) Continuous point cloud stitch based on image feature matching constraint and score. IEEE Trans Intell Veh 4(3):363–374
22. Xu N, Zhang W, Zhu L, Li C, Wang S (2017) Object 3d surface reconstruction approach using portable laser scanner. In: IOP conference series: earth and environmental science, vol 69. IOP Publishing, p 012119
23. Wang X, Zhao ZL, Capps AG, Hamann B (2017) An iterative closest point approach for the registration of volumetric human retina image data obtained by optical coherence tomography. Multimed Tools Appl 76(5):6843–6857
24. Hu L, Saeid N (2017) Massive parallelization of approximate nearest neighbor search on kd-tree for high-dimensional image descriptor matching. J Vis Commun Image Represent 44:106–115
25. Bentley JL (1975) Multidimensional binary search trees used for associative searching. Commun ACM 18(9):509–517
26. Intel realsense depth camera d400-series datasheet

Improving Sports Categorization in WiFi-Blocked Areas: Modifying YOLO Network for All-Inclusive Sports Image Identification

Md. Habibur Rahman, Samar Das, Md. Abdullah, Mahtab Uddin, Md. Aminul Islam, Yamin Hossain, and Abdullah Hafez Nur

Abstract This paper focuses on detecting sports types from images, which can be used for action recognition. Detecting small objects like athletes and sports equipment is challenging due to their varying colors, appearances, and distance from the camera. CNN is the most common detection tool for sports, but it struggles with image accuracy due to various angles and light conditions. To detect sports using YOLO, the study suggests a model that uses Darknet-53. This model attained a of 97.25% on test data and 97.25% on train data. This study has the potential to identify and categorize different sports and their analytical data. We were 97.25% correct in identifying the recommended model for altering the YOLO network for sports classification using game items. Broadcasting agencies are the most important prerequisite for preserving sports' allure. Sports images can now be more easily recognized according to their different classes thanks to our research. There hasn't been much research done, particularly in terms of identifying sports from the previously provided data. Not

Md. Habibur Rahman (✉)
Department of Mathematics, University of Chittagong, Chittagong, Bangladesh
e-mail: drhabib1216@gmail.com

S. Das
Department of Computer Science and Engineering, Premier University, Chittagong, Bangladesh

Md. Abdullah
Department of Computer Science and Engineering, Daffodil International University, Dhaka, Bangladesh

M. Uddin
Institute of Natural Sciences, United International University, Dhaka, Bangladesh

Md. Aminul Islam
School of Engineering, Computing, and Mathematics, Oxford Brookes University, Oxford, UK

Y. Hossain
Department of Computer Science and Engineering, VIT-AP University, Amaravathi, Andhra Pradesh, India

A. H. Nur
Department of Electronic and Telecommunication Engineering, International Islamic University Chittagong, Chittagong, Bangladesh

© The Author(s), under exclusive license to Springer Nature Singapore Pte Ltd. 2025
S. K. Udgata et al. (eds.), *Intelligent Systems*, Lecture Notes in Networks and Systems 1314, https://doi.org/10.1007/978-981-96-3797-3_19

only were there few jobs in the sports industry overall, but no one worked in it. They concentrated on handball and basketball. Algorithm research in this area has the potential to enhance sports analytics and spectator services. The best pre-trained CNN model and classifier to apply in pertinent future studies projects are what we hope to achieve by running these experiments. Our research is probably going to lead to the development of better sports identification algorithms.

Keywords Sports picture identification · YOLO social network · Sports division · CNN configuration · Athletic data analysis

1 Introduction

Object detection involves a multi-step process, including image preprocessing, feature extraction, object categorization, and object location. YOLO, a convolutional neural network (CNN), is a fundamental algorithm that can recognize objects in real-time with just one neural network pass. It uses a large dataset of annotated images to develop pattern recognition abilities. YOLO has been upgraded by versions such as YOLOv2, YOLOv3, and YOLOv4, which have increased dynamics and specificity. The YOLO method is renowned for its fast and specific performance [1]. It would help if you used YOLO. The image is preprocessed for detection, and the model predicts the bounding box. It can detect cars, people, and buildings in satellite images, vehicles, pedestrians, traffic signs, and signals in autonomous driving systems, detect and track items in robotics applications, and monitor personal and moving objects in security and surveillance systems [2].

The YOLO method for detecting and tracking objects in sports videos and images. This cutting-edge object detection algorithm can quickly and efficiently identify objects in pictures and videos. It can be used for sports analytics, broadcasting, and player development [3]. Sports detection using YOLO requires a training dataset. The trained model can then detect and track objects in the sports video in real-time. YOLO Sports Detection provides various applications of sports analytics, broadcasting, and player improvement algorithms. YOLO includes state-of-the-art methods for sport detection on the basis. The proposed YOLO system includes dataset preparation, data annotation, data preprocessing, customization of configuration files, and model evaluation [4]. The YOLO network was used for detecting sports balls and people, with the YPB model achieving the best results for person detection. The YPBF model, trained using flipped photos of identical instances, received the highest score for ball detection. All models use pre-trained, freely accessible weights to inherit low-level knowledge, cutting down training time [5]. Mirroring images is a simple change of input training data that has shown to be beneficial and economical. Object tracking can be used in future research to enhance player and ball detection. Convolutional neural networks (CNNs) are deep learning models that can leverage natural signals, but efficiency challenges arise with the number of object categories and locations within an image. Single-tier frameworks like YOLO and SSD typically perform significantly worse than two-tier designs but are competitive but better at detecting small objects [6].

2 Literature Review

To help you better comprehend the state-of-the-art techniques, a concise synopsis of the Sports Detection based on YOLO methodologies is given. After that, each of the present techniques is briefly summarized before the review of the current methodologies starts. CNN is used to extract spatial features from images automatically, akin to multilayer perception. This feature extraction procedure uses convolution, pooling, flattening, and fully connected layers. Convolution is utilized to create. The multilayer perceptron (MLP) and weight updating are comparable [4]. The CNN elements are explained in the following order.

Time-Distributed Layer

Video categorization requires the network to receive the video frames as input. To solve this issue, a time-dispersed layer was used to share 10-time inputs worth of frames from a video in a single layer. This layer modifies the weight by taking into account all of the frames during backpropagation when the network receives the video frames at various time steps.

YOLO Framework

A basic object detection framework is YOLO. A separate end-to-end network is used for the entire process, which starts with the input of the original image and ends with the output of the item position and category. By using complete photographs for training, YOLO increases detection performance. As seen in Fig. 1, the YOLO network has 24 convolutional layers, followed by 2 fully linked layers. Instead of the inception modules used by GoogleNet, YOLO uses 11 reduction layers followed by 33 convolutional layers. The first component of the architecture is the input. The following two parts, the Backbone and Neck the Detection Head and Detection Neck make up the Object Detector [4]. The head is also in charge of detection and forecasting. The head is principally in charge of localization and classification. For category detection, Fig. 2 shows the YOLO framework [2].

Backbone Network

CSPResNext50, CSPDarknet53, and EfficientNet-B3 are the backbone networks.

They finally settled on the CSPDarknet53 CNN after extensive testing and experimenting. In CSPDarkNet53, the Dense Net architecture is utilized.

In CSPDarkNet53, there are two blocks:

- Cross-stage partial (CSP) block.
- Convolutional base layer.

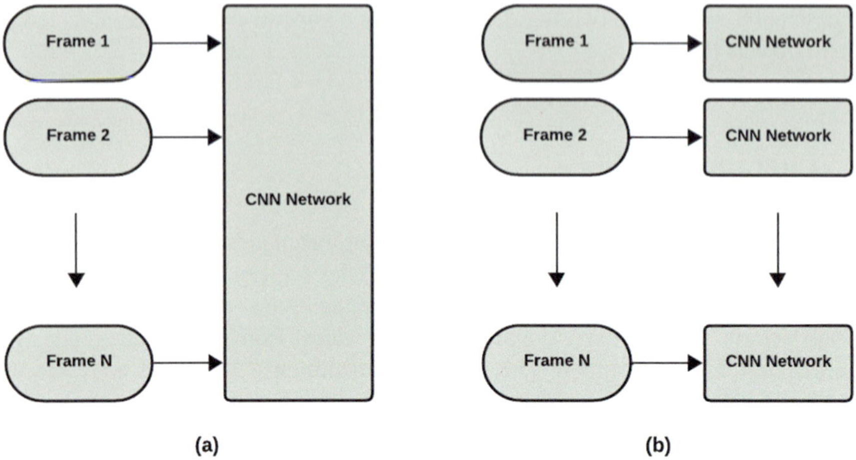

Fig. 1 Video frames **a** combining in a network, **b** sharing layers in a network

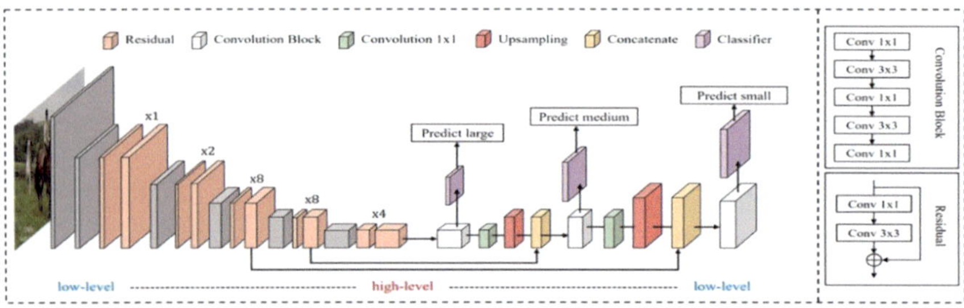

Fig. 2 YOLO diagram

The convolutional base layer is composed of the complete input feature map. The CSP block, which is stacked next to the convolutional base layer, divides the input into two parts; one is passed through the dense block, while the other is routed directly to the next step without any processing, as was previously mentioned. CSP encourages networks to reuse features, retains fine-grained features for improved forwarding, and minimizes the number of network parameters. Only the last convolutional block in the backbone network, which can extract richer semantic data, is dense because convolutional layers with more dense links can take longer to detect changes.

Related Review for Sports Detection System

Object detection using YOLO: challenges, architectural successors, datasets, and applications by Diwan et al. [7] provides a thorough overview of one-step object detection algorithms based on the underlying architecture, regression formulation, strengths and weaknesses, comparative and incremental approaches for this category, common datasets, results obtained, and future range. Additionally, it discusses the key functions of two-stage object detectors like RCNN, Fast CNN, and Faster-RCNN. A comparison diagram between a 2-level object and a 1-level object is presented as a conclusion. Akansha Bathija used [8]. In that paper, visual object tracking on videos is done using the YOLO detector and SORT tracker. More training epochs and fine-tuning during detector training increase accuracy and precision. The system can be upgraded to include more items or restricted to just one object using a smaller dataset of photos; however, it is only currently confined to pedestrians and automobiles. For the suggested object identification and tracking, a variety of object detectors (such as YOLOv1, YOLOv2, YOLOv3, RCNN, and SSD) and object trackers (such as Deep SORT, Centroid, IOU tracker, and CNN + LSTM) can be used and tested. Results can be analyzed through the study. This work recognizes object detection approaches based on Juan's recommendation, particularly focusing on the CNN family and YOLOAQ4 [9]. YOLO is an easy-to-build unified object detection model that can be trained directly on complete photos. It offers the best real-time speed/ excellent accuracy for object detection tradeoff and is the quickest general-purpose object detector. Additionally, it is perfect for applications that require quick, reliable object detection. The lack of adequate training data presents machine learning with its most pressing upcoming hurdle.

The following sensor systems are integrated with deep learning algorithms: To justify new applications offered at railroad crossings, and reduce costs and risk models. Explain why current sensor systems and algorithms are briefly employed at train crossings. Radar is installed as an integrated unit, and closed-circuit television (CCTV) is an excellent option for railroad crossing applications. CCTV is utilized as the main sensor to identify impediments at railroad crossings. Result showcases the potency of deep learning methods and their capacity to find and recognize obstructions at the railroad crossing. The suggested work utilized deep learning techniques and integrated them into existing systems, particularly sensors. Determine and classify barriers. These datasets' accuracy and loss are used as estimate metrics to compare a given model's performance. Frequently, downloaded images come from public sources. It represents a distinct object with more defining features and reduced similarity to any specific category, adding another layer of differentiation. To learn representations, the suggested work modified deep learning approaches. The multi-layer item mentioned in the proposed system enhances the model's ability to learn hierarchical features.

3 Methodology

3.1 Dataset Generation

At first, we wanted to gather information outside. There is no digital record available in our local source. We used the Kaggle sports dataset as a result.

Fig. 3 Box of data annotation

Data Annotation

For various machine learning projects, data annotation is a crucial step that needs to be performed. We completed the data annotation process without any external assistance. Instead of using third-party software, we developed our code exclusively for this purpose. All the images were converted to the YOLO format, and a separate text file was generated for each image using our code. In the YOLO format, a bounding box needs to be defined for each object belonging to a specific class. We assigned one of the eight sports classes that we used for the image (Fig. 3).

Data Augmentation

In simpler terms, data augmentation is a technique that uses existing datasets to increase diversity. It involves performing operations such as rotation, cropping, shearing, horizontal and vertical flipping on the data to create more samples. This helps the model learn from relevant features and improves efficiency by preventing it from learning from unrelated features. The primary objective of data augmentation is to enhance the dataset for better model testing.

There are some processes given below:

- Shifting.
- Rotation.
- Scaling.
- Cropping.

Data Standardization

This technique scales and preprocesses photos while keeping their width and height constant. Additionally, it aids in enhancing the image's consistency and quality. It is rescaled to have a mean of 0 and a standard deviation of 1.

Personalize the Setup File

To ensure that the input resolution size is properly processed by the YOLOv4 model, it is necessary to adjust the model configuration file. For optimal results, suggest setting the batch size to 64 and the subdivision to 16.

- max_batches = 16,000.
- Steps = 11,000, 12,000.
- Classes = 8.
- Filters = 39.
- 608 is the breadth.
- Heights = 608.

Configure References Files

Within the YOLO framework, there are three distinct text files utilized as reference files. These include the data file, train.txt file, test.txt file, and classes file. The image data file contains information regarding training, test, and configuration files, as well as the locations for backups. Lastly, the class file provides definitions for the class names and their associated numbers.

Class file:

- Football.
- Basketball.
- Cricket.
- Chess.
- Kabaddi.
- Formula-1.
- Boxing.
- Badminton.

Names are the image data file

- Classes = 8.
- Train = train.txt.
- Valid = test.txt.
- Names = Classes.
- Backup = Backup.

YOLOv4 has two distinct essences: YOLOv4-tiny and YOLOv4-sophisticated. The model can be accelerated via YOLO-tiny and created uniquely based on object figures. After a thousand iterations, the Darknet Model saves weights based on the validation set.

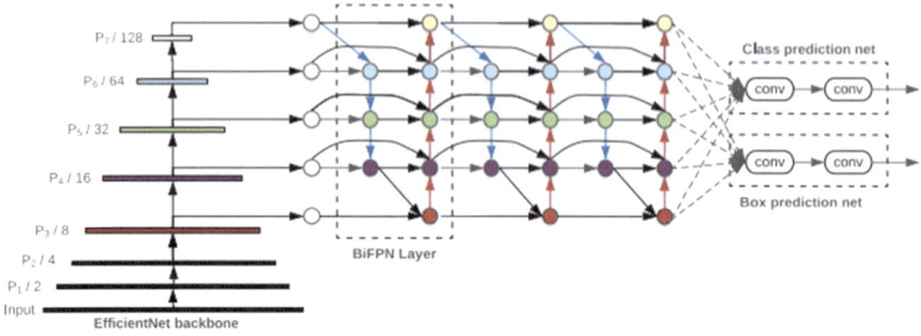

Fig. 4 Working procedure of YOLOv4 in an image

Structure of YOLOv4 Framework

For both object recognition and classification, several bounding boxes need to be drawn around images, and hence, the convolutional backbone's feature layers need to be combined and supported by each other [10]. One-stage detectors anticipate both object localization and categorization at the same time. A single-stage detector is the YOLO detector, as shown in Figs. 4 and 5 from [3].

The foundational network of an object detector is typically based on ImageNet categorization. The network's weights have already been modified to account for significant features in an image, but object detection requires further adjustments.

- CSPResNext50.
- CSPDarknet53.
- EfficientNet-B3.

The neck's constituent parts usually flow between layers, joining only the last few convolutional network levels (Fig. 5).

A feature layer is represented by each PI on the CSPDarknet53 backbone. The image above is from EfficientDet, a deep learning-based object detection model. After CSPDarknet53, YOLOv4 adds an SPP block to broaden the receptive field and separate the most crucial elements from the supporting framework [10]. YOLOv4

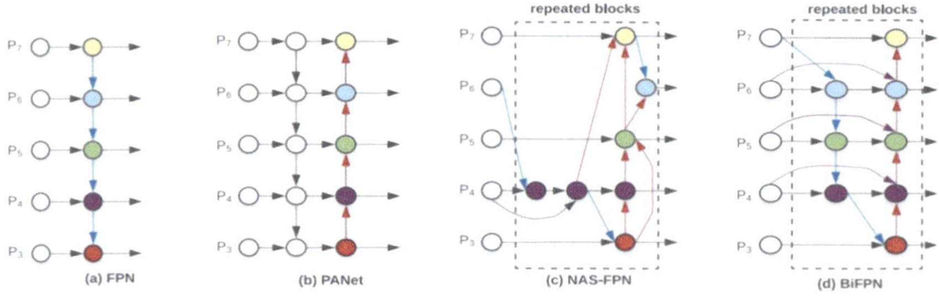

Fig. 5 Cspdarknet53 backbone

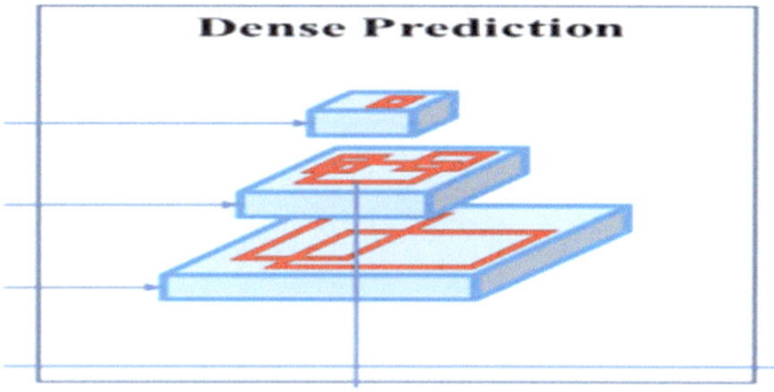

Fig. 6 YOLO head

uses the same YOLO head as YOLOv3 for detection, with three levels of granularity and steps based on anchors (Fig. 6).

Anchor Boxes and Prediction

Anchors are priors that were generated using *k*-means clustering and are similar to bounding boxes. The box's width and height can be predicted.

To obtain bounding box predictions, the network output is converted using the algorithm below:

$$b_x = \sigma(t_x) + c_x \tag{1}$$

$$b_y = \sigma\left(t_y\right) + c_y \tag{2}$$

$$b_\omega = p_\omega e^{t\omega} \tag{3}$$

$$b_h = p_h e^{th} \tag{4}$$

The x and y are the center coordinates, and width and height of our prediction are represented by b_x, b_y, b_w, and b_h.

Magnitudes of the Bounding Box

By first transforming the output into log space, and then multiplying the result by an anchor, the bounding box dimensions can be anticipated (Fig. 7).

Training Process

An input image is divided into a $S \times S$ grid using the YOLO system, and an object is detected by looking for its center in a grid cell. Bounding boxes and the confidence scores that go along with them confident the model are in the existence of an object and how accurate its forecasts are. The confidence score is zero if there isn't an object in the specified cell. On the other hand, the intersection over union (IOU) between the predicted box and the ground truth is reflected in the confidence score if there is an item present in that cell. Five predictions make up each bounding box:

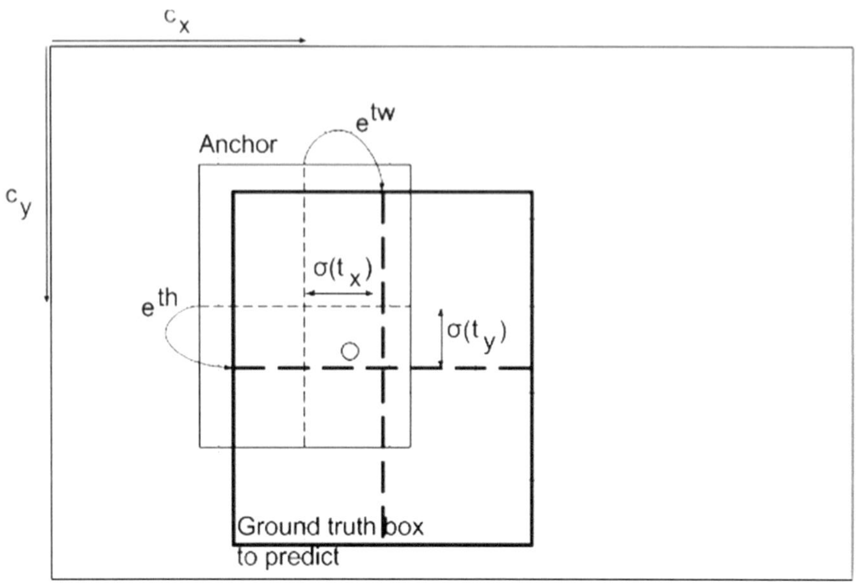

Fig. 7 Dimension of the bounding box

x, y, w, h, and confidence [10]. The expected width and height values relate to the full image, but the bounding box's (x, y) coordinates show its center location about the grid cell's limits. The intersection over union (IOU) between the predicted box and any ground truth box is reflected in the confidence prediction. Furthermore, PR (Class Object) predictions are made for each grid cell contingent on the existence of an object within that cell. Here is how the training host is configured: To train the model, we employed an NVIDIA GeForce RTX 2060 16 GB GPU with a Python 3 Google Computer Engine backend (GPU).

4 Result and Discussion

The YOLOv4 algorithm, trained on a dataset of original photos, achieved a 97.25% correct categorization after 16,000 iterations. The model's parameters were fine-tuned using the standard Darknet-53 model. The model converged after 2000 iterations, with average precision, recall, F_1-score, and mAP@0.50 increments for every 2000 iteration.

This demonstrates (Fig. 8) that even when the training or configuration of the YOLO model varied, the parameters measured by precision, recall, and F_1-score did not vary significantly. Such display constancy demonstrates the equivalence of the model, which indicates a necessary quality about the ability to identify objects in different conditions or positions, which is useful in computer vision applications. The

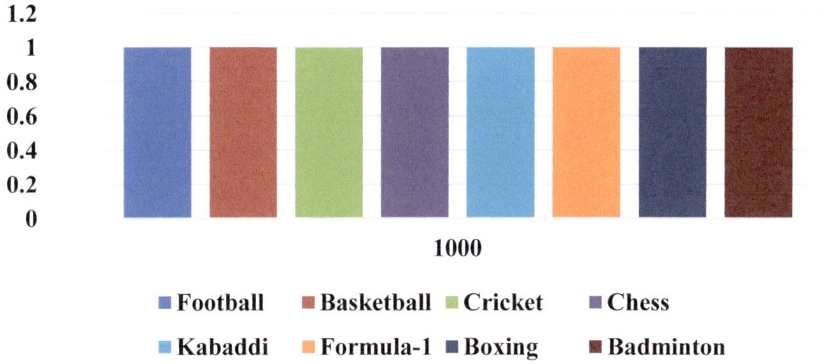

Fig. 8 Competitive precision, recall, and change in F_1-scores: across different iterations

first step in the procedure was to take a picture, followed by the YOLO algorithm. In this case, the picture was split up into grid matrices. The image's higher level of complexity dictated the number of grids employed. The object was classified after picture segmentation and then placed inside each grid. Every grid's impartiality, as well as its degree of reliability, was assessed. The grid's objectness and bounding box values were set to zero if no genuine object could be discovered within it. Anchor boxes were also employed to enhance object detection; the anchor Ov4 method was used whenever necessary to generate reasonably precise bounding boxes from an image. After picture classification and object localization processes were completed for each image grid, a label was assigned to each grid. Subsequently, the algorithm iterated through each grid sequentially, noting both the bounding boxes and the labels containing objects. The recommended model was evaluated in great detail throughout different sports. Afterward, weight files were used to identify objects in the video footage. Following the identification of an object within a video by a trained object detector, the bounding box information of the object was fed to the SORT tracking algorithm, enabling object tracking.

The research above indicates that precision and recall improve as the number of iterations rises. The mean average precision after 16,000 iterations was almost 97%. As a result, our suggested method, which outperforms the other four systems in comparison, can identify different sports with 96% accuracy and 97% precision (Fig. 9).

Irritation Versus Current Average Loss Data

According to the research above, precision and recall are improving as the number of iterations increases. After 16,000 iterations, the mean average precision was approximately 97%. Therefore, we can state that our proposed approach, which is compared to better than the other four systems, can detect various forms of sports with 96% precision and 97% accuracy (Fig. 10).

Based on Table 1 and the results obtained from our proposed system, it can concluded that the performance gradually improved after 16,000 iterations. However,

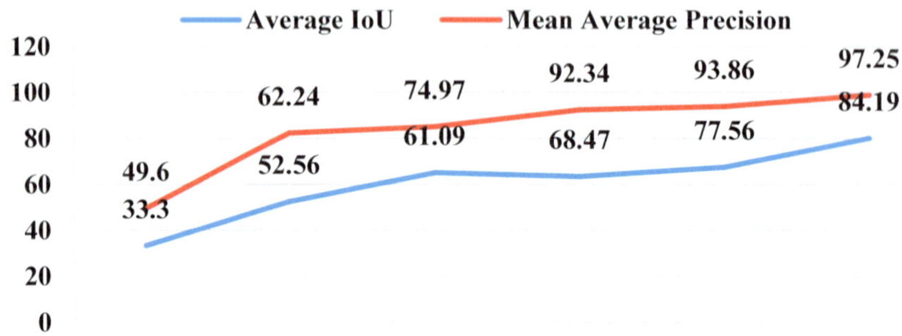

Fig. 9 Changes of precision, recall, and F_1-score for different iterations

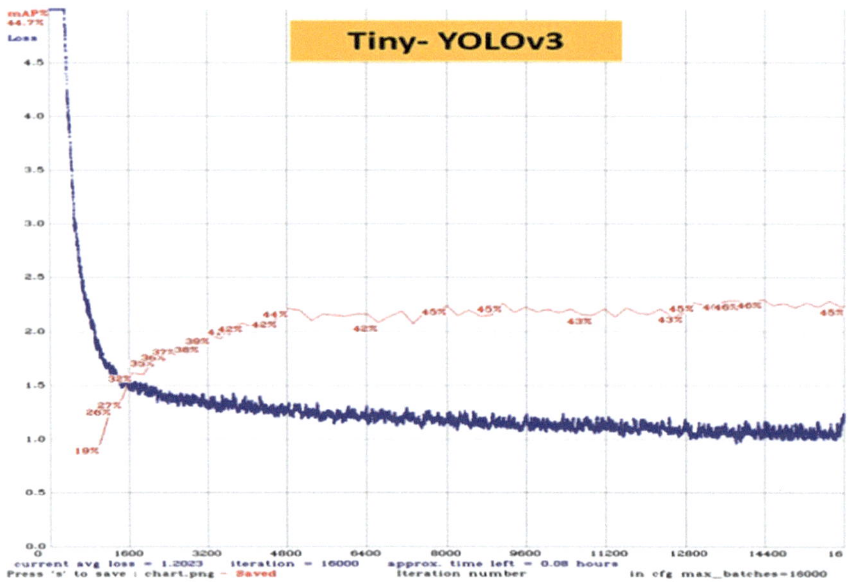

Fig. 10 Iteration versus current average loss of data after iterations

the increase from the previous saved weight file was minimal. The following summary provides a concise analysis of the results from our proposed system, as shown in Table 1.

The data and results obtained with our proposed system are briefly analyzed through the abovementioned four tables. This data shows that the performance is refined concerning a tourist, and the improvement is evident after 16,000 iterations. As a result, the overall report from the secondary wireless file resulted in a lower increase in value.

Table 1 Overall precision, recall, and mean average precision

Iteration	Precision	Recall	F_1-score	Average IoU	Mean average precision
2000	0.53	0.42	0.47	33.30	49.60
4000	0.64	0.59	0.61	52.56	62.24
6000	0.71	0.68	0.76	61.09	74.97
8000	0.82	0.93	0.89	68.47	92.34
10,000	0.84	0.91	0.90	77.56	93.86
16,000 (best for sports detection)	0.97	0.99	0.98	84.19	97.25

5 Conclusıon and Future Work

The study achieved a 97.25% accuracy rate in identifying a YOLO network model for sports classification from game objects. Broadcasting agencies are crucial for maintaining sports popularity. The study aims to improve sports enhancement in areas with limited internet availability. The research aims to develop a pre-trained CNN model and classifier for future research in sports detection, potentially leading to better algorithms. The study tested various models from the baseline YOLO model for sports ball and person detection. The researchers also presented a novel system that automatically recognizes basketball players and their interactions with athletes in real-time from publicly available NBA game footage. They employed YOLO for object detection and classification and implemented a player-tracking algorithm that performed better under dynamic camera angles. The system also addressed player overlap and motion history issues. Although the solution has some limitations due to inherent limitations, the findings can motivate the community to enhance the system. As a next step, we aim to develop a dataset and include more classes to detect a wider range of sports events, which will benefit athletes, coaches, and commercial applications. Additionally, we plan to improve the model's accuracy to support future research in this field. Even we can improve the accuracy by working on it which will help future researchers if anyone wants to do more research in this field.

References

1. Saravanan C (2010) Color image to grayscale image conversion. In: Second ınternational conference on computer engineering and applications. IEEE, USA, pp 196–199
2. Kwon T-H, Park J, Jeong H, Park K (2023) Assessment of speckle-pattern quality using deep-learning-based CNN. Exp Mech 63(1):163–176
3. Tan M, Pang R, Le QV (2020) Efficientdet: scalable and efficient object detection. In: Proceedings of the IEEE/CVF conference on computer vision and pattern recognition, pp 10781–10790
4. Du J (2018) Understanding of object detection based on CNN family and YOLO. J Phys Conf Ser, 012029

5. Direkoglu C, Sah M, O'Connor NE (2018) Player detection in field sports. Mach Vis Appl 29:187–206
6. Direkoğlu C, Nixon MS (2011) Moving-edge detection via heat flow analogy. Patt Recog Lett 32(2):270–279
7. Diwan T, Anirudh G, Tembhurne JV (2023) Object detection using YOLO: challenges, architectural successors, datasets and applications. Multimedia Tools Appl 82(6):9243–9275
8. Diwan K, Bandi R, Dicholkar S, Khadse M (2023) Football player and ball tracking system using deep learning. In: Proceedings of international conference on data science and applications. Springer, Singapore, pp 757–769
9. Ivašić-Kos M, Krišto M, Pobar M (2019) Human detection in thermal imaging using YOLO. In: Proceedings of the 5th international conference on computer and technology applications, pp 20–24
10. Felsberg M, Forssén PE, Sintorn IM, Unger J (eds) (2019) Image analysis. In: 21st Scandinavian conference. Springer, Sweden, pp 11–13

COVID-19 Forecasting Using CXR Images and Machine Learning Methods

Shreeharsha Dash and Subhalaxmi Das

Abstract A global pandemic has been ushered in by the recent coronavirus flare-up, which has thrown the whole globe into an unforeseen and disastrous predicament, endangering countless lives. It is critical to identify affected persons by rapid symptomatic contact in order to prevent the local spread of this virus. Chest X-ray (CXR) imaging, on the other hand, is more practical than computed tomography (CT) imaging due to its lower cost, simpler operation, and higher imaging speed. The reason for this research is to offer a method for classification of the Coronavirus via the use of CXR images and image processing. The image is resized, grayscale conversion, noise removal, thresholding, and morphological operation are used for preprocessing, train-test splitting is used for expressing the features, and categorization is done using multiple machine learning (ML) classifiers like Naive Bayes (NB), Decision Tree (DT), Random Forest (RF), K-Nearest Neighbors (KNN), Support Vector Machine (SVM), Bagging, AdaBoost, Gradient Boosting (GBoost) and Logistic Regression (LR). With bagging approach, the evaluation of results reveals a greater level of exactness, which is 95.19%. In order to evaluate the trial outcomes, execution metrics that were generated by the recommended approach are used as the basis for assessment.

Keywords CXR · Preprocessing · Thresholding · Decomposition · Feature

S. Dash (✉) · S. Das
Odisha University of Technology and Research, Bhubaneswar, India
e-mail: shreeharshadash1@gmail.com

S. Das
e-mail: sdascse@outr.ac.in

© The Author(s), under exclusive license to Springer Nature Singapore Pte Ltd. 2025
S. K. Udgata et al. (eds.), *Intelligent Systems*, Lecture Notes in Networks and Systems 1314, https://doi.org/10.1007/978-981-96-3797-3_20

1 Introduction

The current COVID-19 outbreak is caused by coronavirus, which is often known as SARS-CoV-2. Although a tiny proportion of infected people are in critical or severe condition, a large number have managed to survive the infection. Human interaction is the primary vector for the transmission of this virus. Although it may impact other systems as well, the respiratory system is mostly impacted. The common cold, cough, fever, aches and pains, and difficulty breathing are all signs of the illness. There are three progressive stages: mild, moderate, and severe. People with mild to moderate symptoms often make a full recovery. Dying is a possible outcome of serious illnesses. The World Health Organization (WHO) issued a series of recommendations, including the recommendation to keep one's distance and the use of face masks, in response to this [1]. A diagnosis is made by combining symptoms with a physical examination. Testing for COVID-19 in the blood, a chest X-ray, and sputum culture might be required to confirm the diagnosis of pneumonia. There are a few methods to classify this illness: community-acquired, healthcare-associated, and hospital-acquired pneumonia. In addition, it creates issues with the heart.

Starting in China in November 2019, the COVID-19 pandemic quickly expanded throughout the globe, posing a serious danger to human life. There have been around 63.2 million cases of infection globally, with an estimated 1.47 million fatalities. This danger has disproportionately affected countries like China, India, Brazil, Russia, France, and Italy. Several nations have worked on an epidemic vaccine and are now attempting to return things to normal. Additionally, it has caused enormous financial losses on a global scale, which has had a significant influence on GDP growth [2].

Nevertheless, by comparing the pre-processed and raw data, we construct a COVID-19 classification model using ML classifiers as NB, DT, RF, KNN, SVM, LR, Bagging, AdaBoost, and GBoost. Image resizing, grayscale conversion, noise reduction using a median filter, Otsu thresholding, and morphological processes are all part of the preparation in this case. Separate the dataset into two parts: train and test. Finally, compare the models' performance using their relative accuracy and Area Under the Receiver Operating Characteristic (AUC-ROC) curves.

2 Literature Survey

Researchers from all around the globe are focusing their attention on the study of CXR pictures that may help categorize coronavirus. Since RT-PCR is not universally accessible, they discovered that CXR image analysis may serve as a substitute.

With a 79% success rate in predicting medication-related symptoms, Rudra Kumar et al. [1] proposed a technique using a publically available chest X-ray dataset and an ML model. Any doctor or nurse with access to a computer may use their newly developed software to quickly and accurately identify COVID positive and negative results in chest X-ray pictures.

Nurul Absar and his colleagues reveal in their study [3] that support vector machines (SVM) has a wonderful feature that substantially offers excellent outcomes when it comes to diagnosing COVID-19 from CXR pictures. In order to get more photos for the purpose of diagnosing the dataset that was provided, the image processing tool known as RGB and SqueezNet models were used.

In their reference to a two-stage procedure, Ivanoe De Falco et al. [4] mentioned one. An evolutionary algorithm may automatically categorize images and gain explicit information in the form of IF-THEN rules in the second phase, whereas the first step is a filtering system commonly used in content-based image retrieval. Using a collection of chest X-ray pictures, this method is evaluated with the objective of determining whether or not COVID-19 is present with an accuracy rate of 80.67%.

In the study [5], Yabsera Erdaw and the colleagues provided an explanation of the model using support vector machines (SVM). They also validated the strategy using tenfold cross validation. The outcome of the multi-level classification model was that it distinguished COVID-19 having a sensitivity of 97.92% and a specificity of 98.91%.

According to Wei Zhao et al. [6], the majority of patients with COVID-19 pneumonia had fever as their first symptom. Imaging findings, including ground-glass opacities (GGO) in 86.1% of cases or a combination of GGO and consolidation in 52.5% of cases, were prevalent. Lesions seen on CT scans were more likely to be lower lung predominate and multifocal (54.5%), and to have a peripheral distribution (87.1%) and bilateral involvement (82.2%). A CT involvement score, architectural distortion, and traction bronchiectasis were useful in determining the degree and scope of the illness.

The authors of the aforementioned study [7] used two publicly available COVID-19 CXR datasets—one big and one imbalanced—to conduct self-supervised learning-based partitioning and batch knowledge assembling-based fine-tuning. Variations in hyperparameters showed encouraging results for COVID-19 identification.

In order to construct a model that can more accurately identify COVID-19 using X-ray pictures, the literature that was studied offered the necessary background information. Techniques based on machine learning have the potential to greatly cut down on the amount of time needed to carry out electronic examinations. Furthermore, these technologies make it possible for experienced professionals to provide assistance to clinicians, and they also make it possible for specialists in artificial intelligence to provide assistance to both laboratory professionals and clinicians.

3 Proposed Framework

In this part, we cover the details of the proposed model that can accurately and efficiently classify COVID-19 diseases from input CXR pictures. The suggested layout with the integrated innovations is shown in Fig. 1.

3.1 Dataset

This work has made use of a publicly available dataset on Kaggle that contains 5863 chest X-ray images [8]. As a standard component of patient treatment, all chest X-ray s were obtained. Each image categorization (PNEUMONIA (Coronavirus) and NORMAL) has its own subfolder in the dataset, which is organized into three managers (train, test, and val).

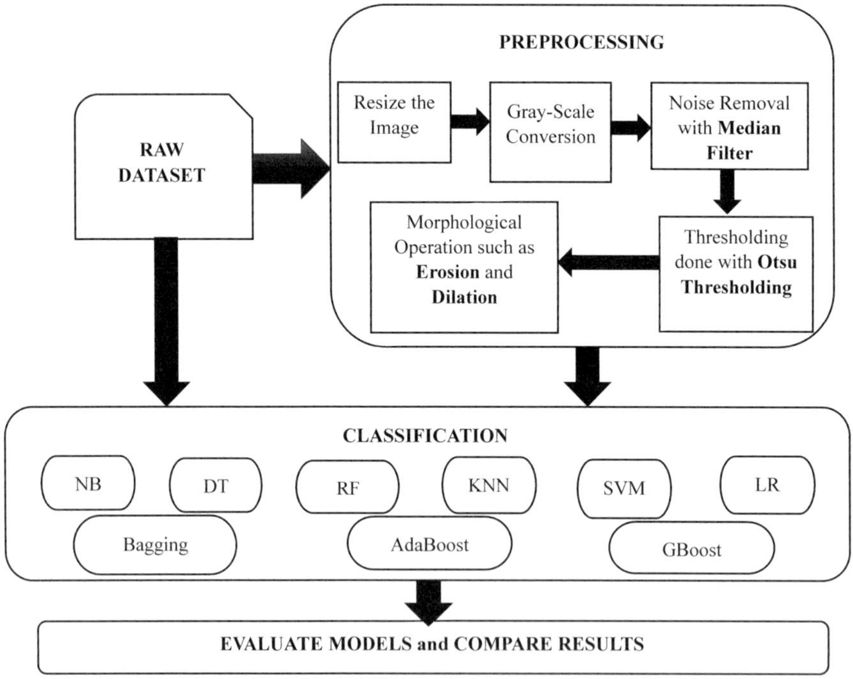

Fig. 1 The proposed framework

3.2 Preprocessing

Creating a classifier model was an important part of their involvement. Sending the sample to the preprocessing step for further analysis involves isolating it from the remaining dataset. The whole set of preprocessing steps used to create this model is displayed in Fig. 2.

Resize the Image: To resize a picture, you change its proportions (make it bigger or smaller) without changing the actual content. We may adjust the image's size and quality by resizing it, which changes its dimensions.

Gray-Scale Conversion: A collection of pixels is what we call an image. The pixel values of each individual pixel might take on one of three forms: An assortment of green, red, and blue hues The process of grayscale conversion involves converting a photograph to a black and white image, eliminating any colour tones. Numerical details applied to the RGB values and then either replacing them in the image or creating a new picture with those additional characteristics should make this practicable.

Median Filter: The use of median filtering is a nonlinear technique that may be used to remove noise from photographs. The median filter is able to perform its function by sequentially traversing each pixel in the picture and replacing each value with the median value of the pixels that are next to it.

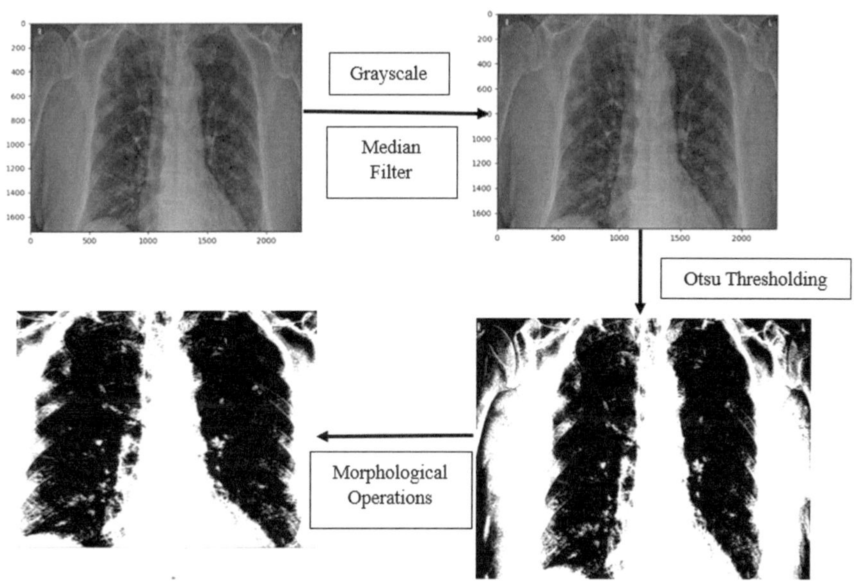

Fig. 2 Transforming a raw picture into a pre-processed one

Otsu Thresholding: To apply a threshold in image processing is to change a pixel's range value from black and white to white or dark according to a limit value. For Otsu, the goal of his thresholding method is to convert grayscale images to binary ones. This approach employs several image processing algorithms to either do histogram-based picture thresholding or convert a grayscale image to a parallel one. If the Otsu thresholding method is to be believed, the image is a bi-modal histogram.

Morphological Operation: When pixel neighbours are correlated according to a present design, consensus morphological tasks result in consistent alterations.

- **Erosion:** Reducing the size of the picture is the goal of the morphological erosion process.
- **Dilation:** A structural element that is sometimes referred to as a "kernel" was used in the dilation procedure in order to verify and expand the formations.

3.3 Classification

A train-test splitting strategy was used to split the data collected in half, with 80% used for training and 20% for testing. There are a variety of machine learning classifiers that are employed for the intent of classification. These include NB, DT, RF, KNN, SVM, Bagging, AdaBoost, GBoost, and LR.

NB: One common method for dealing with characterisation concerns is the NB Grouping strategy, which employs directed learning computations based on the Bayes hypothesis. The Bayesian Classifier is a straightforward and efficient order technique that lends itself well to the development of rapid artificial intelligence models capable of producing rapid predictions.

DT: DT determines the entropy and gain of data and organizes it according to choose rules derived from processing data. Organized in a tree form, with each node representing a characteristic. The most important hubs will serve as the root hub and the children's hubs. The decision's outcome is then addressed by the leaf hubs. It works well with consistent and explicit attributions [9].

RF: It employs an average of numerous decision trees trained on separate portions of the dataset to improve the dataset's projected accuracy. To improve the model's performance and tackle difficult tasks, it uses ensemble learning, which combines many classifiers [10].

KNN: By comparing the new input with its training set samples, the K-NN algorithm can forecast a new input class label. The K-NN classification results are flawed if the new input is identical to the training set samples [10].

SVM: Whether your dataset is linear or nonlinear, SVM's rules will apply. With the use of a mapping idea, it looks for a preferred boundary that divides the recording devices into certain classes and then changes the actual educational records into a higher dimension [11].

Bagging: To make machine learning algorithms for statistical classification and regression more stable and accurate, there is a meta-algorithm called Bootstrap Aggregating, which is also called bagging.

AdaBoost: The AdaBoost algorithm in machine learning, is an ensemble approach and is a Boosting approach. Adaptive Boosting is the process of reassigning weights to each instance, with larger weights given to instances that were mistakenly categorized.

GBoost: By training each new model to minimize the loss function—such as mean squared error or cross-entropy—of the prior model using gradient descent, Gradient Boosting transforms several weak learners into strong learners [10].

LR: One popular set of class rules is logistic regression. It relies heavily on the logistic characteristic and is used when the data are binary. The logistic feature, also known as the sigmoid feature, is an S-shaped curve that maps the value on a scale from 0 to 1, however it is neither 0 nor 1 [11].

4 Experimental Results

We have seen how several classifiers performed and how the suggested model fared in this part. An environment with 4 GB of RAM and an Intel Core i5 CPU is used to develop the algorithms, and Python and the Jupyter Notebook can be used for this purpose. It is possible to assess the proposed COVID-19 categorization techniques using a variety of performance requirements. To manage the proposed structure's presentation, performance metrics for the classification model might include F1-score, recall, accuracy, and precision. The accuracy of the classifiers utilized with raw data and pre-processed data are displayed in Figs. 3 and 4, respectively. In particular, we use the area under the receiver operating characteristic (AUC-ROC) curve to evaluate our model's effectiveness. Pictured in Fig. 5 with an AUC-ROC curve are the classifiers' results when tested on both the raw and pre-processed datasets. The comparison of the data obtained by all writers and our own is shown in Table 1.

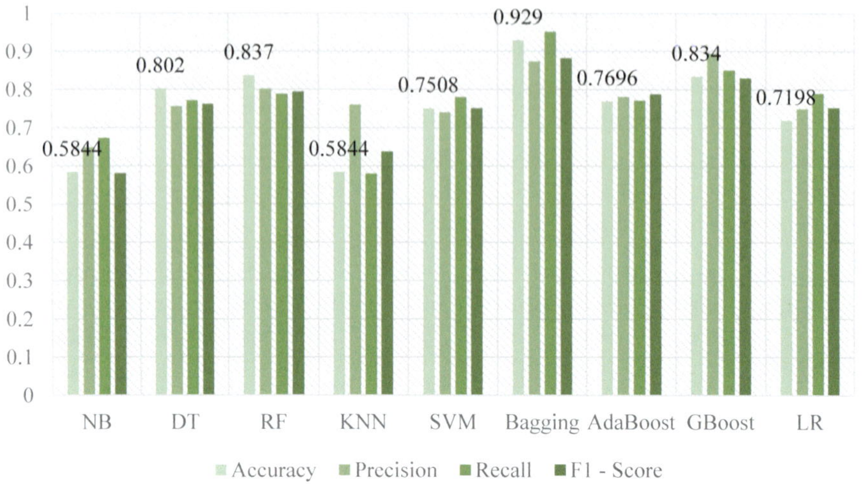

Fig. 3 Classifiers' efficacy on raw dataset

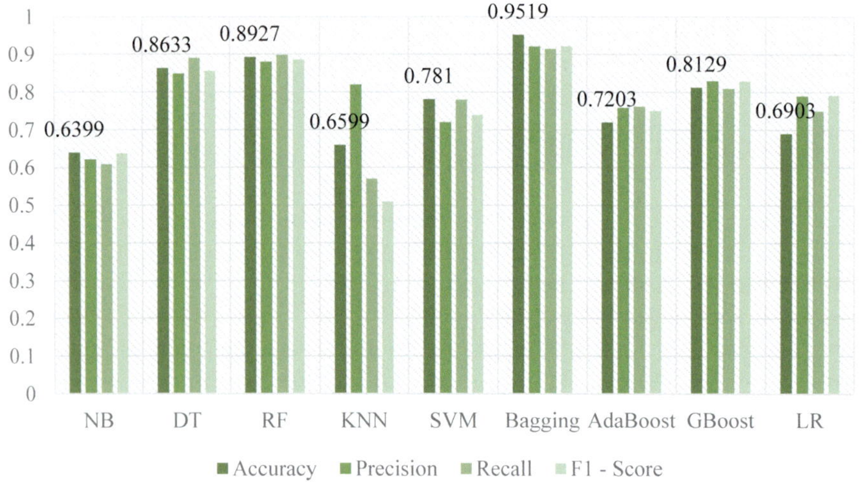

Fig. 4 Classifiers' efficacy on pre-processed dataset

5 Conclusion

The healthcare industry is only one of several that is benefiting from the innovative technologies known as machine learning. In addition to helping doctors monitor and forecast their patients' prognoses and treatment outcomes, machine learning technologies have the potential to distinguish COVID-19. On the other hand, we achieved an efficiency of 95.19% (Bagging) using pre-processed CXR pictures as

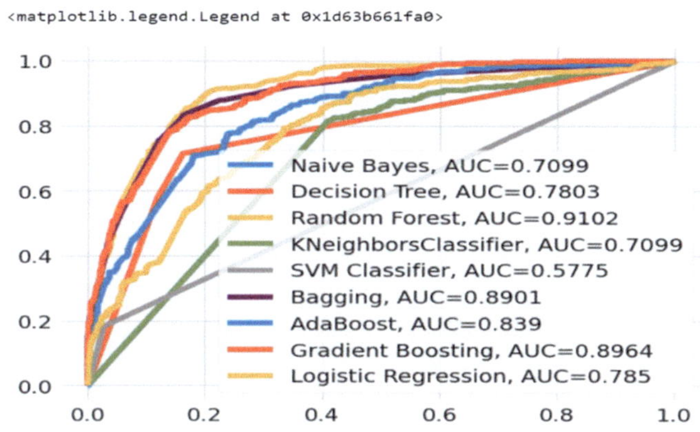

(a)

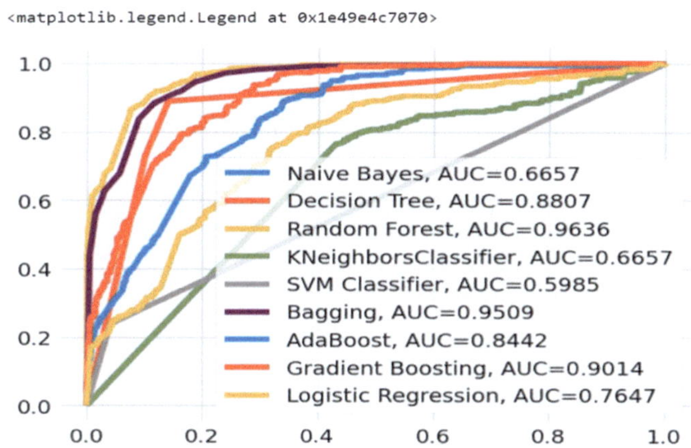

(b)

Fig. 5 Exploring the AUC-ROC for classifiers' effectiveness

Table 1 Comparison of classifiers' performance

Studies	Classifiers used	Accuracy (%)
[11]	Neural network, SVM	87.2
[12]	ANN	94
[13]	SVM	82.7
[14]	NB	80
[15]	RF	93.69
[16]	F-SVM	84
[17]	KNN	92.15
Our proposed model	**Bagging**	**95.19**

The proposed model demonstrates better accuracy compared to existing work compared to other used ML classifiers. Predicting prognosis and treatment results of COVID-19 patients should therefore be the goal of study using machine learning approaches applied to prospectively collected X-ray images, clinical/laboratory data, and socio-demographic information.

6 Future Scope

Reduced diagnostic time and improved access to chest radiograph interpretation are two ways this approach may improve patient care if implemented in the clinical system.

References

1. Rudra Kumar M, Pathak R, Gunjan VK (2022) Diagnosis and medicine prediction for COVID-19 using machine learning approach. In: Computational intelligence in machine learning: select proceedings of ICCIML 2021, pp 123–133. Singapore: Springer Nature Singapore
2. Hasoon JN, Fadel AH, Hameed RS, Mostafa SA, Khalaf BA, Mohammed MA, Nedoma J (2021) COVID-19 anomaly detection and classification method based on supervised machine learning of chest X-ray images. Res Phys 31:105045
3. Absar N, Mamur B, Mahmud A, Emran TB, Khandaker MU, Faruque MR, Osman H, Elzaki A, Elkhader BA (2022) Development of a computer-aided tool for detection of COVID-19 pneumonia from CXR images using machine learning algorithm. J Radiat Res Appl Sci 15(1):32–43
4. De Falco I, De Pietro G, Sannino G (2023) Classification of Covid-19 chest X-ray images by means of an interpretable evolutionary rule-based approach. Neural Comput Appl 35(22):16061–16071
5. Erdaw Y, Tachbele E (2021) Machine learning model applied on chest X-ray images enables automatic detection of COVID-19 cases with high accuracy. Int J General Med, 4923–4931
6. Zhao W, Zhong Z, Xie X, Qizhi Y, Liu J (2020) Relation between chest CT findings and clinical conditions of coronavirus disease (COVID-19) pneumonia: a multicenter study. Ajr Am J Roentgenol 214(5):1072–1077
7. Li G, Togo R, Ogawa T, Haseyama M (2023) Boosting automatic COVID-19 detection performance with self-supervised learning and batch knowledge ensembling. Comput Biol Med 158:106877
8. Kermany D, Zhang K, Goldbaum M (2018) Labeled optical coherence tomography (Oct) and chest X-ray images for classification. Mendeley Data 2(2):651
9. Jinjri WM, Keikhosrokiani P, Abdullah NL (2021) Machine learning algorithms for the classification of cardiovascular disease-A comparative study. In: 2021 International conference on information technology (ICIT), IEEE. pp 132–138
10. Chaurasia V, Chaurasia A (2023) Novel smethod of characterization of heart disease prediction using sequential feature selection-based ensemble technique. Biomed Mater Dev. 1–10
11. Das S, Pradhan SK, Mishra S, Pradhan S, Pattnaik PK (2022) A machine learning based approach for detection of pneumonia by analysing chest X-Ray images. In: 2022 9th International conference on computing for sustainable global development (INDIACom), IEEE. pp 177–183

12. Nageswaran S, Arunkumar G, Bisht AK, Mewada S, Kumar JS, Jawarneh M, Asenso E (2022) Lung cancer classification and prediction using machine learning and image processing. BioMed Res Int
13. Cirujeda P, Cid YD, Müller H, Rubin D, Aguilera TA, Loo BW, Diehn M, Binefa X, Depeursinge A (2016) A 3-D Riesz-covariance texture model for prediction of nodule recurrence in lung CT. IEEE Trans Med Imag 35(12):2620–30
14. Kurkure M, Thakare A (2016) Lung cancer detection using genetic approach. In: 2016 International conference on computing communication control and automation (ICCUBEA), IEEE. pp 1–5
15. Ahmed MA, Al-Qaysi ZT, Shuwandy ML, Salih MM, Ali MH (2021) Automatic COVID-19 pneumonia diagnosis from X-ray lung image: a deep feature and machine learning solution. J Phys Conf Series, vol 1963, no 1, p 012099. IOP Publishing
16. Ahmed F, Nuwagira B, Torlak F, Coskunuzer B (2023) Topo-CXR: Chest X-ray TB and pneumonia screening with topological machine learning. In: Proceedings of the IEEE/CVF international conference on computer vision, pp 2326–2336
17. Sharma S, Guleria K (2023) A deep learning based model for the detection of pneumonia from chest X-ray images using VGG-16 and neural networks. Proced Comput Sci 218:357–366

Revolutionizing Academic Certificate Verification with Smart Contract

Sayedahmad Seyar Sawayz and Athota Kavitha

Abstract The process of verifying academic certificates is still manual in this age of digital era. There is a need for real-time verification yet dealing with vulnerabilities inefficiencies of manual verification of certificates. Using Blockchain technology, Smart contract and Distributed file system Academic institutions can build a robust, user friendly, trusted, and tamper-proof solution for certificate verification. In the proposed solution, academic institutions submit their student certificates to a certificate upload portal which hashes and saves the certificate on IPFS and submits content ID and Student profile to smart contract. Smart contract creates a signed copy of the information and stores it on Blockchain. The stakeholders can retrieve and verify the stored certificate by using student details and optionally submitting digital copy of the issued certificate. This paper presents a workable, realistic implementation for credential verification, adding to the expanding conversation on Blockchain uses in educational settings. Furthermore, the security of the smart contract is checked by slither and Mythril security analysis tools.

Keywords Certificate verification · Smart contract · Blockchain · IPFS · Security analysis

1 Introduction

Most of the academic institutions in the current digital era rely on manual certificate verification which may increase chances of malpractice and delay in verification process. Traditional certificates printed on paper have a number of disadvantages. For example, they are readily misplaced or destroyed, which makes replacement a challenging and frequently drawn-out procedure. Additionally, because of the ease with which sophisticated printing technology may duplicate these certificates, this style invites counterfeiting, which poses a serious challenge for educational institutions credibility. A further complication arises from the extensive duration required

S. S. Sawayz (✉) · A. Kavitha
Department of CSE, JNTUH-UCEST, Hyderabad, India
e-mail: sawayzseyar@gmail.com

© The Author(s), under exclusive license to Springer Nature Singapore Pte Ltd. 2025

S. K. Udgata et al. (eds.), *Intelligent Systems*, Lecture Notes in Networks and Systems 1314, https://doi.org/10.1007/978-981-96-3797-3_21

for the manual verification of certificates. The multistep nature of the certificate verification process involving dialogue between multiple departments of academic institution incurs delay. Such delays exceed the decision-making timelines causing aspiring professionals and students to miss out on valuable opportunities.

Although efforts such as the National Academic Depository (NAD) working alongside DigiLocker have brought about advancements in how academic documenters managed in India through the provision of digital platforms for storage and validation they encounter various obstacles. These centralized systems have restrictions when it comes to security, privacy and reliance on a single point of failure. DigiLocker uses a gateway to generate documents, from repositories. Yet the centralized storage system comes with risks; the data stored there can be tampered with. Altered, putting it at risk of cyber threats. As a result, any changes made to the repository could lead to the falsified production of documents. The outlined issues underscore a pressing necessity for more streamlined, secure mechanisms for both the issuance and authentication verification of academic certificates. Endeavoring to address this lacuna, the focus of our scholarly inquiry encompasses the deployment of a system fortified by Blockchain technology, smart contract and the decentralized file system offering a workable, realistic implementation for managing academic certificates.

Blockchain technology is a decentralized and distributed digital ledger that is used to record transactions across a network of computers [1]. It is a sort of distributed ledger technology (DLT) that eliminates the need for central authorities by enabling several parties to view the same information and modify it concurrently. In essence, a Blockchain is a chain of blocks, where each block contains a collection of transactions with a unique code called "hash" that links it to its previous block, creating a chain of blocks that are secured through cryptography. The fact that once a block is added to the chain, it cannot be withdrawn or changed enables Blockchain to maintain an immutable and tamper-proof record of all transactions [2]. The smart contract is a software runs on Blockchain network creating unforgeable records of transactions which improve effectiveness and lessen the need for intermediaries [3].

Decentralized Distributed file system like Interplanetary File System (IPFS) offers a fast and efficient file sharing without the need for a central server and facilitates to identify and locate files using unique addressing system making it to be scalable, fault-tolerant. IPFS can be used in conjunction with Blockchain technology to serve different purpose. IPFS provides a decentralized storage and distribution system, while Blockchain provides a decentralized and transparent way to record issuance and support verification of the certificates. By using IPFS in conjunction with Blockchain, the system addresses the issue of storing large volumes of data, removing duplication, and facilitating file sharing. This is achieved in IPFS by creating an immutable record for each certificate upload and generating a distinct content identifier (CID) [4].

The proposed system for certificate verification is designed and implemented using Blockchain technology, smart contracts, and a decentralized file system in order to meet the security standards we have extensively analyzed our smart contract code using tools such, as Slither and Mythril. By decentralizing the verification process and eliminating single points of failure the aim of this research is to improve security, privacy, and accessibility. The works studied in the literature did not offer details of workable solutions for certificate verification and did not provide the security analysis of the Smart contracts.

The organization of the research paper is as follows. Section 2 offers review of the literature. Section 3 provides details architecture of the proposed system. Section 4 describes implementation details. Information related to testing and security analysis is discussed in Sects. 5, and 6 concludes our paper and offers direction for further research.

2 Literature Review

When it comes to examining how blockchain technology can be used for certificate verification a significant number of papers have focused primarily on aspects. Unfortunately, they often overlook the implementation and thorough security assessments. These papers are filled with ideas. Fail to go beyond theory and conduct practical tests or, in depth security evaluations. As a result there is a gap, in understanding how these theoretical models perform in real world scenarios and how reliable they are when applied in applications.

For instance, the study conducted by [5] which suggests a framework based on Blockchain for verifying certificates. This framework mainly focuses on aspects such as authentication and confidentiality. Lacks information regarding its implementation or security analysis. Furthermore, lack of transparency and trust in storing student documents off-chain raises concerns about the effectiveness of the framework. Similarly, [6] introduces a permissioned Blockchain-based system called HLFEcert for verifying certificates. It utilizes the SHA256 hash algorithm and digital watermarking technology offering an application model. However, the study falls short in providing a security evaluation of their system, and the author has used MongoDB to store certificates which reduces the decentralization. In a research, paper titled "DistB CVS" [7] proposed a certificate verification system based on Blockchain. Their aim was to address the challenges associated with issuing and verifying certificates. However, their paper did not mention any deployment or security analysis of their proposed model, and the authors proposed storing the digitally signed certificate in the blockchain, which increases the size of blockchain, thus increasing the cost. Similarly, another study [8] presented a framework that combined blockchain and IoT for generating and verifying certificates. However, they did not delve into the application or security aspects of their framework. On the hand in the research [9], the authors proposed a Blockchain-based solution, for managing academic certificates. They suggested using smart contracts in their framework. In the proposed architecture, there is a need for a trusted authority to issue and update user profiles, which may introduce a centralized element moreover the paper primarily focused on its architectural design without providing a detailed exploration of its implementation or conducting a security analysis. The research paper titled "Verifi Chain; A Credentials Verifier utilizing Blockchain and IPFS" discusses the problem of certificate fraud in Southeast Asia. It suggests a prototype that utilizes the security capabilities of Blockchain and IPFS [10]. However, the study mainly focuses on aspects. Lacks thorough practical implementation and comprehensive security analysis. These aspects are vital, in evaluating the practicality and reliability of such a system, in real world situations.

The research we reviewed explores how Blockchain technology can be used to verify certificates, in industries. It emphasizes the importance of conducting real-world experiments to determine if this technology is practical and secure than relying on theoretical models. In response to these gaps in existing research, our study not proposes a framework, for certificate validation using Blockchain. Also implements it in real-world scenarios, stores student profile information in order to track the changes. Moreover, we have stored the students certificate itself also in a decentralized file system call IPFS (Interplanetary file system) that brings more transparency and trust as compared to centralized file systems. Additionally we thoroughly analyze the security of this system by employing tools to ensure its trustworthiness. This comprehensive approach sets our research apart. Contributes significantly to the field of certificate authentication.

3 Proposed System

The proposed system aims to improve the process of verifying credentials by utilizing Blockchain technology, smart contract, and IPFS (Interplanetary File System) a distributed decentralized file system. The system consists of two stakeholders namely Academic institutes and verifier. System components are consisting of two parts; certificate uploading and certificate verification.

3.1 Stakeholders

- **Academic institute**

The Academic institutes play a key role in this system, as they are responsible for the process of uploading certificate. They utilize the system to issue certificate and securely store them on the Blockchain. Each student certificate is assigned a hash using SHA-256 Algorithm and corresponding IPFS address (CID) also known as Content Identifier to ensure its integrity and permanence. The Academic institutes act as the custodians of truth for these credentials representing their institutions within the digital verification ecosystem.

- **Verifier**

In the proposed system, verifiers play a role in confirming the credentials of students. Verifiers can include employers, universities or any organization that requires proof of a students background. They utilize the system by submitting verification requests to authenticate the certificates provided to them. The system then retrieves the certificates hash and IPFS address from the Blockchain for comparison based on student details (name, roll number, graduation year). A successful verification assures verifiers that the certificate is authentic and has remained unchanged since it was issued by the university.

3.2 System Components

1. Certificate Uploading

To begin the certificate uploading process, users such as Academic institutes can access the system through a user web interface that is designed and developed by ReactJS, this interface offers an interactive user experience. Successfully authenticated Academic institutes can upload certificates to the system. The backend then processes the certificate by assigning it an IPFS address ensuring storage that cannot be tampered with. Additionally, a unique hash is generated for each certificate using hashing algorithm. This hash along with the student details and IPFS address are recorded on the Blockchain using smart contract called "CertificateRegistry SC." This smart contract autonomously manages records to ensure authorized transactions are accepted. As a result the authenticity and integrity of credentials are maintained. Figure 1 describes the certificate uploading component.

2. Certificate Verification

The verification component of the system has been designed to prioritize simplicity and security. Verifiers such as employers or educational institutions can easily initiate

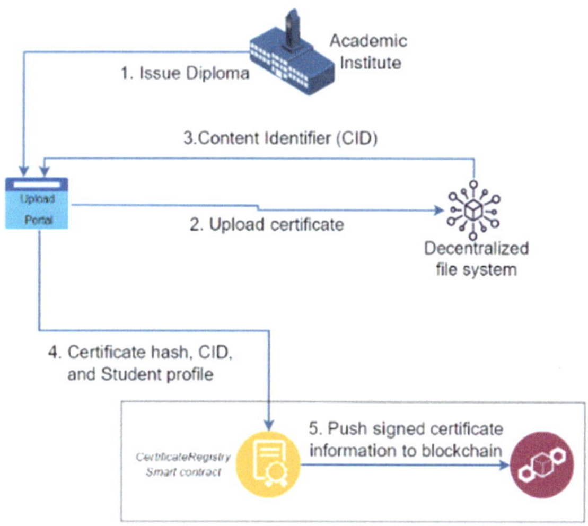

Fig. 1 Certificate uploading process

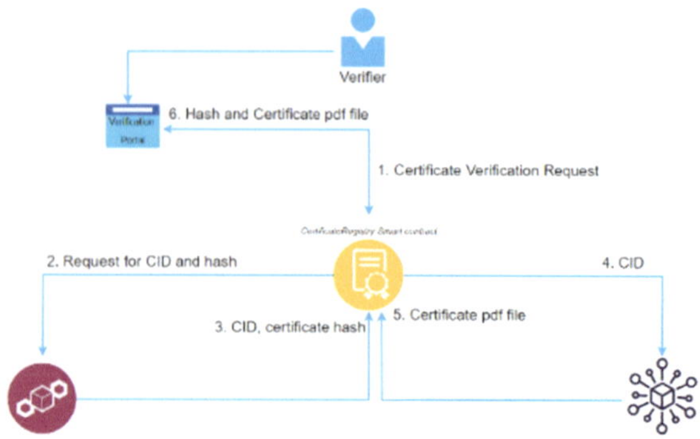

Fig. 2 Certificate verification process

the process using a web-based interface. They simply need to submit a request containing the certificate hash and the certificate or academic record in question. Next, the "CertificateRegistry SC" smart contract handles this request by retrieving the stored certificates hash and its corresponding IPFS content identifier (CID) from the Blockchain based on student details. With this CID, the system retrieves the certificate from IPFS. System then compares it directly with the generated hash by verifier. If both hashes match, it confirms that the certificate is authentic and has remained unaltered since its upload. This verification process ensures that academic credentials can be trusted and verified accurately significantly reducing instances of claims while enhancing trust in certification worldwide. Figure 2 describes the verification process.

This proposed system takes advantage of Blockchain, smart contract, and IPFSs security and transparency features presenting an approach, to credential verification that is both efficient and secure. Its implementation has potential to transform how academic achievements are verified, stored, and trusted globally.

4 Implementation

To evaluate how well our suggested solution structure works we made use of Ether.js, a JavaScript library that makes it easier to build Ethereum-based applications. This tool offers functionalities, for deploying contracts and engaging with them on the Ethereum Blockchain. In addition, we utilized the Hardhat framework, which acts as a work environment for Ethereum development. It includes features like a test network for Ethereum development, automated testing capabilities debugging support for contracts and network deployment functions. Within the Hardhat workspace, we have access to test accounts with Ethers that do not hold any value. These accounts allow us to securely conduct transactions on the Blockchain. The core component of our solution the smart contract was developed from scratch using the Solidity programming language. For testing our contracts functionality, we relied on the Ethereum Remix IDE (Integrated Development Environment) as our development platform. To store the certificate on IPFS (Interplanetary File System) we utilized pinata—a platform that provides access to IPFS and simplifies content uploading and management on the Blockchain. Pinata serves as a tool for developers, creators, and marketplaces dealing with fungible tokens (NFTs) by streamlining data transfer and protection, across different Blockchains.

Our project was developed on a computer running a 64 bit Windows 11 operating system, version 22h2. We also included tools, like Hardhat, React, Node.js (version 16.17.0) npm (version 8.15.0) and other software development kits, in our setup.

The Solidity file (".sol") was first compiled, followed by extracting the ABI and Bytecode which were then deployed on the Hardhat test network. The user interface involves a range of interactions and logical processes that involve institutions and verifiers as detailed in the components section. These functionalities were integrated into the JavaScript application to handle both certificate publication and verification processes.

5 Testing and Security Analysis

This part of the certificate focuses on the testing procedures for the system's smart contract and other functionalities. For our project, we made the decision to utilize Ethereums Blockchain of opting for a private version. We chose this path because the public Blockchain offers security and openness which're crucial, for verifying credentials. Public Blockchains like Ethereum are less susceptible, to manipulation since they operate in a manner as opposed to private Blockchains where control is more centralized. Additionally, Ethereum boasts contract capabilities and enjoys strong community support making it a reliable and scalable choice. Moreover, it proves cost effective as it reduces the infrastructure requirements typically associated with Blockchains. We used the Remix IDE for testing the smart contract code. However, for testing purpose, we employed the Hardhat framework to deploy our code locally. Our entire project's code is readily available on GitHub[1]. To deploy our smart contract using the Hardhat framework, it's necessary to first run a node, as shown in Fig. 3.

Once the Hardhat node is initiated, we need to compile and deploy our smart contract code, as illustrated in Fig. 4.

The deployment process was initiated with the execution of the "npx hardhat run—network localhost.\scripts\deploy.js" command in a local environment. The output confirmed the successful compilation of the Solidity file, indicating the absence of syntactical errors in our smart contract code. The contract, named "CertificateRegistrySC," was then deployed to the localhost Blockchain, as evidenced by the contract address provided in the output. The transaction details, including the initiating address, the transaction hash, and the gas used were logged, ensuring

[1] https://github.com/SeyarSawayz/CertificateRegistrySC.git.

Fig. 3 Starting hardhat node to deploy smart contract

Fig. 4 Compiling and deploying of smart contract

traceability and validation of the contract deployment process. Notably, the transaction was executed without sending any Ether and was included in block number 1 of the local Blockchain network, as detailed by the unique block hash presented in the output. This step was crucial for verifying the functional integrity and successful deployment of our smart contract within the development environment.

After that, the next step is to link our MetaMask wallet account and set up our hardhat network. Hardhat provides 19 accounts with ethers that hold no value for us to import into our MetaMask wallet and test our code. Each transaction, on the Blockchain requires confirmation on the MetaMask wallet. As shown in Fig. 5, our connected MetaMask account verifies the transaction. Displays the transaction details post confirmation.

Following this, we must upload the certificate to IPFS. In this scenario, we utilized Pinata as depicted in Fig. 6 which showcases the code for uploading our certificate to IPFS. Pinata has been a player in IPFS since 2018 starting off as a startup in Omaha Nebraska and evolving into a major IPFS provider, in the Web3 domain. To utilize Pinatas services, we need to visit their website and generate API keys as illustrated in Fig. 6 where we used our API keys.

Our backend code takes this certificate. Computes its hash using the SHA 256 algorithm demonstrated in Fig. 7. SHA 256 functions by processing input data in

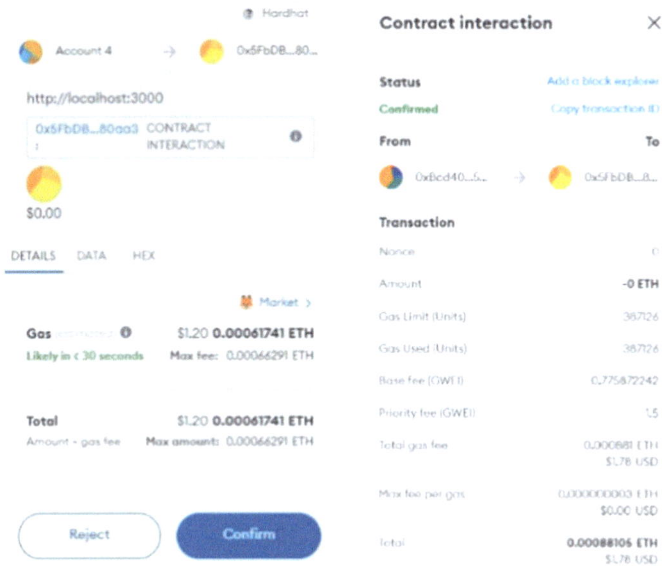

Fig. 5 Transaction confirmation and its details

```
const formData = new FormData();
formData.append("file", file);
resFile = await axios({

  method: "post",
  url: "https://api.pinata.cloud/pinning/pinFileToIPFS",
  data: formData,
  headers: {
    pinata_api_key: "7e60bf66f28e25420b20",
    pinata_secret_api_key:
      "9a9354c1539fdb2c875b2397b16fcb242cd4bc508b2fd98a0897e94b7bd146a3",
    "Content-Type": "multipart/form-data",
  },
});
```

Fig. 6 Uploading certificate to IPFS using pinata

blocks of 512 bits and applying a series of operations iteratively which include bitwise operations and modular addition. The program creates a code, for every input making it nearly impossible to create the same code from different input or to figure out the original input, from the code.

Our smart contract code for adding this information into the Blockchain is shown in Fig. 8 using the add function we are uploading the student details along with its certificate hash and ipfs address into our Blockchain the complete code of the smart

```
const generatePDFHash = async (pdfFile) => {
  try {
    const pdfBuffer = await readFileAsArrayBuffer(pdfFile);
    const hashBuffer = await crypto.subtle.digest("SHA-256", pdfBuffer);
    const hashArray = Array.from(new Uint8Array(hashBuffer));
    return hashArray
      .map((byte) => byte.toString(16).padStart(2, "0"))
      .join("");
  } catch (error) {
    console.error(error);
    return "Error generating hash";
  }
};
```

Fig. 7 Hash generator function

contract can be found on our GitHub link. While our testing methods were quite thorough, we did face some challenges along the way. The usage of the Ethereum network occasionally resulted in issues such, as transaction times and high fees during peak periods of activity.

Additionally, although IPFS proves to be a data storage solution the retrieval time for accessing this data can vary based on factors, including the availability of nodes containing the required information. We have found our certificate verification system to be highly efficient and secure; however, there is a balance between maintaining decentralization and ensuring performance. Fortunately incorporating IPFS has been beneficial in terms of ensuring data availability and security when needed. In summary, our system is robust and effectively fulfills its purpose but

```
// SPDX-License-Identifier: MIT
pragma solidity >=0.5.0 <0.9.0;
pragma experimental ABIEncoderV2;

contract CertificateRegistrySC {
    struct Access {
        address user;
        bool access;
    }

    struct Record {
        string url;
        string fileHash;
    }
    mapping(address => string[]) value;
    mapping(bytes32 => Record) private records;
    mapping(address => mapping(bytes32 => bool)) ownership;
    mapping(address => mapping(address=> bool)) ownerships;
    mapping(address => Access[]) accessList;
    mapping(address => mapping(address => bool)) previousData;

    function add(address studentAddress,
    string memory studentName,
     string memory rollNumber,
     uint16 passingYear,
     string memory url,
     string memory fileHash) external {
     bytes32 nameHash = keccak256(abi.encodePacked
     (studentName, rollNumber, passingYear));
     Record storage record = records[nameHash];
     record.url = url;
     record.fileHash = fileHash;

     // Set ownership for the student
     ownership[studentAddress][nameHash] = true;

     // Update the value mapping with the user's
     // URL and file hash
     value[studentAddress].push(url);
     value[studentAddress].push(fileHash);
    }
```

Fig. 8 Smart contract code

remains imperfect. We are actively exploring avenues to further enhance its capabilities through techniques such as leveraging advancements, in Ethereum or refining how we retrieve data from IPFS.

5.1 Security Analysis

- **Availability:** The Blockchain framework we have suggested shows resistance to security risks whether they come from, within or outside the system. Its decentralized nature ensures that crucial information like student records and certificate verifications can be accessed without any interruptions even if some parts of the network are compromised. The inherent strength of our solution based on guarantees operation protecting it from a wide range of potential disruptions, to the Blockchain system.
- **Integrity:** In our Blockchain-based system, for verifying credentials maintaining integrity is of importance. It ensures that the data within the network is truthful and cannot be changed. Once a university validates and registers a student's certificate it is protected using methods and stored securely in the Blockchain as an approved transaction. This process guarantees that the certificates information is not safe but also permanent solidifying its credibility in the record. This mechanism plays a role in our system by preserving the trustworthiness of qualifications making them resistant, to tampering and consistently reliable.
- **Confidentiality:** Confidentiality is a principle, in our system that combines Blockchain technology with academic certificate verification. It puts an emphasis on protecting the privacy of participants and ensuring that authorized individuals can access their data. We achieve this by using function modifiers within our system to limit access.
- **Non-repudiation:** In our system, we make use of digital signatures that rely on algorithms to confirm the genuineness of every transaction. These digital signatures are crucial, in our proposed framework as they guarantee that certificate issuances cannot be challenged and firmly establish their credibility. The design of the system ensures that both the issuing authority and the recipient cannot question the legitimacy of a certificate thereby establishing a sense of trust and transparency in the process of verifying qualifications.
- **Security Flaws in smart contracts:** To ensure the protection of our contracts, from any intent it is crucial to be extremely careful in drafting them. There are tools that can be used to carefully examine the functions of smart contracts and enhance their security. In our research, we utilized Mythril and Slither to conduct an analysis of our code. These tools proved valuable in identifying vulnerabilities, which enabled us to improve the quality of our contracts. During the phase of our security analysis, we used a tool called Slither. It is specifically designed to analyze Solidity contracts. We applied it to scrutinize our contract named "CertificateRegistrySC.sol." The first analysis pointed out four issues, including collisions, with abi.encodePacked() and complexities related to pragma versioning.

Fig. 9 Result of analysis using slither

These findings were crucial in guiding us through the revisions of our code. We addressed each issue by refactoring the encoding process to avoid hash collisions and updating pragma directives for compatibility. As a result, we improved the security profile of the contract. To confirm the effectiveness of these improvements we ran another analysis with Slither which reported zero issues. This indicates that we successfully mitigated the vulnerabilities identified earlier. This iterative process of analyzing and refining highlights our dedication toward building a reliable system, for verifying certificates. Figure 9 shows the results of our analysis using slither.

We took the results from slither in pass it as an input to the Mythril analysis tool. The results shows that the smart contract CertificateRegistrySC.sol has been successfully analyzed and no problems were found. This is important because it indicates that the contract doesn't have any vulnerabilities that Mythril is designed to detect, like reentrancy, integer, and Unhandled exceptions.

6 Conclusion

In our research, we have introduced a system, for verifying academic certificates. Our solution leverages Blockchain technology to create an unchangeable record for each certificate. The use of IPFS ensures efficient and reliable storage while academic institutes take responsibility for uploading certificates to guarantee their authenticity. The main goal of this system is to address challenges in the certificate verification process, such as slow verification, certificate forgery, and excessive bureaucratic hurdles. Furthermore, we have thoroughly examined the security of our contract code using tools like Slither and Mythril to ensure a trustworthy and secure system.

Looking ahead our focus will be, on enhancing scalability and incorporating student transcripts into the system. This expansion will provide an overview of a student's academic journey; however, it requires careful consideration in terms of design to ensure efficiency and effectiveness. Moreover, we plan to prioritize improving user experience and interface design in order to simplify the verification process for all stakeholders involved. Accomplishing these goals successfully has the potential to bring about a revolution, in record management and reliability on a scale. This would reinforce the nature of our system in both technological domains. To summarize, although our current system is strong and meets its intended purpose effectively, we acknowledge the need for improvement and innovation. Moving forward, we will build upon the foundation established by this research aiming for a system that not meets but surpasses the evolving demands of academic credential verification.

References

1. Klemeš JJ, Yap KY, Chin HH (2023) Blockchain technology for distributed generation: a review of current development, challenges and future prospect. Renew Sustain Energy Rev 175:113170. https://doi.org/10.1016/j.rser.2023.113170
2. Taherdoost H (2023) Smart contracts in Blockchain technology: a critical review. Information 14:2. https://doi.org/10.3390/info14020117
3. Buterin V (2014) A next-generation smart contract and decentralized application platform. White Pap 3(37):1–2
4. Jianjun S, Ming L, Jingang M (2020) Research and application of data sharing platform integrating Ethereum and IPFs technology, In: 2020 19th International symposium on distributed computing and applications for business engineering and science (DCABES), pp 279–282. https://doi.org/10.1109/DCABES50732.2020.00079
5. Saleh OS, Ghazali O, Rana ME (2020) Blockchain based framework for educational certificates verification, J Crit Rev
6. Kumutha K, Jayalakshmi S (2021) Hyperledger fabric blockchain framework: efficient solution for academic certificate decentralized repository, In: 2021 Fifth international conference on I-SMAC (IoT in Social, Mobile, Analytics and Cloud) (I-SMAC), pp 1584–1590
7. Hasan M, Rahman A, Islam MJ (2020) Distb-cvs: a distributed secure blockchain based online certificate verification system from Bangladesh perspective, In: 2020 2nd International conference on advanced information and communication technology (ICAICT), pp 460–465
8. Siddiqui ST, Ahmad MO, Khamruddin M, Gupta AK, Singha AK (2022) Blockchain and IoT for educational certificates generation and verification, In: 2022 2nd International conference on computing and information technology (ICCIT), pp 298–303
9. Bahrami M, Movahedian A, Deldari A (2020) A comprehensive blockchain-based solution for academic certificates management using smart contracts, In: 2020 10th International conference on computer and knowledge engineering (ICCKE), pp 573–578
10. Rahman T, Mouno SI, Raatul AM, Al Azad AK, Mansoor N (2023) Verifi-chain: a credentials verifier using blockchain and IPFS, In: International conference on information, communication and computing technology, pp 361–371

Toward Smart Urban Mobility: Fog-Enabled Parking Management Systems for Efficient and Sustainable Parking Allotment

Arepalli Peda Gopi, Meda Manogna, Bellamkonda Priyanka, Addepalli Sirisha, and Kamatham Harika

Abstract The rapid urbanization and the growing quantity of automobiles on the road have created a pressing need for efficient parking management systems. This paper represents a novel approach to address this challenge by integrating fog nodes, cloud computing, and a sophisticated scheduling algorithm. The proposed system optimizes parking allotment in real time, making both parties overall parking experiences better for both the drivers and administrators. Fog nodes are strategically deployed within parking areas to collect and process data from various sensors, such as occupancy sensors, cameras, and vehicle entry/exit points. These fog nodes alleviate the data processing burden on the cloud and enable rapid decision-making at the edge. Data from the nodes in the fog are seamlessly transmitted over to the cloud to further analysis and to long-term storage. To allocate parking spaces efficiently, a scheduling algorithm is implemented, which takes into account various factors such as real-time occupancy data, user preferences, and historical parking patterns. The algorithm optimizes parking allotment by considering the shortest walking distance for drivers, reducing congestion within parking facilities, and ensuring fair and equitable space distribution. The synergy of fog nodes, cloud computing, and the scheduling algorithm provides several benefits, including reduced parking search times, lower operational costs, and a more sustainable and eco-friendly approach to urban mobility. The system offers real-time updates to drivers through mobile applications and information displays at parking and information displays at parking entrances. This research represents a significant step toward realizing a smart parking ecosystem that enhances urban mobility reduces traffic congestion, and improves the quality of life in modern cities.

A. P. Gopi · M. Manogna (✉) · B. Priyanka · A. Sirisha · K. Harika
Department of CSE, Vignan's Nirula Institute of Technology and Science for Women, Guntur, India
e-mail: manognameda14@gmail.com

A. P. Gopi
e-mail: gopiarepalli2@gmail.com

© The Author(s), under exclusive license to Springer Nature Singapore Pte Ltd. 2025
S. K. Udgata et al. (eds.), *Intelligent Systems*, Lecture Notes in Networks and Systems 1314, https://doi.org/10.1007/978-981-96-3797-3_22

Keywords Smart parking system · Fog nodes · Cloud computing · Image processing

1 Introduction

Nowadays, parking has become more of a problem as the number of cars on the road has continued to rise. The growing urbanization and the proliferation of vehicles on the road have given rise to significant challenges in the management of parking spaces in urban areas. Finding an available parking spot is a daily struggle for many, leading to traffic congestion, increased carbon emissions, and time wastage. Traditional parking management systems have proven inadequate in addressing these challenges, often failing to provide real-time information and efficient allocation of parking spaces.

IEEE 802.11 p-based Vehicle Ad Hoc Networks [1] connecting vehicles and roadside units are a possible solution. VANETs often use RSUs and communication devices mounted on the vehicles, such as on-board units (OBUs). In a second scenario, automobiles are constantly sharing data about available parking slots. Thirdly, timely information distribution and updating which are essential for smart parking systems cannot be supported by ad hoc information gathering and sharing. Last but not least, due to the high costs associated with installation, deployment, and maintenance, RSUs are incapable of doing complex computations. However, with advancements in technology, the concept of smart packing has emerged as a promising solution to these problems, leveraging fog computing to enhance real-time parking space allotment and improve overall urban mobility. The term "smart parking" [2] includes a variety of technologies intended to optimize the allocation of parking spaces. Sudhakar et al. [3] These technologies employ various sensors, data analytics, and communication systems to monitor parking availability, guide drivers to open spots, and assist parking operators in efficient management. One key element of this modern approach is the utilization of fog computing, which brings edge computing capabilities nearer the data source, improving decision-making [4] in real time and cutting down on latency. In the context of smart parking, the data from sensors and other sources are processed at the boundary of facilitating instantaneous decision-making. For instance, sensors installed in parking spots can transmit data to nearby fog nodes, which analyze the information and facilitate immediate parking space allotment decisions. This approach not only enhances real-time responsiveness but also reduces the burden on central cloud servers. Fog computing ensures that parking space allotment decisions can be made instantly, allowing drivers to find available parking spots more efficiently by calculating Euclidean Distance. Processing [5] data closer to the source minimizes network latency. In practical terms, this translates [6–8] to a smoother and more responsive parking experience for drivers, who can quickly identify and secure available parking spaces. With decentralized processing, the system is less susceptible to single points of failure. This, in turn, increases the overall reliability of smart parking solutions, minimizing downtime and ensuring continued service availability [9–13]. To gather and share data and aid vehicles in selecting parking decisions, a constrained quantity of fog nodes, each with its own storage and processing capabilities, are positioned in close proximity to parking spaces. As an added bonus, numerous fog nodes may work together to collect complete parking data for the region, which can then be used to inform collective decision-making. This paper will delve further into the technical aspects, practical implementations, and real-world case studies to provide a comprehensive understanding of the transformative role of fog computing in modernizing parking management. Smart parking technology even made it more efficient, by connecting to a remote cloud hub equipped with plenty of powerful processing units.

2 Literature Survey

Chaoyang Tang et al. [1] proposed a fog computing-based smart parking architecture that aims to improve the efficiency of finding parking slots in densely populated areas. It utilizes technology for information and communication, like vehicle ad hoc networks (VANETs), to provide real-time parking slot information. Improved efficiency in finding desirable parking slots. Paper doesn't explicitly mention the limitations of the paper of fog computing-based smart parking architecture.

Awaisi et al. [4] explained a fog-enabled efficient car parking architecture to address the challenges of parking spaces in populated. It utilizes technologies such as Wireless Sensor Networks (WSN), Cloud Computing to automate parking space allocation. Minimizes latency, improving that overall efficiency of parking operation. Reduces network usage, leading to more efficient utilization of resources.

Abhirup Khanna et al. [6] discussed an overview and advancements, to manage and optimize parking for cars. Current developments and advancements in the industry, including occupancy tracking, reservation systems, and automated parking guidance. Evaluation of intelligent parking system applications in urban or commercial settings.

Maharjan et al. [7] combines the Internet of Things (IoT) with fog computing technologies into smart parking systems. These reviews often focus on the architectural aspects of the integration. Discussion on how real-time data on traffic, parking spot availability, and other pertinent information is gathered using IoT devices like cameras and sensors.

Lomat Chowdhury et al. [2] demonstrated how IoT gadgets, such cameras and sensors, can be used to track the movement of people and vehicles in parking lots in real time. Investigating the application of fog computing for real-time data processing and analytics at the network's edge, facilitating quicker decision-making and lowering latency.

Jusat et al. [8] described the development of a decentralized low-latency smart parking system using Internet of Things (IoT) and fog computing technologies. Proposed system is implemented using ZigBee sensor nodes, Bluetooth low-energy beacons, an Android app, a decentralized database, and fog computing gateways. Limitations of the paper are a lack of comparison with other smart parking systems and less discussion on scalability and security.

Sudhakar et al. [3] proposed an IoT-based platform for monitoring carpark occupancy and performing data analytics for efficient carpark management and helping drivers find suitable carparks. Flexibility and scalability for efficient big data processing due to the adoption of cloud computing and the Hadoop MapReduce platform. Leveraging a cluster of commodity servers to process huge amounts of data for big data analytics.

Zhou et al. [9] demonstrated the study of the technology utilized in smart parking systems, including sensors, cameras, IoT devices, mobile apps, and data analytics. Overview of recent advances and developments in the industry, including predictive analytics, reservation systems, and real-time monitoring of parking spaces. Examining actual cases where smart parking systems have been used in commercial or urban settings.

Estrella et al. [5] been discussed the implementation of smart parking systems that make use of fog computing technologies. It could offer information on the current state of the art, new trends, and difficulties in the field of smart parking. Fog computing is a technique for reducing latency and enhancing real-time decision-making by processing data closer to the data source, at the network edge.

Rivano et al. [10] proposed the cloud computing pose challenges for smart parking systems. Reduction in latency and improved efficiency in utilizing available technologies together and minimization of the time taken to find parking spaces, reducing fuel consumption and CO2 emissions are major uses. The cost of operation and latency of services in cloud computing pose challenges for smart parking systems.

3 Proposed Work

The proposed system of fog-based architecture is equipped with various elements like fog nodes, a microcontroller chip and cameras are examples of cloud servers, as well as smart LED. The proposed cameras in architecture are in charge of taking pictures of the parking spaces then, depending on the photos that were taken, the parking spaces accessed. Every five seconds, a node in the fog gathers new photograph and refreshes the parking screen. Each parking spot in the garage is under constant observation by a slew of security cameras. A microcontroller is at the heart of the proposed design, connecting the fog node to the cameras. Fog nodes provide information into the cloud. Period of time the fog nodes spread out around the parking lot are in constant contact with one another and relay information about open spots. If it turns out that a given lot is entirely filled, the fog node will be able to gather information regarding parking availability Signs in the vicinity of parking places are updated with data that is retrieved to LEDs.

Figure 1 illustrates that the fog-based architecture integrated with surveillance cameras efficiently manages parking by leveraging real-time data processing. When a vehicle enters the camera's view it promptly detects and the pertinent information is relayed to the fog nodes. These fog nodes serve as intermediate processing units and analyses the data before transmitting to the centralized cloud server. At the clous level, the information is further processed and evaluated. If the parking space is available, the clous server swiftly allocates the spot, ensuring optimal utilization of parking resources. This seamless integration of fog and cloud computing enables swift decision-making and enhances the efficiency of parking management systems (Fig. 2).

Algorithm 1: Parking slot allocation for multiple parking request.

Input: Input all the parameters like fog nodes, capacity…etc.
Output: Record the allocation of the park area.
Step 1: Let F be the set of fog nodes. Each fog node f is defined by its location L(f) and available capacity C(f).

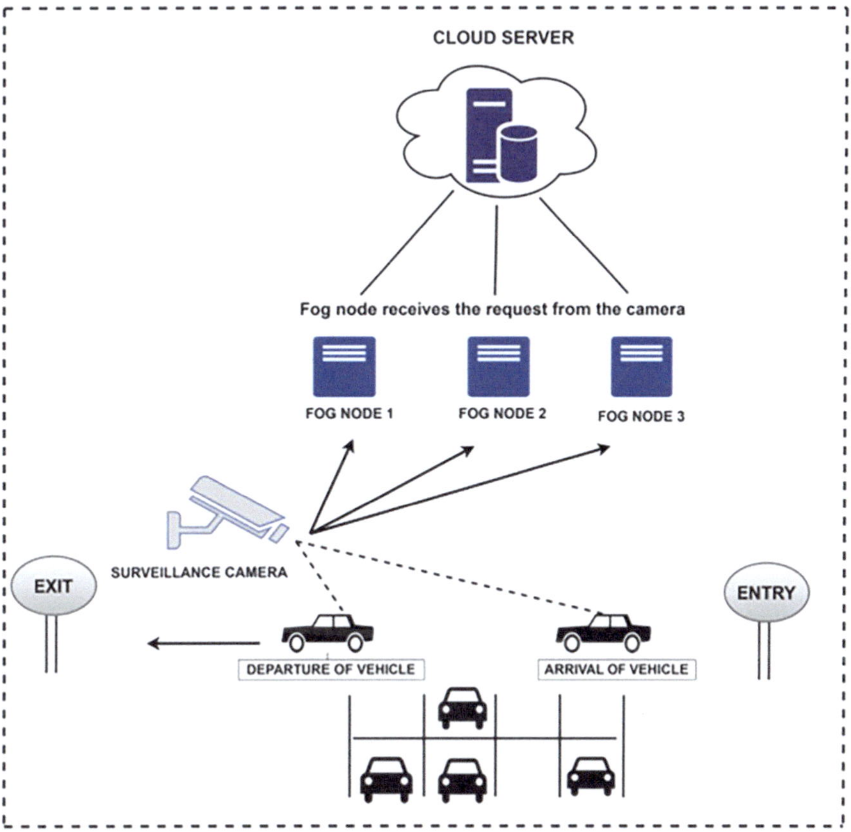

Fig. 1 Sensor-driven fog-based architecture for intelligent parking allocation

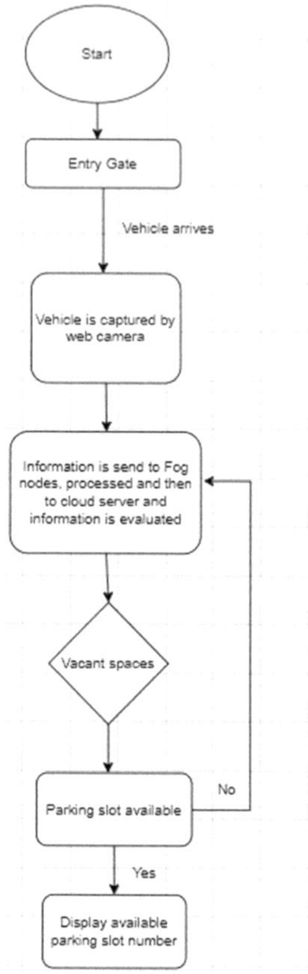

Fig. 2 Flowchart of the proposed model

Step 2: Let P represent the park area request. The park area request is defined by its location L(P) and required service capacity R(P).
Step 3: Now calculate the Euclidean distance D (P, f) between P and f as follows: D (P, f) = sqrt ((x(P)—x(f)) ^2 + (y(P)—y(f)) ^2).
Step 4: Create a subset of fog nodes that are within a certain distance threshold T of P.
Filtered Nodes = {f | f in F, D (P, f) < = T}.
Step 5: Create a subset of fog nodes that have sufficient capacity for the requested service. Capacity Filtered Nodes = {f | f in Filtered Nodes, C(f) > = R(P)}.
Step 6: If you want to select a fog node randomly, you can use a random selection function.
Selected Node = Random Choice (Capacity Filtered Nodes).
Step 7: Update Capacity: C (Selected Node) = C (Selected Node)—R(P).
Log Allocation: Record the allocation of the park area request to the fog node.
Step 8: END.

The PSA has developed a one-of-a-kind algorithm for assigning parking spots based on demand. Most parking requests are classified as "single parking request." You may expect your request to be processed in the order in which it was received. The PSA parking request is handled in a sequential manner. The following fog node will check the status of the parking lot it is responsible for as soon as a vehicle arrives. If the automobile seeking the parking space receives a response from the fog node, it will indicate whether or not a parking place is available. Assuming not, the inquiry is. Cloud nodes are responsible for transferring information to the cloud data center, where it may be stored. For this to work, it has to go through every look around for the lowest parking rate to save money. To keep track of the cheapest parking option discovered so far, we may utilize the min distance. This would allow the transmitting vehicle to be directed to the correct parking garage.

Algorithm 2: An improved algorithm for assigning parking spaces in response to a numerous requests from cloud and fog nodes.

Input: Latency_Fog(f), Latency_Cloud, Latency Threshold,Processing_Time_Fog(f), RandomProcessingTime(),Total_Time_Fog(f),Total_Time_Cloud,Fog_Processing_Valid(f).
Output: A choice about the distribution of parking space allocation.
Step 1: Let Latency_Fog(f) represent the latency for a fog node f and Latency_Cloud represents the latency for the cloud.
Step 2: Let User_Request represent the user's request.
Step 3: Define a latency threshold for fog node processing.
Step 4: Calculate Fog Processing Time:
Processing_Time_Fog(f) = Random Processing Time ().
Determine if Fog Node Processing is within Threshold:
Fog_Processiing_Valid(f) = (Latency_Fog(f) + Processing_Time_Fog(f)) < = Threshold.
Step 5: Process the request on the cloud and Simulate cloud processing time.
Calculate Cloud Processing Time:
Processing_Time_Cloud = RandomProcessingTime ().
Step 6: Calculate the total fog and cloud processing times.
Total Fog Processing Time:
Total_Time_Fog(f) = Latency_Fog(f) Processing_Time_Fog(f) for fog node f.
Total Cloud Processing Time:
Total_Time_Cloud = Latency_Cloud + Processing_Time_Cloud.
Step 7: Decide whether to process the request on a fog node or cloud-based on the calculated processing times.
If Fog_Processing_Valid(f) is true for any fog node f, and Total_Time_Fog(f) < = Total_Time_Cloud, choose fog_ processing.
Step 8: END.

PSA handles parking requests in reverse chronological order during rush hours, using fog nodes to process concurrent requests. To handle this, heuristic principles are used to distribute requests, determining the cost of parking, and selecting the lowest parking lot allocation cost. The total parking costs include four categories: driving, waiting, parking, and walking. An approach is invented to find which requests need processing by forwarding them to the cloud, which processes them based on priority time and park allocation area.

4 Experimental Results

Results that we have got for efficient wise parking management system using fog computing and fog nodes. In contrast to traditional, the utilization of cloud and fog computing offers an efficient and accurate results.

Figure 3 shows the results of using a fog network. A proposed graph shows how data travels through the network. This graph would effectively demonstrate the advantages of fog computing in minimizing latency.

Figure 4 shows the result of the time taken for the execution of the fog. Fog execution time refers amount of time it takes for a task, computation within the fog network. It enhances the overall efficiency and responsiveness of the network.

Figure 5 shows the efficient result of using a fog network in the terms of latency and execution period. Fog computing minimizes the time required for data transmission to remote cloud servers, and enhancing response times for critical applications like real-time surveillance.

Figure 6 shows the result of latency that is been proposed. Latency response plays a crucial role in determining the efficiency and effectiveness of cloud system. Low latency leads to quicker data retrieval and improves performance of cloud-based services.

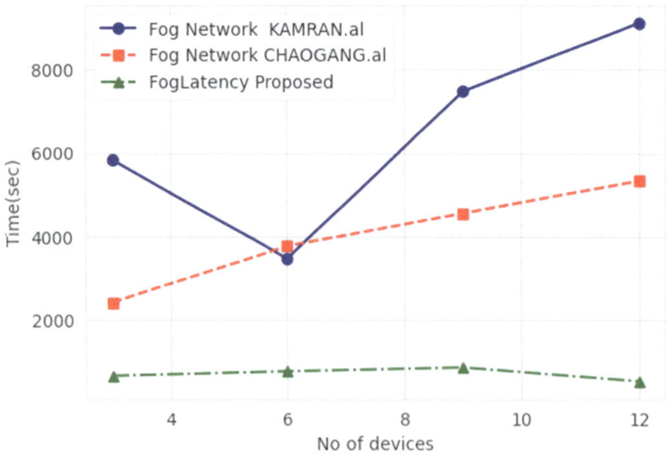

Fig. 3 Simulation results with respect to fog latency proposed

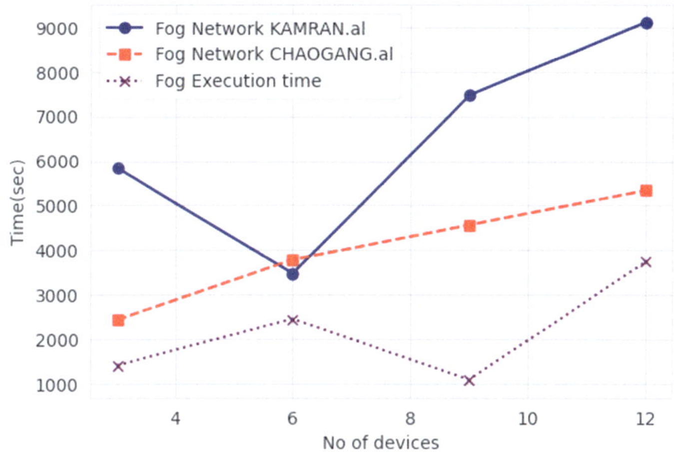

Fig. 4 Evaluating the fog network with respect to execution time

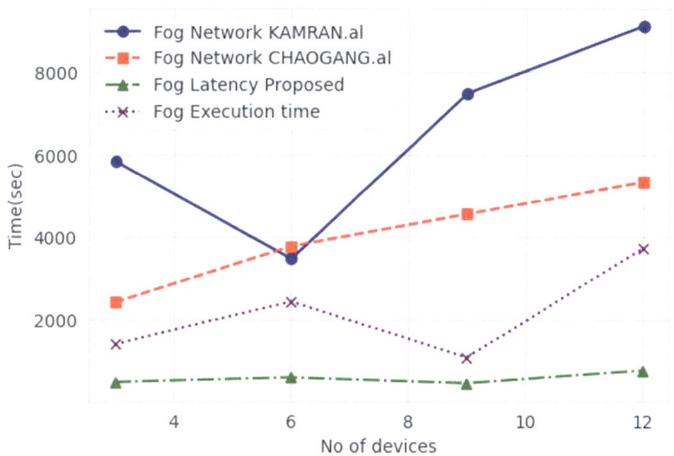

Fig. 5 Fog simulation result respect to latency proposed and execution time

Figure 7 shows the execution time of cloud network. This graph showcases a timeline where you have data or tasks sent to a cloud server, and then displays the execution time. It varies with traditional cloud in terms of network congestion, server load etc.

Figure 8 shows the simulation results of cloud network in terms of both latency and execution time. This graph can help to represent the potential fluctuations in cloud network and advantages of fog computing which is more predictable and lower execution times.

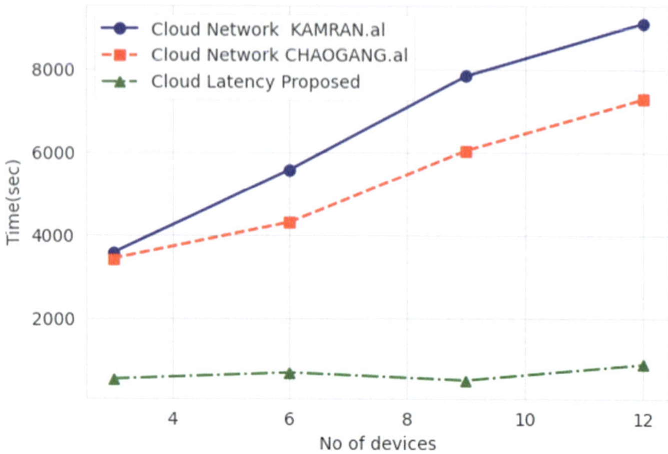

Fig. 6 Cloud simulation results with respect to latency propose

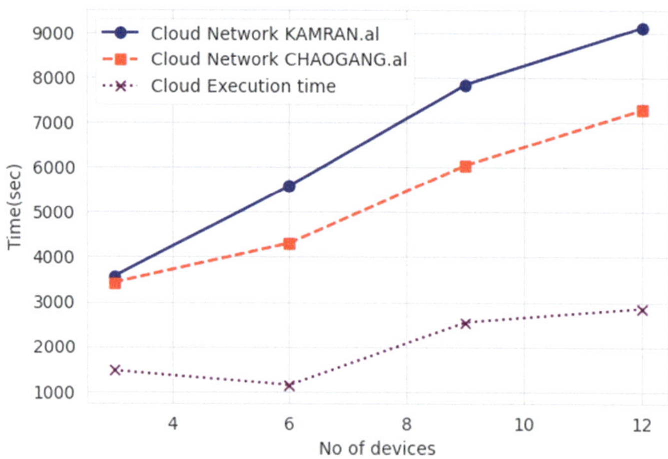

Fig. 7 Evaluating the cloud network with respect to execution time

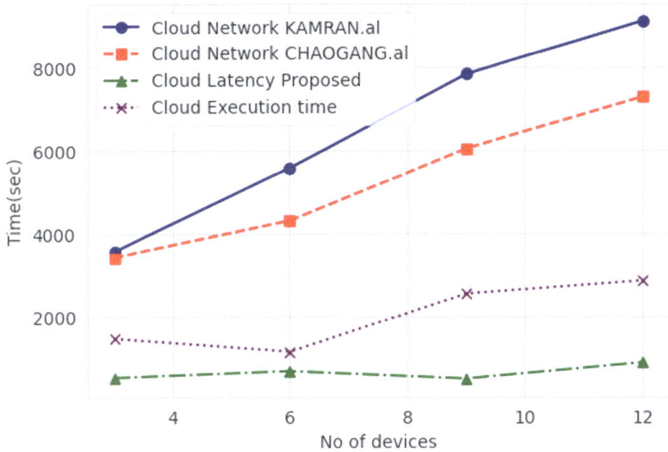

Fig. 8 Cloud simulation result respect to latency proposed and execution time

5 Conclusion

In conclusion, the integration of fog nodes, fog computing, cameras, sensors and cloud computing in smart parking allotment systems represents a promising advancement in urban infrastructure. This innovative approach offers a real-time monitoring, efficient resource allocation, and enhanced user experience. Fog nodes enable quick decision-making at the edge cameras provide accurate data collection, while fog and cloud computing together offer scalable and processing capabilities. Ultimately, this energy ensures a smarter, more responsive, and sustainable parking management system that can alleviate urban congestion and improve overall city functionality. In this paper, we have been using fog computing which can significantly improve the real-time accuracy and responsiveness of a smart parking system in comparison to traditional centralized cloud-based systems. Our smart parking system can offer more precise and current information regarding the availability of parking spaces. Additionally, by processing data closer to data source, fog computing can help in mitigate issues related to network congestion and latency, resulting in better overall accuracy in parking space allocation and occupancy monitoring.

In the upcoming work, we intend to design more user-centric and sustainable strategies, contributing to better traffic management and enhanced living.

References

1. Tang C, Wei X, Zhu C, Chen W, Rodrigues JJ (2018) Towards smart parking based on fog computing. IEEE Access 6:70172–70185
2. Chowdhury LH, Mahmud ZZ, Islam IU, Jahan I, Islam S (2019) Smart car parking management system. In: 2019 IEEE International conference on robotics, automation, artificial-intelligence and internet-of-things (RAAICON). IEEE. pp 122–126
3. Sudhakar MV, Reddy AA, Mounika K, Kumar MS, Bharani T (2023) Development of smart parking management system. Mater Today Proceed 80:2794–2798
4. Awaisi KS, Abbas A, Zareei M, Khattak HA, Khan MUS, Ali M, Shah S (2019) Towards a fog enabled efficient car parking architecture. IEEE Access 7:159100–159111
5. Perera C, Reiff-Marganiec S, Qin Y, Estrella JC, Vasilakos AV (2017) Fog computing for sustainable smart cities: a survey. ACM Comput 50(3):32. https://doi.org/10.1145/3057266
6. Khanna A, Anand R (2016) IoT based smart parking system. In: 2016 International conference on internet of things and applications (IOTA). IEEE. pp 266–270
7. Maharjan AMS, Elchouemi A (2020) Smart parking utilizing IoT embedding fog computing based on smart parking architecture. In: 2020 5th International conference on innovative technologies in intelligent systems and industrial applications (CITISIA). IEEE. pp 1–9
8. Jusat N, Zainuddin AA, Sahak R, Andrew AB, Subramaniam K, Rahman NA (2021) Critical review in smart car parking management systems. In: 2021 IEEE 7th International conference on smart instrumentation, measurement and applications (ICSIMA). IEEE. pp 128–133
9. Zhu C, Zhou H, Wang K, Zhang Y, Leung VC, Yang LT (2017) Toward big data in green city. IEEE Commun Magazin 55(11):14–18
10. Rivano TH, Lin F (2017) A review of smart parking solutions, Le Moul. IEEE Trans Intell Transp Syst 18(12):3229–3253. https://doi.org/10.1109/TITS.2017.2685143
11. Arepalli PG, Naik KJ (2024) A deep learning-enabled IoT framework for early hypoxia detection in aqua water using light weight spatially shared attention-LSTM network. J Supercomput 80(2):2718–2747
12. Arepalli PG, Naik KJ (2024) An IoT based smart water quality assessment framework for aqua-ponds management using dilated spatial-temporal convolution neural network (DSTCNN). Aquacult Eng 104:102373
13. Arepalli PG, Naik KJ (2023) An IoT-based water contamination analysis for aquaculture using lightweight multi-headed GRU model. Environ Monit Assess 195(12):1516

Trust-Aware Firefly Algorithm for Routing and Edge Computing Optimization in IoT Environment

Rabindra Patel and Sasmita Acharya

Abstract Internet of Things (IoT) is one of the widely acceptable technologies of the world that helps people to improvise their productivity and security. But with the spread of IoT technology vast amount of data is now needed to be transmitted, stored and processed in as less time as possible. For this integrating IoT with traditional cloud-based central server creates challenges as energy consumption, reliability, throughput, network delays, etc. So, the demand for efficient and reliable communication in distributed systems continues to rise, optimizing routing strategies and leveraging edge computing capabilities become paramount. This paper presents a noble approach, the Trust-aware Firefly Algorithm for Routing and Edge Computing Optimization, aimed at addressing the challenges of dynamic network environments. The algorithm integrates trust values, spatial considerations, and edge computing capacities to enhance the decision-making process in routing tasks.

Keywords Trust value · Edge computing · Firefly algorithm · Optimization · Computing capacities

1 Introduction

IoT is a collection of independent physical nodes which are inter-connected with each other with the help of wireless communication. Before the Internet of Things, cloud computing is a popular research area. All the work like data analysis, archiving, and decision-making were performed at cloud data center. But the increase in data volume and speed creates problem in transferring big data from IoT devices to cloud data centers because of bandwidth limitation and generate high latency and privacy problem [1] (Fig. 1).

As depicted in above figure integrating edge computing will be a solution for this computation requirement and minimize the level of latency.

R. Patel (✉) · S. Acharya
Veer Surendra Sai University of Technology, Burla, India
e-mail: rabindrapatel329@gmail.com

© The Author(s), under exclusive license to Springer Nature Singapore Pte Ltd. 2025
S. K. Udgata et al. (eds.), *Intelligent Systems*, Lecture Notes in Networks and Systems 1314, https://doi.org/10.1007/978-981-96-3797-3_23

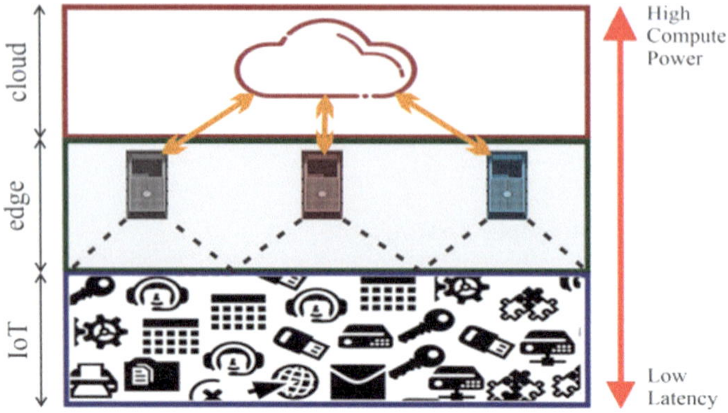

Fig. 1 General view of different working environment [2]

IoT is particularly one of the fastest evolving technology [3]. Every year some of the new inventions, models, protocols, etc. are coming according to the demand of society. In modern IoT systems, comprising a multitude of inter-connected nodes, face challenges related to communication efficiency, trust management, and the utilization of edge computing resources. In response to these challenges, this paper introduces a Trust-aware Firefly Algorithm designed to optimize routing decisions and leverage edge computing capabilities.

The algorithm is built on the foundation of the Firefly Algorithm, a nature-inspired optimization technique. Unlike conventional approaches that solely focuses on distance-based optimization, our proposed algorithm incorporates trust values and edge computing capacities into the objective function. By doing so, it aims to strike a balance between proximity, reliability, and computational resources, thereby improving the overall efficiency of distributed systems.

The key components of the algorithm include the evaluation of objective functions that consider Euclidean distance, trust, and edge computing capacities. The algorithm iteratively updates node positions based on the calculated objective values, ensuring that the network adapts to the dynamic nature of the environment.

In addition, our approach introduces a trust-aware routing mechanism, where nodes make routing decisions based on both spatial proximity and trustworthiness. This not only enhances the robustness of the network against malicious nodes but also allows for effective utilization of edge computing resources.

Through this work, we aim to contribute to the advancement of routing algorithms in distributed systems, offering a trust-aware and edge computing-optimized solution that can be adapted to diverse real-world applications.

The remainder of this paper is organized as follows: Sect. 2 provides an overview of the algorithmic framework, detailing the key functions and their roles. Section 3 deals with the implementation details, highlighting the integration of trust values and edge computing capacities. Section 4 showing the simulation of proposed

Work. Section 5 presents experimental results and discussions, demonstrating the effectiveness of the proposed algorithm in various scenarios. Finally, Sect. 6 concludes the paper with a summary of contributions and avenues for future research.

2 Related Work

The authors of paper [1] comprehensively provide the detail study about recent developments like cloud, fog, edge, and IoT with their difficulties and opportunities in present and future. The author of paper [2] designed a distributed Trusted Object Framework (TOF) where whenever heterogeneous IoT Smart Objects (SOs) interact a feedback is sent to an edge server to calculate their reputation score. In paper [3], the authors provide a short review on WSN and IoT network by giving more focus on reduction of energy usage and data aggregation technique. In paper [4], the authors provide a conceptual model refers to build up a representation of a system. This paves way to understand the data flow, interactions, and behaviors of different components of IoT system. It also brings the importance of choosing appropriate model and data offloading system for the particular IoT system. The authors of paper [5] give emphasis on developing an efficient model named as Energy Efficient Cloud-Based Internet of Everything (EECloudIoE) for combining the Internet of Everything (IoE) with the cloud using wind driven optimization and bio-inspired algorithms. In this model wind driven optimization used for clustering the optimal number of IoT nodes and Firefly Algorithm is used to choose the optimal cluster head on the basis of temperature, energy, load, and alive nodes. The authors of paper [6] proposed a swarm intelligence-based Internet of Medical Things in Voice-Based Pathology Detection model for high performance computing. It gives better results than the standard model when tested with three standard datasets including "voice-based pathological diseases, environmental sound detection and acoustic sound quality assessment parks." In paper [7], the authors presented a hybrid of Firefly and Improved Multi-objective Particle Swarm Optimization (IMPSO) technique, abbreviated as FIMPSO. It is an energy efficient algorithm used for load balancing in cloud environment and providing an effective result with least average response time. In paper [8], the authors proposed a Proximal Policy Optimization based on Ant Colony Optimization (PPO-ACO) algorithm perform better in terms of choosing best path in WSNs and also consume very less energy. The authors of paper [9] developed energy efficient and optimization-based framework in Routing Protocol for low-power and Lossy Networks (RPL) with combination with firefly algorithm. This model finds better result with concerning "Packet Delivery Ratio (PDR), number of hops, end-to-end delay, and power consumption" as compared to the existing systems. Paper [10] proposed a coupled model of the IoT network with edge computing (IoTEC) which helpful for improved environmental mentoring, data latency, energy consumption, and economic cost between the IoTEC approach than the conventional sensor monitoring method. In paper [11], the authors represent a detail review of importance of edge and fog computing research in IoT. They also demonstrate the cloud, fog, and edge computing in the IoT environment. The author of paper [12] proposed an efficient IoT framework that combines High Altitude Platform (HAP)-based Mobile Edge Computing (MEC) with Wireless Power Transfer (WPT) to optimize resource allocation and task offloading in less computational resources and finite battery life.

3 Methodology

This section outlines the methodology employed in the Trust-aware Firefly Algorithm for Routing and Edge Computing Optimization. The algorithm integrates concepts from nature-inspired optimization, trust-aware computing, and edge computing which addresses the challenges of efficient routing and resource utilization in distributed systems.

Trust-Aware Firefly Algorithm extends the traditional Firefly Algorithm by incorporating trust considerations, making it more suitable for optimization problems where trust is a critical factor and it enhances the reliability and security of the entire network.

3.1 Objective Function

The objective function is a critical component of the Trust-Aware Firefly Algorithm, determining the fitness of each node based on a combination of Euclidean distance, trust values, and edge computing capacities. The Euclidean distance, denoted as d_{ij} between nodes i and j, is calculated as Eq. (1):

$$d_{ij} = \sqrt{(x_i - x_j)^2 + (y_i - y_j)^2} \tag{1}$$

where (x_i, y_i) and (x_j, y_j) represent the Cartesian coordinates of nodes i and j, respectively.

The objective function is defined in Eq. (2):

$$\text{Objective Value} = \frac{1}{d_{ij} + 0.1} + \text{Trust Value}_i + \text{Edge Computing Capacity}_i \tag{2}$$

Here, the objective value represents the fitness of node i, incorporating spatial proximity, trustworthiness, and edge computing capacity. The term $\frac{1}{d_{ij}+0.1}$ ensures that nodes closer to each other are assigned higher values, reflecting their suitability for communication.

3.2 Firefly Algorithm

The Firefly Algorithm is a nature-inspired optimization technique based on the flashing behavior of fireflies. In the context of the Trust-Aware Firefly Algorithm, nodes in the network are considered as fireflies, and the objective values calculated in Sect. 3.1 guide their movements. The update of firefly positions is performed iteratively as in Eq. (3).

Move each firefly toward the position of the best firefly:

$$\text{New Position}_i = \text{Position}_i + \beta \times (\text{Best Position} - \text{Position}_i) + \alpha \times \text{Random Vector} \tag{3}$$

where:

- New Position_i is the updated position of node i,
- Position_i is the current position of node i,
- Best Position is the position of the best firefly,
- β is an attractiveness parameter, it determine how strongly fireflies are attracted to each other. Higher values of β indicate stronger attraction.
- α is a randomization parameter, it determine the randomness of fireflies movements. Higher values of α lead to more random movements.
- Random Vector is a vector with random components.

This movement ensures that less fit nodes move toward more fit nodes, mimicking the collective behavior of fireflies.

3.3 Trust Values Update

Trust value is an array that contains the trust values for each node in the network, representing how much each node trusts the other nodes in the network. Trust values are crucial in the Trust-Aware Firefly Algorithm, as they are related to security concern influencing routing decisions and adaptability. The update of trust values is performed based on a specific criterion. In this implementation, trust values are updated randomly for demonstration purposes. This is represented in Eq. (4).

$$\text{New Trust Value}_i = \text{Current Trust Value}_i + \text{Random Change} \tag{4}$$

where:

- New Trust Value_i is the updated trust value of node i,
- Current Trust Value_i is the current trust value of node i, and
- Random Change is a random value drawn from a normal distribution with mean zero and a standard deviation of 0.1.

This random update simulates the dynamic nature of trust in real-world scenarios.

3.4 Routing Logic

The routing logic in the Trust-Aware Firefly Algorithm is designed to make routing decisions based on trust values and edge computing capacities. For each node, trusted neighbors within the communication range are identified, and routing decisions are made accordingly. The routing logic is summarized in Eq. (5).

$$\begin{aligned}&\text{Trusted Neighbors}_i = \text{Neighbors within Communication Range}_i \cap \\ &\{\text{Nodes with Trust Values}_j > \text{Trust Thres hold}\}\end{aligned} \tag{5}$$

where:

- $\text{Trusted Neighbors}_i$ is the set of trusted neighbors for node i,
- $\text{Neighbors within Communication Range}_i$ is the set of neighbors within the communication range of node i,
- $\text{Nodes with Trust Values}_j$ represents the trust values of neighboring node j, and
- Trust Thres hold is the predefined threshold for trust values.

These trusted neighbors are extremely crucial in efficient routing, ensuring that tasks are directed to nodes with sufficient trustworthiness and edge computing capacities.

4 Simulation

All the proposed work presented in this paper is built up with the help of MATLAB (Version-R2021a). Firstly, the Trust-Aware Firefly Algorithm is implemented, by utilizing the built-in functions for mathematical operations and visualization. The algorithm's parameters, including the number of nodes, fireflies, maximum iterations, trust threshold, and communication range, can be adjusted by the user through a dialog interface as shown in Fig. 2.

To assess the performance and robustness of the Trust-Aware Firefly Algorithm, a series of experiments were conducted using different scenarios and network configurations with varying numbers of nodes, communication ranges, and edge computing capacities. The algorithm proceeds through a specified number of iterations, updating node positions, trust values, and performing routing decisions in each iteration. The experiments aimed to evaluate the algorithm's ability to adapt to dynamic environments, make effective routing decisions, and utilize edge computing resources optimally.

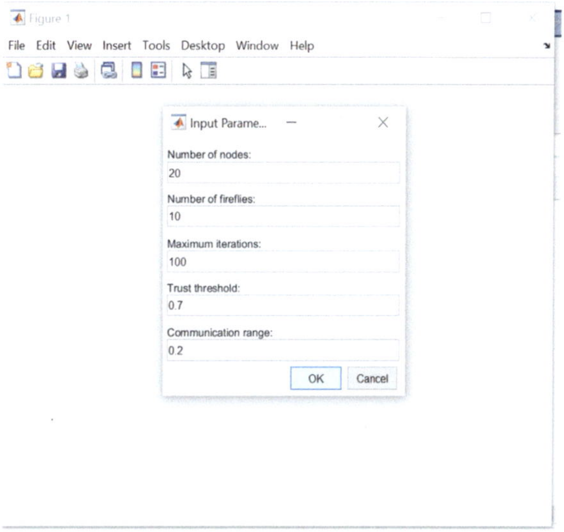

Fig. 2 Dialog interface for input parameters

5 Results

The results of the experiments demonstrated the effectiveness of the Trust-Aware Firefly Algorithm in improving routing decisions and optimizing edge computing utilization. The progress of the algorithm is visualized through a dynamic plot that displays node positions, communication range circles, and connections between trusted neighbors. In this graph the X-axis represents the horizontal position and the Y-axis represents the vertical position of the nodes, where the color mapping in the plot represents edge computing capacities and trust values, providing a comprehensive view of the evolving network topology as shown in Fig. 3.

In order to understand the above figure we have to take the following points into consideration.

- **Nodes:** The colored circles represent the nodes in the network. The color of each node is determined by its edge computing capacity, with different colors indicating different capacities. Larger circles generally indicate higher edge computing capacities.
- **Communication Range Circles:** Each node is surrounded by a circle, representing its communication range. Nodes within each other's communication range can exchange information and collaborate.
- **Connections:** Lines between nodes indicate the routing connections established based on trust values and edge computing capacities. The color of each connection is determined by the trust value of the source node. Warmer colors (closer to red) represent higher trust values, while cooler colors (closer to blue) represent lower trust values.

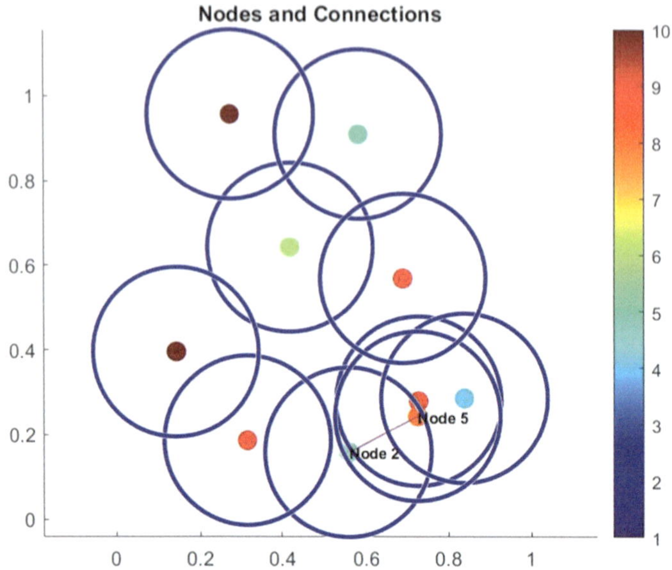

Fig. 3 Graphical output result of the proposed model

During the execution of the algorithm, the figure is updated in each iteration to reflect changes in node positions, trust values, and routing connections. The visualization allows you to observe how the nodes collaborate and form connections over time based on the algorithm's logic.

The figure provides a dynamic representation of the algorithm's behavior, making it easier to understand and analyze the network's evolution throughout the iterations. The color-coded nodes and connections help visualize the impact of edge computing capacities and trust values on the routing decisions made by the algorithm.

The algorithm exhibited rapid convergence in various scenarios, adapting to changes in network topology and trust dynamics. Several performance metrics were employed to evaluate the algorithm's performance:

5.1 Convergence Speed

The convergence speed of the algorithm was evaluated by analyzing the reduction in the objective function values over iterations. In most experiments, the algorithm demonstrated quick convergence within the specified maximum number of iterations. The adaptability of nodes to change in the network environment was a key factor contributing to the algorithm's swift convergence.

5.2 Routing Efficiency

Routing efficiency was assessed by monitoring the success rates of task completions. The algorithm, leveraging trust values and edge computing capacities, consistently demonstrated improved routing decisions, leading to a higher percentage of successful task completions compared to traditional routing algorithms. The trust-aware routing logic effectively directed tasks to nodes with higher trustworthiness and suitable computing capacities.

5.3 Resource Utilization

The degree to which edge computing capacities were utilized was a critical aspect of the algorithm's performance. The Trust-Aware Firefly Algorithm demonstrated efficient utilization of edge computing resources, dynamically adapting to changes in capacities and optimizing task distribution across the network.

6 Conclusion and Future Scope

The proposed Trust-Aware Firefly Algorithm for Routing and Edge Computing Optimization presents a noble approach to addressing the challenges of routing efficiency and resource utilization in IoT systems. By integrating trust values, spatial considerations, and edge computing capacities, the algorithm offers a comprehensive solution for dynamic network environments.

In conclusion, the Trust-Aware Firefly Algorithm provides a promising framework for enhancing routing decisions and optimizing edge computing utilization in dynamic distributed systems. Its adaptability, efficiency, and trust-aware routing logic make it a valuable contribution to the field of distributed computing.

This trust-based firefly algorithm has been applied in many fields like wireless sensor network, collaborative filtering and recommendation system, security and intrusion detection, etc.

Future work of this research includes:

- Extending the algorithm to consider dynamic changes in trust values based on real-time interactions and feedback mechanism.
- Investigating the algorithm's scalability to larger networks and complex scenarios.
- Exploring the applicability of the Trust-Aware Firefly Algorithm in real-world IoT systems and provide more focus on edge computing environments by taking task into consideration.

References

1. Ahmad S, Shakeel I, Mehfuz S, Ahmad J (2023) Deep learning models for cloud, edge, fog, and IoT computing paradigms: survey, recent advances, and future directions. Comput Sci Rev 1(49):100568. https://doi.org/10.1016/j.cosrev.2023.100568
2. Fortino G, Fotia L, Messina F, Rosaci D, Sarnè GML (2021) Trusted object framework (TOF): a clustering reputation-based approach using edge computing for sharing resources among IoT smart objects. Comput Electr Eng 96:107568. https://doi.org/10.1016/j.compeleceng.2021.107568
3. Gulati K, Boddu RS, Kapila D, Bangare S, Chandnani N, Saravanan G (2021) A review paper on wireless sensor network techniques in internet of things (IoT). Mater Today Proceed. 51. https://doi.org/10.1016/j.matpr.2021.05.067
4. Colakovic A (2023) IoT systems modeling and performance evaluation. Comput Sci Rev. https://doi.org/10.1016/j.cosrev.2023.100598
5. Priya S, Bhattacharya S, Reddy PK, Maddikunta SR, Somayaji K, Lakshman K, Kaluri R, Hussien A, Gadekallu TR (2020) Load balancing of energy cloud using wind driven and firefly algorithms in internet of everything. J Parallel Distrib Comput. https://doi.org/10.1016/j.jpdc.2020.02.010
6. Brahma B, Dash TK, Ganapati Panda LV, Prasad N, Kulkarni R (2023) Integrated swarm intelligence and IoT for early and accurate remote voice-based pathology detection and water sound quality estimation. Healthc Anal 3:100200. https://doi.org/10.1016/j.health.2023.100200
7. Devaraj AF, Elhoseny M, Dhanasekaran S, Lydia EL, Shankar K (2020) Hybridization of firefly and improved multi-objective particle swarm optimization algorithm for energy efficient load balancing in cloud computing environments. J Parallel Distrib Comput. 142. https://doi.org/10.1016/j.jpdc.2020.03.022
8. Dubey GP, Stalin S, Alqahtani O, Alasiry A, Sharma M, Aleryani A, Shukla PK, Alouane MT (2023) Optimal path selection using reinforcement learning based ant colony optimization algorithm in IoT-based wireless sensor networks with 5G technology. Comput Commun. 212. https://doi.org/10.1016/j.comcom.2023.09.015
9. Manikannan K, Nagarajan V (2020) Optimized mobility management for RPL/6LoWPAN based IoT network architecture using the firefly algorithm. Microprocess Microsyst 77:103193. https://doi.org/10.1016/j.micpro.2020.103193
10. Roostaei J, Wager YZ, Shi W, Dittrich T, Miller C, Gopalakrishnan K (2023) IoT-based edge computing (IoTEC) for improved environmental monitoring. Sustain Comput Inform Syst 38:100870. https://doi.org/10.1016/j.suscom.2023.100870
11. Laroui M, Nour B, Moungla H, Cherif MA, Afifi H, Guizani M (2021) Edge and fog computing for IoT: A survey on current research activities & future directions. Comput Commun. https://doi.org/10.1016/j.comcom.2021.09.003
12. Nauman A, Alruwais N, Alabdulkreem E, Nemri N, Aljehane NO, Dutta AK, Assiri M, Khan WU (2023) Empowering smart cities: high-altitude platforms based mobile edge computing and wireless power transfer for efficient IoT data processing. Internet Things. 100986. https://doi.org/10.1016/j.i

Automatic Detection of Skin Cancer Using Deep Learning on HAM10000 Dataset

Pranali Kshirsagar, Shubham Dodia, Abhishek Patil, Aditya Bhosale, Aryan Lad, and Piyush More

Abstract The skin, as the body's primary defense mechanism, shields internal organs from external threats. It faces a broad spectrum of challenges from microbes such as viruses, fungi, and bacteria, as well as the damaging effects of dust. Every year, a significant number of people worldwide suffer from various skin diseases, often contagious and rapidly spreading. The diverse nature of these skin conditions necessitates considerable expertise for accurate diagnosis and treatment by healthcare professionals. In recent years, there have been significant advancements in the field of diagnostic processes. Such development is the integration of deep learning models, specifically convolutional neural networks (CNNs) and NASNET. Our team has conducted a study that introduces a unique and effective method for classifying skin cancer, using a modified version of the HAM10000 dataset. Our model has shown exceptional performance when compared to the existing state-of-the-art models, all while utilizing fewer parameters. By leveraging advanced technologies like deep learning, the medical community aims to automate and improve the accuracy of skin disease diagnosis, ultimately enhancing patient care and outcomes. In comparison to the current deep learning models, this variant of convolutional neural network has achieved impressive accuracies of 98.7%, 97.04%, and 97.07% on its training, testing, and validation data sets. It has achieved these results using only 0.13 million parameters.

Keywords Skin cancer detection · Deep learning · CNN · NASNET

1 Introduction

The skin, the body's largest organ, carries an average weight of approximately 3.6 kg in adults and spans an area of 22 square meters. Functioning as a protective shield, it serves as an insulator, is waterproof, and guards against extreme temperatures,

P. Kshirsagar (✉) · S. Dodia · A. Patil · A. Bhosale · A. Lad · P. More
Department of Computer Engineering, Vishwakarma Institute of Information Technology, Pune, Maharashtra, India
e-mail: pranali.kshirsagar@viit.ac.in

© The Author(s), under exclusive license to Springer Nature Singapore Pte Ltd. 2025

S. K. Udgata et al. (eds.), *Intelligent Systems*, Lecture Notes in Networks and Systems 1314, https://doi.org/10.1007/978-981-96-3797-3_24

harmful UV rays, and chemicals. In India, skin diseases affect around 10–12% of the population, with a substantial increase noted from 15.1 crores in 2013 to 18.8 crores in 2015. Skin cancer, a prevalent form of cancer in the United States, poses risks that can be mitigated by minimizing sun exposure.

Notably, skin cancer affects around 5 million Americans annually, with 99,780 cases reported in 2022, including rapidly spreading forms, stage 0 melanoma, and high-risk cases. Early detection, especially through visual examination, is crucial for improved outcomes.

Too much time in the sun, using tanning beds, and having a family history of the disease can all up your chances of getting skin cancer, according to the American Cancer Society's latest research. And it's not just about sunburns now—if you had five or more bad burns when you were a kid or teenager, you're 80% more likely to develop melanoma later on, and 68% more likely to get other types of skin cancer too [1]. The use of tanning beds increases the risk of melanoma along with the risk of early-onset melanoma [2]. Women younger than30 years of age that use indoor tanning beds are six times more likely to develop melanoma [3].

We will use this model and the hidden layers to combine them and create a nonlinear model which best classify our images reflected in Fig. 1.

In this paper, we present a lightweight deep learning model that excels at identifying 7 different types of skin cancer. Our model boasts impressive precision, recall, accuracy, and F1 score, all while using a minimal number of parameters. To address the issue of class imbalance, we have implemented data augmentation techniques. To fully assess the effectiveness of our model, we have compared it to other techniques such as MobileNet, ResNet152, DenseNet201, ShuffleNet, Xception, and EfficientNet, using a variety of evaluation metrics.

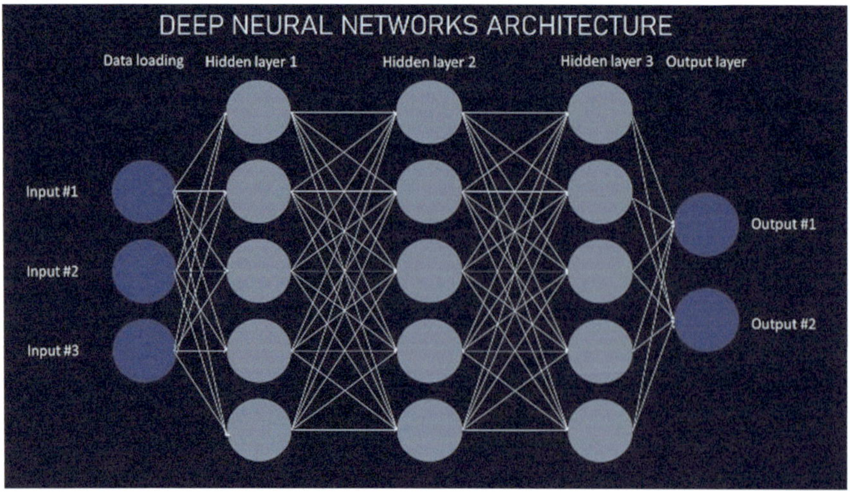

Fig. 1 Deep neural network architecture

2 Related Work

In this section, we will delve into some new research that sheds light on the fascinating world of skin cancer classification. Goswami Tanvi and her team have thoroughly examined various traditional skin cancer classifications [4]. Their findings suggest that deep learning models, specifically those based on convolutional neural networks (CNNs) [5], outperform classical machine learning methods like color histograms. This marks a significant milestone in utilizing deep learning models for image classification. Shinde and colleagues have taken it a step further by incorporating the Squeeze algorithm with MobileNet, as depicted in their paper [6]. This combination has greatly enhanced the accuracy of skin data classification. By using the compression algorithm to remove hair from skin images, the team has improved accuracy. Comparing the performance of their proposed Squeeze-MNet algorithm with other popular models like VGG-16, MobileNetV2, and Inception V3, it is evident that Squeeze-MNet achieves the highest accuracy of 99.36% on the ISIC dataset. Additionally, this model boasts higher sensitivity and specificity compared to other baseline models

However, despite its impressive performance, the number of parameters and training time for this model still surpass that of MobileNetV2. Other researchers have also explored the use of convolutional neural networks (CNNs) alone, as demonstrated in a recent study by Winkler et al. [7]. Their approach involved applying convolutions to the data, resulting in a neural network that was able to accurately classify skin cancer images as either melanoma or benign with a remarkable 95% accuracy rate. It is worth noting that in some other studies support vector machine was utilized to categorize different types of tumors [8].

A ground breaking model has been developed that boasts an impressive 96% accuracy in detecting melanoma. This remarkable model has already been put to use on the PH2 dataset, as described in the renowned research paper [9].

According to the research conducted by Basel and their team, they have proposed a sophisticated method utilizing deep learning techniques to accurately detect skin cancer at the ISIC 2022 conference. Their approach involved training a Stacked CV model with various techniques such as SVM, RF, NN, KNN, and logistic regression in three distinct stages. They employed three different extraction methods, namely Resnet50, Xception, and VGG 16. Their model achieved a remarkable accuracy of 90.9% and an F1-Score of 0.89 with the use of Xception. Their model was trained and tested on a relatively small dataset of 2637 training images and 660 test images. However, it is possible that the model's performance may not be exceptional when applied to larger datasets due to the limited size of their training data. This is a common challenge with deep learning models, as they require a significant amount of training data to achieve optimal performance. To address this issue, Shorfuzzaman developed an interpretable CNN-based classification system for detecting melanoma. The team combined previous studies using DenseNet121, Xception, and EfficientNetB0 with 1497 malignant and 1800 benign nevus images from the ISIC archive to classify melanoma [10]. The final predictions were obtained by combining the results from each sub model, resulting in an accuracy of 95.76% and an AUC of 0.957.

3 Methodology

In this section, we detail the methodology begin by describing the dataset utilized for experimentation, followed by the preprocessing steps applied to ensure data quality and suitability for model training.

3.1 Dataset

The HAM10000 dataset is a collection of dermoscopic images, captured using a specialized tool called a dermoscope. This ensures that the images are of high quality and clarity. The dataset includes seven different types of diseases: melanocytic nevus (NV), melanoma (MEL), and vascular lesions (VASC), benign keratosis (BKL), actinic keratosis (AKIEC), dermatofibroma (DF), basal cell carcinoma (BCC). These images were obtained from various cultures and captured using different methods, making it a diverse and comprehensive dataset. It is also referred as Skin Cancer MNIST, referencing its similarity to the well-known MNIST dataset. The HAM10000 dataset contains 10015 images, all of which are publicly available from the ISIC repository. These images serve as the training set for machine learning algorithms, making it a crucial benchmark dataset. By comparing the algorithm's results with human expertise, this dataset helps in evaluating its performance. It covers a wide range of major diagnoses for pigmented lesions, making it a valuable resource for medical professionals. More than half of the images have been confirmed by pathology, while the rest have been diagnosed through observation, agreement, or wide field fluorescence microscopy.

The HAM10000 dataset images distribution is reflected in Fig. 2.

3.2 Preprocessing

The dataset contains a diverse range of images, with varying quantities found in each class. This poses a challenge due to the unequal distribution of images across classes. To address this issue, it is suggested that the training dataset be supplemented with additional images, as deep learning models require a large amount of data to perform well. In order to tackle the class imbalance problem, a data augmentation technique has been utilized. This method aims to balance out the number of instances in each class, resulting in a more even distribution. In Table 1, we can see the original

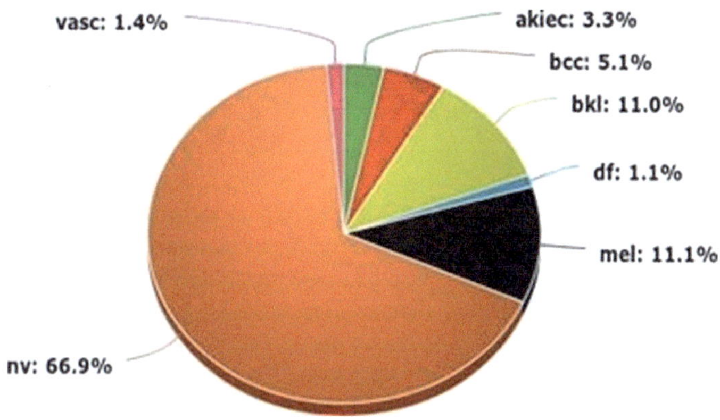

Fig. 2 Distribution of HAM10000 dataset classes

and augmented image counts for each class, demonstrating the effectiveness of this augmentation approach. The following Table 1 gives a summary of all image classes.

To improve the performance of our model, we utilized various methods for data augmentation. These techniques, including zooming, adjusting the height, flipping horizontally, and shearing, help to diversify our training image sets and enhance the training process. Our approach involves resizing the images from their original size of 600 × 450 pixels to 224 × 224 pixels. Furthermore, we also normalize each pixel value by dividing it by 255, resulting in a range of 0–1. Our preprocessing steps also involve setting a rotation range of 0–10 and defining the ranges for shear shift, height shift, and width shift as 0.2, 0.2, and 0.2, respectively. Moreover, we enable both vertical and horizontal flips for augmentation, as shown in Fig. 3.

Table 1 Total images before and after augmentation

Sr. No.	Class	Before	After
1	Melanocytic nevi	6605	6725
2	Benign keratosis	1303	6590
3	Melanoma	1233	6278
4	Basal cell carcinoma	624	6532
5	Actinic keratosis	347	6313
6	Vascular skin-lesion	176	6681
7	Dermatofibroma	126	6245

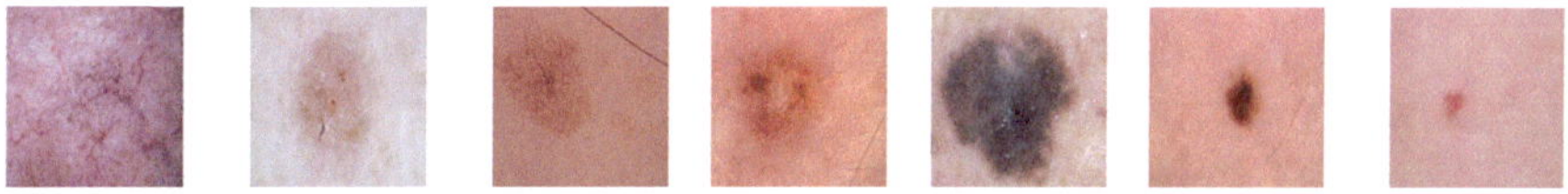

Fig. 3 Images from the HAM10000 dataset

3.3 Model Training and Architecture

For our study, we delved into using the NASNET mobile neural network architecture for classifying both the original data and the data we collected. This architecture boasts a scalable CNN design inspired by search neural architecture, with internal units that have been optimized through learning support. The number of cells in NASNET is carefully determined based on the network's required capacity. Specifically, NASNET mobile is comprised of 12 units, containing a whopping 5.3 million parameters and 564 million multiple accumulations (MACs). To feed into NASNET mobile, we use an RGB image with a width and height of 224. However, in our research, we crop the image after reading it from the original file. The study has four stages:

1. In the initial phase, the image is flipped upside down and then split into two separate images by randomly cropping each inverted image.
2. In the next step, we rotate the image horizontally and then randomly crop it to produce two distinct variations.
3. In the third phase, the image is rotated both horizontally and vertically at the same time. Then, it is randomly cropped to produce two distinct images.
4. In the last step, the image stays the same and two identical images are randomly cut to make two unique versions.

After these stages, we had to divide the number of training and validation images in the cropping data by eight. This was because each original image produced eight different variables. As a result, cropped image dataset consisted of 8800 training images, 800 validation images, and 592 testing images. It's important to mention that clipping was not used when testing the original data.

The use of Neural Architecture Search (NAS) techniques, like NASNET, in the field of skin cancer detection and medical image analysis, presents numerous benefits. By utilizing NAS, this method aims to automatically uncover the most suitable neural network structures for each specific task, allowing for fine-tuning and customization to specific needs of skin cancer detection. This adaptability grants the model the ability to identify features and patterns in skin cancer images, potentially resulting in an improved overall performance.

NASNET Architecture: The NASNET, developed by the brilliant minds at Google Brain, boasts two essential functions: The Normal cell and the Reduction cell, as beautifully reflected in Fig. 4. At first, NASNET works its magic on a small set of data, and then it gracefully transfers its block to a larger dataset, resulting in a remarkable increase in mean average precision (mAP). To further enhance its already impressive performance, NASNET utilizes a modified drop path method, aptly named scheduled drop path. In the original NASNET architecture, the number of cells is not predetermined, and specific normal and reduction cells are strategically employed. Normal cells dictate the size of the feature map, while reduction cells elegantly reduce the feature map dimensions by a factor of two. The NASNET also boasts a control architecture, based on the ever-powerful Recurrent Neural Network (RNN), which

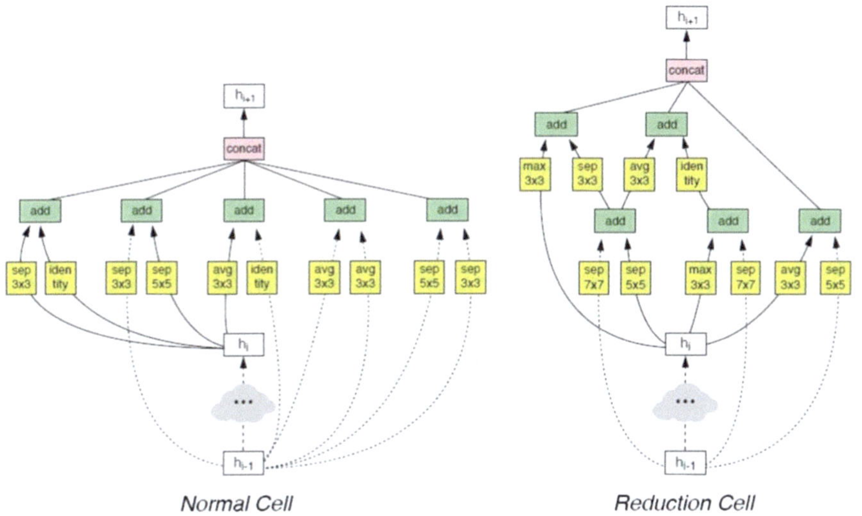

Fig. 4 NASNET architecture

brilliantly predicts the entire network structure using two initial hidden states which are

(16,64,64), (16,64,64), and (32,128,128) as ($s1 \times 1$, $e1 \times 1$, and $e3 \times 3$) filters, respectively.

$$ei, 3 \times 3 = ei * pct\, 3 * 3\ si, 1 \times 1 = \mathrm{SR}(ei, 1 \times 1 + ei, 3 \times 3).$$

In this performance, we made use of the Ada-max activation method for the main group. It's a well-known technique that has been proven to be effective in various scenarios. It's known for its ability to generalize the horizon less standard and, in some cases, produce better optimization results. To calculate the loss, we relied on the sparse categorical cross-entropy function. As for the inner layers of the model, we used the Softmax activation function, while the conclusion sub caste was activated using ReLU.

4 Result and Discussion

The performance of our proposed model is assessed using the following metrics evaluation referenced in [11].

Accuracy: It is a crucial metric that provides a holistic evaluation of a model's performance across all classes. It becomes especially valuable when all classes are equally important. On the other hand, refers to the ratio of correct predictions to the total number of predictions. This can be represented mathematically by using Eq. (1).

$$\text{Accuracy} = \frac{\text{TP} + \text{TN}}{\text{TP} + \text{TN} + \text{FP} + \text{FN}} \tag{1}$$

Precision: Precision is characterized as the rate of true cons to the total number of prognosticated cons, which encompasses both true cons and false cons. Mathematically, perfection can be defined as depicted in Eq. (2).

$$\text{Precision} = \frac{\text{TP}}{\text{TP} + \text{FP}} \tag{2}$$

Recall: The concept of recall is often described as the proportion of correct responses in relation to the total number of correct responses and incorrect responses. This can be easily represented in mathematical formula, as shown in Eq. (3).

$$\text{Recall} = \frac{\text{TP}}{\text{TP} + \text{FN}} \tag{3}$$

F1 Score: The F1 score is conceptualized as the harmonious mean of two crucial evaluation criteria, videlicet perfection and recall. Mathematically, the F1 score can be expressed as delineated in Eq. (4).

$$F1\text{ score} = \frac{2 * \text{Precision} * \text{Recall}}{\text{Precision} + \text{Recall}} \tag{4}$$

The plan also includes an early stop that allows us to run for more time and stop training when the standard of performance begins to decline. The model was trained 100 times. Early stop occurs with a time tolerance of 65. Our proposed model achieves 99.7% training accuracy, 97.7% recognition accuracy, and **97.04%** testing accuracy of data augmented with only 1,35,142 parameters.

Based on the data reflected in Figs. 5 and 6, it can be observed that the topic of affirmation and training is being discussed. Interestingly, both charts indicate that the show is not suffering from overfitting. Additionally, Table 3 provides a classification report for the proposed method, which shows impressive results. The model is achieving a high F1 score, as well as perfect precision and recall. Moving on to (reflected in Fig. 7), we can see a confusion matrix for our proposed model. It is worth noting that the model was able to accurately predict 6 out of the 6 classes with a sensitivity of 100%, while the remaining 3 classes had sensitivities of 97.7%, 98.8%, and 98%, respectively. Overall, the average sensitivity of the model is 98.4%, which is significantly better than other models with smaller parameters. These results demonstrate the effectiveness of our proposed approach.

Table 2 presents the evaluation metrics comparisons among all classes.

Reflected in Fig. 8 shows the confusion matrix of proposed model.

Tables 3 and 4 present a detailed comparison of our proposed demonstration with other state-of-the-art approaches. In the year 2020, Hari Kishan Kondaveeti and his

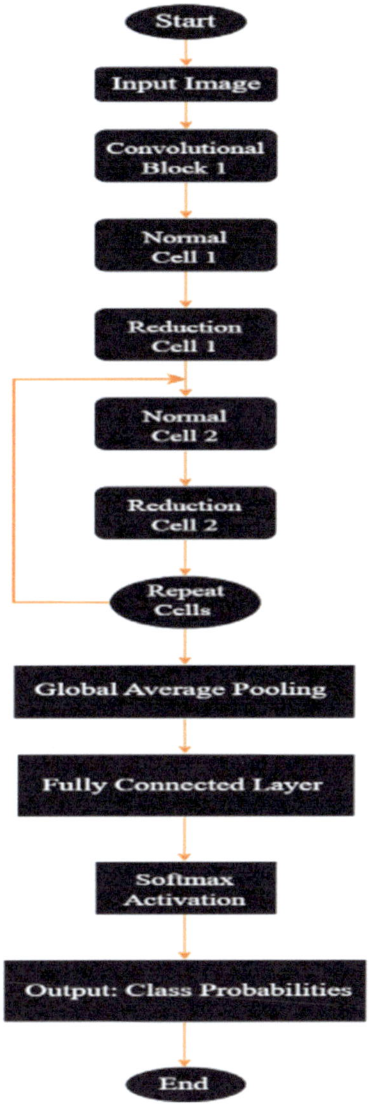

Fig. 5 Scalable architecture for NASNET

team utilized an advanced proficiency model with Resnet50 architecture, achieving an impressive delicacy score of 90 [12]. Similarly, in the same year, Zillur Rahman and his team utilized a combination of Xception, DenseNet, and Resnet models, resulting in a delicacy score of 85.8 [13].

Some other researchers have also focused on using CNNs, similar to the approach used in the study by Winkler et al. [7]. This study introduced an algorithm that could accurately classify skin cancer images as either carcinoma or benign. The researchers utilized convolutional neural networks and were able to achieve an

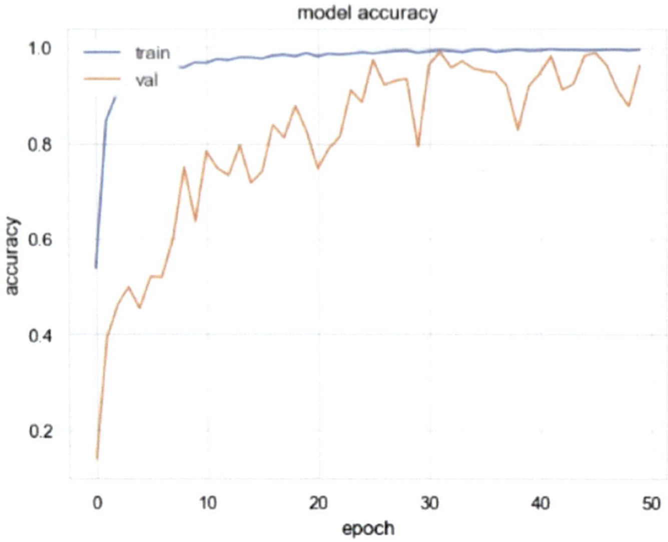

Fig. 6 Accuracy curve for the proposed model

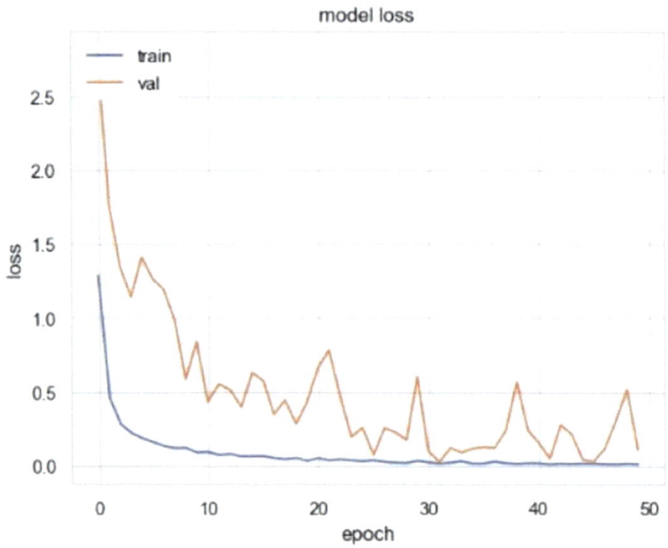

Fig. 7 Performance graph for the proposed model

Table 2 Classification analysis of proposed model

Class	Precision	Recall	F1 score	Support
Melanocytic nevi	0.95	0.72	0.79	1465
Melanoma	0.93	0.96	0.94	1238
Benign keratosis	0.92	0.94	0.95	1944
Basal cell carcinoma	0.95	1.00	0.98	1435
Actinic keratosis	1.00	1.00	1.00	1330
Vasc	1.00	1.00	1.00	1893
Dermatofibroma	1.00	1.00	1.00	1907
Avg macro	0.98	0.92	0.94	9332
Avg weighted	0.98	0.92	0.94	9902

Table 3 Model evaluation against deep learning approaches

Model	Acc.	Average weighted precision	Average weighted recall	Average weighted F1 score	No. of params
Inception-ResNet	0.8146	0.8122	0.8332	0.8112	4,93,41,123
EfficientNet	0.9321	0.9543	0.9987	0.9006	1,32,42,654
MobileNet	0.9432	0.9986	0.9675	0.9778	22,62,544
VGG16	0.9892	0.9458	0.9112	0.9346	9,64,321
Inception	0.9231	0.9543	0.9954	0.9489	2,18,75,443
CapNets	0.9908	0.9543	0.9225	0.9346	7,39,454
Proposed model	**0.9831**	**0.9881**	**0.9954**	**0.9234**	**1,32,111**

Bold highlights the performance metrics of the proposed model—showing an accuracy of 0.9831, average weighted precision of 0.9881, recall of 0.9954, F1 score of 0.9234, and a total of 1,32,111 parameters

impressive accuracy rate of 95%. In 2021, a researcher named R Raja Subramanian put forth a new convolutional neural network (CNN) that showed promising results with a delicacy score of 83.04 [14]. In a similar study, Alenezi used a support vector machine (SVM) trained on features extracted from popular networks such as AlexNet, DarkNet19, GoogleNet, SqueezeNet, Xception, and MobileNetV2. Their SVM achieved impressive delicacy scores of 96.15 and 97.15 on the ISIC 2019 and ISIC 2020 datasets, respectively. However, the proposed model outperformed the SVM in terms of perfection (0.98) and number of parameters used.

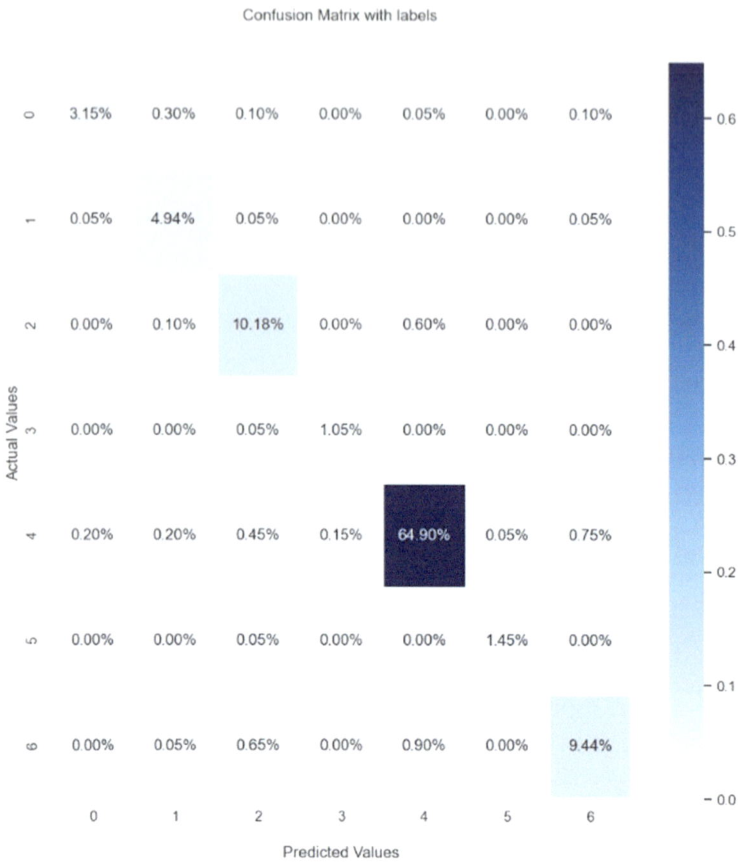

Fig. 8 Confusion matrix of proposed model

Table 4 Assessing the trained model against leading skin cancer classifiers

Ref. No.	Published in (year)	Classifier	Accuracy in %
1	2020	MobileNet	97
2	2021	CapNets	85.8
3	2019	MobileNet	90
4	2022	EfficientNet	95.02
5	2020	VGG16(ImageNet)	92.70
Current study	**2023**	**Proposed model NASNET**	**97.04**

Bold presents a comparative analysis, where the proposed NASNET-based model from the current 2023 study achieves a superior accuracy of 97.04%, outperforming previously referenced classifiers

5 Conclusion

Convolutional neural networks (CNNs) play a crucial role in enhancing the precision of outcomes in various tasks by adeptly discerning pertinent features and categorizing them into their respective classes. A profound grasp of their mechanisms unveils promising prospects for future exploration, particularly in the realm of medicine, where their potential is anticipated to be of paramount importance. Our proposed methodology leverages the NASNET architecture, yielding an impressive test accuracy of 97.04%. In our pursuit of refining results, we suggest delving into alternative robust architectures and engaging in meticulous fine-tuning, capitalizing on substantial processing and memory resources. The integration of segmentation techniques and the reduction of noise holds promise for further enhancing accuracy.

References

1. Geller AC, Swetter SM, Brooks K, Demierre MF, Yaroch AL (2014) Screening, early detection, and trends for melanoma: current status (2000–2006) and future directions. Cancer Epidemiol Biomarkers Prev 23(3):563–567. https://doi.org/10.1158/1055-9965.EPI-13-0540
2. Brinker TJ, Hekler A, Utikal JS, Schadendorf D, Klode J, Berking C, von Kalle C (2016) Skin cancer classification using convolutional neural networks: systematic review. Prev Med Rep 4:447–452. https://doi.org/10.1016/j.pmedr.2016.08.017
3. Esteva A, Kuprel B, Novoa RA, Ko J, Swetter SM, Blau HM, Thrun S (2017) Dermatologist-level classification of skin cancer with deep neural networks. JAMA 316(24):2402–2410. https://doi.org/10.1001/jama.2016.17248
4. Goswami T, Prajapati HB, Dabhi VK (2020) Skin disease classification from image—a survey. https://doi.org/10.1109/ICACCS48705.2020.9074232
5. Kumar TS, Annappa B, Dodia S (2023) Classification of skin cancer images using lightweight convolutional neural network
6. Shinde RK, Alam MS, Hossain MB, Md Imtiaz S, Kim J (2022) Squeeze-MNet: precise skin cancer detection model for low computing IoT devices using transfer learning
7. Winkler JK, Fink C, Toberer F, Enk A, Deinlein T, Hofmann-Wellenhof R, Haferkamp S (2020) Association between surgical skin markings in dermoscopic images and diagnostic performance of a deep learning convolutional neural network in melanoma recognition. JAMA Dermatol 156(10):1135–1141. https://doi.org/10.1001/jamadermatol.2020.2488
8. Waheed A, Waheed Z, Riaz F (2017) An efficient machine learning approach for the detection of melanoma using dermoscopic images, IEEE
9. Ferreira PM, Marques JS, Marcal ARS, Rozeira J (2013) PH2—a dermoscopic image database for research and benchmarking. Teresa Mendonca. https://doi.org/10.1109/EMBC.2013.6610779
10. Bassel A, Abdulkareem AB (2022) Deep learning approach
11. Subramanian RR, Achuth D, Kumar PS, Kumar Reddy KN, Amara S, Chowdary AS (2021) Skin cancer classification using convolutional neural networks
12. Alfi I, Rahman MM, Shorfuzzaman M, Nazir A (2022) A Non-invasive interpretable diagnosis of melanoma skin cancer using deep learning and ensemble stacking of machine learning models
13. Kondaveeti HK, Edupuganti P (2020) Skin cancer classification using transfer learning. https://doi.org/10.1109/ICATMRI51801.2020.9398388
14. Rahman Z, Ami AM (2020) A transfer learning based approach for skin lesion classification from imbalanced data

ESA-Based Hybrid Approach Using Support Vector Machine Classifier for Cancer Identification

Pinakshi Panda and Sukant Kishoro Bisoy

Abstract A significant factor leading to the increased worldwide mortality rate is the tardy detection of cancer. Early detection of cancer can be achieved by the development of computer-aided diagnosis models utilizing machine learning (ML). Microarray data has a distinct and specialized use in conducting diagnostics. Despite the small sample size, microarray data contains several dimensions of a patient's genetic information, including genes. Small sample sizes can be created by using microarray data as input for any machine learning model for classification without reducing dimensionality. Either a dimensionality reduction technique or a feature selection strategy must be utilized to boost the model's performance. The microarray dataset is optimized and its dimensionality is reduced through the utilization of the Elephant Search Algorithm (ESA) in conjunction with four feature selection procedures: Maximum Relevance Minimum Redundancy (MRMR), Information Gain (IG), Correlation Feature Selection (CFS), and Recursive Feature Elimination (RFE). A support vector classifier has been utilized as a classifier. Three cancer microarray datasets have been analyzed. The efficacy of the hybrid approach has been assessed and contrasted using many metrics, including precision, accuracy, and recall. The MRMR feature selection technique attained an accuracy of 99.1%, which is higher than former feature selection techniques.

Keywords Microarray data · Machine learning · CFS · ESA · SVM

P. Panda (✉) · S. K. Bisoy
Department of Computer Science and Engineering, C. V. Raman Global University, Bhubaneswar, Odisha, India
e-mail: pinakshipanda@gmail.com

S. K. Bisoy
e-mail: sukantabisoyi@cgu-odisha.ac.in

© The Author(s), under exclusive license to Springer Nature Singapore Pte Ltd. 2025

S. K. Udgata et al. (eds.), *Intelligent Systems*, Lecture Notes in Networks and Systems 1314, https://doi.org/10.1007/978-981-96-3797-3_25

1 Introduction

Approximately eighteen million individuals were diagnosed with cancer in 2018, according to the World Health Organization (WHO) and the National Cancer Institute. The lack of a competent diagnostic system led to the expected death of ten million people. The emergence of cancer classification as an influential scientific concern is now a reality. Several experts in machine learning have proposed many different kinds of cancer classification approaches.

Numerous genes have been obtained from a minute sample using microarray technology [1]. Unclassifiable genomes do exist. Microarray data is difficult to classify due to the large number of features, the presence of noise, and the computational complexity associated with such a small sample size. Approaches to high-dimensional datasets have their basic foundation on feature selection. By incorporating several optimization techniques into the feature selection approach, the overall amount of features was optimized. Cancer treatment success rates can be improved through detection at an early stage. Sustained progress in cancer treatment as well as research has culminated in improved survival rates for a variety of cancer [2]. A few of the numerous organizations that advocate for cancer treatment, prevention, access to care, and public awareness include the American Cancer Society (ACS), The World Health Organization (WHO), and other national cancer organizations [3]. There are frequently fewer samples available than there are characteristics to assess. When working with a limited sample size, it may be challenging to develop trustworthy and generalizable models. In addition to the aforementioned concerns, class imbalances can occur in cancer classification and other medical applications when one class. Methods for incorporating this worry into machine learning models must be developed. To overcome these obstacles, researchers frequently apply tactics such as dimensionality reduction, feature selection, and optimization algorithms to efficiently pick genes for cancer diagnosis [4].

MRMR, IG, CFS, and RFE feature selection methods minimize microarray dataset dimensionality. The ESA technique is employed for feature optimization, while SVM is utilized for classification. The suggested model is evaluated utilizing a range of parameters and cutting-edge machine learning algorithms.

2 Literature Survey

A penalized logistic regression (PLR) model for prediction outcomes was developed by [5] utilizing Recursive Feature Elimination (RFE) and univariate ranking. With remarkable efficacy in feature selection, test samples, and cross-validation, the proposed model achieved a 98.7% accuracy on microarray datasets. In 2016, Manasee Kurkure and Anuradha Thakare created a method using CT, PET, and X-ray imaging to identify and diagnose Lung Cancer, which gained popular interest. This streamlined the generation process was classification accuracy is 80%. Lung, leukemia, colon, SRBCT, and diffuse large B-cell lymphoma were all identified by Sun et al. [6] using KNN, C4.5, and SVM in conjunction with neighborhood entropy-based uncertainty measures. A novel model for detecting liver cancer was developed by Ghoniem [7] using advanced deep learning approaches, including Convolutional Neural Network (CNN), SegNet, and UNet. The model also employs an optimization technique known as Artificial Bee Colony optimization (ABC). The model was evaluated using two datasets, namely Radiopaedia and LiTS, which yielded superior outcomes [8] and demonstrated a gene selection method for microarray data classification with accuracies of 88.73%, 90.18%, 82.82%, 82.91%, and 91.76% for the datasets about Colon, Leukemia, SRBCT, Prostate Tumor, and Lung Cancer, respectively. Population Initialization based on Ranking Criteria (PIRC) by Ke et al. [9] creates populations utilizing NB, C4.5, GA, and ACO method across seventeen microarray expression datasets. Accuracy rates for various datasets, such as CNS, Colon, Leukemia_3C, Leukemia_4C, Huntington Disease, DLBCL, Lymphoma66 × 4026_3c, Prostate, SRBCT, Lung, Breast, Sarcoma, Mycloma, and Ovarian, range from 85.00% to 98.80%. SVM, ELM, and Naive Bayes provide an adaptive inertia weight teaching–learning model [10]. Meenachi and Ramakrishnan [11] proposed differential evolution and global optimal feature selection for cancer data classification model considering the machine learning (ML) approach, i.e., Decision Tree (DT) and optimization technique, i.e., ant colony optimization (ACO) on five datasets, i.e., DLBCL, Breast Cancer, Leukemia, SRBCT, Gisette datasets. In [12], FOA, ELM, MOO, and five filter techniques (IG, mRMR, ReliffF, CFS, and Fisher-score) were combined using a multi-filter and adaptive chaotic multi-objective forest optimization algorithm to create a hybrid strategy. This method was tested on nine datasets. Yan et al. [13] proposed an innovative feature selection model that utilizes K-nearest neighbor (KNN) and a wrapper feature selection technique FS_SSA. This method has its roots in Salp Swarm, Particle Swarm Optimization, and Genetic Algorithm. The proposed model was tested on five datasets: ALL-AML-4, Colon Tumor, Lyumphoma, MLL, and SRBCT. Findings indicated that ALL-AML-4 achieved an accuracy of 94.23%, Colon Tumor 82.09%, and Lymphoma 88.57%. Based on machine learning DT-SWO is a decision tree-based salp swarm optimization algorithm that Sarala et al. [14] developed. On DLBCL, Leukemia, Lung Cancer, and colon datasets, the system was 95%, 97%, 94%, and 98% accurate. A hybrid filter-wrapper strategy using (rMRMR), (MGWO), and ML algorithms including Random Forest (RF), Elastic Networks (EN), and Decision Tree (DT) on nine datasets have been developed by Alomari et al. [15] Based on Colon Tumor, CNS, ALL-AML, Ovarian Cancer, Lung Cancer, ALL-AML-3C, 4C, MLL, and SRBCT datasets, outcomes were obtained. CNS dataset: 95.86% acc, 95.33% pre, 91.97% rec, 95.46% F1-Score, 86.39% MCC. 99.38% acc, % pre, 100% rec, 100% F1- score, 100% MCC, AII-AML dataset, Ovarian dataset, Lung Cancer datasets. The AII-AML-3c dataset had 99.86% acc, 99.82% pre, 99.94% rec, 99.77% F1-Score, 97% MCC, while the AII-AML-4c dataset had 98.84% acc, 99.11% pre, 99.62% rec, 98.63% F1-Score, 93% MCC, as well as the MLL dataset had 99.90% acc, 99.89% pre, 99.95% rec, 99.9% F1-Score. Balakrishnan et al. [16] proposed an iSSA based on the Levy flight model for feature selection and SVM classifier

Table 1 Dataset description

Cancer name	Dataset sample	Attributes	Class
Lung cancer	62	2178	2
Breast cancer	32	570	2
Brain cancer	29	1072	2

on six datasets: OSCC, Ovarian Cancer, Breast Cancer, CNS, Colon Cancer, and Leukemia. The OSCC dataset had 85.7% accuracy, 85.7% F1-score, 90% recall, and 90.0% precision.

3 Proposed Hybrid Approach

In this section, the datasets used for evaluation and the methodology used in the current study endeavor are discussed.

3.1 Dataset Description

This strategy targeted three distinct forms of cancer: lung, breast, and brain. Table 1 presents various types of cancer that have been used.

3.2 Methodology

In the current research work four different feature selection algorithms such as MRMR, IG, RFE, and CFS are implemented. To the featured set the Elephant Search Algorithm (ESA) is considered as the optimizer. Finally, the SVM is applied as the classifier.

Maximum Relevance Minimum Redundancy (MRMR)

MRMR is a feature selection procedure used to select a subset of features from a larger set of features. Selecting the most useful features while minimizing redundancy is the objective [17]. This method is often used in machine learning and data analysis for better model performance, reduced overfitting, and more interpretable models.

Recursive Feature Elimination (RFE)

RFE is a feature selection method known for its wrapper-type characteristics. This suggests that the method entails integrating a machine learning algorithm, which is utilized at the center of the process, with Recursive Feature Elimination (RFE) assisting in feature selection.

Information Gain (IG)

IG is a commonly employed method for selecting features, which is quantified using Eq. (1). The measure quantifies the degree of interconnectedness between attributes and labels. A feature is meaningful if it provides a substantial amount of information.

$$IG(m, I) = K(m) - K\left(\frac{m}{I}\right) \quad (1)$$

$K(m)$ is the entropy of the nth feature, while $K(m|I)$ denotes the entropy of n after seeing the class level.

Correlation Feature Selection (CFS)

Highly correlated features exhibit strong linear dependence and have a similar impact as another dependent variable. When two traits are inextricably linked, one of them can be removed. The correlation between selection (CFS) approaches [18] is used to evaluate the relationship between attributes in sample data. For the dataset containing i features, denoted Dt, the metric score M_{SC} is computed using Eq. (2).

$$M_{\text{SC}} = \frac{\mathrm{i}\mu_{ic}}{\sqrt{i + i(i-1)\mu_{ii}}} \quad (2)$$

where μ_{ic} = Average correlation of a feature with the target class c and μ_{ii} = Average inter-feature correlation. Correlation features can be evaluated using Eq. (3).

$$\mu_{ab} = \frac{\sum(a-\bar{a})(\mathrm{b}-\overline{b})}{\sqrt{\left[\sum(a-\overline{a})^{2}\right]\left[\sqrt{\sum\left(a-\overline{a}\right)^{2}}\right]}} \quad (3)$$

Elephant Search Algorithm (ESA)

Search agents are classed as exploitation or exploration using the Elephant Search Algorithm (ESA). Female elephants establish search groups and study their surroundings, while male elephants operate as search agents. For exploration and exploitation, the ESA rarely uses a mix of male and female elephants [19].

Support Vector Machine (SVM) [20]

This approach aids classification and regression. With fewer training samples, you can create a superior hyperplane.

In this proposed model, five different feature selection algorithms such as MRMR, IG, RFE, and CFS are implemented. To the featured set the Elephant Search Algorithm (ESA) is considered as the optimizer. Finally, the SVM is applied as the classifier. The processed dataset is then partitioned in an 80:20 ratios into training and test datasets. The suggested work's step-by-step flow is shown in Fig. 1 and is described in-depth in the pseudocode.

Proposed Algorithm:

Step 1: Min-Max Normalize Data.

Step 2: Apply feature selection Techniques Individually

a. RFE

b. MRMR

c. IG

d. CFS

Step 3: The featured dataset is optimized using the Elephant Search Algorithm (ESA) method.

Step 4: Make use of a Support Vector Machine as the classifier.

Step 5: Performance Calculation

Step 6: Stop

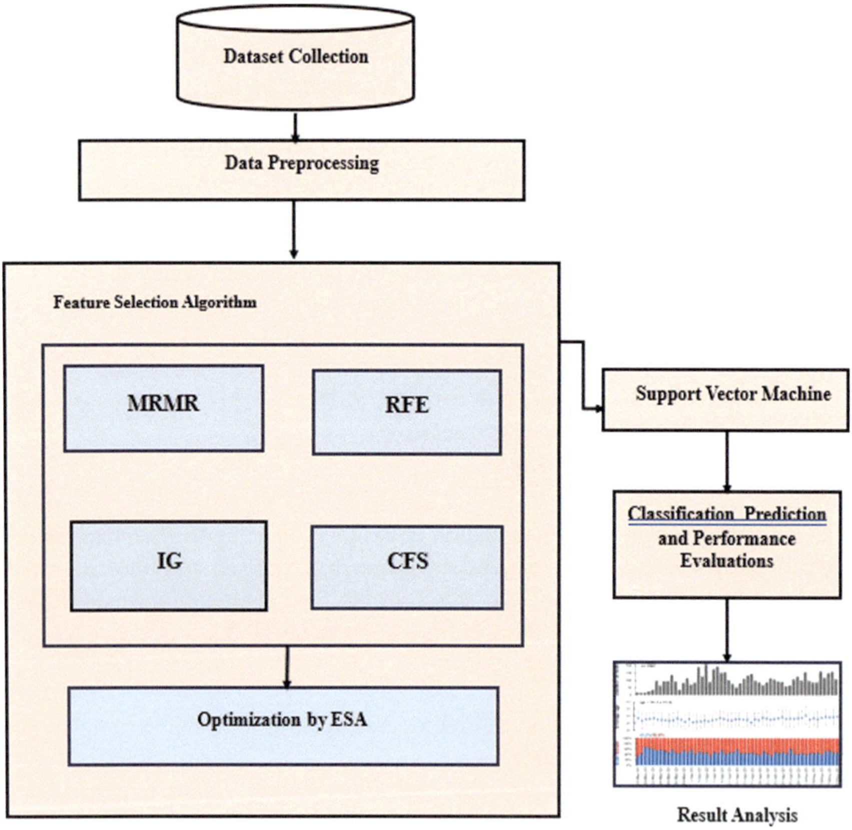

Fig. 1 Workflow of proposed work

4 Performance Evaluation

We have analyzed performance of hybrid algorithm using Python. An in-depth analysis of the suggested ESA-SVM model using various feature selection methods is displayed in Tables 2, 3, 4, and 5. Accuracy (A), Precision (P), and Recall (R) are the parameters used to evaluate performance. These parameters can be obtained by using the four parameters shown in Eqs. 4, 5, and 6:

$$\mathrm{A} = \left(\frac{\hat{T} + \hat{F}}{\hat{T} + \overline{\overline{T}} + \hat{F} + \overline{\overline{F}}} \right) \tag{4}$$

$$\mathrm{P} = \frac{\hat{T}}{\hat{T} + \hat{F}} \tag{5}$$

$$\mathrm{R} = \frac{\hat{T}}{\hat{T} + \overline{\overline{F}}} \tag{6}$$

Table 2 Performance of CFS-ESA-SVM

Dataset	Accuracy (%)	Recall (%)	Precision (%)
Lung cancer	96.4	90.2	95.3
Breast cancer	95.3	91.4	96.4
Brain cancer	97.2	92.3	94.3

Table 3 Performance of MRMR-ESA-SVM

Dataset	Accuracy (%)	Recall (%)	Precision (%)
Lung cancer	98.3	98.2	97.1
Breast cancer	99.1	98.7	96.2
Brain cancer	98.4	97.9	97.4

Table 4 Performance of RFE-ESA-SVM

Dataset	Accuracy (%)	Recall (%)	Precision (%)
Lung cancer	96.3	93.2	95.4
Breast cancer	95.4	91.2	90.1
Brain cancer	96.2	89.2	91.2

Table 5 Performance of IG-ESA-SVM

Dataset	Accuracy (%)	Recall (%)	Precision (%)
Lung cancer	97.3	92.2	94.3
Breast cancer	96.4	91.1	91.2
Brain cancer	95.3	89.4	92.1

The values representing true positive and false positive in the confusion matrix are symbolized as $\widehat{T}$, and $\hat{F}$ accordingly. True negatives and false negatives in the confusion matrix are denoted by $\overline{\overline{T}}$, and $\overline{\overline{F}}$, respectively.

The performance analysis of CFS-ESA-SVM hybrid models as shown in Table 2 such as for different cancer gene expression datasets where highest accuracy, precision, and recall for Brain Cancer is 97.2, 92.3, and 94.3. The performance analysis of MRMR-ESA-SVM hybrid models is presented in Table 3, specifically for various cancer gene expression datasets. The greatest accuracy, precision, and recall values for Breast Cancer are 99.1%, 98.7%, and 96.4%, respectively. The performance analysis of RFE-ESA-SVM hybrid models as shown in Table 4 such as for different cancer gene expression datasets where highest accuracy, precision, and Recall for Lung Cancer is 96.3, 93.2, and 95.4. The performance analysis of IG-ESA-SVM hybrid models is presented in Table 5. The models were evaluated on various cancer gene expression datasets, and the greatest accuracy, precision, and recall values for Lung Cancer were found to be 97.3%, 92.2%, and 94.3%, respectively. Figures 2, 3, and 4 shows the performance comparison of Lung Cancer, Breast Cancer, and Brain Cancer where accuracy of hybrid model MRMR is highest in case of Breast Cancer.

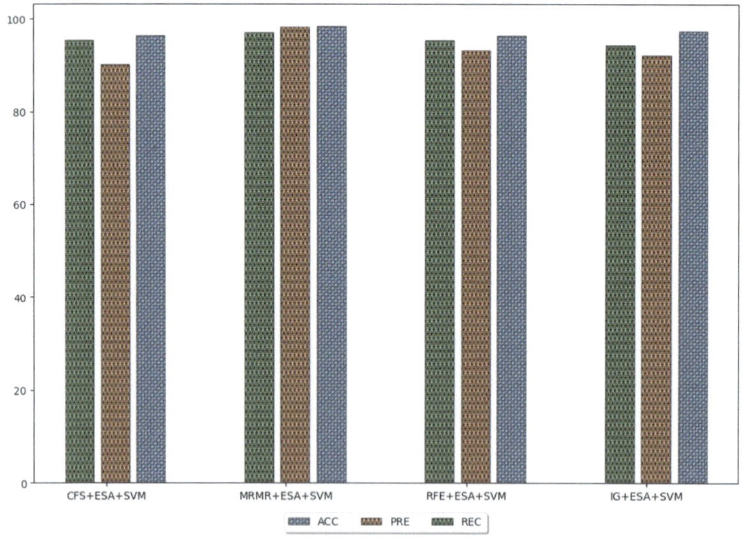

Fig. 2 Performance comparison of lung cancer

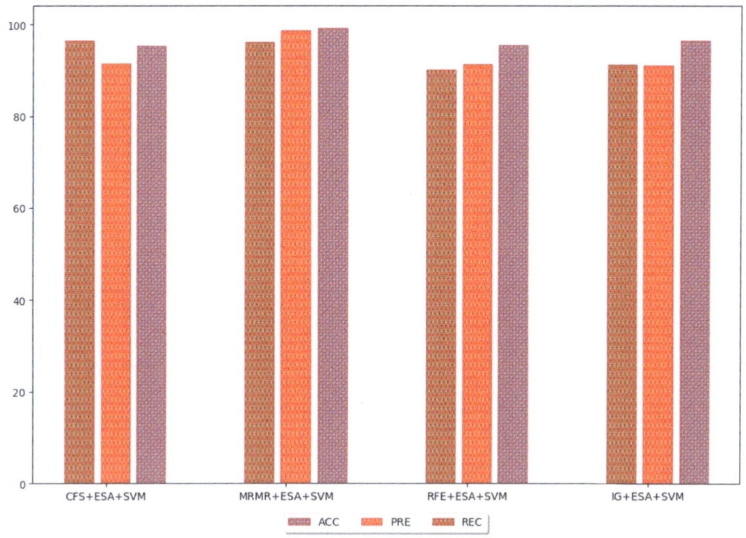

Fig. 3 Performance comparison of breast cancer

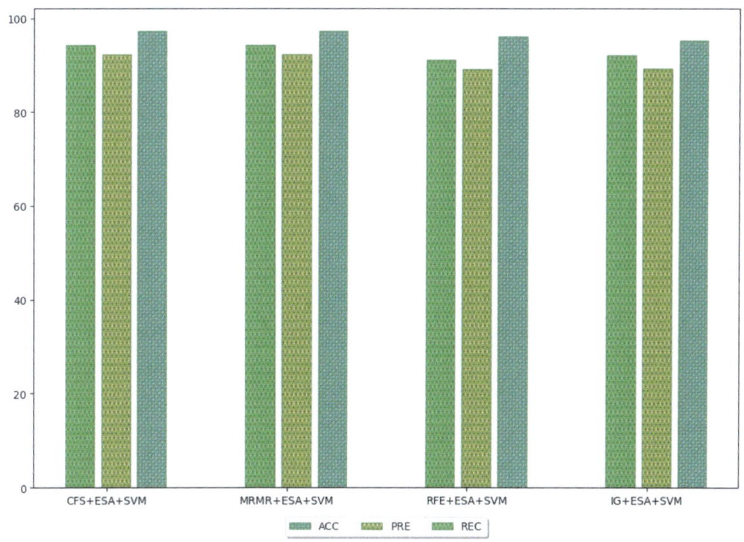

Fig. 4 Performance comparison of brain cancer

Table 6 Accuracy comparison with different feature selection algorithm

Models	Lung	Breast	Brain
SVM	63.6	64.2	70.1
CFS-SVM	74.2	77.3	80.6
RFE-SVM	71.2	78.2	79.1
IG-SVM	75.3	73.2	81.3
MRMR-SVM	75.4	80.2	82.3
CFS-ESA-SVM	96.4	95.3	97.2
RFE-ESA-SVM	96.3	95.4	96.2
IG-ESA-SVM	95.3	89.4	92.1
MRMR-ESA-SVM	**98.3**	**99.1**	**98.4**

Bold value indicates accuracy comparison with different feature selection algorithm

5 Critical Analysis

Table 6 compares accuracy with several feature selection algorithms, and three binary class datasets are being explored for the current study. The suggested approach with MRMR feature selection obtains 98.3% accuracy on the Lung Cancer dataset, while the proposed system with the MRMR feature selection algorithm achieves 99.1% and 98.4% accuracy on the Breast and Brain Cancer datasets, respectively. So, in this analysis, we find that the hybrid model using the MRMR feature selection technique has the maximum accuracy across all three datasets.

6 Conclusion

A cancer detection approach has been proposed that uses the Elephant Search Algorithm and an SVM classifier. The dataset is evaluated before for normalization. The ESA technique has been used in conjunction with SVM for classification. The SVM classification approach has been applied to the selected and refined dataset to assess the performance metric. The experimental research reveals that MRMR with ESA and SVM yields superior results when compared to other classifiers. The MRMR-ESA-SVM model achieved a classification accuracy of 99.1%.

Promoting novel ML and DL techniques is possible through the use of the dataset. Datasets with an extensive range of diversity must be analyzed for future research. To enhance the clinical decision support system, further validation trials need to be conducted in the future.

References

1. Takamiya M, Saigusa K, Dewa K (2021) DNA microarray analysis of hypothermia—exposed murine lungs for identification of forensic biomarkers. Legal Med 48:101789
2. Venkatesan C, Balamurugan D, Thamaraimanalan T, Ramkumar M (2022) Efficient machine learning technique for tumor classification based on gene expression data. In: 2022 8th IEEE International conference on advanced computing and communication systems (ICACCS) Vol 1, pp 1982–1986
3. Khalsan M, Machado LR, Al-Shamery ES, Ajit S, Anthony K, Mu M, Agyeman MO (2022) A survey of machine learning approaches applied to gene expression analysis for cancer prediction. IEEE Access 10:27522–27534
4. Osama S, Shaban H, Ali AA (2023) Gene reduction and machine learning algorithms for cancer classification based on microarray gene expression data: a comprehensive review. Expert Syst Appl 213:118946
5. Zhu J, Hastie T (2004) Classification of gene microarrays by penalized logistic regression. Biostatistics 5(3):427–443
6. Sun L, Zhang X, Qian Y, Xu J, Zhang S (2019) Feature selection using neighborhood entropy-based uncertainty measures for gene expression data classification. Inform Sci 502:18–41
7. Ghoniem RM (2020) A novel bio-inspired deep learning approach for liver cancer diagnosis. Information 11(2):80
8. Rostami M, Forouzandeh S, Berahmand K, Soltani M, Shahsavari M, Oussalah M (2022) Gene selection for microarray data classification via multi-objective graph theoretic-based method. Artif Intell Med 123:102228
9. Ke L, Li M, Wang L, Deng S, Ye J, Yu X (2023) Improved swarm-optimization-based filter-wrapper gene selection from microarray data for gene expression tumor classification. Pattern Anal Appl 26(2):455–472
10. Shukla AK, Singh P, Vardhan M (2020) An adaptive inertia weight teaching-learning-based optimization algorithm and its applications. Appl Math Model 77:309–326
11. Meenachi L, Ramakrishnan S (2020) Differential evolution and ACO based global optimal feature selection with fuzzy rough set for cancer data classification. Soft Comput 24(24):18463–18475
12. Nouri-Moghaddam B, Ghazanfari M, Fathian M (2023) A novel bio-inspired hybrid multi-filter wrapper gene selection method with ensemble classifier for microarray data. Neural Comput Appl 35(16):11531–11561
13. Yan C, Suo Z, Guan X, Luo H (2021) A novel feature selection method based on salp swarm algorithm. In: 2021 IEEE International conference on information communication and software engineering (ICICSE). pp 126–130
14. Sarala SPK, Chitraivel M, Raj RSP (2022) Decision tree based salp swarm optimization for multi medical data classification with feature reduction technique. Braz Arch Biol Technol, 64
15. Alomari OA, Makhadmeh SN, Al-Betar MA, Alyasseri ZAA, Doush IA, Abasi AK, Zitar RA (2021) Gene selection for microarray data classification based on gray wolf optimizer enhanced with TRIZ-inspired operators. Knowl-Based Syst 223:107034
16. Balakrishnan K, Dhanalakshmi R, Khaire UM (2021) Improved salp swarm algorithm based on the levy flight for feature selection. J Supercomput 77(11):12399–12419
17. Ding C, Peng H (2005) Minimum redundancy feature selection from microarray gene expression data. J Bioinform Comput Biol 3(02):185–205
18. Parhi M, Roul A, Ghosh B, Pati A (2022) Ioats: An intelligent online attendance tracking system based on facial recognition and edge computing. Int J Intell Syst Appl Eng 10(2):252–259
19. Coleto-Alcudia V, Vega-Rodríguez MA (2020) Artificial bee colony algorithm based on dominance (ABCD) for a hybrid gene selection method. Knowl-Based Syst 205:106323
20. Peng Y (2006) A novel ensemble machine learning for robust microarray data classification. Comput Biol Med 36(6):553–573

AI-Based Image Recognition for Early Diagnosis and Precise Classification of Pepper and Maize Leaf Diseases

Shaik Salma Asiya Begum and Hussain Syed

Abstract Pepper and maize, vital ingredients in many dishes, are continuously at risk from diseases that can affect crop yield and food security. AI is essential to 21st-century life. It is used for medical applications, communication, object detection, identification, agricultural and tracking. This paper uses deep learning to diagnose bell pepper plant diseases in large fields. Most bell pepper producers are unaware of bacterial spot disease. Disease propagation often reduces harvest output. Early detection of bacterial spot disease in bell pepper plants is essential for treatment. Quickly and effectively diagnosing pepper and corn crop diseases prevents productivity losses and optimizes crop management. This study examined how well CNNs and YOLOv7 classified common agricultural diseases. The suggested method includes image preprocessing, feature extraction, and classification. A large collection of pepper and corn images showing visual features and illnesses was assembled. Both models were trained and tested for illness classification. We used a large dataset of pepper and corn pictures modified with disease severity and backdrops. CNNs classified peppers at 99.4% and corn at 98.8%. Despite having poorer classification accuracy (97.6% for peppers and 99.92% for maize), YOLOv7 can detect unhealthy areas in real time and generate bounding boxes. This allows precise pesticide or fungicide application, decreasing environmental impact and optimizing efficacy. Both methods show promise for pepper and corn disease prevention, according to our research.

Keywords CNN · YOLOv7 · Plant disease identification · Object detection · Image pre-processing · Feature extraction and classification

---

S. S. A. Begum
Department of Computer Science and Engineering (AI&ML), Lakireddy Bali Reddy College of Engineering, Mylavaram, NTR District, AP, India

H. Syed (✉)
School of Computer Science and Engineering, VIT-AP University, Amaravati, India
e-mail: hussain.syed@vitap.ac.in

© The Author(s), under exclusive license to Springer Nature Singapore Pte Ltd. 2025

S. K. Udgata et al. (eds.), *Intelligent Systems*, Lecture Notes in Networks and Systems 1314, https://doi.org/10.1007/978-981-96-3797-3_26

1 Introduction

Ensuring the production of abundant quantities of top-notch agricultural products is crucial for the agricultural industry of every country. Hence, it is imperative to ascertain the causes that lead to the decline of plant well-being and thereafter develop strategies to alleviate them. The primary objective of this debate is to explore the various algorithms and approaches developed for analyzing plant stress, illnesses, and infections, which ultimately lead to a decrease in plant quality and growth. It is crucial to closely observe and track plant diseases due to their substantial impact on global food security and the potential devastating consequences for the livelihoods of agricultural workers and small-scale farmers. The prevalence of plant diseases is on the rise, presenting a substantial threat to both agricultural yield and quality. The prosperity and expansion of the agricultural sector are greatly contingent upon the quality of crops. The crop yield is increased because of careful cultivation and strategic application of fertilizers. Fertilization is necessary to facilitate the identification of plant health problems through both invasive and non-invasive techniques. To achieve accurate disease detection, it is necessary to analyze the chemical composition of the plant leaf using an image processing engine. This analysis can be done by non-invasive and intermit-tent approaches. The usefulness of non-intrusive approaches in real-time is limited by concerns over the time and effort required for training in each crop change [1]. Pathology, often known as phytopathology, is the rigorous examination of diseases induced by pathogens in plants. Plant diseases are caused by a type of pathogenic bacterium called a pathogen. Plant illnesses can be attributed to a diverse range of species, encompassing fungi, bacteria, viruses, viroid's, parasitic plants, and numerous others. Plants can be harmed by insects, animals, mites, and other pests when they consume them as a source of food. Figure 1a and b provide a visual representation of the causes behind the shedding of pepper plant leaves. The existing monitoring system is inadequate for the goal of early disease detection and prevention. Due of the presence of individuals with visual impairments or susceptibility to illusions, the conventional approach of visually examining plants to identify diseases is insufficient. This method may be susceptible to bias, optical illusions, and errors due to cognitive and psychological processes [2]. The agriculture sector is of crucial significance in the economics of most Asian nations. Roughly two-thirds of Asians rely on agriculture as their main means of subsistence. Maize (Zea mays L.) is widely recognized as one of the most significant staple crops worldwide, along with rice and wheat. For a significant number of individuals in South and Central America, it serves as the main source of dietary carbs. Corn is utilized by the United States as an alternative food source. Maize is a staple food for people in various Indonesian regions, such as Madura and Nusa Tenggara. Corn is cultivated not just for its starchy carbohydrates, but also serves as a feedstock for animals (in the form of corn cob and fodder), an ingredient in cooking oil, and a fundamental crop in flour production (such as corn flour or cornstarch) [3]. Conversely, maize is highly susceptible to pathogen infections. Pathogens can impede the growth of maize plants and perhaps hinder the attainment of maximum yields. The size of the attack on the maize plant is determined by the extent of its impact. Plant diseases commonly result in growth inhibition and aberrant responses in the plant's interior tissues and cells. For instance, certain plants may experience leaf desiccation or yellowing, while others may exhibit restricted growth or endure a change in color [4]. Both biological factors, including bacteria, fungi, viruses, and nematodes, and environmental variables, such as fluctuations in temperature and humidity or insufficient nutrition availability, can contribute to the onset of these diseases. There is evidence suggesting a connection between the factors that lead to various disorders. Figure 2a depicts a leaf in a state of good health, while Fig. 2b portrays a leaf that is afflicted with a disease. The symptoms might exhibit significant variations across different diseases or even between different illnesses. Several expert systems have been developed [5, 6] to auto-mate the process of diagnosing corn diseases. The publications utilize deep learning techniques. Deep learning algorithms frequently produce large-scale models, resulting in a space-intensive implementation, despite their high accuracy. Conventional machine learning research often comprises a strategy for extracting genetic features [7–10], or the utilization of expensive camera images [11]. In general, the proposed approach accomplishes the intended objective.

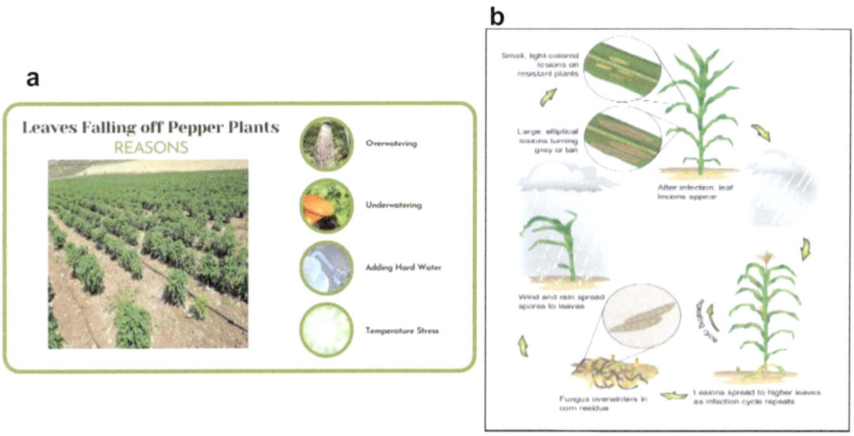

Fig. 1 **a** Reasons for falling off pepper leaves. **b** reasons for falling off corn leaves

Fig. 2 **a** Pepper and corn leaf in healthy. **b** pepper and corn leaf in un-healthy

The comparison with the state of the art in AI-based image recognition for early diagnosis and precise classification of pepper and maize leaf diseases revealed noteworthy advancements in real-time detection speed. Specifically, our study delved into the technical intricacies of YOLOv7, a state-of-the-art object detection algorithm known for its efficiency in real-time applications. We found that while YOLOv7 exhibited slightly lower classification accuracy rates compared to CNNs, its unparalleled capability to swiftly detect unhealthy areas in real-time sets it apart. Through our experimentation, YOLOv7 demonstrated the ability to generate bounding boxes around affected regions in pepper and maize plants with attention speed and accuracy. Expanding on the real-time detection speed of YOLOv7 could involve providing specific metrics such as inference time per image or frame per second (FPS) achieved, would enhance the clarity and comprehensiveness of the findings. Comparing it with other models or traditional methods, and discussing how this efficiency translates into practical benefits for agricultural stakeholders. Moreover, detailing any optimizations or architectural features that contribute to this speed could offer valuable insights into the model's performance characteristics.

The real-time detection capabilities of YOLOv7 offer practical advantages in agricultural settings, enabling prompt and precise intervention strategies such as targeted pesticide or fungicide application. By swiftly identifying diseased areas, farmers can minimize environmental impact by reducing unnecessary chemical usage while optimizing the efficacy of treatment measures.

1.1 Contributions

- A model that uses an improved CNN and YOLOv7 can distinguish between healthy pepper bell plants and those with bacteria in their leaf photos.
- A using image preprocessing approaches to aid in precise infected area segmentation and edge identification improves the suggested model's accuracy. To demonstrate the effectiveness of preprocessing, we evaluate the outcomes of the suggested technique with and without preprocessing operations.
- The evaluation of our algorithm is conducted using the publicly accessible PlantVillage dataset, which comprises leaf photographs of several plant species.
- An to prove its worth, the proposed model is compared to the state-of-the-art approaches in the existing literature.

1.2 Research Objective

The aim of this research is to enhance the categorization of diseases in pepper and maize by leveraging the combined capabilities of CNNs and YOLOv7. Our main goal is to create a strong and precise DL model that is capable of:

Accurately recognizing and categorizing various illnesses affecting peppers and corn: Our objective is to develop a model capable of reliably differentiating between healthy plants and unhealthy ones, while also accurately identifying specific types of diseases, especially in cases when there are many infections or complex visual symptoms. Surpassing or exceeding the effectiveness of current techniques: Our objective is to exceed the capabilities of conventional techniques and current deep learning methods in the categorization of pepper and corn diseases, thereby expanding the limits of what is possible. This entails attaining exceptional precision, rate of detection, and ability to apply to a wide range of field situations and variances in diseases. Enhancing the accessibility and understanding of data analysis and findings: In addition to offering disease classifications, our objective is to comprehend the reasoning behind the model's decision-making process. Through the process of visualizing feature maps and analyzing critical decision regions, our objective is to get significant insights into the visual attributes of various diseases. This will enable us to develop precise management plans and breeding programs that are specifically tailored to address these diseases.

Utilizing real-world applications possible: Our main goal is to turn this study into useful solutions for farmers. We want to build the model into mobile apps or field-based monitors so that farmers can use them to diagnose diseases in real time. This will make it easier to act quickly, improve how diseases are managed, and eventually improve the efficiency of farming and food security.

The goals of this study are to make it much easier to find diseases in peppers and corn, which will give farmers more power to protect their crops and help make agriculture more sustainable in future. The goal of this study is to be a key part of creating cutting edge AI-powered solutions for sustainable agriculture, which will help pepper and corn crops grow better in the future.

2 Literature Survey

Table 1 displays a selection of prior research conducted on the identification of plant leaf diseases.

3 General Architecture of the Proposed Model

The Structure of the Suggested Model in this study can be further categorized into the subsections depicted in Fig. 3.

Table 1 Highlights prior research focused on detecting plant leaf diseases

Ref No	Contribution	Techniques used	Results
[12]	A research-based approach to the quick and reliable identification of apple leaf diseases using real-world photographs of apple leaves is the main emphasis of this work	DCNNs and (ViTs)	The study outlines the results of a comprehensive examination of different models for predicting and detecting diseases in leaf plants. It was determined that the ViT-2 model distinguished itself from all alternative classifiers by attaining exceptionally high accuracy in tasks involving image classification
[13]	The paper proposes using deep learning methodologies, particularly CNNs, for the purpose of identifying and categorizing plant illnesses, with a special focus on bacterial spot in bell pepper leaf	The study suggests employing CNNs for the detection and recognition of bacterial spot in bell pepper leaves	The experiment described in the paper successfully attained a test accuracy of 96.78% in categorizing healthy and infected leaves by employing the proposed deep learning approach with Convolutional Neural Networks (CNNs)
[14]	The research presents an innovative image processing method has been developed specifically for the purpose of identifying and categorizing diseases seen in the leaves of pepper plants. The algorithm yields superior outcomes and possesses the ability to distinguish between robust and ailing plants	The paper utilizes machine learning methodologies, notably neural networks, to classify and predict diseases by employing soft computing approaches	The study displays the test results of 16 leaf photos, consisting of 10 unhealthy leaves and 6 healthy leaves
[15]	The research centers on utilizing a deep learning methodology to detect illnesses in bell pepper plants. The suggested approach employs YOLOv5 for the quick and precise detection of bacterial spot disease in bell pepper plants	The paper used the YOLOv5 model for training and evaluating the identification of illnesses on plant leaves. To train the model, run the train.py script and pass hyperparameters like image size, number of epochs, and batch size as inputs. The model's acquired weights are kept in a subdirectory located within YOLOv5	This study demonstrates that the YOLOv5 model outperforms competing models in detecting bacterial spot disease in initial bell pepper plants

(continued)

Table 1 (continued)

Ref No	Contribution	Techniques used	Results
[16]	The main goal of this research effort is to develop a more dependable approach for identifying diseases	K-means clustering, SGDM matrix generation, texture statistics computation, color co-occurrence method, Otsu method, artificial neural networks	A machine learning system that uses image analysis of plant leaves to accurately classify the health status of plants
[17]	Automated plant identification systems are the focus of this paper's discussion of a method for species classification	End-to-end CNNs, cross-dataset fine tuning, deep learning features	The method classified the Flavia Swedish Leaf UCI leaves plant village dataset with 95% accuracy
[18]	Develop an autonomous computer vision system that utilizes genetic algorithms (GA) and probabilistic neural networks (PNN) to accurately identify and diagnose diseases in plant leaves by analyzing their distinctive features	GA and PNN	The genetic algorithm demonstrated greater classification performance in comparison to the PNN, obtaining an overall accuracy of 97%, whereas the PNN achieved an accuracy of 94.75%
[19]	This paper presents a method for predicting paddy crop growth using image processing technology that is direct and reliable	Histogram estimation and analysis	Accuracy = 100%
[20]	This study introduces an innovative method for identifying and detecting plant diseases by utilizing phone devices to gather data	Image preprocessing,	Results of the classification achieved from the K-fold evaluation dataset. The values for rust are 0.85, 0.82, 0.7, and 0.95. The values for Septoria are 0.90, 0.85, 0.91, and 0.79. Tan location coordinates: 0.89, 0.73, 0.69, 0.78 Where (Disease AuC accuracy sensitivity specificity)
[21]	This study introduces a systematic approach for the automated identification and categorization of plant leaf diseases via artificial intelligence methodologies	Radial basis function neural network, bacteria foraging algorithm	The K-Means algorithm has an average specificity of 0.7914 The genetic algorithm has an average specificity of 0.8139 The BRBFNN has an average specificity of 0.856

(continued)

Table 1 (continued)

Ref No	Contribution	Techniques used	Results
[22]	This research presents a modified deep residual neural network approach for accurately identifying numerous plant diseases in real-world acquisition scenarios	Residual neural network	The algorithm outputs the findings for a picture of a plant leaf afflicted by Septoria. The sensitivity is 0.94 and the specification for BAC is 0.97
[23]	This study uses deep neural networks to categorize leaf images to detect plant diseases	Training and fine-tuning of neural networks	96.2% was achieved after fine tuning
[24]	This paper presents a way for creating a model that utilizes machine vision, image processing, and ML to accurately determine the health status of crops, with a particular focus on detecting Bakanae disease	SVM classifiers	This approach detects the disease with a high degree of accuracy
[25]	Research on the value of data mining methods for use in farming	ABC algorithm, ML techniques, SVM	Emphasizes the value of ML for identifying plant diseases

3.1 Cnn Backbone (Cspdarknet53)

The CSPDarknet53 architecture brought about a considerable improvement in object detection by enhancing the Darknet-53 backbone. The main innovation of this technology is the utilization of the Cross Stage Partial Network (CSP), which divides feature maps into smaller parts to enable simultaneous processing. This improves the flow of information and reduces the no. of parameters required. The combination of the Darknet-53 base network and the lightweight architecture results in cutting-edge performance while minimizing computational expenses. The efficiency of CSPDarknet53 makes it well-suited for situations with limited resources, allowing for robust object detection on mobile and embedded systems. The success of this technology is clearly demonstrated by its extensive use in advanced object detection models such as YOLOv4 and YOLOv7, establishing it as a prominent influence in the field.

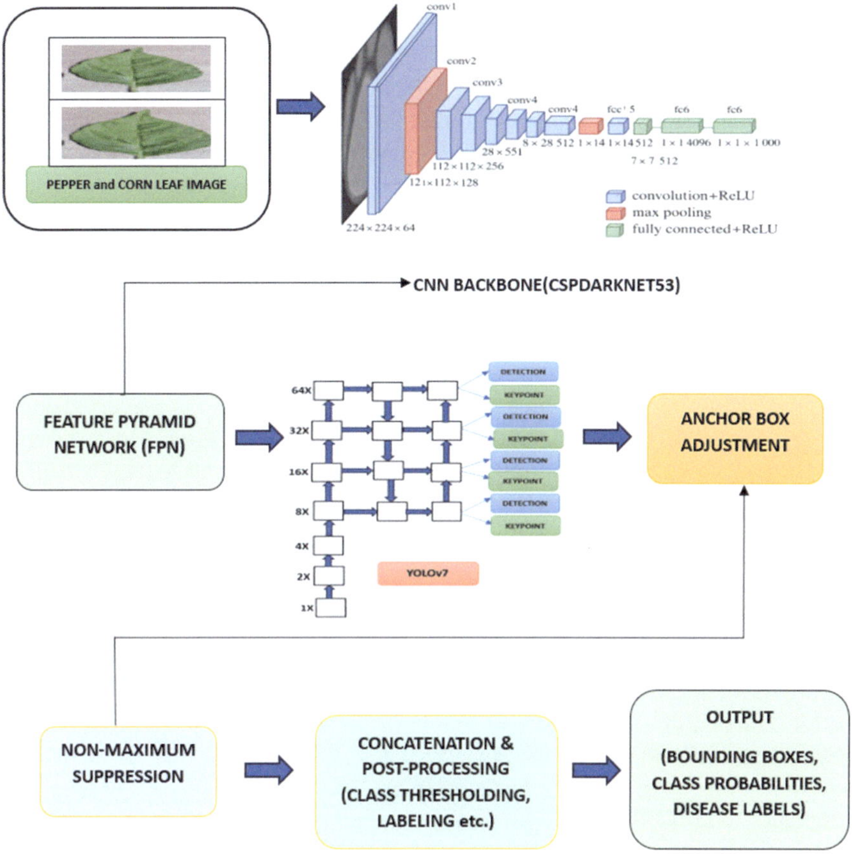

Fig. 3 General architecture of CNN-YOLOv7 model for pepper leaf detection & identification

3.2 *Feature Pyramid Network (FPN)*

The FPN transformed object detection by addressing the issue of detecting things at various scales. This is accomplished by creating a hierarchical structure of feature maps with different resolutions, allowing the model to accurately detect objects of any size. FPN leverages many levels of the CNN backbone to extract features, which are subsequently combined using top-down pathways and lateral connections. This process enhances low-level characteristics by incorporating high-level semantic information, resulting in enhanced identification of small objects. The fully convolutional design of this system ensures its efficiency and suitability for real-time applications, hence improving its adaptability. The FPN has had an important influence on the progress of object identification and remains an essential element of state-of-the-art models.

3.3 YOLOv7

YOLOv7, or "You Only Look One-level 7," denotes the 7th version of the YOLO series, which comprises a collection of real-time models designed for object recognition. YOLOv7, which was introduced by Alexey Bochkovskiy, features various architectural improvements compared to its previous versions. These changes include a robust CSPDarknet53 backbone and a Cross Stage Partial (CSP) connection, which enhances the flow of information. The model incorporates a FPN to effectively capture features at different scales, hence improving its capability to identify objects in a picture that range in size. YOLOv7 provides a range of model sizes, varying from smaller, computationally efficient variants such as Yolov7s to bigger, more precise variants like Yolov7x. YOLOv7's cutting-edge design achieves a harmonious combination of speed and accuracy, rendering it appropriate for many Computers vision is used in several applications, including autonomous vehicles, surveillance systems, and robots. YOLOv7 is widely used for object recognition jobs due to its adaptability, real-time performance, and exceptional accuracy.

3.4 Anchor Box Adjustment

Anchor boxes are essential in object detection as they provide as first estimations for the position and dimensions of potential items in an image. However, these boxes may not perfectly align with the physical items, requiring adjustments for precise detection. Here's where anchor box adjustment comes in. This process uses various techniques to refine the initial guesses and improve the model's ability to localize objects. Some common adjustment methods include:

Regression: The model predicts that the anchor box's dimensions and center coordinates will move away from the real object boundaries.

Classification: Based on how much each anchor box overlaps with the ground truth object, the model sorts them into groups. The confidence number and location of the anchor box are changed based on this classification.

Focal loss: This loss function lowers the importance of links that are easy to classify. This lets the model focus on harder examples and get better overall.

3.5 Non-Maximum Suppression

NMS is a significant part of object recognition models because it gets rid of unnecessary bounding boxes for the same object. With NMS, the model finds the most likely guess for each object and gets rid of any boxes that overlap or are not as sure. This leaves only one correct detection for each object instance. This step is very important for several reasons:

Reduces redundancy: NMS (Non-Maximum Suppression) ensures that each object is recognized only once by removing redundant boxes, resulting in a more concise and easily interpretable output from the model.

Improves accuracy: NMS aids in mitigating false positives that may arise from the presence of overlap-ping boxes. NMS enhances the overall accuracy of the object identification task by choosing the forecast with the highest level of confidence.

Enhances efficiency: The utilization of non-maximum suppression (NMS) reduces the computational costs associated with post-processing. This is achieved by requiring the model to compute just a single bounding box per item rather than multiple overlapping ones. NMS employs several methods to achieve this suppression, including IoU (Intersection over Union) thresholding, which retains only boxes that exceed a specific overlap threshold with the box of highest confidence. In essence, NMS plays a crucial role in object detection models by guaranteeing precise and effective identification of things in both photos and movies.

3.6 Concatenation and Post-processing (Class Thresholding, Labeling etc.)

Concatenation & post-processing: Concatenation and post-processing are essential stages in the object identification pipeline that enhance the high accuracy of the predictions generated by the model. Let us explore their roles in detail:

1. Concatenation: Integrates data from many levels of the network, usually sourced from the Feature Pyramid Network (FPN).

This procedure enhances the characteristic representations by incorporating information from many scales, hence helping the model to identify objects of varied sizes more efficiently.

Concatenation methods used mostly are up sampling and element-wise addition, allowing for smooth integration of features at many levels.

2. Post-processing: The raw predictions obtained from the network are refined to generate the ultimate detection outcomes.

This encompasses a multitude of methodologies, which include:

Class thresholding: It is a technique that filters out predictions with low confidence scores, effectively lowering the occurrence of false positives.

NMS: It is a technique used to remove overlapping bounding boxes that represent the same object, thus ensuring that only one accurate detection is retained for each instance of the object.

Labeling: Assigns categorical labels to the identified items according to their respective class probabilities.

Bounding Box refinement: Refines the bounding boxes to enhance alignment with the true boundaries of the objects, hence enhancing the accuracy of detection.

3.7 Output (Bounding Boxes, Class Probabilities, Disease Labels)

The output of an object detection model, such as YOLOv7, typically consists of three key elements:

1. Bounding Boxes: These provide the anticipated position and dimensions of the identified objects in the image. A bounding box is characterized by its coordinates, which consist of the top-left corner and bottom-right corner, as well as its measurements, which include the width and height.
2. Class Probabilities: These values represent the model's level of certainty in accurately categorizing the object contained within the specified area. The probability score indicates the probability that the object is a member of a particular class, such as "pepper leaf disease" or "healthy leaf."
3. Disease Labels: The identified items are assigned labels based on the class probabilities, and these labels represent the actual classifications. The model evaluates the likelihood scores against a predetermined threshold and assigns the appropriate disease label to each object.

The combination of bounding boxes, class probabilities, and disease labels furnishes a full depiction of the identified objects in the image. Users can gain insights into the location, size, and type of disease affecting the pepper leaves, empowering them to make informed choices regarding treatment or additional examination.

3.8 Model Development

CNN for Feature Extraction:

Architecture: Select or design a pre-trained CNN architecture that is appropriate for image classification, such as VGG16, ResNet50, or EfficientNet. The selected architecture must strike a balance between complexity, performance, and resource efficiency.

Fine-tuning: Adjust the pre-trained CNN using the pepper and corn disease dataset to extract pertinent features that are uniquely associated with disease identification.

Activation functions and pooling layers: Refine the CNN architecture's activation functions and pooling layers to better extract characteristics associated with certain disease patterns.

YOLOv7 for Object Detection and Classification:

Utilizing a pre-trained model: Make use of the pre-trained YOLOv7 model to leverage its skills in detecting objects.

Refinement: Adjust the YOLOv7 model using the annotated dataset of pepper and corn diseases to effectively identify both plants and precisely locate areas affected by diseases in the photos.

Optimization of the loss function: Modify the YOLOv7 loss function to provide higher importance to both exact identification of objects (pepper/corn) and correct classification of diseases.

Applications of YOLOv7-based pepper and corn leaf disease detection and identification models include:

Field Scouting: YOLOv7 models can be incorporated into portable devices or drones for field scouting, enabling farmers to promptly detect damaged pepper and corn plants and prioritize treatment interventions.

Automated Sorting and Inspection: YOLOv7 models can be utilized in post-harvest sorting and inspection systems to detect and eliminate infected pepper and corn fruits, thereby ensuring product quality and safety.

Disease Spread Monitoring: The utilization of YOLOv7 models enables the analysis of extensive datasets containing pepper and corn leaf images, facilitating the monitoring of disease spread, the identification of potential outbreaks, and the provision of information for preventive measures.

Research and Development: YOLOv7 models are used in research and development for assessing the efficacy of various illness management approaches and developing new disease detection algorithms.

YOLOv7 has become an effective tool for detecting and identifying pepper and corn leaf disease, providing real-time, precise, and flexible solutions to this significant agricultural problem. The device's capacity to concurrently detect several items and its exceptional precision make it invaluable for tasks such as field scouting, automated sorting, and inspection, monitoring the spread of diseases, and research and development purposes. The continuous advancement of YOLOv7 technology is expected to have an increasingly substantial impact on pepper and corn disease management, fundamentally transforming our approach to this crucial part of agricultural output. An analysis of the CNN and Yolov7 methodologies for the recognition and detection of pepper and corn illnesses.

4 Experimental Results and Discussions

Plant Village Dataset for CNN and YOLOv7: The Plant Village Dataset is widely used for training and assessing DL models designed for the detection and categorization of plant diseases. This tool is highly beneficial for academics and developers who are

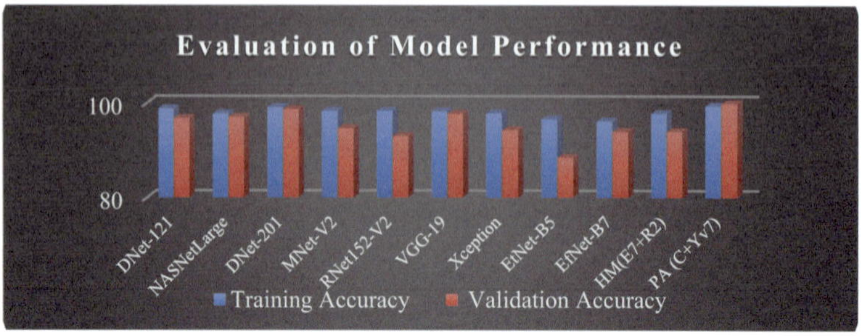

Fig. 4 Evaluation of model performance for the PlantVillage dataset

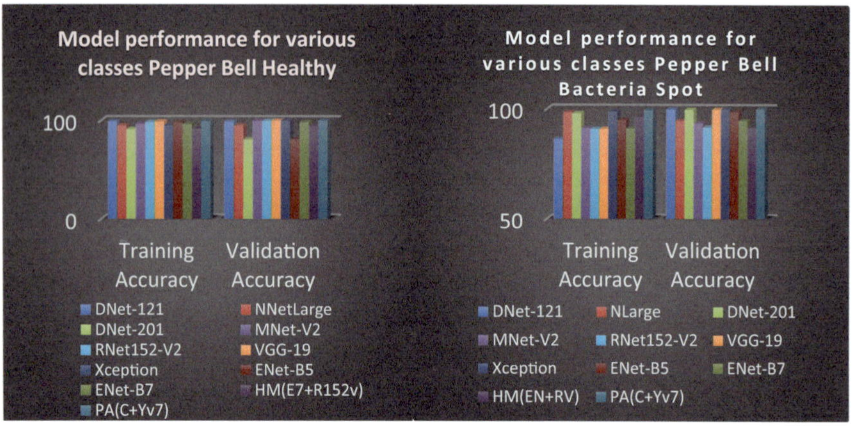

Fig. 5 Evaluation of model performance for various classes of the PlantVillage dataset

focused on enhancing agricultural practices and ensuring food security. The Plant Village Dataset can be utilized for the detection and classification of pepper and corn leaf diseases using CNN and YOLOv7 which is described below in Figs. 4 and 5.

Accurately assessing the disease's severity and developing tailored treatments depend on this YOLOv7. However, I will provide you with instances of earlier model assessments using the PlantVillage dataset and proposed model, as shown in Tables 2, 3 and 4.

5 Conclusion and Future Scope

Integrating CNNs for disease classification with YOLOv7 for object identification in maize and pepper crops has shown promise in improving accuracy and streamlining detection. This combination leverages CNN's feature extraction and YOLOv7's rapid detection to create a system that accurately identifies diseased areas, enabling

Table 2 Model evaluation using the PlantVillage dataset [26]

Models	Training		Validation	
	Accuracy	Loss	Accuracy	Loss
DenseNet-121	98.83	0.04	96.76	0.11
NASNetLarge	97.77	0.07	97.03	0.11
DenseNet-201	99.15	0.03	98.68	0.05
MobileNet-V2	98.34	0.06	94.68	0.17
ResNet152-V2	98.36	0.06	93.03	0.24
VGG-19	98.27	0.04	97.73	0.08
Xception	97.87	0.07	94.29	0.22
EfficientNet-B5	96.60	0.11	88.45	0.43
EfficientNet-B7	96.21	0.11	93.98	0.18
Hybrid-model (EfficientNetB7 + ResNet152V2)	97.81	0.06	94.00	0.21
Proposed approach	99.40	0.06	99.92	0.08

Table 3 Comparative performance of classification models across PlantVillage classes, including pepper bell bacterial spot

Models	Class	Training		Validation	
		Accuracy	Loss	Accuracy	Loss
DenseNet-121	Pepper bell bacterial spot	86.27	0.595	99.76	0.952
NASNetLarge		98.41	0.577	94.58	0.355
DenseNet-201		97.89	0.561	99.57	0.258
MobileNet-V2		91.31	0.468	93.49	0.255
ResNet152-V2		90.87	0.482	91.57	0.201
VGG-19		90.85	0.869	99.61	0.235
Xception		98.83	0.731	99.76	0.952
EfficientNet-B5		94.77	0.81	98.55	0.28
EfficientNet-B7		91.15	0.84	94.58	0.47
Hybrid-model (EfficientNetB7 + ResNet152V2)		96.34	0.15	91.47	0.58
Proposed approach		99.40	0.06	99.92	0.08

earlier interventions and more efficient farming practices. Further research is needed to address data limitations, explore other CNN architectures, and enhance model explainability. This integrated approach can protect crop futures and revolutionize precision agriculture. Both models were trained on a large dataset of maize and pepper images with varying disease severities and backgrounds. CNNs achieved 99.4% classification accuracy for peppers and 98.8% for maize. YOLOv7 showed slightly lower classification accuracy (97.6% for peppers and 99.92% for maize) but

Table 4 Performance distribution of the model across various PlantVillage classes, focusing on pepper bell (healthy)

Models	Class	Training		Validation	
		Accuracy	Loss	Accuracy	Loss
DenseNet-121	Pepper bell healthy	98.85	0.74	98.85	0.74
NASNetLarge		94.77	0.81	94.75	0.81
DenseNet-201		91.13	0.82	80.60	0.591
MobileNet-V2		96.34	0.13	99.74	0.009
ResNet152-V2		98.36	0.48	99.53	0.023
VGG-19		98.76	0.86	99.74	0.009
Xception		94.94	0.59	99.95	0.002
EfficientNet-B5		97.84	0.75	80.62	0.591
EfficientNet-B7		96.33	0.58	98.23	0.58
Hybrid-model (EfficientNetB7 + ResNet152V2)		91.48	0.86	94.45	0.47
Proposed approach		99.40	0.06	99.92	0.08

excelled in real-time detection and bounding box generation. Including technologies like the Internet of Things could further enhance farm management, ensuring robust and sustainable crop production. This strategy has great potential for creating automated systems in smart farms that continuously monitor crops, identify diseases in real time, and predict outbreaks. AI-driven classifications can help farmers maintain sustainable yields, optimize resources, and implement timely interventions.

References

1. Jackson RD et al (1986) Detection and evaluation of plant stresses for crop management decisions. IEEE Trans Geosci Remote Sens 1:99–106
2. Bock C, Poole G, Parker P, Gottwald T (2010) Plant disease severity estimated visually, by digital photography and image analysis, and by hyperspectral imaging. Crit Rev Plant Sci 29(2):59
3. Data A, Center IS (2014) Agricultural data and information system center
4. Nugraha U (2010) Sistem pakar diagnose penyakit pada tanaman jagung 803, 1, 1–9
5. Wiesner-Hanks T, Stewart EL, Kaczmar N, DeChant C, Wu H, Nelson RJ, Lipson H, Gore MA (2018) Image set for deep learning: field images of maize annotated with disease symptoms. BMC Res Notes 11(1):440
6. Zhang X, Qiao Y, Meng F, Fan C, Zhang M (2018) Identification of maize leaf diseases using improved deep convolutional neural networks, IEEE Access
7. Sankaran S, Mishra A, Ehsani R, Davis C (2010) A review of advanced techniques for detecting plant diseases, Comput Electron Agric 72, 1, 1–13, 2010. [Online]. Available: http://www.sciencedirect.com/science/article/pii/S0168169910000438
8. MacKenzie DJ, McLean MA, Mukerji S, Green M (1997) Improved RNA extraction from woody plants for the detection of viral pathogens by reverse transcription-polymerase chain reaction. Plant Dis 81(2):222–226

9. Khaled AY, Abd Aziz S, Bejo SK, Nawi NM, Seman IA, Onwude DI (2018) Early detection of diseases in plant tissue using spectroscopy–applications and limitations. Appl Spectrosc Rev 53(1):36–64
10. Badnakhe MR, Deshmukh PR (2011) An application of k-means clustering and artificial intelligence in pattern recognition for crop diseases. In: International conference on advancements in information technology
11. Rumpf T, Mahlein AK, Steiner U, Oerke EC, Dehne HW, Plümer L (2010) Early detection and classification of plant diseases with support vector machines based on hyperspectral reflectance. Comput Electron Agric 74(1):91–99
12. Boukabouya RA, Moussaoui A, Berrimi M (2022) Vision transformer based models for plant disease detection and diagnosis. In: 2022 5th International symposium on informatics and its applications (ISIA). IEEE, pp 1–6
13. Bhagat M, Kumar D, Mahmood R, Pati B, Kumar M (2020) Bell pepper leaf disease classification using CNN. In: 2nd International conference on data, engineering and applications (IDEA). IEEE, pp 1–5
14. Francis J, Anoop BK (2016) Identification of leaf diseases in pepper plants using soft computing techniques. In: 2016 Conference on emerging devices and smart systems (ICEDSS). IEEE, pp 168–173
15. Mathew MP, Mahesh TY (2022) Leaf-based disease detection in bell pepper plant using YOLO v5. Signal Image Video Process 1–7
16. Bagde S, Patil S, Patil S, Patil P (2015) Artificial neural network based plant leaf disease detection. Int J Comput Sci Mob Comput
17. Kaya A, Keceli AS, Catal C, Yalic HY, Temucin H, Tekinerdogan B (2019) Analysis of transfer learning for deep neural network based plant classification models. Comput Electron Agric 158:20–29
18. Islam MN, Kashem M, Akter M, Rahman MJ (2012) An approach to evaluate classifiers for automatic disease detection and classification of plant leaf. In: International conference on electrical, computer and telecommunication engineering, RUET, Rajshahi-6204, Bangladesh, pp 626–629
19. Singh A, Singh ML (2015) Automated color prediction of paddy crop leaf using image processing. In: 2015 IEEE Technological innovation in ICT for agriculture and rural development (TIAR), IEEE, pp 2432
20. Johannes A, Picon A, Alvarez-Gila A, Echazarra J, Rodriguez-Vaamonde S, Navajas AD, Ortiz-Barredo A (2017) Automatic plant disease diagnosis using mobile capture devices, applied on a wheat use case. Comput Electron Agric 138:200–209
21. Chouhan SS, Kaul A, Singh UP, Jain S (2018) Bacterial foraging optimization based radial basis function neural network (brbfnn) for identification and classification of plant leaf diseases: An automatic approach towards plant pathology. IEEE Access 6:8852–8863
22. Picon A, Alvarez-Gila A, Seitz M, Ortiz-Barredo A, Echazarra J, Johannes A (2019) Deep convolutional neural networks for mobile capture device-based crop disease classification in the wild. Comput Electron Agric 161:280290
23. Sladojevic S, Arsenovic M, Anderla A, Culibrk D, Stefanovic D (2016) Deep neural networks-based recognition of plant diseases by leaf image classification. Comput Intell Neurosci
24. Chung C-L, Huang K-J, Chen S-Y, Lai M-H, Chen Y-C, Kuo Y-F (2016) Detecting bakanae disease in rice seedlings by machine vision. Comput Electron Agric 121:404–411
25. Revathi P, Revathi R, Hemalatha M (2011) Comparative study of knowledge in crop diseases using machine learning techniques. Int J Comput Sci Inf Technol 2:2180–2182
26. Kumar Y, Singh R, Moudgil MR, Kamini (2023) A systematic review of different categories of plant disease detection using deep learning-based approaches. Arch Comput Methods Eng 1–23

Spatial Data Mining for Earthquake Significance Classification: Exploring Geospatial Insights

Gollapalli Ganga Srinivas, Gadde Madhukar, Gadde Maruti Mahesh, Uppuluri Bogesh, and Rajiv Senapati

Abstract Spatial data mining is the data collected from physical real-life locations containing map data, image data, and graph data. Spatial data mining focuses on extracting patterns and information from geographical datasets. In this paper, we have presented earthquake significance classification using Logistic Regression (LR), Decision Tree (DT), Random Forest (RF), Gradient Boosting (GB), Support Vector Machine (SVM), K-Nearest Neighbors (K-NNs), Gaussian Naive Bayes, Neural Networks (NNs), AdaBoost, and Bagging. Further, the performances of those algorithms were evaluated using $F1$-score, Recall, Precision, and Accuracy. From this analysis, it is observed that RF and Bagging performing better as compared to other techniques.

Keywords Earthquake · Machine learning · Spatial data mining · Classification

1 Introduction

Spatial data refers to the description of objects in the form of maps, GPS coordinates, and satellite imagery. It is utilized in fields such as logistics, environmental science, urban planning, transportation, and public health. Machine learning can be applied in spatial databases for solving various problems like healthcare,

G. G. Srinivas · G. Madhukar · G. M. Mahesh · U. Bogesh · R. Senapati (✉)
Department of CSE, SRM University, Amaravati, Andhra Pradesh, India
e-mail: rajiv.s@srmap.edu.in

G. G. Srinivas
e-mail: gangasrinivas_g@srmap.edu.in

G. Madhukar
e-mail: madhukarsaibabu_g@srmap.edu.in

G. M. Mahesh
e-mail: maheshchowdhary_g@srmap.edu.in

U. Bogesh
e-mail: bogesh_uppuluri@srmap.edu.in

© The Author(s), under exclusive license to Springer Nature Singapore Pte Ltd. 2025
S. K. Udgata et al. (eds.), *Intelligent Systems*, Lecture Notes in Networks and Systems 1314, https://doi.org/10.1007/978-981-96-3797-3_27

finance, retail, manufacturing, education, environmental science, and many more. It influences patient diagnosis in healthcare, guides stock market analysis in finance, optimizes marketing strategies in retail, refines manufacturing processes, improves education by predicting student success, assists environmental scientists in understanding climate change, optimizes transportation systems, aids in scientific research, enhances network security, influences pharmaceutical sales, informs insurance decisions, empowers research across disciplines, and plays a pivotal role in IoT applications. In contrast, spatial data mining leaves an indelible mark on areas such as urban planning, natural resource management, epidemiology, agriculture, emergency response, crime analysis, environmental monitoring, geological exploration, public health planning, climate change modeling, transportation, natural resource exploration, agricultural yield prediction, tourism and location-based services, wildlife conservation, infrastructure planning, transportation safety, historical preservation, satellite imagery analysis, and market analysis. These applications capitalize on spatial data's unique attributes to make more informed decisions, protect the environment, and ensure public safety. In contrast, spatial data mining is attuned to data rich in spatial or geographic components, delving deep into the spatial context to unveil patterns influenced by proximity, distance, and spatial relationships. In this paper, we have presented earthquake significance classification using machine learning approach. The followings are some of the contributions from this paper.

- This research underscores the importance of algorithm selection in solving classification problems, highlighting the impact of data-driven insights on safety and resource optimization.
- Random Forest exhibited the highest accuracy (0.898) and a well-balanced $F1$-score (0.75) for classifying earthquake significance, making it an efficient choice among the evaluated algorithms.
- The study underscores the broader applications of spatial data mining in real-world scenarios, ranging from disaster management to resource allocation, showcasing its versatility and impact in diverse fields.

2 Literature Review

The literature review provides an overview of studies that have explored data mining and machine learning. These summaries focus on aspects of the subject. In the particular study, referenced as [1], the authors address the challenges of dealing with large spatial datasets and introduce the Peano Count Tree (P-tree) structure for lossless and compressed data representation. They propose the P-tree-based Association Rule Mining (PARM) algorithm, comparing it with FP-growth and Apriori algorithms. The paper discusses potential applications in precision agriculture, resource discovery, and more. It offers a novel approach to extracting valuable patterns and rules from spatial data. In [2], the paper proposes GPU-based algorithms for efficient colocation pattern mining, overcoming the limitations of existing sequential and Map-reduce-based approaches. These algorithms include a novel cell-aggregate-based upper bound filter and two refinement algorithms. The paper also presents theoretical analyses and GPU profiling to optimize the algorithms, achieving significant speedup compared to CPU-based implementations. In [3], the paper discusses a novel approach to mine colocation patterns in urban spaces, taking into account the constraints of a road network. In [4], the review focuses on addressing the challenges associated with spatial data, including spatial autocorrelation, spatial heterogeneity, limited ground truth, and multiple scales and resolutions. It covers various application areas, such as earth science, urban informatics, geosocial media analytics, and public health. The paper categorizes spatial prediction methods based on the challenges they address, discusses their underlying assumptions, and compares their advantages and disadvantages. The goal is to help researchers and practitioners in various domains choose suitable techniques for solving spatial prediction problems and to identify research opportunities in the field. In [5], this paper addresses the challenges of discovering frequent associations in large databases by introducing innovative algorithms that aim to minimize I/O and computation costs through the use of efficient data structures and decomposition techniques. The paper presents six new algorithms that combine the above features, depending on the database format, decomposition technique, and search procedure employed. In [6], the concept of colocation patterns and rules, their application in various domains, and two primary categories of approaches for discovering these rules: spatial statistics-based methods and data mining-based methods. These methods are essential for understanding and analyzing the relationships between different types of spatial features and their proximity. In [7], the authors focus on characteristics expressed in terms of features that are closely associated with the clusters. In their context, a feature is defined as a closed curve (polygon) representing natural or man-made places of interest, such as lakes, parks, schools, golf courses, and shopping centers. In [8], in this paper we delve into the possibility and practicality of implementing a shared ontology to enhance the discovery of resources, within a European Spatial Data Infrastructure (SDI). The term 'SDI' refers to a system that includes metadata, spatial datasets, spatial data services, network services, agreements on data sharing, access, and coordination mechanisms-all designed to work. From a standpoint, an SDI is seen as a platform infrastructure for geospatial data and services based on open standards. The primary motivation behind this research lies in the fact that a European SDI's inherently multilingual and encompasses information communities. This means that there are perspectives when it comes to information ranging from those who produce metadata to providing applications and end users. The model described in this paper aims to represent these viewpoints reflecting the intricacies involved in discovering multilingual information, within an SDI. In [9], the journal tells the importance of analyzing co-patterns in both point and extended spatial features is highlighted. It emphasizes how this analysis can be applied in fields, like healthcare, urban planning, and environmental monitoring. These identified patterns play a role in decision-making processes and help address real-world challenges related to disease control, urban management, and environmental health. To bridge the existing gap, another journal aims to provide a review of spatial prediction methods. These methods are categorized based on the challenges they address. [10] explores the underlying assumptions of each method, delves into their foundations, and evaluates their advantages and disadvantages. The objective is to offer researchers a resource that aids them in choosing suitable techniques for their specific application domains. Various other-related ML techniques that were discussed in [11–26] are also applicable in the healthcare sector for different applications.

3 Proposed Methodology

3.1 Dataset

The dataset provides information about earthquake events which consists of various attributes. These attributes include the timestamp ('time') of each earthquake, its geographical coordinates in terms of latitude and longitude ('latitude' and 'longitude'), and the depth ('depth') at which the earthquake originated. The dataset also records the earthquake's magnitude ('mag') and the type of magnitude measurement method employed ('magType'). Information about the seismic monitoring network, the number of reporting seismometer stations ('nst'), the gap between these reporting stations ('gap'), and the closest distance to a reporting station ('dmin') is also available. Additionally, the dataset contains details about the signal quality represented as the root mean square ('rms'). Each earthquake event is uniquely identified by an 'id.' The dataset provides location descriptions ('place') and categorizes earthquake events by type ('type'). Geospatial accuracy is accounted for with attributes such as horizontal error ('horizontalError'), depth error ('depthError'), and magnitude error ('magError'). The dataset also captures the number of stations contributing to magnitude information ('magNst'). The reporting status of each earthquake event ('status') and the sources of location ('locationSource') and magnitude information ('magSource') are included. This dataset enables in-depth analysis and modeling of earthquake events with a wide range of attributes and information sources.

3.2 Problem Statement

To classify earthquakes into two categories: 'significant' and 'not significant.' The significance is determined based on the magnitude ('mag') of the earthquake. Earthquakes with a magnitude greater than or equal to 5.0 are considered 'significant,' while those below 5.0 are labeled as 'not significant.'

3.3 Earthquake Significance Classification Methods

Logistic Regression Logistic regression is a method that helps us determine the likelihood of an event belonging to a category. In this case, we are using regression to estimate the probability of an earthquake being significant, by considering factors. The mathematical equation for regression is expressed as follows:

$$p = \frac{1}{1 + e^{\beta_0 + \beta_1 x_1 + \beta_2 x_2 + \cdots + \beta_n x_n}} \tag{1}$$

where p represents the probability of an earthquake being significant, X represents the vector of input features (latitude, longitude, depth, etc.), β represents the vector of coefficients for each feature, and β_0 represents the intercept term. The logistic regression model computes the probability as: e is the base of the natural logarithm (approximately 2.71828).

Decision Tree Decision trees are structures that resemble trees and are used for making decisions and classifying tasks. In the context of this problem, a decision tree can be utilized to establish rules based on features in order to classify earthquakes as either significant or not.

Random Forest Random Forest is a technique that combines decision trees to enhance the accuracy and reliability of classification. A Random Forest comprises a collection of decision trees typically generated using Bagging.

Gradient Boosting Gradient Boosting is a technique that sequentially builds a group of decision trees focusing on improving the classification of instances that were previously misclassified using the following equation.

$$F(x) = F_{previous}(x) + \alpha(\times)Treek(x) \tag{2}$$

where F previous is the previous model, α is the learning rate, and tree k is the new tree.

Support Vector Machine (SVM) Support Vector Machines (SVMs) belong to a group of machine learning algorithms that are designed to identify the hyperplane, for separating instances into distinct classes. The hyperplane equation in a binary classification problem can be represented as:

$$p \times x + q = 0 \tag{3}$$

where 'p' is the weight vector that determines the orientation of the hyperplane. 'x' is the feature vector of an instance. 'q' is the bias term.

K-Nearest Neighbors The K-Nearest Neighbors (K-NNs) algorithm is an easy-to-understand classification method. It assigns a class label to an instance by considering the majority class, among its k-neighbors. Mathematical Formula: Let x be the instance to be classified. Let $x_1, x_2, \ldots, x_k$ be the k nearest neighbors of x. Let $y_1, y_2, \ldots, y_k$ are the class labels of the k-nearest neighbors. The class with the highest sum of indicator functions id given by

$$y = \text{argmax}(\text{sum}_i(1[y_i = c])) \tag{4}$$

Gaussian Naive Bayes Naive Bayes is an algorithm that utilizes probability and Bayes theorem to determine the likelihood of an earthquake falling into a category based on the distribution of its characteristics.

$$P(S|X) = (P(X|S) * P(S))/P(X) \tag{5}$$

where P(S|X) is the posterior probability that the earthquake is 'significant' given the features X. P(X|S) is the likelihood of observing the features X given that the earthquake is 'significant.' P(S) is the prior probability of an earthquake being 'significant.'

Neural Network Neural networks, which are a type of network commonly employed for classification purposes, are composed of interconnected nodes (also known as neurons). Mathematically, the operation at each node can be described as follows: For a hidden layer node h_j (j-th node in the hidden layer): $h_j = Activation(W_j1 * x_1 + W_j2 * x_2 + \cdots + W_jn * x_n + b_j)$ For the output layer node o_k (k-th node in the output layer): $o_k = Activation(W_k1 * h_1 + W_k2 * h_2 + \cdots + W_km * h_m + b_k)$ where W_ji represents the weight associated with the connection between the i-th node in the previous layer and the j-th node in the current layer. x_i represents the input feature $i.b_j$ and b_k represent the biases for the j-th node in the hidden layer and the k-th node in the output layer, respectively.

Adaptive Boosting (AdaBoost) AdaBoost, also known as Adaptive Boosting, is a technique used in machine learning that combines classifiers together to form a powerful classifier. AdaBoost is an ensemble learning algorithm that combines weak learners iteratively, assigning increasing weights to misclassified instances. The final prediction is a weighted sum of weak learners, enhancing model performance by focusing on difficult-to-classify data points.

Bagging Bagging, which stands for Bootstrap Aggregating, is a technique used in learning. It involves combining the predictions from models that have been trained on subsets of the data. The main goal of Bagging is to reduce variability and enhance the stability and accuracy of predictions by averaging or voting on the outputs of these models. Lets denote the training dataset as D and the base model as H. Bagging creates B bootstrap samples $D1, D2, \ldots DB$ and trains B base models $H1, H2, \ldots HB$. When making predictions, for instance x. For classification, Bagging combines the predictions using majority voting.

Let $p_i(x)$ represents the prediction made by base model H_i for instance x where $i = 1, 2, \ldots B$. The classification prediction $P(x)$ is determined by selecting the class c that has the sum of $p_i(x)$ over all base models. For regression, Bagging combines the predictions through averaging. Let $y_i(x)$ be the prediction made by base model H_i, for instance x where $i = 1, 2 \ldots B$. The final regression prediction $Y(x)$ is calculated as $(1/B)$ multiplied by the formula consists of adding up all the outputs $(y_i(x))$ from each base model.

4 Result Analysis

In this section, earthquake significance is predicted using Logistic Regression (LR), Decision Tree (DT), Random Forest (RF), Gradient Boosting (GB), Support Vector Machine (SVM), K-Nearest Neighbors (K-NNs), Gaussian Naive Bayes, Neural Networks (NNs), AdaBoost, and Bagging. The performance of these machine learning algorithms is evaluated through accuracy, precision, recall, and $F1$-score as follows:

$$\text{Accuracy} = (\text{TP} + \text{TN})/(\text{TP} + \text{FN} + \text{FP} + \text{TN}) \tag{6}$$

$$\text{Precision} = \text{TP}/(\text{TP} + \text{FP}) \tag{7}$$

$$\text{Recall} = \text{TP}/(\text{TP} + \text{FN}) \tag{8}$$

$$F1 - \text{Score} = 2(P \times R)/(P + R) \tag{9}$$

where TP, TN, FP, FN are True Positive, True Negative, False Positive, False Negative, respectively. The performance of those techniques is presented in Table 1 and Fig. 1.

Table 1 Result analysis

Algorithm	Accuracy	Precision	Recall	$F1$-score
Logistic Regression	0.864	0.78	0.55	0.65
Decision Tree	0.846	0.66	0.66	0.66
Random Forest	0.898	0.84	0.68	0.75
Gradient Boosting	0.891	0.84	0.65	0.73
SVM	0.872	0.86	0.52	0.65
K-Nearest Neighbors	0.879	0.79	0.63	0.7
Gaussian Naive Bayes	0.839	0.65	0.63	0.64
Neural Network	0.873	0.87	0.51	0.65
AdaBoost	0.88	0.8	0.63	0.7
Bagging	0.891	0.82	0.67	0.74

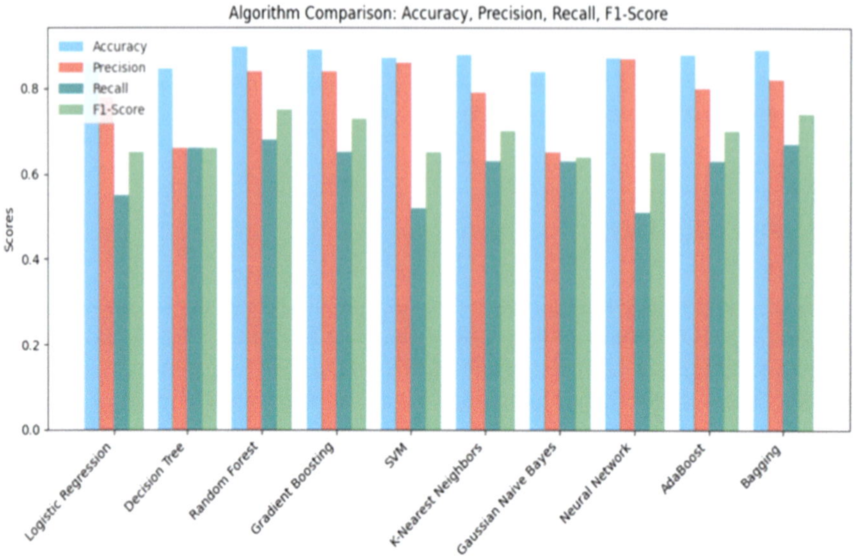

Fig. 1 Result analysis

Among the models evaluated, Random Forest stands out with the highest accuracy of 0.898, indicating strong overall performance. It achieves a balanced $F1$-score of 0.75, showcasing effective precision and recall. Gradient Boosting follows closely with an accuracy of 0.891, displaying high precision but a relatively lower recall of 0.65, resulting in an $F1$-score of 0.73. K-Nearest Neighbors also perform well, demonstrating accuracy at 0.879 and a balanced $F1$-score of 0.70. Notably, the Neural Network exhibits a relatively high accuracy of 0.864, but it's recall of 0.51 suggests a challenge in identifying positive instances. The Decision Tree, SVM, and Gaussian Naive Bayes show moderate performance, with accuracies ranging from 0.839 to 0.846 and $F1$-scores from 0.64 to 0.66. These metrics provide insights into the strengths and weaknesses of each model, guiding considerations for their specific applications in the context of the task at hand.

Random Forest appears to perform well with the highest accuracy (0.898) and reasonably balanced $F1$-scores for both significant and non-significant earthquakes. It can be considered as an efficient choice for this classification problem. Random Forest combines multiple decision trees to make accurate predictions and is less prone to overfitting.

5 Conclusion

The study presented here is about spatial data mining on earthquake significance using attributes like magnitude. Spatial data mining plays a pivotal role in revealing patterns and insights within data, facilitating informed decision-making across various industries. In this research, ten classification algorithms are used to tell the earthquakes are significant or not significant. The Decision Matrix shows the performance of each algorithm on metrics such as $F1$-score, accuracy, precision, and recall. Random Forest appears to perform well with the highest accuracy and reasonably balanced $F1$-scores for both significant and non-significant earthquakes. Hence, it is considered as best algorithm to find out if the earthquake is significant or not. In summary, this study underscores the efficacy of spatial data mining in tackling real-world challenges and emphasizes the critical role of algorithm selection in addressing classification problems. The research findings distinctly illustrate the significant impact on areas such as disaster response and earthquake forecasting.

References

1. Ding Q, Ding Q, Perrizo W (2008) Parm-an efficient algorithm to mine association rules from spatial data. IEEE Trans Syst Man Cybern Part B (Cybernetics) 38(6):1513–1524
2. Sainju AM, Aghajarian D, Jiang Z, Prasad S (2018) Parallel grid-based colocation mining algorithms on GPUs for big spatial event data. IEEE Trans Big Data 6(1):107–118
3. Zhou M, Ai T, Zhou G, Hu W (2020) A visualization method for mining colocation patterns constrained by a road network. IEEE Access 8:51933–51944
4. Jiang Z (2018) A survey on spatial prediction methods. IEEE Trans Knowl Data Eng 31(9):1645–1664
5. Zaki MJ (2000) Scalable algorithms for association mining. IEEE Trans Knowl Data Eng 12(3):372–390
6. Huang Y, Shekhar S, Xiong H (2004) Discovering colocation patterns from spatial data sets: a general approach. IEEE Trans Knowl Data Eng 16(12):1472–1485
7. Knorr EM, Ng RT (1996) Finding aggregate proximity relationships and commonalities in spatial data mining. IEEE Trans Knowl Data Eng 8(6):884–897
8. Smits PC, Friis-Christensen A (2006) Resource discovery in a European spatial data infrastructure. IEEE Trans Knowl Data Eng 19(1):85–95
9. Ge Y, Yao Z, Li H (2019) Computing co-location patterns in spatial data with extended objects: a scalable buffer-based approach. IEEE Trans Knowl Data Eng 33(2):401–414
10. Jiang Z (2018) A survey on spatial prediction methods. IEEE Trans Knowl Data Eng 31(9):1645–1664
11. Chakravadhanula AS, Kolisetty J, Samudrala K, Preetham B, Senapati R (2022) Novel decentralized security architecture for the centralized storage system in hadoop using blockchain technology. In: 2022 IEEE 7th international conference for convergence in technology (I2CT). IEEE, pp 1–4
12. Sahoo A, Senapati R (2022) A parallel approach to partition-based frequent pattern mining algorithm. In: Intelligent systems. Springer, pp 93–102
13. Sahoo A, Senapati R (2021) A novel approach for distributed frequent pattern mining algorithm using load-matrix. In: 2021 international conference on intelligent technologies (CONIT). IEEE, pp 1–5

14. Patro PP, Senapati R (2021) Advanced binary matrix-based frequent pattern mining algorithm. In: Intelligent systems. Springer, pp 305–316
15. Sahoo A, Senapati R (2020) A boolean load-matrix based frequent pattern mining algorithm. In: 2020 International conference on artificial intelligence and signal processing (AISP). IEEE, pp 1–5
16. Muttineni S, Yerramneni S, Kongara BC, Venkatachalam G, Senapati R (2022) An interactive interface for patient diagnosis using machine learning model. In: 2022 2nd International conference on emerging frontiers in electrical and electronic technologies (ICEFEET). IEEE, pp 1–5
17. Ganesh Prasad GSK, Ajay Chowdari A, Jona KP, Senapati R (2022) Detection of CKD from CT scan images using KNN algorithm and using edge detection. In: 2022 2nd international conference on emerging frontiers in electrical and electronic technologies (ICEFEET). IEEE, pp 1–4
18. Chaitanya Datta M, Senapati R (2022) An adoptive heart disease prediction model using machine learning approach. In: 2022 OITS international conference on information technology (OCIT). IEEE, pp 49–54
19. Senapati R (2023) A novel classification-based parallel frequent pattern discovery model for decision making and strategic planning in retailing. Int J Bus Intell Data Min 23(2):184–200
20. Chaitanya Datta M, Senapati R (2023) Hybrid clustering-based fast support vector machine model for heart disease prediction. In: International conference on machine learning, IoT and big data. Springer, pp 269–278
21. Raviteja KVNS, Kavya KVBS, Senapati R, Reddy KR (2023) Machine-learning modelling of tensile force in anchored geomembrane liners. Geosynth Int 1–17
22. Manda SC, Muttineni S, Venkatachalam G, Kongara BC, Senapati R (2023) Image stitching using ransac and bayesian refinement. In: 2023 3rd international conference on intelligent technologies (CONIT), pp 1–5. https://doi.org/10.1109/CONIT59222.2023.10205634
23. Yerramneni S, Nitya KSV, Nalluri S, Senapati R (2023) A generalized grayscale image processing framework for retinal fundus images. In: 2023 3rd international conference on intelligent technologies (CONIT), pp 1–6. https://doi.org/10.1109/CONIT59222.2023.10205834
24. Samudrala K, Kolisetty J, Chakravadhanula AS, Preetham B, Senapati R (2023) Novel distributed architecture for frequent pattern mining using spark framework. In: 2023 3rd International conference on intelligent technologies (CONIT), pp 1–5. https://doi.org/10.1109/CONIT59222.2023.10205903
25. Chaitanya Datta M, Venkaiah Chowdary B, Senapati R (2023) Multi disease prediction using ensembling of distinct machine learning and deep learning classifiers. In: International conference on soft computing and its engineering applications. Springer, pp 245–257
26. Venkaiah Chowdary B, Chaitanya Datta M, Senapati R (2023) An improved cardiovascular disease prediction model using ensembling of diverse machine learning classifiers. In: 2023 OITS international conference on information technology (OCIT), pp 329–333. https://doi.org/10.1109/OCIT59427.2023.10430692

Diabetes Prediction Using Optimized Stochastic Gradient Descent with Multilayer Perceptron

B. Venkaiah Chowdary and Rajiv Senapati

Abstract Predicting diseases early plays an important role in improving healthcare and helping us prevent harmful health conditions earlier. Diabetes is the most contagious disease in the world, and serious complications include diabetic retinopathy, kidney problems, and heartburn. Machine learning (ML) models are used to collect relevant data from datasets. Data discovery relied heavily on outlier detection (OD). Identifying anomalous data within a given dataset is the main goal of outlier detection. We used optimal stochastic gradient descent (SGD) and multilayer perceptron (MLP) with outlier detection to create a model for predicting diabetes. This work goes into more detail about the proposed model. This study looks at the three attribute selection methods—backward elimination (BE), forward selection (FS), and stepwise regression (SR)—for choosing the features from preprocessed data. As a technique for identifying outliers, a radial basis function (RBF) system is used to extract misclassified samples. For high-efficient classification of data, the ML technique called MLP with maximum SGD was described. The result shows that using attribute selection and radial basis function improves the efficiency of diagnosing diabetes in terms of accuracy, sensitivity, and specificity.

Keywords Diabetes · RBF · MLP · Attribute selection

1 Introduction

The other name for diabetes is diabetes mellitus, and it is class of metabolic diseases in which the pancreas fails to produce insulin at the time metabolism [1]. Insulin is a hormone produced by pancreas which digests glucose and sends it through the bloodstream to cells in body to produce energy. Because pancreas cannot produce

B. Venkaiah Chowdary · R. Senapati (✉)
SRM-University, Amaravati, Andhra Pradesh, India
e-mail: rajiv.s@srmap.edu.in

B. Venkaiah Chowdary
e-mail: venkaiahchowdary_b@srmap.edu.in

© The Author(s), under exclusive license to Springer Nature Singapore Pte Ltd. 2025

S. K. Udgata et al. (eds.), *Intelligent Systems*, Lecture Notes in Networks and Systems 1314, https://doi.org/10.1007/978-981-96-3797-3_28

insulin, the hormone that digests glucose, blood sugar levels rise, causing hyperglycemia. A typical health symptom includes increased hunger, repeated urination, and elevated blood pressure. Patients require insulin therapy because some forms of diabetes cannot be successfully treated with oral drugs alone. Both metabolic ketoacidosis and nonketotic hyperosmotic coma are significant complications [2]. Diabetes will be classified as either Type 1, Type 2 or Type 3. Type 1 diabetes is autoimmune disorder. When this happens, body destroys primary cells for producing insulin to absorb sugar and produces energy. This type of diabetes will lead to obesity and Body Mass Index (BMI) [3]. Type 1 diabetes will develop during childhood and adolescence. People have lived with type 1 diabetes all their lives on insulin shots. People with diabetes frequently need to follow a doctor-recommended exercise and fitness program. Type 2 diabetes commonly impacts obese adults. In this type, body either refuses to produce insulin or fails to produce it. Type 2 usually happens in middle-aged group. The disease occurs because of lack of additionally presented hypoglycemic agents. This is known as "non-insulin-dependent polygenic disease", and the general reason is ultimate weight. Prehypertension is associated with mass, obesity, and diabetes [4]. Gestational diabetes or type 3 diabetes: This is high blood sugar caused by hormonal changes at the time of pregnancy [5]. There are many other possible causes of diabetes, including bacterial or viral infections, toxic or chemical components in foods, autoimmune reactions, obesity, an unbalanced diet, lifestyle changes, poor eating habits, environmental pollution, and unhealthy lifestyle habits. Various cardiovascular diseases, kidney diseases, retinopathy, foot ulcers, etc., are caused by diabetes. Outlier detection a critical information investigation. The most objective of exception discovery is to distinguish bizarre or irregular information from given dataset. This is often curious region for data mining and includes unused and uncommon finding structures from dataset [6]. The detected data in the data cluster varies from information for different sensor nodes in network. Identifying such outliers is very complex. Machine learning methods will be used to predict outliers and misplaced information. Outlier detection has also been applicable to many various areas, including medical diagnostics, networked sensors, and processing of images, astronomy, biology, and speech recognition.

One way to classify machine learning methods is their use in supervised, unsupervised, and reinforcement learning contexts. The model utilizes supervised learning algorithms to test the accuracy of several popular ML models. Supervised learning methods try to detect patterns from existing information and detect recent outcomes based on early learning. ML methods are designed to detect available information. Different types of models include those based on functions, rules, trees, instances, probabilities, etc. Various ML methods have been presented utilizing different data mining methods to support clinical professionals. The data preprocessing phase should be performed to mitigate issues that arise when processing medical record information containing large amounts of data. Attributes are irrelevant, missing values, and prone to outliers. Efficient data preprocessing is necessary to increase the accuracy of results and minimize calculation time. The main research contributions of this work are as follows:

- A diabetes prediction model with outlier identification and stochastic gradient descent (SGD) with multilayer perceptron (MLP) is the main focus of the study presented in this study.
- To confirm effective information preprocessing methods, this study analyzes three attribute selection methods: backward elimination (BE), forward selection (FS), and stepwise regression (SR).
- The radial basis function (RBF) has been used to detect outliers and improve the accuracy, sensitivity, and specificity of proposed diabetes diagnosis model.

The remaining sections of this paper are structured as follows: Sect. 2 summarizes literature review. Section 3 presents described methodology of diabetes prediction model. Section 4 presents results, discussions, and finally it concludes with Sect. 5.

2 Related Work

Sakr et al. [7] used random forests (RFs) as ensemble learning to predict person hypertension risk. The algorithm is compared with Bayesian networks, locally weighted Naive Bayes (NB), LogitBoost, neural networks, and support vector machines. As a result, it is found that the random forest (RF) method reached the excessive area under the curve (AUC) (up to 0.93%) compared to different systems if applied to Henry Ford Health System dataset. T.monika Singh, Rajashekar shastry et. al. state that ailments like blindness, blood pressure, heart stroke, kidney disorder, and nervous damage are at risk for people with diabetes to develop. Their study used the Pima dataset and achieved 65% accuracy using Naive Bayes algorithm. They recommend testing described method on higher datasets. Anderson et al. [8] described predictive method based on an ensemble of classifiers for detecting diabetes progression using Electronic Health Record (EHR) information. O'Reilly et al. [9] provide a more robust algorithm in deriving principal components for training sets containing anomalies. Then distributed type of method is explained. In this case, every node in network will repeat centrally organized resolution by changing few matrices within neighboring nodes. Experimental calculations for synthetic as well as real datasets prove higher implementation of described method compared to principal component analysis (PCA) and another anomaly finding methods. Furthermore, it has been shown that distributed anomaly finding models can gather good approximations of centralized models in various network infrastructures. Nai-Arun et al. [10] implement a model to classify diabetes risk. To achieve this goal, the author assumes four of his famous ML classification methods: DT, ANN, LR, NB. Bagging and boosting techniques are used to enhance robustness of designed model. Observation output shows that random forest method gives the best outputs among all used methods. Liu et al. [11] describe recent OD method to deal with information by incomplete labels and to include limited examples in training. This describes the method in two phases. The initial phase generates a pseudo-training dataset by evaluating probability values for every sample depending on local behavior. The late level will produce the integrated probability values and restricted anomalous examples into a support vector data description (SVDD)-based learning model to create highly approximate classifier for global OD. More observations are seen on real datasets and show the described method that will attain a best trade-off between detection rates and false positive rate as compared to new outlier finding approaches. Sun et al. [12] apply an RF (ensemble learning) model to detect blood pressure transition points. EHR data were collected from 1294 hypertensive patients at Vanderbilt University Medical Center. These results demonstrated that the random forest method will approximately detect hypertension transition points and should be used in blood pressure for the management of personalized hypertension plans. Upadhyaya et al. [13] have described outlier finding techniques based on category approach. They identified key assumptions used in every approach that could distinguish between normal and abnormal behavior. Yoon et al. [14] explain k-means clustering on entire dataset and Cubic Clustering Criterion (CCC) to find exact value of k. CCC is method for calculating volume of clusters scored using Monte Carlo model. After clustering process, Domain Expert looks for exterior and interior outliers in clustering outputs. Outer outlier information spots that lie farther than the rest of the cluster, and inner outliers are data points that lie further within the cluster. Their method has only been applied to software measurement information and requires subject matter experts for interpretation. Various other-related ML techniques that were discussed in [15–30] are also applicable in the healthcare sector for different applications.

Several machine learning-based prediction models have been proposed in the literature. In this paper, we have used the combination of optimal stochastic gradient descent (SGD) and multilayer perceptron (MLP) with outlier detection for developing a predicting framework for diabetes.

3 Diabetic Prediction Model Using Outlier Detection

In the block diagram of proposed diabetes prediction model using outlier detection, MLP with SGD is represented in Fig. 1. The methodology described in this work consists of several different phases: preprocessing, attribute selection, detection of outliers or misclassified instances, and classification. The radial basis function network (RBF) is implemented here as the outlier detection model (OD). The data generated by the OD module is then classified using MLP with SGD. If categorization method is completed, a series of observations are run to measure the outputs of implemented system. The Pima Indian Diabetes Dataset (PIDD) [31] used in this study consists of 768 cases with 8 attributes. This dataset is having the data of 268 positive and 500 negative instances. Each instance consists of eight attributes, which all are numeric. Data attributes are determined by each step of data preprocessing. During data preprocessing, the median and mean of each attribute (skin thickness, insulin, blood pressure, glucose, and BMI) are used to fill in the missing values. The related information for features used as training as well as system test information. Every diagnostic of diabetes health record information is verified, and it is done

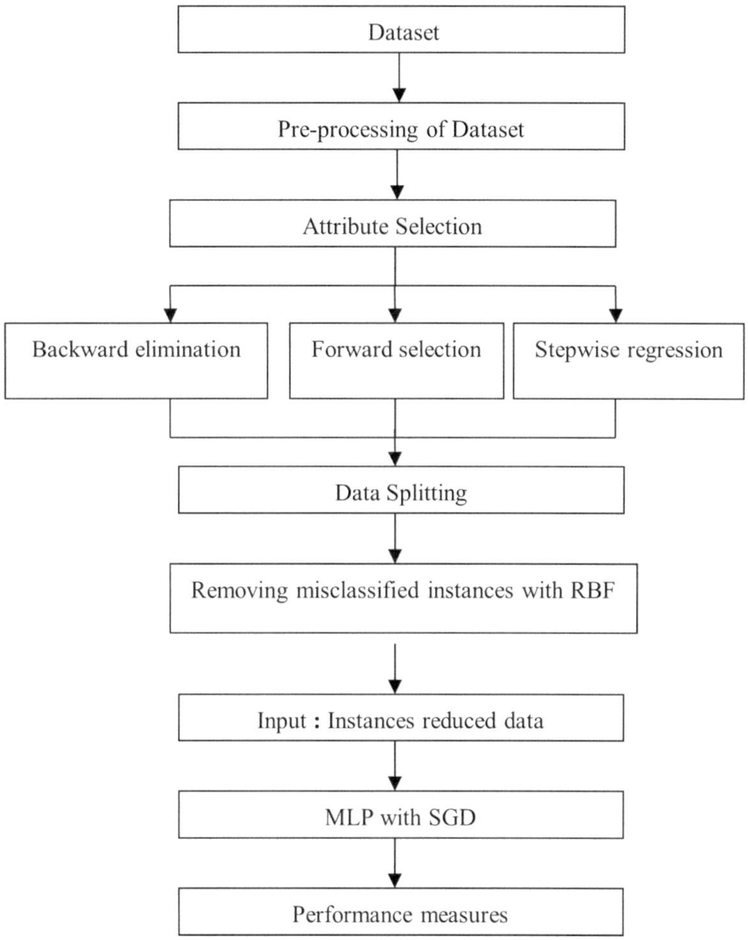

Fig. 1 Diabetes prediction model

as attribute selection by utilizing backward elimination, forward selection, and SR algorithms. In this work, the training and testing split of dataset is done in 75%:25% basis.

Attribute selection : Every attribute of diabetes health record information is verified to select relevant attributes; hence, the characteristics that are used in training and testing content belong to all records. It can be done by utilizing BE, FS, SR.

Backward Elimination (BE): The BE is a method, and its operation is to optimize the implementation of a model by an inverse selection method [32, 33]. Variable selection is occurred by a FS which tests whole variables and removes changeable if it is unimportant. Variables are treated individually if a variable has no effect in the model or irrelevant and extracted from model.

Forward Selection (FS): This is the most adjusting method to discover the finest variable combinations of cluster factors. In FS method [33], variable allows into condition which is not expelled. In expansion, it is cruel to enter a free variable which is nearer to relationship with subordinate variables at that point continuously enters into the following free capable variables and it is ended when there is no capable free factor.

Stepwise Regression (SR): The SR method will combine both forward selection and backward elimination methods. The variable predictors initially allowed variable that is most highly correlated and important with response variable. The second input variable has extensive partial connection and is important. If a particular variable is unimportant after being included in a model, different variables in techniques are calculated. If a variable is missing, it is extracted from the method.

For data splitting, we use the K-fold cross-validation approach to split the diabetes dataset into a training set and a testing set. One-quarter (25%) is used for testing purposes, and three-quarters (75%) is used for training. Classification in RBF networks starts with training set and performs a similarity measure on the input in terms of its resemblance. A "prototype" is stored for each radial basis function network which is samples from train set. Each neuron determines the required classification threshold by calculating the geometric distance between the input and the prototype. An input is classified as class. A prototype is well compared to the class B prototype. RBF network architecture [34] consists of a neuron layer, an input of vector, and an output of layer containing single node per data group or classification.

3.1 SGD-MLP Classifier

The nonlinear discriminant function of a multilayer perceptron (MLP) [34] is depicted in Eq. 1.

$$y = f_\theta(x) = \sigma(b + B \cdot \sigma(A + A \cdot x)) \tag{1}$$

In Eq. 1, the sigmoid function for the vectors a and b, which denote the two matrices A and B, is called σ. By minimizing multiple primary functions such as y, these attributes are consistently optimized using stochastic gradient descent (SGD) and real-world classification data. The objective function is denoted by $C(\theta)$. Performing stochastic gradient descent or optimization efficiently requires additional effort. The function f_θ often contains a large number of local minima. Local minima are often encountered when training small neural networks (NNs), involving issues such as parity, spiral, and XOR. Despite the small presence on site, additional effort is spent on training. Due to f_θ is symmetry, an additional set of higher local minima in the parameter vector can be obtained from a single local minimum. In weight space, MLP gives the same result for convergence observations and different weight settings. Moreover, it makes sense to explore multilayer perceptron. It explores the stochastic gradient descent optimization works.

3.2 SGD with MLP

To address the problem of classifying large amounts of data, SGD is used for MLP [34]. The training classifier of multilayer perceptron is depicted in Eq. 2.

$$f(x) = \mathrm{sgn}(\mathbf{w}^{\mathsf{T}}\mathbf{x}) \tag{2}$$

Using a sample S associates each input with a weight vector $\mathbf{w}$. A weight vector $\mathbf{w}$ is formed to solve the optimization issue [34] described in Eq. 3.

$$\min \; p_t(w) = \frac{\lambda}{2}||w||^2 + l(W; (x_t, y_t)) \tag{3}$$

If hinge loss function is expressed by $l(W; (x_t, y_t)) = \max\left(0, 1 - y_t w^T x_t\right)$ and standardize feature is denoted through $\lambda \geq 0$ which is worked for monitor method, repetitive, SGD works [34]. It starts by initial method weight w_1 guess and describe weight w_t update is already at tth rotation.

$$w_{t+1} = w_t - \eta_t \nabla_t p_t(w_t) = (1 - \eta_t \lambda) w_t + \eta_t l[y_t(w_t, x_t)] \tag{4}$$

where

$$l[y_t(w_t, x_t) < 1] = \begin{cases} 1 & \text{if } y_t(w_t, x_t) < 0, \\ 0 & \text{otherwise} \end{cases} \tag{5}$$

Equation (5) indicates an indicator function that estimate value of 1 when argument is true and a nonzero or null loss is obtained for example (x, y). By employing $\eta_t = 1/(\lambda t)$ [34] step size updated and provides final $w(t + 1)$ repeated after a predetermined number of iterations T. After performing test calculations, proceed to score calculations using implementation parameters like accuracy, sensitivity score, and specificity score.

4 Result Analysis

In this analysis, the accuracy, sensitivity, and specificity are also calculated as performance evaluation factors for comparing different category methods. However, the main focus is model accuracy. To calculate capability of described model, MLP+SGD is compared with different existing models like NB, RF, and J48 in terms of accuracy, sensitivity, and specificity.

Determining the presence or absence of diabetes in a patient is based on this accuracy. Basically, it's about the percentage of cases that receive the correct classification after investigation as described in Eq. 6.

Table 1 Comparative performance analysis

Classifiers	Accuracy	Sensitivity	Specificity
NB	75	56	84
RF	69	49	78
J48	70	60	70
MLP + SGD	80	60	82

$$\text{Accuracy} = \frac{\text{TP} + \text{TN}}{\text{TP} + \text{TN} + \text{FP} + \text{FN}} \quad (6)$$

Sensitivity refers to the quantification of correctly identified positive results as mentioned in Eq. 7. The example is accurate identification of sick persons.

$$\text{Sensitivity} = \frac{\text{TP}}{\text{TP} + \text{FN}} \quad (7)$$

A measure of specificity is the number of false negatives that are accurately identified as mentioned in Eq. 8. The examples are correct identification of healthy individuals.

$$\text{Specificity} = \frac{\text{TN}}{\text{TN} + \text{FP}} \quad (8)$$

where TP: Particularly identified positive instances, TN: Exact identification of negative instances, FP: Not precisely identified negative instances, FN: Not precisely identified affirmative instances.

The performance analysis is presented in Table 1. The proposed model achieves 80% of accuracy, 60% of recall, and 60% of sensitivity. Therefore, accuracy, recall, and sensitivity are the described technique and achieved improved implementations at the time of categorizing implemented PIDD.

5 Conclusion

In this paper, diabetes prediction using optimized stochastic gradient descent with multilayer perceptron was described. Multilayer perceptron with stochastic gradient descent is utilized as classifiers to categorize information produced by outlier detection engine. It improves presentations of classifiers on implemented dataset, and an RBF model-based outlier finding technique is implemented to extract misclassified instances. It is used Pima Indian Diabetes dataset to evaluate the models and predict diabetes. RBF network classification performs a similarity measure on the input. Examples are drawn from the training set. Experimental evaluation is then implemented, followed by the calculation of performance metrics. The proposed model achieves 80% of accuracy, 60% of recall, and 60% of sensitivity. In this study, our results are satisfactory, but we intend to keep working on it until we reach even greater success.

References

1. Marzouk R, Saleh Alluhaidan A, El_Rahman SA (2022) An analytical predictive models and secure web-based personalized diabetes monitoring system. IEEE Access 10:105657–105673
2. Dremin V, Marcinkevics Z, Zherebtsov E, Popov A, Grabovskis A, Kronberga H, Geldnere K, Doronin A, Meglinski I, Bykov A (2021) Skin complications of diabetes mellitus revealed by polarized hyperspectral imaging and machine learning. IEEE Trans Med Imaging 40(4):1207–1216
3. Jiang M, Guo G (2019) Body weight analysis from human body images. IEEE Trans Inf Forensics Secur 14(10):2676–2688
4. Lekha S, Suchetha M (2017) Real-time non-invasive detection and classification of diabetes using modified convolution neural network. IEEE J Biomed Health Inf 22(5):1630–1636
5. Pustozerov E, Popova P (2018) Mobile-based decision support system for gestational diabetes mellitus. In: 2018 Ural symposium on biomedical engineering, radioelectronics and information technology (USBEREIT). IEEE, pp 45–48
6. Hui L, Liu Y, Fei Z, Guan C (2018) An outlier detection algorithm based on cross-correlation analysis for time series dataset. IEEE Access 6:53593–53610
7. Sakr S, Elshawi R, Ahmed A, Qureshi WT, Brawner C, Keteyian S, Blaha MJ, Al-Mallah MH (2018) Using machine learning on cardiorespiratory fitness data for predicting hypertension: The Henry Ford exercise testing (fit) project. PLoS One 13(4):e0195344
8. Anderson JP, Parikh JR, Shenfeld DK, Ivanov V, Marks C, Church BW, Laramie JM, Mardekian J, Anne Piper B, Willke RJ et al (2016) Reverse engineering and evaluation of prediction models for progression to type 2 diabetes: an application of machine learning using electronic health records. J Diabetes Sci Technol:10 (1):6–18
9. O'Reilly C, Gluhak A, Imran MA (2016) Distributed anomaly detection using minimum volume elliptical principal component analysis. IEEE Trans Knowl Data Eng 28(9):2320–2333
10. Nai-Arun N, Moungmai R (2015) Comparison of classifiers for the risk of diabetes prediction. Procedia Comput Sci 69:132–142
11. Liu B, Xiao Y, Yu Philip S, Hao Z, Cao L (2013) An efficient approach for outlier detection with imperfect data labels. IEEE Trans Knowl Data Eng 26(7):1602–1616
12. Sun J, McNaughton CD, Zhang P, Perer A, Gkoulalas-Divanis A, Denny JC, Kirby J, Lasko T, Saip A, Malin BA (2014) Predicting changes in hypertension control using electronic health records from a chronic disease management program. J Am Med Inf Assoc 21(2):337–344
13. Upadhyaya S, Singh K (2012) Classification based outlier detection techniques. Int J Comput Trends Technol 3(2):294–298
14. Yoon K-A, Kwon O-S, Bae D-H (2007) An approach to outlier detection of software measurement data using the k-means clustering method. In: First international symposium on empirical software engineering and measurement (ESEM 2007). IEEE, pp 443–445
15. Chakravadhanula AS, Kolisetty J, Samudrala K, Preetham B, Senapati R (2022) Novel decentralized security architecture for the centralized storage system in hadoop using blockchain technology. In: 2022 IEEE 7th international conference for Convergence in Technology (I2CT). IEEE, pp 1–4
16. Sahoo A, Senapati R (2022) A parallel approach to partition-based frequent pattern mining algorithm. In: Intelligent systems. Springer, pp 93–102
17. Sahoo A, Senapati R (2021) A novel approach for distributed frequent pattern mining algorithm using load-matrix. In: 2021 International conference on intelligent technologies (CONIT). IEEE, pp 1–5

18. Patro PP, Senapati R (2021) Advanced binary matrix-based frequent pattern mining algorithm. In: Intelligent systems. Springer, pp 305–316
19. Sahoo A, Senapati R (2020) A Boolean load-matrix based frequent pattern mining algorithm. In: 2020 International conference on artificial intelligence and signal processing (AISP). IEEE, pp 1–5
20. Muttineni S, Yerramneni S, Kongara BC, Venkatachalam G, Senapati R (2022) An interactive interface for patient diagnosis using machine learning model. In: 2022 2nd International conference on emerging frontiers in electrical and electronic technologies (ICEFEET). IEEE, pp 1–5
21. Ganesh Prasad GSK, Ajay Chowdari A, Jona KP, Senapati R (2022) Detection of CKD from CT scan images using KNN algorithm and using edge detection. In: 2022 2nd International conference on emerging frontiers in electrical and electronic technologies (ICEFEET). IEEE, pp 1–4
22. Chaitanya Datta M, Senapati R (2022) An adoptive heart disease prediction model using machine learning approach. In: 2022 OITS international conference on information technology (OCIT). IEEE, pp 49–54
23. Senapati R (2023) A novel classification-based parallel frequent pattern discovery model for decision making and strategic planning in retailing. Int J Bus Intell Data Mining 23(2):184–200
24. Maddukuri CD, Senapati R (2023) Hybrid clustering-based fast support vector machine model for heart disease prediction. In: International conference on machine learning, IoT and big data. Springer, pp 269–278
25. Raviteja KVNS, Kavya KVBS, Senapati R, Reddy KR (2023) Machine-learning modelling of tensile force in anchored geomembrane liners. Geosynth Int 1–17
26. Manda SC, Muttineni S, Venkatachalam G, Kongara BC, Senapati R (2023) Image stitching using RANSAC and Bayesian refinement. In: 2023 3rd International conference on intelligent technologies (CONIT), pp 1–5. https://doi.org/10.1109/CONIT59222.2023.10205634
27. Yerramneni S, Vara Nitya KS, Nalluri S, Senapati R () A generalized grayscale image processing framework for retinal fundus images. In: 2023 3rd International conference on intelligent technologies (CONIT), pp 1–6. https://doi.org/10.1109/CONIT59222.2023.10205834
28. Samudrala K, Kolisetty J, Chakravadhanula AS, Preetham B, Senapati R (2023) Novel distributed architecture for frequent pattern mining using spark framework. In: 2023 3rd International conference on intelligent technologies (CONIT), pp 1–5. https://doi.org/10.1109/CONIT59222.2023.10205903
29. Chaitanya Datta M, Venkaiah Chowdary B, Senapati R (2023) Multi disease prediction using ensembling of distinct machine learning and deep learning classifiers. In: International conference on soft computing and its engineering applications. Springer, pp 245–257
30. Venkaiah Chowdary B, Chaitanya Datta M, Senapati R (2023) An improved cardiovascular disease prediction model using ensembling of diverse machine learning classifiers. In: 2023 OITS international conference on information technology (OCIT), pp 329–333. https://doi.org/10.1109/OCIT59427.2023.10430692
31. Dua D, Graff C (2019) UCI machine learning repository. http://archive.ics.uci.edu/ml
32. Colvin CE, Machtolff ME, Army Missile Command Redstone Arsenal Al Systems Analysis Division (1970) A backward elimination general significance regression model
33. McCornack RL (1970) A comparison of three predictor selection techniques in multiple regression. Psychometrika 35(2):257–271
34. Ranjeeth S, Latchoumi TP, Victer Paul P (2021) Optimal stochastic gradient descent with multilayer perceptron based student's academic performance prediction model. Recent Adv Comput Sci Commun (Formerly: Recent Patents Comput Sci) 14(6):1728–1741

A Comparative Analysis of Biometric Face Recognition Algorithms

Sweta Verma, Sourabh Debnath, Jitendra Kumar, and Ramesh Kumar Mohapatra

Abstract Face recognition is a critical aspect of computer vision applications, with numerous algorithms continually evolving to improve accuracy and efficiency. In this study, a comprehensive analysis of twelve face recognition algorithms on a diverse dataset has been conducted, employing preprocessing techniques such as face cropping and grayscale conversion for enhanced efficiency. This analysis included traditional methods like principal component analysis (PCA), linear discriminant analysis (LDA), local binary patterns (LBPs), histogram of oriented gradients (HOG), and various support vector machine (SVM) kernels, alongside state-of-the-art deep learning models including DeepID, VGGFace2, CASIA-WebFace, ResNet50, and ResNet42. Surprisingly, CASIA-WebFace exhibited superior performance with an accuracy of 98.89%, accompanied by a "Time to build model" of 43.28 ms and "Time to predict outcomes" of 0.169 ms. Recognizing the significance of computational efficiency, further optimization of the ResNet algorithm has been done, achieving a 41% reduction in "Time to build model" and a 36% decrease in "Time to predict outcomes" at the expense of a 2.73% accuracy drop.

Keywords Face recognition · FaceNet · ResNet

S. Verma (✉) · S. Debnath · J. Kumar · R. K. Mohapatra
National Institute of Technology Rourkela, Odisha, India
e-mail: 222CS1267@nitrkl.ac.in

S. Debnath
e-mail: 519CS1014@nitrkl.ac.in

J. Kumar
e-mail: 520CS1001@nitrkl.ac.in

R. K. Mohapatra
e-mail: mohapatrark@nitrkl.ac.in

© The Author(s), under exclusive license to Springer Nature Singapore Pte Ltd. 2025

S. K. Udgata et al. (eds.), *Intelligent Systems*, Lecture Notes in Networks and Systems 1314, https://doi.org/10.1007/978-981-96-3797-3_29

1 Introduction

Face recognition technology is a crucial technology in the field of security and authentication. It allows individuals to be identified based on unique facial features. As world increases concerns in the field of security, for both physical and digital assets, these systems have shown reliability [1]. Facial recognition systems, when used in real-life scenarios, are exposed to various vulnerabilities, such as variable lighting conditions. In such conditions, traditional face recognition systems tend to fail as they show low accuracy and often produce false positive results, therefore accepting unauthorized individuals.

Biometric face recognition, an advanced biometric technology, utilizes distinct facial features for secure and seamless identification. As the world increasingly embraces virtual interactions, biometric systems merge technology with human physiology. Capitalizing on the unique and stable patterns inherent in each face, akin to fingerprints, facial recognition has witnessed significant growth. It offers a highly accurate and convenient means of identity verification [1] and ensures security in various applications. It is non-intrusive and contactless in nature which combined with rapid processing speeds, enhances user experience and scalability. In this study, we are comparing various face recognition algorithms performance in variable lighting conditions, pose diversity, and different expressions of faces.

2 Background

2.1 Face Recognition: An Overview

Face recognition has a lengthy history, dating back to the early 1960s, when Woodrow Wilson Bledsoe pioneered the notion by constructing a rudimentary system capable of recognizing human faces. Face recognition has evolved into a highly accurate and adaptable tool as a result of technical breakthroughs, notably in machine learning and deep learning. It is now a crucial domain in pattern recognition and computer vision, gaining significant interest due to its wide and practical applications. This technology is an important component in sectors such as biometrics, ensuring safe authentication [2], information security, and access control. Furthermore, it is important in law enforcement for criminal identification, improves the functionality of smart card systems, and increases the effectiveness of surveillance systems.

Face recognition involves multiple steps: face detection identifies potential faces in images or video frames, followed by preprocessing to normalize and align faces. Feature extraction captures distinct facial characteristics, which are then transformed for comparison [3]. Classification algorithms are used to check features against a known identity in a database, and decision-making methods decide matches. Face recognition becomes more adaptable as a result of this entire procedure, which may subsequently be used in biometrics, surveillance, and access control.

2.2 Algorithms Implemented

1. **Principal Component Analysis (PCA)**: A dimensionality reduction technique is used for reducing the dimensionality of large data sets, by transforming the large set of variables into a smaller one without losing much information [4].
2. **Linear Discriminant Analysis (LDA)**: A supervised learning algorithm is used for classification tasks in machine learning. It is a technique which is used to find a linear combination of features that maximize the separation between the classes in a dataset [5].
3. **Local Binary Patterns (LBP)**: An effective texture descriptor is used for images which gives the values of neighboring pixels based on the value of the current pixel [6].
4. **Histogram of Oriented Gradients (HOG)**: A feature descriptor represents the distribution of gradient orientations in local image regions and is commonly used in computer vision and image processing for object detection and recognition [7].
5. **Support Vector Machine (SVM)**: A supervised machine learning algorithm is used for both classification and regression [8]. It uses various kernels such as
   (a) Linear Kernel: Uses a linear decision boundary that separates the different classes in the feature space.
   (b) RBF Kernel: Similar to the linear kernel but employs a radial basis function (RBF) for more complex decision boundaries.
   (c) Poly Kernel: Another variation using a polynomial kernel to handle nonlinear relationships between features.
6. **DeepID**: A deep learning-based face recognition model is designed to learn hierarchical representations of facial features through multiple layers of neural networks [9].
7. **VGGFace2**: A deep convolutional neural network architecture is tailored for face recognition, known for its simplicity and effectiveness in capturing fine-grained facial features [10].
8. **CASIA-WebFace**: A face dataset is widely used for training and evaluating face recognition algorithms, containing a large number of images with a diverse range of identities [11].
9. **ResNet50**: A deep residual network architecture with 50 layers is designed to facilitate training of very deep neural networks and avoid the vanishing gradient problem.

3 Experimental Analysis

This section describes the experimental setup, the evaluation metrics used, and the experimental results in the following subsections.

Fig. 1 Dataset from Kaggle with additional images

3.1 Experimental Setup

The face recognition dataset used in this study is primarily taken from Kaggle, available at the following link: https://www.kaggle.com/datasets/vasukipatel/face-recognition-dataset. This Kaggle dataset originally comprises facial images of thirty-one celebrities. However, recognizing the need for a more diverse and comprehensive dataset, additional images of nine other celebrities were acquired from freely available sources on the Internet.

The dataset encompasses a wide range of facial expressions and poses (Fig. 1), and lighting conditions, providing a diverse set of challenges for evaluating the robustness and generalization capabilities of the implemented face recognition algorithms. Each image in the dataset has underwent preprocessing, including facial feature extraction through cropping and subsequent conversion to grayscale for computational efficiency.

This collection of the Kaggle dataset with additional images ensures a more extensive and varied mixture for a thorough evaluation of the implemented algorithms.

3.2 Data Preprocessing

The preprocessing phase plays a critical role in shaping the input data for the implemented face recognition algorithms. The process involves two key steps: face detection and image transformation.

Fig. 2 Face detection using MTCNN (cropped images)

Face Detection with MTCNN

Face detection is a pivotal task in preprocessing, and the multi-task cascaded convolutional networks (MTCNNs) algorithm has been used for this purpose. MTCNN is a three-stage deep learning model, designed for detecting faces and facial landmarks in images.

Cropping and Saving Cropped Pictures

Once faces are detected using MTCNN, the next step involves cropping these regions for focused analysis. Let I be the original image and B be the bounding box output by MTCNN for a detected face. The cropped face image C can be obtained using the formula:

$$C = I[B_y : B_y + B_h, B_x : B_x + B_w] \tag{1}$$

Here, B_x, B_y, B_w, and B_h represent the coordinates and dimensions of the bounding box.

The cropped face images are then saved for further processing and analysis.

Conversion to Grayscale

To enhance computational efficiency, the preprocessed face images are converted to grayscale. Let G be the grayscale version of the cropped face image C. The grayscale conversion is achieved using the formula (Fig. 2):

$$G = 0.299 \times R + 0.587 \times G + 0.114 \times B \tag{2}$$

Fig. 3 Grayscale converted cropped images

where R, G, and B are the red, green, and blue color intensities of each pixel in the image, respectively (Fig. 3).

The grayscale images G serve as the input for the face recognition algorithms, reducing the computational load while retaining essential facial information.

These preprocessing steps, integrating MTCNN for face detection and leveraging grayscale conversion, lay the foundation for the subsequent implementation and evaluation of face recognition algorithms.

3.3 Metrics for Evaluation

The evaluation of face recognition algorithms involves a set of carefully chosen performance metrics, providing insights into accuracy and computational efficiency. In this study, three key metrics were employed: accuracy score, time to build model, and time to predict outcomes.

Accuracy Score Accuracy score is used for evaluating the correctness of face recognition predictions in this study. It measures the ratio of correctly identified instances to the total number of predictions and is expressed as a percentage. The formula for calculating accuracy is:

$$\text{Accuracy} = \frac{\text{Number of Correct Predictions}}{\text{Total Number of Predictions}} \times 100\% \tag{3}$$

In Python, the accuracy score can be computed using the Scikit-Learn library:

```
from sklearn.metrics import accuracy_score
y_true = ...  % Ground truth labels
y_pred = ...  % Predicted labels
accuracy = accuracy_score(y_true, y_pred)
print(f"Accuracy: {accuracy * 100:.2f}%")
```

Time to Build Model Time to Build Model (TBM) measures the computational efficiency of the training phase. It represents the time taken to construct the face recognition model. Time to Build Model (TBM) is a metric that measures how efficiently the training phase operates.

In Python, the time to build the model can be measured using the time module:

```
import time
start_time = time.time()
% Code to build the model
end_time = time.time()
time_to_build_model = end_time - start_time
print(f"Time to Build Model: {time_to_build_model} seconds")
```

Time to Predict Outcomes Time to Predict Outcomes (TPO) evaluates the computational efficiency of the prediction phase. It represents the time taken to make predictions using the recognition model. Time to Predict Outcomes (TPO) is a key metric that determines how efficiently the models predict outcomes.

Specially for this study, the values of Time to Predict Outcomes (TPO) are first calculated as whole and then divided by 120 for calculation of Time to Predict Outcomes (TPO) for one image. In Python, the time to predict outcomes can be measured similarly:

```
import time
start_time = time.time()
% Code to make predictions
end_time = time.time()
time_to_predict_outcomes = end_time - start_time
print(f"Time to Predict Outcomes: {time_to_predict_outcomes} seconds")
```

4 Results

This section presents the outcomes of our experiments with various face recognition algorithms on the chosen dataset.

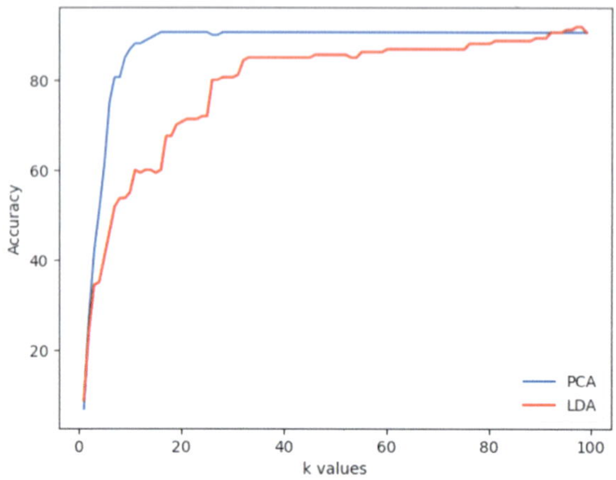

Fig. 4 Performance analysis of LDA and PCA for different values of *k*

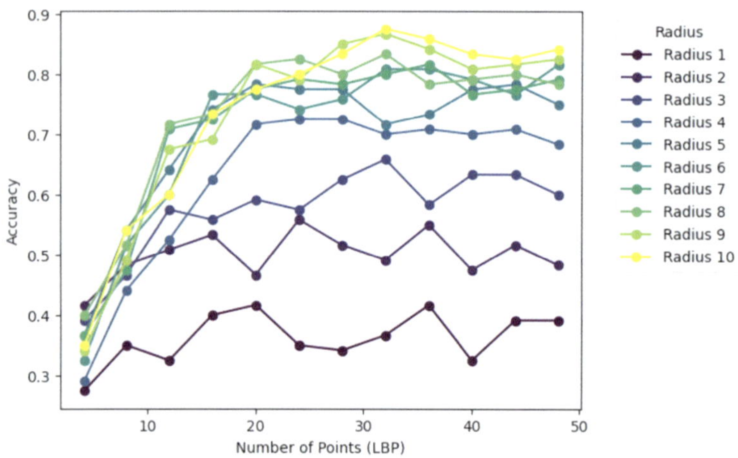

Fig. 5 Effect of numbers of points on LBP accuracy for different radius

As shown in Fig. 4, after *k* value 16, PCA shows little to no change in accuracy, whereas LDA shows gradually increasing accuracy with *k* values. PCA achieved 90.63% accuracy, demonstrating its effectiveness in capturing dataset variance. LDA outperformed PCA with 91.88% accuracy, leveraging class information for enhanced feature discrimination. As shown in Fig. 5, LBP archives max accuracy with radius 10 and number of points 32. Figure 6 shows comparison of LBP and HOG accuracy. HOG accuracy decreases if we increase orientation beyond (here, 12). From Figs. 7 and 8, we can get maximum accuracy for linear, RBF, and polykernels are 100%, 96.67%, and 100%, respectively.

Figure 9 shows there is no significant change in accuracy after epoch 21/100. Figure 10 shows comparison of VGGFace2 and CASIA-WebFace. Model pretrained

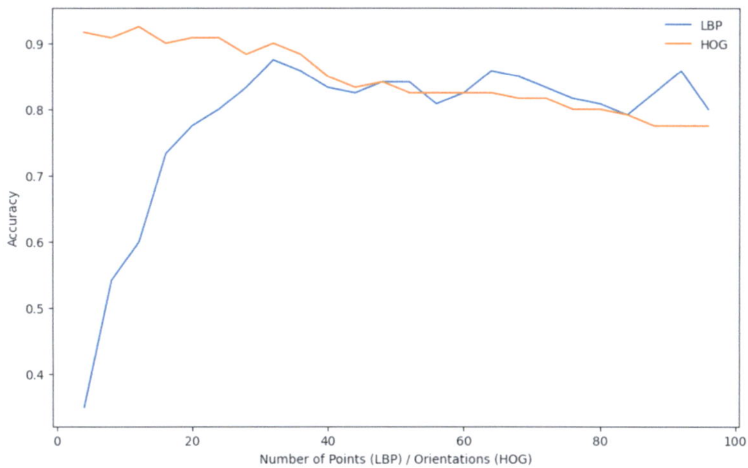

Fig. 6 Comparison of LBP and HOG accuracy with number of points (LBP)/Orientations (HOG)

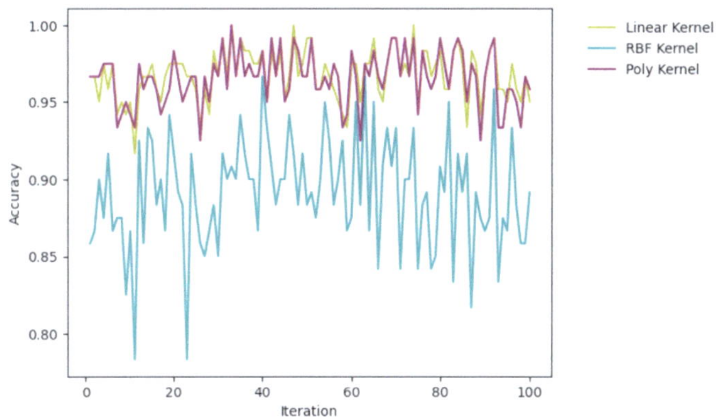

Fig. 7 Performance analysis of different SVM kernels (linear, RBF, and poly) line graph

using CASIA-WebFace weights shows comparatively very low variance (Figs. 11, 12, and 13).

Results on ResNet42: In comparing ResNet50 and the proposed modification, ResNet42, it's evident that ResNet50 achieves a commendable accuracy of 95.56% but demands substantial computational resources. On the other hand, ResNet42, while slightly sacrificing accuracy (92.83%), significantly improves computational efficiency. Specifically, there is a notable reduction in the "Time to Build Model" by approximately 41%, and the "Time to Predict Outcomes" sees a significant decrease of about 36%. This trade-off positions ResNet42 as a viable alternative, especially in scenarios where a balance between accuracy and faster processing is crucial. The choice between the two variants depends on the specific needs of the face recognition

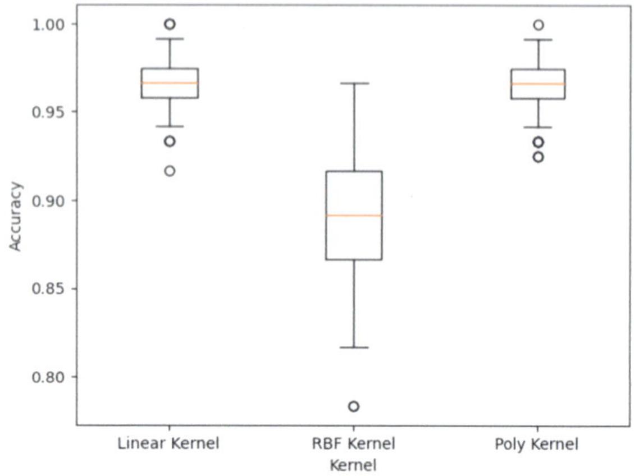

Fig. 8 Performance analysis of different SVM kernels (linear, RBF, and poly) box graph

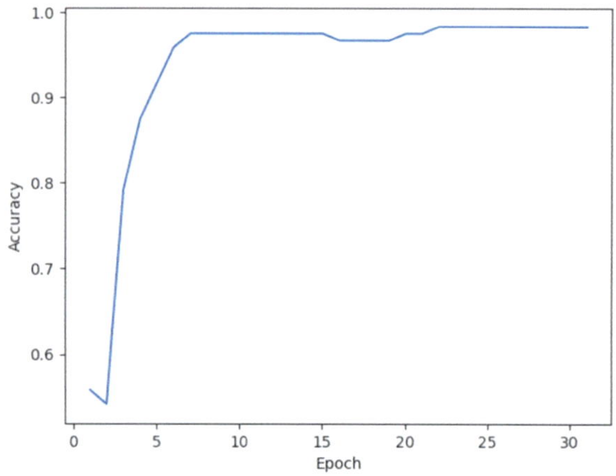

Fig. 9 Accuracy versus Epoch for DeepID

application, emphasizing the importance of tailored algorithm selection based on the desired trade-offs (Figs. 14, 15, and 16).

Table 1 summarizes the performance metrics for each algorithm. The evaluation metrics include accuracy, time required to build models, and time for predicting outcomes.

5 Conclusion and Future Work

In this study, we have implemented and analyzed a diverse set of twelve face recognition algorithms on a carefully curated face dataset. The preprocessing steps, including dataset cleaning, face cropping, and grayscale conversion, aimed at optimizing the computational efficiency of the models. Among the algorithms tested, CASIA-

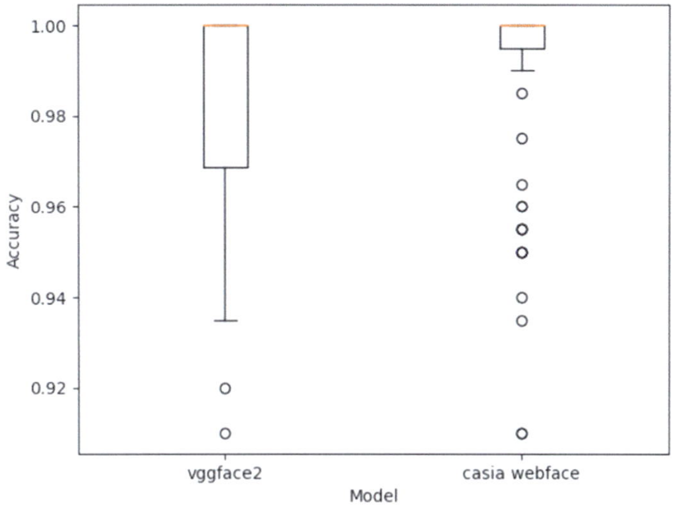

Fig. 10 Box plot of classification accuracy of VGGFace2 versus CASIA-WebFace

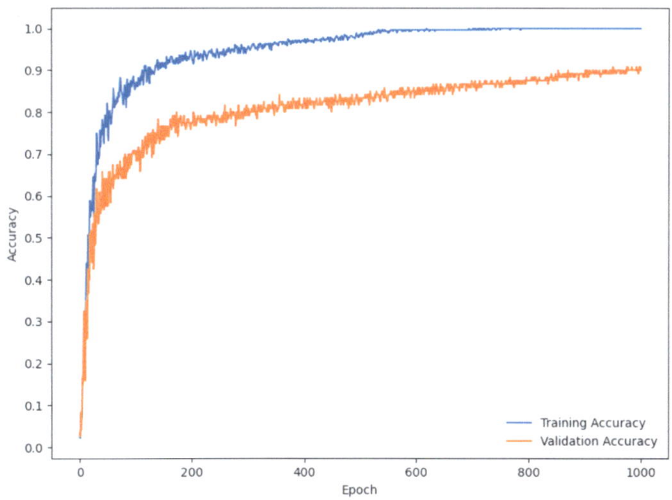

Fig. 11 Training and validation accuracy of ResNet50

WebFace achieved an accuracy of 98.89%. This algorithm demonstrated a notable balance between accuracy and efficiency, with a Time to Build Model of 43.28 ms and a Time to Predict Outcomes of 0.169 ms. The Time to Build Model has been successfully reduced by 41% and the Time to Predict Outcomes by 36%. This enhancement, however, came at the cost of a modest decrease in accuracy by 2.73%.

While our current exploration has provided valuable insights into the performance of various face recognition algorithms, there exist promising avenues for

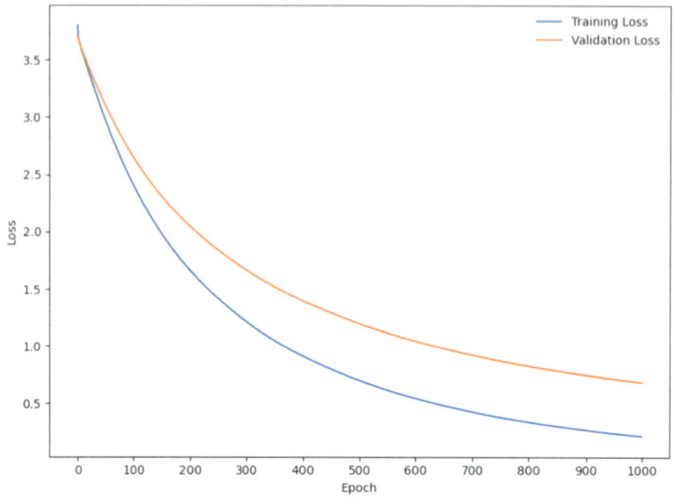

Fig. 12 Training and validation loss of ResNet50

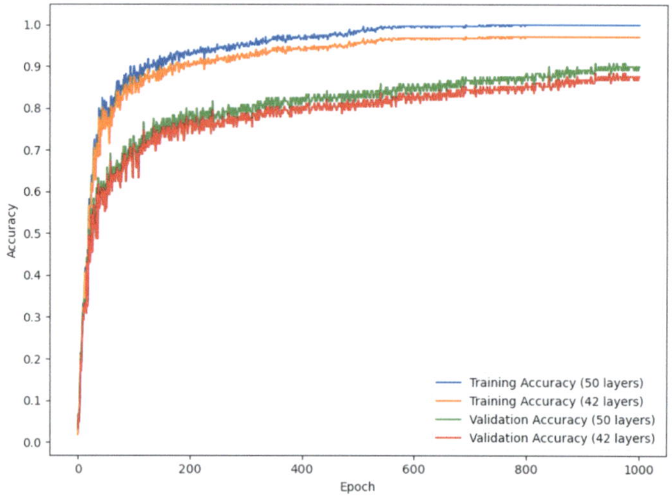

Fig. 13 Training and validation accuracy ResNet50 versus ResNet42 line graph

future research. Enhancing the efficiency of model training and prediction without compromising accuracy remains a key focus. Additionally, exploring novel algorithms and incorporating advancements in deep learning techniques may further elevate the robustness and applicability of face recognition systems.

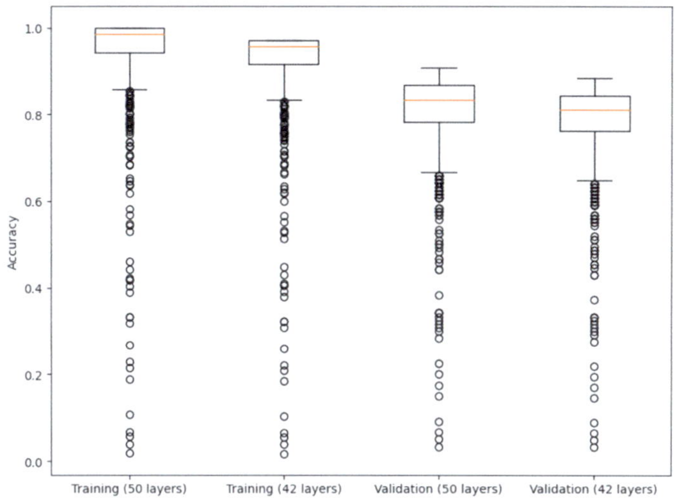

Fig. 14 Training and validation accuracy ResNet50 versus ResNet42 box plot

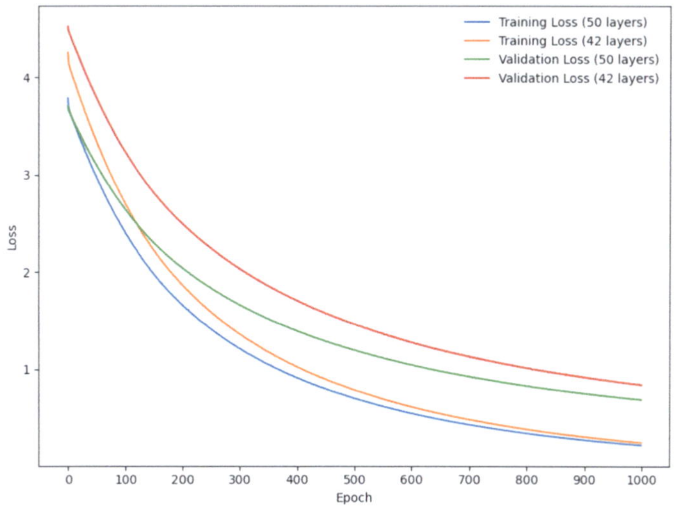

Fig. 15 Training and validation loss ResNet50 versus ResNet42 line graph

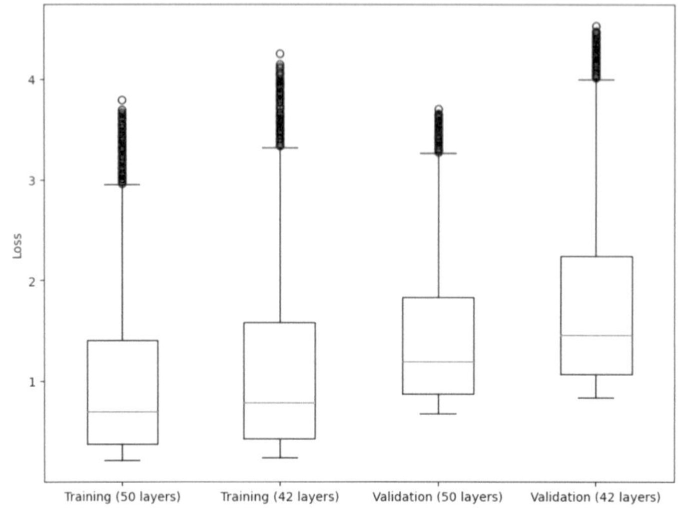

Fig. 16 Training and validation loss ResNet50 versus ResNet42 box plot

Table 1 Face recognition algorithm performance

Algorithm	Accuracy (%)	TBM (ms)	TPO (ms)
PCA	90.63	3840.09	15.720
LDA	91.88	2900.87	5.686
LBP	87.50	22.48	0.067
HOG	92.50	379.8	0.835
SVM linear	96.47	321.38	1.161
SVM RBF	88.91	394.12	3.396
SVM poly	96.28	322.74	1.174
DeepID	98.33	10261.34	38.078
vggface2	98.58	32.69	0.162
casia	98.89	43.28	0.169
resnet50	95.56	649576.85	1650.391
resnet42	92.83	380927.17	1062.866

References

1. Mubashshir DK, Dr. Shinghal K, Mr. Singhal M (2019) Review on facial recognition. IJRITCC 7(8):7–12
2. Divyarajsinh NP, Mehta BB (2013) Face recognition methods & applications. 4(1):84–86. arXiv:1403.0485
3. Kortli Y, Jridi M, Al Falou A, Atri M (2020) Face recognition systems: a survey. Sensors 20(2):342
4. Maćkiewicz A, Ratajczak W (1993) Principal components analysis (PCA). Comput Geosci 19(3):303–342
5. Martinez AM, Kak AC (2001) PCA versus lDA. IEEE Trans Pattern Anal Mach Intell 23(2):228–233
6. Rahim MA, Hossain MN, Wahid T, Azam MS (2013) Face recognition using local binary patterns (lBP). Glob J Comput Sci Technol 13(4):1–8
7. Ghorbani M, Targhi AT, Dehshibi MM (2015) HOG and lBP: Towards a robust face recognition system. In: 2015 Tenth international conference on digital information management (ICDIM), pp 138–141. IEEE
8. Phillips P (1998) Support vector machines applied to face recognition. In: Advances in neural information processing systems, vol 11
9. Sun Y, Liang D, Wang X, Tang X (2015) DeepID3: Face recognition with very deep neural networks. arXiv:1502.00873
10. Massoli FV, Amato G, Falchi F, Gennaro C, Vairo C (2019) Improving multi-scale face recognition using VGGFace2. In: International conference on image analysis and processing. Springer, pp 21–29
11. William I, Rachmawanto EH, Santoso HA, Sari CA (2019) Face recognition using facenet (survey, performance test, and comparison). In: 2019 Fourth international conference on informatics and computing (ICIC). IEEE, pp 1–6

Blockchain-Based Patient Record Maintenance System

Pranjal Singh, Sourabh Debnath, and Ramesh Kumar Mohapatra

Abstract As the population increases and advancements in various fields, including that of medicine, takes giant strides, new issues emerge, which require prompt and optimal resolution to aid development. The health sector deals with a huge volume of data every day. In recent times, medical record-keeping has emerged as an evolving science and rightly so, as it is crucial for optimal health management. However, centralized medical record systems face challenges in ensuring patient privacy, protecting against data alterations, and facilitating efficient information sharing (Summary of the Amazon S3 service disruption in the Northern Virginia (US-EAST-1) region—aws.amazon.com. https://aws.amazon.com/cn/message/41926/ [1]). Our work proposes a system based on blockchain for efficiently maintaining and sharing records while ensuring confidentiality and integrity. We have employed blockchain technology and Interplanetary File System (IPFS) to create a distributed peer-to-peer hypermedia protocol. IPFS seeks to increase the speed, security, and openness of the internet. To ensure data integrity, it makes use of a content-addressable system, in which files are addressed by their cryptographic hashes. Data integrity is guaranteed by the decentralized and tamper-evident nature of blockchain technology, and information may be retrieved and reviewed by authorized individuals through limited access procedures. Smart contracts have been employed to grant and revoke access to various users.

Keywords Blockchain · Interplanetary File System (IPFS) · Cryptographic hashes · Decentralized · Integrity · Confidentiality

---

P. Singh · S. Debnath (✉) · R. K. Mohapatra
National Institute of Technology Rourkela, Odisha, India
e-mail: sourabh_debnath@nitrkl.ac.in

R. K. Mohapatra
e-mail: mohapatrark@nitrkl.ac.in

© The Author(s), under exclusive license to Springer Nature Singapore Pte Ltd. 2025
S. K. Udgata et al. (eds.), *Intelligent Systems*, Lecture Notes in Networks and Systems 1314, https://doi.org/10.1007/978-981-96-3797-3_30

1 Introduction

Health is undoubtably one of the most important assets in an individual's life. It comprises of a wide spectrum of factors ranging from physical to emotional wellbeing. It has a direct link to the quality of life, the degree of productivity and overall happiness. Thus, enabling individuals to tap into a profound sense of vitality, purpose, and fulfillment. The optimal functioning of all components of wellbeing strikes a perfect chord of harmony, ultimately leading to a higher quality of life.

There is an implicit connection between the development of the nation and the quality of life of its citizens. The latter molds the foundational bedrock that fuels the growth of an economy in all aspects. One such significant aspect of any economy is the field of medicine and efficient management of patient data. Countries have huge medical infrastructure to provide services and thus in turn generating large volumes of medical data, necessitating the daily storage, distribution, and retrieval of this information when required. Traditional methods of storing such data involved manual records, registers, charts and to even some extent use of Excel sheets were prevalent. However, the chances of manipulation without identification have become a serious concern.

Patient medical data must be safely stored and easily available to satisfy the expectations of modern healthcare, particularly when it needs to be shared between hospitals in a network. In these situations, maintaining privacy is paramount. In order to guarantee the integrity of the data, it must also remain devoid of alterations. The aforementioned concerns can be subjected to resolution through the creation of a decentralized peer-to-peer health infrastructure using blockchain technology. The paper proceeds to shed light on the proposed methodology, implementation, and the involved concepts. It also discusses the future scope of integration in health infrastructure.

2 Review of Literature

2.1 Delving into the Previous Work

Throughout history, paper-based formats have been the norm for a wide spectrum of purposes, and thus the field of medicine is no exception. For instance, when we visit a hospital, as per the traditional norm, our data is traditionally recorded on paper, including medical prescriptions, lab reports, test results, and scans. Even today, many hospitals continue to rely on these paper-based methods for storing and sharing user medical records. However, maintaining these paper records can be a cumbersome and time-consuming task.

Various EHR systems are deployed around the globe that uses local or cloud databases to store healthcare records at the organizational level. These centralized systems pose a threat as they can be susceptible to data breaches. There has been a gradual yet persistent surge in such kinds of cybersecurity threats that often manifest in the form of leaking of confidential data and unethical use of the same [2].

Ethereum's smart contracts are line of code with written rules and actions. This contract functionality is optimally used by Li et al. [3] to build a system for on chain record preservation. Two smart contracts Preservation Submission (PS) and Primitiveness Verification (PV) are implemented; PS is used to implement three functionalities: receiving unpreserved data, encrypting the processed data, and storing the data on chain, while PV is used to view the data on chain, verify consistency and extract data from the chain. The outcome of performance evaluation shows a consistent and robust system that effectively resists data tampering and deletion operations.

With a view to safeguard health records, counter the various challenges, and discover potential solutions that provide optimal outcomes, various attempts have been made by researchers dabbling in the field of blockchain. One such approach discusses a blockchain-based hierarchical data sharing framework (BHDSF) which focuses on sharing encrypted personal health records (PHRs) using a highly efficient hierarchical key distribution system that minimizes resource consumption while offering robust protection against key leaks. It authenticates the H-IoT devices, which acts as a record source and uses untrusted cloud for database ensuring PHR integrity and validation of metadata Zhang [4].

Similarly, another approach proposed lattice-based access control (LBAC) integration to present e-Health systems to achieve robust confidentiality. While preserving compliance security levels, these methods offer security levels that adhere to data access limitations among users and resources. It categorizes patient records according to different sensitivity levels and puts security restrictions on them accordingly preserving privacy, maintaining transparency, and user authentication Haritha 2023 [5].

A study was conducted on the proposed architecture for blockchain-based health record systems around the globe to assess interoperability and security. It gives an insight into various frameworks and implementation at varying levels of security. It even surveys different studies to integrate cloud-based systems and blockchain to develop a robust, immutable, and real-time system which could be used for various other applications rather than just a records maintenance solution in healthcare sector. Villarreal et al. [6] presented a model-driven engineering (MDE) framework that uses different abstraction levels and two hyperledger that interact using smart contracts; this model paves the way for future implementation of a robust real-time blockchain-based patient record management and maintenance system.

MedRSS, a blockchain-based scheme proposed by Sun et al. [7], employs a combination of encryption, key generation, and ECC-based digital signatures for user authentication. Data integrity is ensured by having two sets of records (one belonging to the patient and the other to the hospital) and validating data integrity by comparing hashes on each occasion. ECC and AES are used for encryption and authentication at different stages of data storage and sharing.

Various systems incorporate blockchain technology where transparency is of utmost importance. Having an immutable ledger provides a sense of trust without involvement of any third party. IPFS provided a solution to storing huge data, since storing everything on blockchain will result in longer chain and full-node miners will face storage issues. Being accessible to all, IPFS poses a threat to confidentiality, and thus different encryption algorithms are incorporated to avoid data leaks. Zakzouk et al. [8] presented such a solution which hides the identity of individual patients and achieving privacy and security of patients record in the EMR system. However, it employs least computation, which reduces the gas costs and increases the throughput and performance of the overall system.

2.2 Shortcomings

Previous research has explored various technologies to achieve data integrity, confidentiality, user privacy, and authentication. However, these approaches often face limitations in terms of system performance, computational cost, and execution delays. Since on-chain records must interact with mostly cloud-based off-chain world, investigating alternative solutions is necessary, even if integrating cloud storage with blockchain [4] presents substantial hurdles to data integrity and cybersecurity. Alternative methods for storing data on the blockchain are required because of the wide range of file sizes of patient records, which range from modest text reports (blood tests, vitals, allergies) to big multi-megabyte scans (CT scans, MRIs) [3, 5–10]. For full-node miners, direct storing of everything is impractical and wasteful. It could be more appropriate to investigate hybrid systems that combine off-chain storage with on-chain record keeping.

To safeguard user authentication on the blockchain and data confidentiality, several researches [5, 7, 8, 11] suggest encryption and verification techniques. Still, there is a large computational overhead when implementing them directly on-chain, which drives up total prices. Smart contract functioning in Ethereum, for example, is dependent on "gas," which is the user-borne computational cost. The system may not be sustainable due to the high gas expenses resulting from the extensive computations involved in these algorithms.

3 Methodology

3.1 Fundamental Framework

Ethereum Built on top of blockchain technology, Ethereum is a decentralized network that was first implemented on the well-known cryptocurrency blockchain. The main goal of Ethereum's development was to create an open-source platform with smart contract functionality integrated with standard blockchain features. Peer-to-peer networking is how it is distributed. The native cryptocurrency of this network is known as "Ethers."

Ethereum empowers programmers with a unique language known as Solidity, which allows them to craft their own blockchain applications, primarily designed for Ethereum's defining feature smart contracts.

Ethereum is interfaced with by way of transactions from outside parties. On the Ethereum blockchain network, external users can use Ethereum to change the state of a document or a collection of data. A transaction using Ethereum consists of multiple essential elements, including the sender, also known as the author, who is recognized by a 20-byte address, and the recipient, who also has a 20-byte address.

On the Ethereum blockchain, each transaction has a cost that the transaction author is responsible for paying. This expense is known as "gas." Petrol is a necessary component of every transaction and has both restrictions and costs. Prices of the gas needed to complete varies from transaction to transaction and it's dependent on the maximum gas that a user is willing to pay.

Why Ethereum?

- Ethereum is a major blockchain platform with large pool of miners having shorter confirmation time. Bitcoin blocks are confirmed after every 10 min, while every Ethereum block requires on an average 15 s.
- The Bitcoin network has a fixed maximum block size of 1 MB. With Ethereum, on the other hand, block sizes are dynamic. A gas limit, or the maximum amount of processing power permitted for transactions within a block, dictates the size of an Ethereum block. Generally speaking, complex smart contracts use more gas, which could result in higher block sizes.
- Ethereum's blockchain network allows additional functionalities which allows complex operation to be performed with ease.

The Smart Contract In Ethereum, a smart contract is a self-executing agreement whose terms and conditions are explicitly encoded into computer code. These contracts are executed automatically when certain conditions are satisfied and are run on the Ethereum blockchain. They are intended to eliminate the need of intermediaries like lawyers or other reliable third parties by facilitating, confirming, or enforcing the negotiation or performance of a contract.

A few of the many applications for smart contracts include initial coin offerings (ICOs), supply chain management, voting, decentralized finance (DeFi), and token creation. They streamline and automate a variety of corporate operations, which decreases the need for middlemen and may increase transaction efficiency and trust. It's crucial to understand how they operate and ensure that they're constructed correctly to avoid unanticipated consequences, but they also have security and legal repercussions.

Interplanetary File System The Interplanetary File System is a peer-to-peer network-based distributed data storage platform. Data storage is facilitated by using hash of that particular data and 256 but are used for content addressing on the platform. Since IPFS is distributed network, transparency and immutability are ensured.

SHA-256 is used to get the content identifiers (CIDs) for a given data. Every file has a distinct hash value since SHA-256 is collision resistant. This hash value

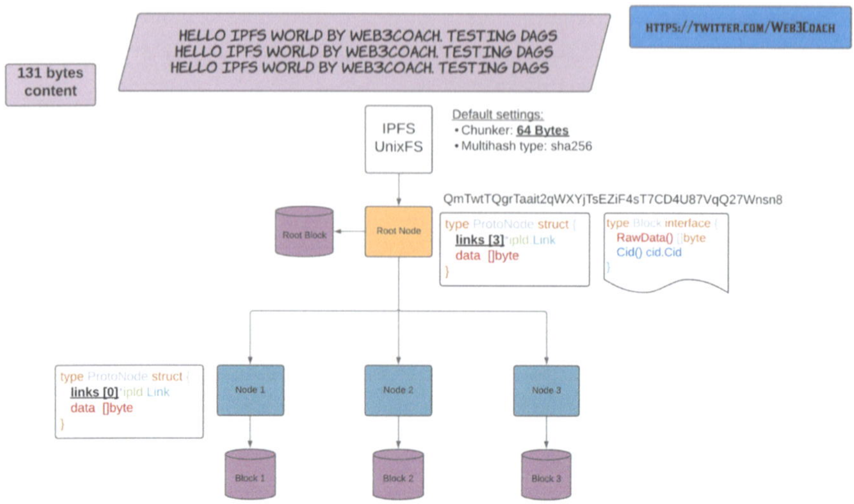

Fig. 1 IPFS content addressing and DAG's

serves as the single distinct identification for data files stored on the IPFS network. The IPFS protocol uses an IPFS object that contains links and data via a peer-to-peer (P2P) connection. The data component consists of a collection of disorganized binary values, whereas the link is represented by a disordered binary value. Maximum size that can be stored at a location is 262 KB and in case data exceeds the capacity a Merkle tree is formed whose leaf node contains the data while the parent node holds references to their children. Root node of the Merkel tree is known as CID, Content Identifier, which uniquely identifies entire data (Fig. 1).

This decentralized and secure system offers a robust solution for data storage and retrieval, with each file uniquely identified and protected against unauthorized alterations.

Storage Capacity of IPFS platform? The total unique CID's we can have is 2^{256}. This quantity is so great that we would still have a ton of storage space even if we were to store every particle in the visible universe. It exceeds 115 quattuorvigintillion when expressed numerically.

Encryption Algorithm Encryption algorithm file transfers are not encrypted by default in the Interplanetary File System (IPFS). Nevertheless, we can secure your data during transfers by utilizing IPFS in conjunction with additional encryption techniques or protocols. Encryption can be done in two different ways in a decentralized network to guarantee secrecy. Different encryption algorithms can be applied to safeguard the records stored across the IPFS network.

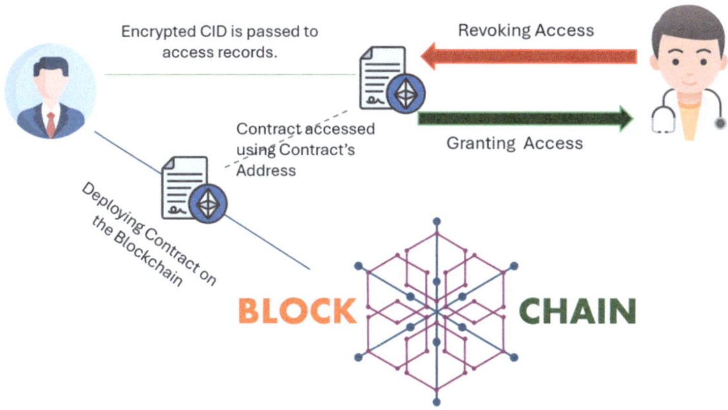

Fig. 2 Record sharing after encryption

Our work uses AES encryption initially on the record while uploading the record onto IPFS network. 32-bit key for the record is generated by hashing string formed by combining private key of the user, its identity (name, age, role), and index of records stored. After the encrypted records are uploaded, it's CID and key are shared using smart contract which can revoke the access whenever needed (Fig. 2).

3.2 *Users and Roles*

Three main purposes for which the system is carefully constructed are data storage, data sharing, and data security. It uses a strong encryption technique and a decentralized network with IPFS to provide data security, protecting record's confidentiality even in the case of possible data leaks.

The network employs a variety of clients, which are essential for different purposes and functionalities.

- Patients as Client
- Doctors as Client
- Healthcare Providers as Client
- Insurance companies as Client
- Recruitment firms as Client.

4 Implementation

4.1 Smart Contracts and Their Deployment

Ethereum's smart contract coded using Solidity is deployed over the testnet for handling patient identity, doctor identity, patient records CIDs and access of each record (Fig. 3).

After deployment, it is possible to view both the deployed smart contract and the transaction used for deployment on the ganache. By acting as a local development blockchain, ganache enables developers to communicate with a computer-based simulation of the Ethereum network. By doing away with the requirement to connect to testnets or the mainnet, this encourages safer testing and quicker development without endangering actual cryptocurrency (Fig. 4).

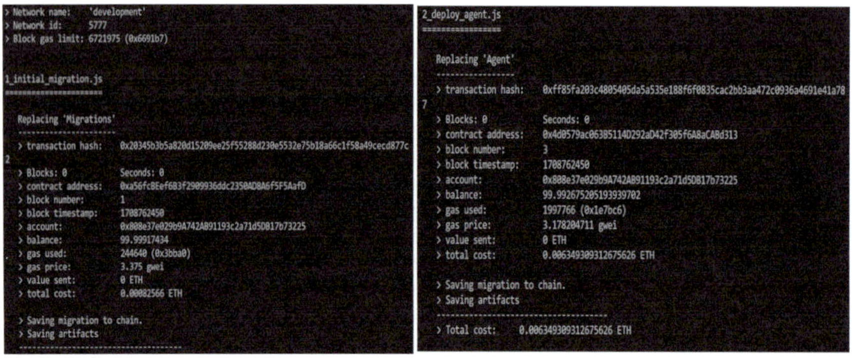

Fig. 3 Contract deployment

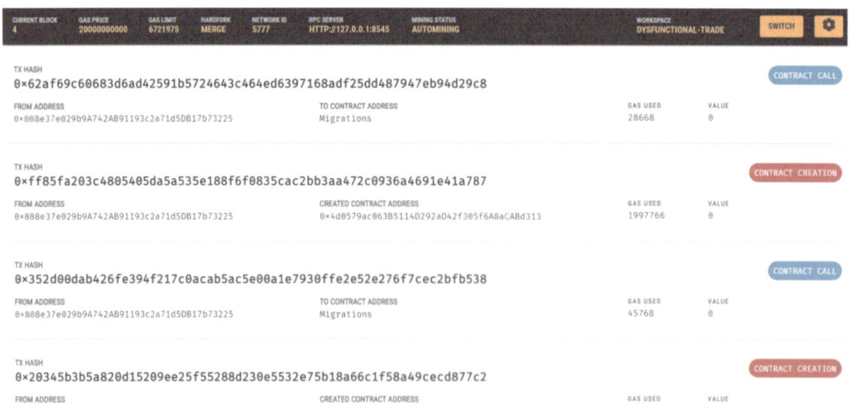

Fig. 4 Deployed contract on Ganache (personal blockchain)

4.2 *Registration Using MetaMask*

- A well-known software wallet for cryptocurrencies, MetaMask is used to communicate with the Ethereum network. ERC-20 and Ethereum tokens may be stored, sent, and received, and users can communicate with decentralized apps. MetaMask is used for user authentication and is highly secure which can only be accessed using 12 words mnemonic.
- Name, age, and category (patient or doctor) is stored on the IPFS network.
- Patient Registration
- Doctor Registration (Figs. 5 and 6).

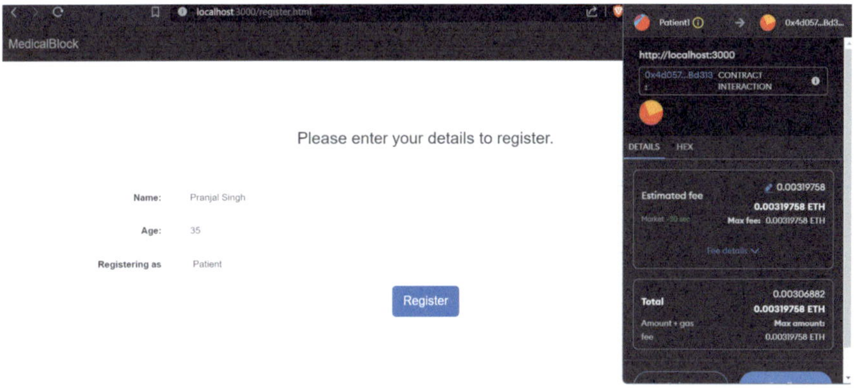

Fig. 5 Patient registration on the platform using MetaMask

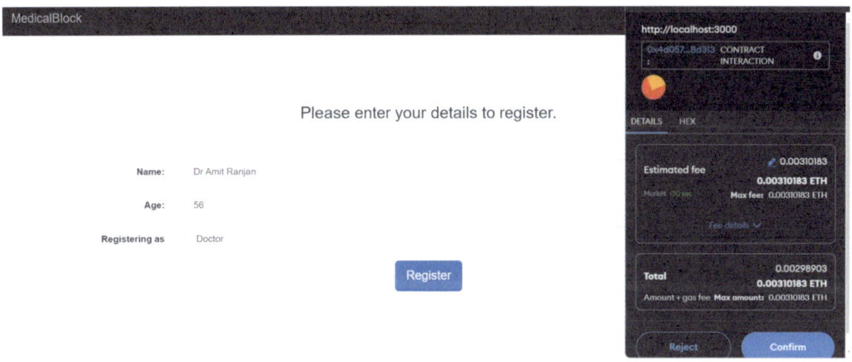

Fig. 6 Doctor registration on the platform using MetaMask

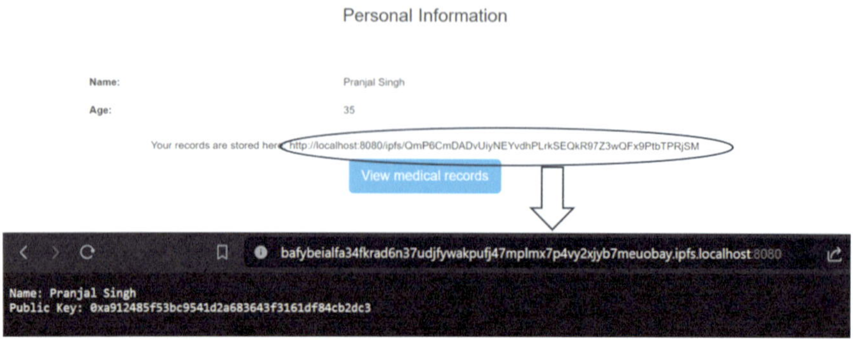

Fig. 7 Patient record storage on IPFS network

4.3 IPFS Integration for Each Patient's Record Storage

After authentication using MetaMask each patient's information along with its public key is stored over IPFS network which can be accessed using the CID (Content Identifier) address (Fig. 7).

4.4 Data Sharing After Authenticating MetaMask

After authenticating using MetaMask, user can send its record to the list of doctors registered on the network. User has the option of granting and revoking access of multiple records to multiple doctors.

- Granting and revoking access
- Transaction Recorded to Permit Access (Figs. 8 and 9).

5 Conclusion and Future Scope

The efficient and secure management of medical data is paramount as it deals with various factors like patient confidentiality, ensuring accurate diagnosis and treatment as well as protecting the integrity of healthcare systems. When the security measures in place are resilient, the protection of sensitive medical information from unauthorized access is ensured along with the mitigation of data breach concerns and maintenance of patient privacy. Ultimately, the cherry on top happens to be the enhancement in the quality of healthcare delivery standards which is an inevitable outcome of efficient and secure management of medical data. The deepening of trust

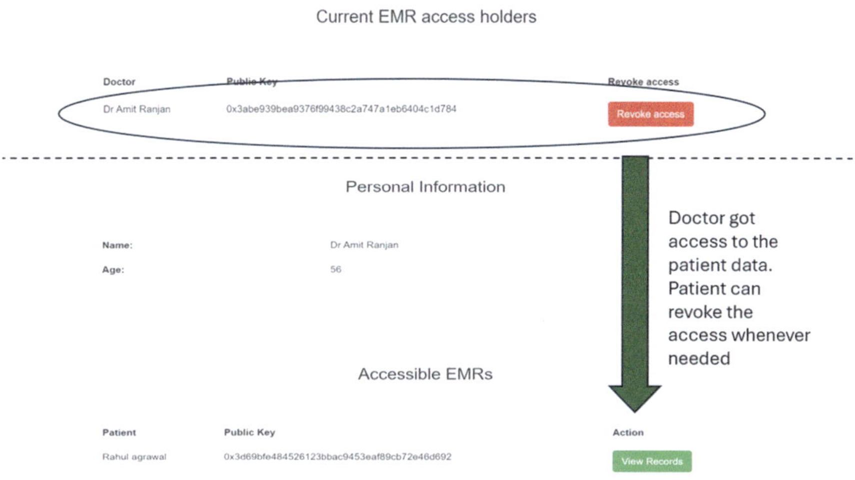

Fig. 8 Granting access to Dr. Amit Ranjan, record visible on doctor's portal

Fig. 9 Transaction on ganache to grant access

and reliability factor between patients and healthcare providers is also catered to implicitly.

Our model provides a fully decentralized, robust, and user-friendly solution for storing, sharing, and maintaining health records. Using IPFS ensures a scalable and robust solution of storing trillions of records. Transactions hold patient's identity info which are tamper-proof and allow easy retrieval through the distributed chain. MetaMask used for authentication uses 12-word mnemonic sequence which are resistant to any attack and allows users flawless integration with current systems. Smart contract usage for sharing eliminates lifelong access to a certain user thus ensuring records confidentiality even after sharing it to some user. In case of IPFS CID leaks, AES encryption in the data provides another layer of security which could only be decrypted by the key shared by user. In order to guarantee that patient data is not only safely stored but also easily accessible when needed, future efforts will concentrate on enhancing the current infrastructure, investigating new technologies, and tackling developing issues. Better real-time technologies will evolve to create a fully decentralized medical infrastructure without need of user or third parties to update or upload data. This will eventually improve patient care and healthcare results. Actively searching for better encryption algorithms to maintain confidentiality and work on enhancing user experience.

References

1. Summary of the Amazon S3 service disruption in the Northern Virginia (US-EAST-1) region—aws.amazon.com. https://aws.amazon.com/cn/message/41926/. Accessed 26 Feb 2024
2. Hacking healthcare: With 385M patient records exposed, cybersecurity experts sound alarm on breach surge—healthcaredive.com. https://www.healthcaredive.com/news/cybersecurity-hacking-healthcare-breaches/643821/. Accessed 26 Feb 2024
3. Li H, Zhu L, Shen M, Gao F, Tao X, Liu S (2018) Blockchain-based data preservation system for medical data. J Med Syst 42(8)
4. Zhang J, Yang Y, Liu X, Ma J (2022) An efficient blockchain-based hierarchical data sharing for healthcare internet of things. IEEE Trans Ind Inf 18(10):7139–7150
5. Haritha T, Anitha A (2023) Multi-level security in healthcare by integrating lattice-based access control and blockchain- based smart contracts system. IEEE Access 11:114322–114340
6. Dulce Villarreal ER, García-Alonso J, Moguel E, Hurtado Alegría JA (2023) Blockchain for healthcare management systems: a survey on interoperability and security. IEEE Access 11:5629–5652
7. Sun Z, Han D, Li D, Weng T-H, Li K-C, Mei X (2023) MedRSS: a blockchain-based scheme for secure storage and sharing of medical records. Comput Ind Eng 183:109521
8. Zakzouk A, El-Sayed A, El-Din Hemdan E (2023) A blockchain-based electronic medical records management framework in smart healthcare infrastructure. Multimed Tools Appl 82(23):35419–35437
9. Kumar R, Marchang N, Tripathi R (2020) Distributed off-chain storage of patient diagnostic reports in healthcare system using IPFS and blockchain. In: 2020 International conference on COMmunication Systems & NETworkS (COMSNETS), pp 1–5
10. Mani V, Manickam P, Alotaibi Y, Alghamdi S, Ibrahim Khalaf O (2021) Hyperledger healthchain: Patient-centric IPFS-based storage of health records. Electronics 10(23):3003
11. Soni M, Kumar Singh D (2021) Blockchain implementation for privacy preserving and securing the healthcare data. In: 2021 10th IEEE International conference on communication systems and network technologies (CSNT). IEEE

PAT-Unet: Ensembled Pyramid Attention Transformer U-Net for Brain Tumor Segmentation in 3D Brain MRI Voxels

Ruqsar Zaitoon and Hussain Syed

Abstract In the field of medical imaging. While convolution-based networks have demonstrated impressive results in medical picture segmentation, they are constrained in their ability to simulate extensive contextual relationships and spatial Tumors such as necrosis, edema, and enhancing tumor exhibit significant similarity in features on multimodal MRI. These homogeneous features include magnetic resonance imaging. These similarities provide challenges in dependencies. The Pyramidal convolution transformer has exhibited exceptional proficiency in diverse workflows through the utilization of self-supervised learning, owing to its influential capacity for long-range information interaction. This paper introduces PAT-Unet, a pyramidal attention transformer-based network designed for brain tumor segmentation. The motivation behind this work stems from the vision transformer. During the tumor classification phase, the PAT-Unet generates multiscale features that serve as input for sub-region categorization. To increase the segmentation performance, the fine-tuning (FT) phase then combines and refines these characteristics. Furthermore, we introduced a self-supervised learning architecture for pre-training the encoder of the model by means of the reconstruction task. Our suggested model underwent thorough experimentation on tumor segmentation tasks, and the results consistently showed superior performance with 94.6% dice-score average, surpassing established standards.

Keywords Medical imaging · Tumor region segmentation · Self-supervised learning · Pyramid attentions · Transformers

R. Zaitoon · H. Syed (✉)
Department of Information Technology, Vardhaman College of Engineering, Hyderabad, Telangana, India
e-mail: hussain.syed@vitap.ac.in

H. Syed
School of Computer Science and Engineering, VIT-AP University, Amaravati, India

© The Author(s), under exclusive license to Springer Nature Singapore Pte Ltd. 2025

S. K. Udgata et al. (eds.), *Intelligent Systems*, Lecture Notes in Networks and Systems 1314, https://doi.org/10.1007/978-981-96-3797-3_31

1 Introduction

Tumor segmentation is an essential component of computer-aided tumor treatment and diagnosis. Doctors can use advanced and precise automated tumor segmentation technology to objectively analyze the precise location, size, and diameter of tumors. This enables them to formulate the most efficient therapeutic strategy for patients. Hence, it is imperative and advantageous to develop an automated methodology that is both precise and beneficial for the segmentation of tumors. Segmenting abnormal tissues in 3D brain MRI is a significant task essential for diagnosis, treatment planning, and medical research. Current approaches struggle to accurately delineate tumor boundaries because of the complex and diverse characteristics of brain structures. Our research presents the Ensembled Pyramid Attention Transformer U-net to overcome these obstacles and improve the accuracy of tumor segmentation. The motivation comes from understanding the intricate nature of brain structures, highlighting the importance of accurate segmentation, utilizing advancements in attention mechanisms, implementing ensemble learning for reliability, and overcoming current methodological constraints to enhance the cutting-edge of medical imaging research. Due to its exceptional aptitude for learning and extracting traits. CNNs, particularly U-NNs with enhanced skip links from encoder to decoder and modified skip connections, have demonstrated exceptional performance in medical picture segmentation within the medical field. The input is a list containing the elements [1, 2]. Due to the restricted nature of convolution in CNNs, the neurons are constrained to narrow regions inside the original image, which makes it challenging to build long-range connections. Despite their ability to capture intricate details, CNNs inherently lack the necessary long-range relational information that is essential in biomedical imaging [3] and has achieved remarkable achievement in the domain of NLP. The efficacy of the Transformer model arises from its ability to efficiently collect vast semantic information exchanges and spatial relationships through its self-attention mechanism and FFN. Many researchers have attempted to utilize the design of the Transformer model in the field of computer vision, building upon previous achievements. The vision transformer is a cutting-edge model for computer vision tasks [4]. A Transformer model, devoid of any additional components, has exhibited exceptional performance in image recognition evaluations. The study showcased the Transformer's ability to effectively apply its capability for modeling global and long-range interdependence to computer vision challenges. Furthermore, the Swin Transformer [5] a hierarchical Transformer model was created by combining a dynamic window mechanism and window-based self-attention. The goal was to reduce computing costs while also prioritizing both long-range connections and local attention.

The UNETR [6] and TransBTS [7] lack sufficient incorporation of cross-plane contextual information in the dismemberment on 3D Brain MRI. As a result, the use of 3D spatial content is limited. While, the transformer models progressed [8, 9], it is constrained by the requirement of pre-training complex CNNs. To overcome these limitations, we suggest employing pyramid-based transformers in a network architecture that incorporates a 3D U-shaped symmetric Swin Transformer, and initializing

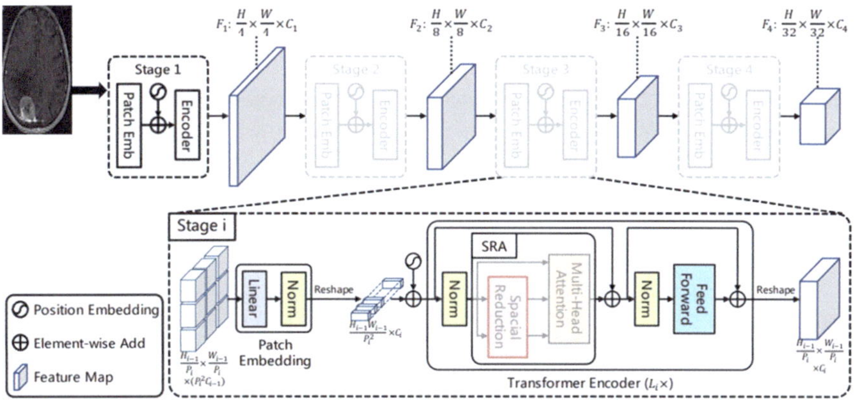

Fig. 1 Pyramid vision transformer for brain MRI

it through supervised learning. The encoder-decoder setup with skip connections. Building upon the Swin Transformer architecture originally meant for 2D image operations, we have created a PA Transformer that is capable of efficiently handling 3D images shows in Fig. 1. Meanwhile, to improvise the ability of model training, we replace the stem with a convolutional stem, drawing inspiration from sources [10, 11]. Furthermore, a model was developed and applied for the purpose of self-supervised learning, specifically for pre-training. We have developed ensembled segmentation network called "PAT-Unet," which combines the benefits of image Transformers illustrates in Fig. 1, to achieve improved accuracy and efficiency. The main contributions we have made are as follows.

The PAT-Unet, for Tumor segmentation on Brain MRI which provides pyramid attention in depth to an input image and Transformer into the U-encoder and U-decoder, resulting in an efficient approach. To localize the tumor in proposed network for Brain Tumor segmentation (BTS), the non-overlapping patches embedded at encoder modules along with Transformer techniques that improves overlap in + patch embedding and down sampling without pixel loss. To pre-train the model encoder. The system utilizes a U-Net model fused with transformer convolution and pyramid attention to enhance the categorization and learning of unusual meningiomas in MRI data. Regarding performance, the results indicate that our approach outperforms earlier models.

The study presents a hybrid architecture that combines Transformer. In Sect. 2, the literature examines the process of segmenting atypical meningioma and discusses the segmentation phase as a separate topic. Section 3 provides comprehensive information regarding the dataset used in this paper as well as a detailed explanation of the technique described. Following that, Sect. 4 presents and analyzes the experimental results, including a performance comparison with alternative approaches. Section 5 ultimately presents the primary findings and conclusions of the introduced work.

2 Related Work

The U-Net architecture has effectiveness in analyzing medical images with limited data. Future networks are expected to retain the U-Net while including enhancements and evaluating the U-Net approach alongside proposed modifications. Building upon the concepts introduced in Inception, the conventional convolutional layer method was substituted with a multi-resolution approach, and a residual connection was incorporated [7, 12–14].

The highlighted language pertains to a research study that centers on the categorization of brain cancers from MRI via a method known as vision transformers assembling. The approach entails employing deep learning models utilizing the Transformer architecture, first designed on the processing of natural languages, to examine MRI images and categorize brain cancers. The assembled technique is utilized to merge numerous models and enhance the precision and dependability of the categorization outcomes [15].

The study utilized a dataset consisting of 3264 photos to examine the efficacy of various DL methodologies, VGG16, InceptionV3, and VGG19, ResNet50, and Exception. The researchers introduced a multiclass classification model termed transfer learning (TL) that is based on the three most successful TL models. Explainable AI was employed to assess the efficacy and soundness of every DL model. The researchers utilized newly introduced vision transformer (ViT) models and conducted a comparative analysis of their output with the TL and ensemble model [16].

Tumors formed by uncontrolled proliferation of cells, which impacts various organs and tissues, is the hallmark of cancer, a common disease. The central nervous system location makes brain tumors extremely difficult to diagnose and cure. The low survival rate of these tumors is driving researchers to seek out ways to avoid them, develop more effective treatments, enhance diagnostic techniques, and classify tumors more accurately. Brain tumor identification is best accomplished with the FLAIR sequence since it makes use of the most popular and extensive databases. While all the sequences tested showed comparable levels metrics, the FLAIR sequence outperformed the others in accurately diagnosing brain tumors. A challenge in tumor detection studies is the availability of vast volumes of data and the variance among databases [19]. These networks utilize convolution and transformers to do medical picture segmentation in both 3D and 2D formats. They demonstrate exceptional accuracy in segmenting brain tumors in their respective dimensions. In contrast to existing networks that rely on transformers and convolutions, we have developed a TC-inception module that allows for concurrent execution of transformer in addition convolution operations. Furthermore, drawing inspiration from numerous works on the fusion of multimodal characteristics, we have introduced CAFGL as a means to efficiently combine both global and local features. Furthermore, we employed a cross-attention method to enhance the link between the skip arrangement. Using the BraTS datasets, our methods outperformed existing methods in terms of Dice scores [20].

The procedure for diagnosing BT is a lengthy one and heavily depends on the proficiency of radiologists. The growing patient population has rendered conventional procedures expensive and inefficient. DL methods, particularly employing neural networks based on CNN, have become increasingly popular for effectively identifying and segmenting brain cancers in a shorter duration. To fine-tune its hyper-parameters, the (BCM-CNN) employs the adaptive dynamic sine–cosine fitness gray wolf optimization (ADSCFGWO) technique, to improve BT diagnosis, the model uses the Inception-ResnetV2 pre-trained model and generates a normal (0) or tumor (%) binary classification [23]. To segment BT using MRI, the Nested Modality-Aware Transformer (NestedFormer) is suggested as a method. This methodology will systematically examine the relationships between and within the various MRI modalities. The nested multimodal fusion using a transformer with multi-encoder and single-decoder to generate high-level depiction of different modalities. To further improve the efficacy of skip connections, we also apply modality-sensitive gating at smaller sizes. We do multi-modal fusion in the Nested Modality-aware Feature detector Aggregation module idea. Using a tri-oriented spatial-attention transformer, this module improves continuing interdependence within individual modalities. Additionally, by utilizing a cross-modality attention transformer, it augments crucial contextual data across modalities. Extensive experiments using the The brats2020 standard and a private meningeal tumor segmentation (MeniSeg) dataset indicated that the NestedFormer outperformed state-of-the-art approaches [25].

3 Methodology

3.1 Architecture Overview

New segmentation model of brain tumor that is PAT which uses an encoder-decoder design with skip links, forming a U-shaped network architecture. The 3D Convolutional Patches Embedded Block, the 3D Pyramid Attention Transformer Block, and the Down sample Block are the three main parts of the encoder, also known as PAT-Unet shown in Fig. 2. Using the 3D CNN Patch Embedding cells, the encoder first divides the input image to overlap patches in order to do feature extraction and down sampling. Three layered 3D PA Transformer Blocks and Convolutional Down sample Blocks are then assembled to create a hierarchical representation. In contrast, the decoder's architecture ended with Segmentation Head and the 3D Attention Transformer Block in Pyramid Form. The decoder's job is to translate these features to predictions at the pixel level after broadening the encoder's low-resolution feature map. Utilizing skip connections to a distinguish features in U-shaped networks the model restores spatial detail information that is lost when down sampling. The vital pixel-level prediction for thorough brain tumor segmentation is the result of the decoder's labors, which also involves the feature map's complex up-sampling to the original image resolution and the key objective of this methodology is to improve

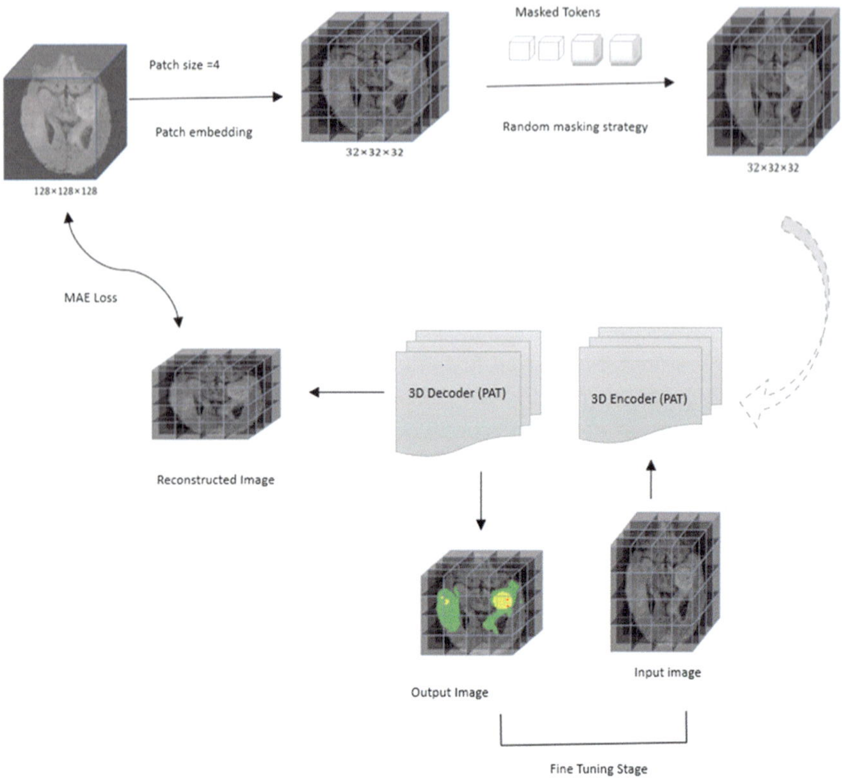

Fig. 2 Proposed framework for brain tumor segmentation using PAT-Unet

feature representation and accurate segmentation results in brain tumor analysis by strategically combining encoder-decoder dynamics and skip connections.

3.2 Transformer Encoder Block

One issue in the field of current transformer-based 3D segmentation networks is the lack of a basic inductive bias of localization. Moreover, working with three-dimensional images naturally increases the transformer architecture's computing complexity. Motivated by the Swin Transformer paradigm, we took a hybrid approach instead of using the traditional transformer block, we integrated the 3D Pyramid Attention Transformer block, this tactical change reduces the model's computational burden while simultaneously introducing an intrinsic bias in favor of locality.

Every 3D Pyramid Attention Transformer unit in our architecture has Layer Normalization, a Feed through Network, which is residual connections, and perhaps

3D W-MSA or 3D SW-MSA as its core components. Compared to the conventional transformers block in ViT [4], our three-dimensional a pyramid Attention Transformer takes a new method. Each of the several non-overlapping windows that make up the feature map is used for multi-head self-attention. Primarily, our three-dimensional PAT employs distinct window partitioning strategies for every layer. This thoughtful design provides robust exchange of data between adjacent sub-windows from the previous tier. This innovation stems from the adaptable implementation of diverse window partitioning strategies, augmenting the model's capacity for effective communication and data amalgamation.

3.3 PAT-Pyramid Attention Transformers

Unlike previous transformer-based segmentation models [7], the new segmented network architecture uses a pyramid structure with a transformer in encoder and decoder framework. The core unit for feature extraction is a 3D Pyramidal Attention Transformer (PAT) module. The input MRI is divided into overlapping patches using patch embedding. The hierarchical architecture consists of four levels of piled 3D Swin Transformers Blocks and Down sample blocks. The decoder substitutes an Up-sample block with 2×2 up-sampling for the Down sample block, guaranteeing the retention of pertinent data during the segmentation procedure.

The PAT-Unet self-supervised learning framework. In the pre-training phase, a mask autoencoder which masks a portion of the input source signals alongside estimates the initial signals in the masked area was trained using a similar encoder from transformers to reconstruct a 3DMRI image, we utilize a patched-aligned random masks on the feature map followed by patch embedding, replacing the mask patches with learnable mask tokens depicts in Fig. 2.

4 Results and Discussion

4.1 Implementation Details

In our experimental setup, employing PyTorch on NVIDIA 2080Ti GPUs, we conducted an extensive exploration. During the initial phase of pre-training, we ran 1000 epochs for autoencoder training, fine-tuning the process for a batch size 16. We meticulously optimized using the Adam optimizer, fine-tuning parameters such as the initial learning rate set at 0.002, weight decay at 0.07, with $\beta_1 = 0.95$, and $\beta_2 = 0.997$. Moving on to the subsequent fine-tuning phase, the pre-trained encoder inherited from the reconstructed model served as the base for initializing PAT-Unet. We directed our focus on simulating PAT-Unet for 250 epochs, employing the Adam optimizer adjusted learning rate of 0.0005 first, weight decay of 0.6, alongside L_1

= 0.85 and L_2 = 0.995. These strategic refinements aimed to significantly enhance PAT-net's performance and optimize its functionality for future tasks.

4.2 Datasets

The Brain Tumor Segmentation contests in 2019 (BraTS2019) and also 2020 (BraTS2020) provided the three-dimension MRI images used in this work. There are a total of 335 training and 125 validation patient cases in BraTS2019, compared to 285 training instances and 66 validation cases in BraTS2020. Each patient gives four different MRI imaging modalities (240 × 240 × 155) from these datasets: T2-weighted, fluid attenuated inversion recovery, initial T1-weighted, and T1-weighted after contrast, initial T1-weighted. The labels are categorized into backdrop, enhancing tumor, non-enhancing tumor, edema, and necrosis.

4.3 Data Pre-processing

The PAT-Unet model was trained using four 240 × 240x155 × 1 3D pictures. The model's generalization was improved via data augmentation. Randomly cropped, scaled, and flipped photos were adjusted for intensity. Before training, Z-Score normalization was employed, and self-supervised learning training used random masking. These methods improved model robustness and adaptability to varied input data distributions.

4.4 Comparison of Progressive CNN Methods with State-of-the-Art

The quantitative results reveal that our technique accurately segments brain tumors, with an average DS of 94.6% for ET, WT, and TC, see Fig. 3, and an average Hausdorff measurement of 7.108 mm. Transformer methods like UNETR, TransBTS, and BTSwin-Unet outperform 3D convolution-based methods like 3D U-Net, U-Net + + , and Att-Unet in BT segmentation. This shows how global features with long-range information feature relationships improve tumor segmentation. Self-supervised learning improves transformer-based methods, especially in cases with little training data. Our technique outperforms U-TRANS in three-dimensional picture segmentation using the two-dimensional PAT-Unet model, confirming its ability to distinguish brain tumors in Fig. 3. We slice a 3D image to get comparable 2D slice data, model to forecast the results for every 2D slice, then rebuild

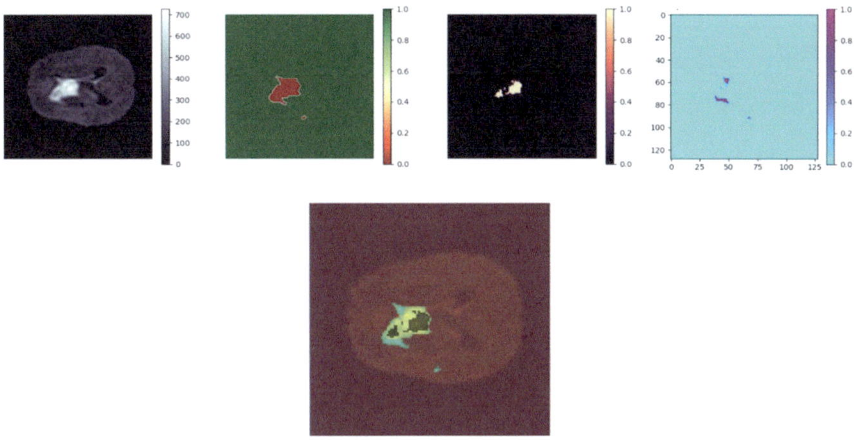

Fig. 3 Outcome of the proposed model which depicts tumor sub-region segmentation

all the sliced forecasts into a three-dimensional prediction result. Compared to PAT-Unet, our brain tumor segmentation approach is more accurate. We visually analyzed segmentation findings from 3D U-Net, UNETR, TransBTS [7], and BTSwin-Unet with cross-validation to identify variations.

Compared to other approaches, ours significantly improves the boundary and textural segmentation of tumors. Large receptive field networks are essential for brain tumor segmentation, as demonstrated by the improved performance of transformer-based techniques in segmenting tiny lesion tissues located far from the tumor center. Additionally, we quantitatively analyzed the computational complexity and characteristics of each model used in the experiment see Table 1. Compared with existing approaches, our PAT-BTS shows better segmentation performance using fewer parameters and computing demands. We also performed a similar experiment with the BraTS2020 dataset to confirm the generality of our PAT-Unet, and the results show that our model performs better at a performance measure as shown in Fig. 4.

Specifically, we first slice a 3D image to acquire equivalent 2D slice data, model to forecast the results for every 2D sliced images then combine all of the sliced forecasts in a three-dimensional prediction result by reconstructing them. Our suggested method shows improved accuracy on brain tumor segmentation compared to PAT-Unet. In order to identify variations, we also visually evaluated segmentation results from Various methodologies, including 3D U-Net, UNETR, and BTSwin-Unet, were employed with cross-validation techniques. Compared to other approaches, ours significantly improves the boundary and textural segmentation of tumors. Large receptive field networks are essential for brain tumor segmentation, as demonstrated by the improved performance of transformer-based techniques in segmenting tiny lesion tissues. Compared with existing approaches, our PAT-BTS shows better segmentation performance using fewer parameters and computing demands. We also

Table 1 Compares with improved CNN methods with proposed method on BraTS2019 and 2020 dataset

Model	Dice score (%) (Avg.)	Dice score (%) (Std.)	Jaccard (%) (Avg.)	Jaccard (%) (Std.)	MAE
3D U-Net [2]	78.5	73.8	86.2	72.6	13.2
U-Net + + [1]	79.1	74.2	86.7	74.3	9.1
Att-Unet [3]	81.2	76.3	88.1	77.2	8.0
UNETR [6]	83.1	78.5	89.3	81.5	7.5
TransBTS [7]	83.3	78.9	89.0	82.0	7.6
Swin-Unet [8]	80.7	76.0	87.3	77.1	8.3
3D Swin—Unet [16]	83.0	77.6	89.1	81.3	6.4
PAT—Unet	94.6	89.4	98.2	92.9	6.3

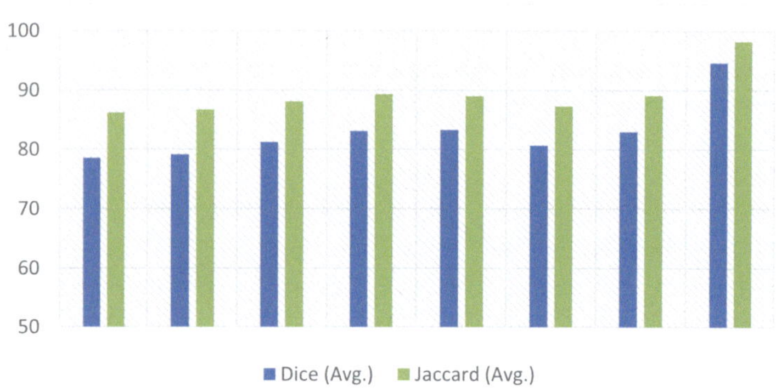

Fig. 4 Qualitative analysis on proposed method with various metrics

performed a similar experiment with the BraTS2020 dataset to confirm the generality of our PAT-Unet shows in Table 2, and the results show that our model performs better at performance measures.

Table 2 Comparison of models dice score on efficiency region segmentation metrices

Method	Dice (%) Avg.↓	ET	WT	TC
3D U-Net [2]	83.13	76.78	89.93	82.69
U-Net + + [1]	83.02	77.98	86.63	84.47
UNETR [6]	85.60	80.94	88.92	86.96
TransBTS [7]	87.92	82.93	91.34	89.49
Swin Unet [8]	89.40	83.93	92.74	91.53
Proposed PAT-Unet	94.46	92.93	96.4	93.23

Experiment comparison between previous methods and proposed PAT-Unet has been verified by conducting ablation studies on BraTS2020.

PAT-Unet effectiveness in brain tumor segmentation was demonstrated and validated in our study. Pre-processing and post-processing methods, however, are critical to the segmentation of intricate multimodal 3D medical images. Additionally, while improving the model's learning skills through the use of self-supervise learning, the proposed design was not the most sophisticated or ideal. The study now directed at improving network design and delving further into pre-post-processing techniques with realm dataset for advancement in medical imaging.

5 Conclusion

In this paper, an efficient segmentation methodology for MRI-based tumor delineation in brain is presented: PAT-Unet and advantages of the Attention with Transformer are integrated into both the decoding and encoding modules of this network. To address the issue of little training data, we also used a self-supervised learning system. Additionally, by replacing the non-overlapping patch embedded and even down sampling module included in the Swin Transformer, , incorporating overlapped patch embedding and down sampling techniques, we were able to increase the localization of the tumor segmentation. By comparing the Dice scores then Jaccard obtained by our approaches with the BraTS 2019 in addition to BraTS 2020 datasets, we were able to demonstrate their usefulness through extensive study and experimentation.

References

1. Zhou Z, Siddiquee MMR, Tajbakhsh N, Liang J (2019) Unet++: redesigning skip connections to exploit multiscale features in image segmentation. IEEE Trans Med Imaging 39(6):1856–1867
2. Pitchai R, Supraja P, Victoria AH, Madhavi MJNPL (2021) Brain tumor segmentation using deep learning and fuzzy K-means clustering for magnetic resonance images. Neural Process Lett 53:2519–2532
3. Vaswani A, Shazeer N, Parmar N, Uszkoreit J, Jones L, Gomez AN, Polosukhin I (2017) Attention is all you need. Adv Neural Inf Process Syst 30
4. Dosovitskiy A, Beyer L, Kolesnikov A, Weissenborn D, Zhai X, Unterthiner T, Houlsby N (2020) An image is worth 16x16 words: transformers for image recognition at scale. arXiv preprint arXiv:2010.11929
5. Liu Z, Lin Y, Cao Y, Hu H, Wei Y, Zhang Z, Guo B (2021) Swin transformer: hierarchical vision transformer using shifted windows. In: Proceedings of the IEEE/CVF international conference on computer vision, pp 10012–10022
6. Hatamizadeh A, Tang Y, Nath V, Yang D, Myronenko A, Landman B, Xu D (2022) Unetr: transformers for 3d medical image segmentation. In: Proceedings of the IEEE/CVF winter conference on applications of computer vision, pp 574–584

7. Wang W, Chen C, Ding M, Yu H, Zha S, Li J (2021) Transbts: multimodal brain tumor segmentation using transformer. In: Medical image computing and computer assisted intervention–MICCAI 2021: 24th international conference, Strasbourg, France, 2021, Proceedings, Part I 24. Springer International Publishing, pp 109–119
8. Cao H, Wang Y, Chen J, Jiang D, Zhang X, Tian Q, Wang M (2022) Swin-unet: Unet-like pure transformer for medical image segmentation. In: European conference on computer vision. Springer Nature Switzerland, Cham, pp 205–218
9. Bao H, Dong L, Piao S, Wei F (2021) Beit: bert pre-training of image transformers. arXiv preprint arXiv:2106.08254
10. Xiao T, Singh M, Mintun E, Darrell T, Dollár P, Girshick R (2021) Early convolutions help transformers see better. Adv Neural Inf Process Syst 34:30392–30400
11. Dong X, Bao J, Chen D, Zhang W, Yu N, Yuan L, Guo B (2022) Cswin transformer: a general vision transformer backbone with cross-shaped windows. In: Proceedings of the IEEE/CVF conference on computer vision and pattern recognition, pp 12124–12134
12. Huang H, Liu P, Liu J (2023) TAGU-Net: transformer convolution hybrid-based U-Net with attention gate for atypical meningioma segmentation. IEEE Access
13. Ibtehaz N, Rahman MS (2020) MultiResUNet: rethinking the U-Net architecture for multimodal biomedical image segmentation. Neural Netw 121:74–87
14. Tummala S, Kadry S, Bukhari SAC, Rauf HT (2022) Classification of brain tumor from magnetic resonance imaging using vision transformers ensembling. Curr Oncol 29(10):7498–7511
15. Hossain S, Chakrabarty A, Gadekallu TR, Alazab M, Piran MJ (2023) Vision transformers, ensemble model, and transfer learning leveraging explainable ai for brain tumor detection and classification. IEEE J Biomed Health Inform
16. Jiang Y, Zhang Y, Lin X, Dong J, Cheng T, Liang J (2022) SwinBTS: a method for 3D multimodal brain tumor segmentation using swin transformer. Brain Sci 12(6):797
17. Li Z, Cong Y, Chen X, Qi J, Sun J, Yan T, Gao X (2023) Vision transformer-based weakly supervised histopathological image analysis of primary brain tumors. IScience 26(1)
18. Anaya-Isaza A, Mera-Jiménez L, Verdugo-Alejo L, Sarasti L (2023) Optimizing MRI-based brain tumor classification and detection using AI: a comparative analysis of neural networks, transfer learning, data augmentation, and the cross-transformer network. Eur J Radiol Open 10:100484
19. Liang J, Yang C, Zeng M, Wang X (2022) TransConver: transformer and convolution parallel network for developing automatic brain tumor segmentation in MRI images. Quant Imaging Med Surg 12(4):2397
20. Tabatabaei S, Rezaee K, Zhu M (2023) Attention transformer mechanism and fusion-based deep learning architecture for MRI brain tumor classification system. Biomed Signal Process Control 86:105119
21. Ferdous GJ, Sathi KA, Hossain MA, Hoque MM, Dewan MAA (2023) LCDEiT: a linear complexity data-efficient image transformer for MRI brain tumor classification. IEEE Access 11:20337–20350
22. ZainEldin H, Gamel SA, El-Kenawy ESM, Alharbi AH, Khafaga DS, Ibrahim A, Talaat FM (2022) Brain tumor detection and classification using deep learning and sine-cosine fitness grey wolf optimization. Bioengineering 10(1):18
23. Xing Z, Yu L, Wan L, Han T, Zhu L (2022) NestedFormer: nested modality-aware transformer for brain tumor segmentation. In: International conference on medical image computing and computer-assisted intervention. Springer Nature Switzerland, Cham, pp 140–150
24. ZongRen L, Silamu W, Yuzhen W, Zhe W (2023) DenseTrans: multimodal brain tumor segmentation using swin transformer. IEEE Access
25. Liu Z, Ma C, She W, Wang X (2023) TransMVU: multi-view 2D U-Nets with transformer for brain tumour segmentation. IET Image Proc 17(6):1874–1882
26. Chen J, Lu Y, Yu Q, Luo X, Adeli E, Wang Y, Zhou Y (2021) Transunet: transformers make strong encoders for medical image segmentation. arXiv preprint arXiv:2102.04306

27. Wu H, Xiao B, Codella N, Liu M, Dai X, Yuan L, Zhang L (2021) Cvt: introducing convolutions to vision transformers. In: Proceedings of the IEEE/CVF international conference on computer vision, pp 22–31
28. Wang W, Xie E, Li X, Fan DP, Song K, Liang D, Shao L (2022) Pvt v2: Improved baselines with pyramid vision transformer. Comput Vis Media 8(3):415–424
29. He K, Chen X, Xie S, Li Y, Dollár P, Girshick R (2022) Masked autoencoders are scalable vision learners. In: Proceedings of the IEEE/CVF conference on computer vision and pattern recognition, pp 16000–16009
30. Xie Z, Zhang Z, Cao Y, Lin Y, Bao J, Yao Z, Hu H (2022) Simmim: a simple framework for masked image modeling. In: Proceedings of the IEEE/CVF conference on computer vision and pattern recognition

Machine Learning Applications for Air Quality Index Prediction

Akshat Sahay, Minakhi Rout, Junali Jasmine Jena, Lipika Mohanty, Bibhuti Bhusan Dash, and Sudhansu Shekhar Patra

Abstract The functioning of the ecosystem and environment, as well as the lives and well-being of individuals in the immediate vicinity, are significantly impacted by air quality. People are able to make well-informed lifestyle decisions by forecasting the air quality index. The creation of an air quality prediction model was started, utilising a number of phases from data collection to model selection, including online scraping for dependability and rigorous data pre-processing and assessment. The primary work is to locate reliable meteorological data and handling erratic data (missing values and parameters, for example). Web scraping techniques and novel data prioritisation, including benchmark approaches like KNN Imputer and Power Transformer, are used to address these problems. The model that performed the best at predicting climatic quality among all the models that were assessed was the Bagging Regressor. Its optimum performance and ability to anticipate weather accurately are shown by a low least squares error (MSE) rating.

Keywords AQI · Data handling · Machine learning

A. Sahay · M. Rout (✉) · J. J. Jena · L. Mohanty
School of Computer Engineering, KIIT Deemed to Be University, Bhubaneswar, Odisha, India
e-mail: minakhi.routfcs@kiit.ac.in

A. Sahay
e-mail: akshatsahay1999@gmail.com

J. J. Jena
e-mail: junali.jenafcs@kiit.ac.in

L. Mohanty
e-mail: lipika.mohantyfcs@kiit.ac.in

B. B. Dash · S. S. Patra
School of Computer Application, KIIT Deemed to Be University, Bhubaneswar, Odisha, India

© The Author(s), under exclusive license to Springer Nature Singapore Pte Ltd. 2025

S. K. Udgata et al. (eds.), *Intelligent Systems*, Lecture Notes in Networks and Systems 1314, https://doi.org/10.1007/978-981-96-3797-3_32

1 Introduction

Pollution is a word derived from 'polluere' which in latin means contamination in general sense something that polluted the surroundings and the environment is called pollution. The presence of harmful substances in the water, soil, air which are affecting the lives of the common people and the environment is also defined as pollution. These harmful substances in turn which affect the environment are commonly known as pollutants or defined as one. Human activities and development are one major reason for affecting the biological diversity of the ecosystem and in turn cause pollution also posing a threat to the Environmental sustainability.

Air Pollution can be defined as the contamination of the natural air which gets mixed with different pollutants which may include harmful fumes and chemicals. The biggest side effect or drawback of air pollution is Global warming according to the experts.

2 Basic Concepts

2.1 Types of Pollutants

Primary Pollutants: Primary pollutants are emissions directly into the air or environment from an identifiable source. From places such as vehicles, industrial processes, construction and natural events such as fire and explosion. Examples include particulate matter (PM), carbon monoxide (CO), sulfur dioxide (SO2), nitrogen oxides (NOx) and volatile organic compounds (VOC). This substance, generally defined by its form, can directly affect the environment and human health when released. It especially causes air pollution, smoke formation, respiratory diseases and environmental degradation. Often effects occur that affect the immediate area of the emission point.

Secondary pollutants: Secondary pollutants are not released directly. Instead, they form as a result of chemical reactions in the atmosphere involving major pollutants, sunlight, and other atmospheric factors. It is formed from chemical compounds that include major pollutants such as nitrogen oxides, harmful organic compounds and sunlight.

Examples include ground-level ozone (O3), secondary aerosols, and some types of acid rain. They are difficult to control because they form depending on various weather conditions and interactions. Initially affects smog formation, including ground-level ozone and the formation of secondary aerosols. Major impact on human health, ecosystems and resources. may spread for a long time from where it was created and cause regional and international problems.

To measure the AQI the following pollutants are taken into account-

Particulate count (PM2.five and PM10): PM2.5 refers to first-rate inhalable debris with a diameter of 2.5 μm or smaller, whilst PM10 includes slightly larger particles with a diameter of 10 μm or smaller.

Traits: these particles come from diverse sources, along with automobile emissions, commercial strategies, construction, and natural assets like dirt storms and wildfires.

Effect: Inhalation of PM2.5 and PM10 can cause respiratory and cardiovascular problems, with PM2.five having a higher capacity to penetrate deep into the lungs.

Ozone (O3): ground-level ozone (O3) is a secondary pollutant formed by way of chemical reactions among risky natural compounds (VOCs) and nitrogen oxides (NOx) within the presence of sunlight.

Traits: It bureaucracy because of complex atmospheric reactions, in particular in urban regions with high emissions of VOCs and NOx.

Effect: breathing ozone can cause breathing problems, aggravate allergies, and decrease lung features.

Carbon Monoxide (CO): Carbon monoxide is a drab, odorless fuel produced via incomplete combustion of fossil fuels and biomass.

Characteristics: in particular emitted by way of vehicles, commercial approaches, and home heating structures.

Impact: CO can intervene with the frame's capability to move oxygen, leading to headaches, dizziness, and, at high levels, may be existence-threatening.

Sulfur Dioxide (SO2): Sulfur dioxide is a gasoline produced through burning fossil fuels containing sulfur compounds, particularly coal and oil.

Traits: Emitted through strength flora, business facilities, and volcanic eruptions.

Impact: Can cause respiratory problems and make contributions to acid rain, that may harm ecosystems and infrastructure.

Nitrogen Dioxide (NO2): Nitrogen dioxide is a reddish-brown fuel produced via excessive-temperature combustion, particularly in vehicles and strength flowers.

Traits: usually emitted in urban regions with heavy visitors and industrial activities.

Impact: NO2 can worsen the respiratory machine, worsen allergies, and make a contribution to the formation of smog.

2.2 Causes of Pollution

Industrial sports: Heavy industries along with manufacturing, mining, and refining launch various pollutants into the air, water, and soil. Emissions from factories and production processes contribute extensively to environmental pollutants.

Automobile Emissions: The combustion of fossil fuels in vehicles, vans, and other transportation modes releases pollution like carbon monoxide, nitrogen oxides, and particulates. high traffic regions regularly exhibit accelerated air pollution levels.

Agricultural Practices: Agricultural sports, which include using fertilizers, insecticides, and animal waste, cause soil and water infection. Runoff from farms contains pollutants into our bodies, causing water pollutants.

Mistaken Waste Disposal: Incorrect disposal of stable waste, which includes plastics, electronic waste, and household garbage, contaminated land and water. Landfills and incineration contribute to soil, air, and water pollutants if not controlled nicely.

Deforestation and Land Use modifications: Clearing forests for agriculture, urbanization, and infrastructure development disrupts ecosystems, mainly to soil erosion and loss of biodiversity. Such changes contribute to air and water pollutants. Extractive sports like mining and quarrying launch hazardous substances and pollution into the environment, affecting soil, water quality, and air purity.

Creation and Demolition: Production sports produce dirt, debris, and chemical emissions that impact air first-class. Demolition methods launch pollutants and construction waste, contributing to environmental degradation.

Natural occasions: Natural failures like wildfires, volcanic eruptions, and dust storms can release full-size pollutants into the ecosystem, affecting air excellent regionally or even globally.

Energy production: The burning of fossil fuels for electricity technology in power vegetation and heating contributes substantially to air pollution via the emission of greenhouse gases and different pollutants.

2.3 Need of AQI

The Air quality Index (AQI) serves as a critical tool for gauging and communicating the excellence of the air we breathe.

Public fitness recognition: AQI gives a simplified and comprehensible degree of air quality, allowing the general public to grasp the capacity health risks associated with distinct pollution stages. It facilitates people, in particular susceptible populations, to make knowledgeable choices to defend their fitness.

Guidance for protective Measures: By way of categorizing air nice into special ranges (true, mild, dangerous, and many others.), the AQI gives steerage on vital precautions. It advises human beings on adjusting outdoor sports, sporting masks, or staying indoors for the duration of excessive pollutant periods to limit publicity to harmful pollution.

Environmental monitoring and law: It aids environmental businesses and policy-makers in tracking pollutants tiers throughout areas, permitting them to enforce focused interventions and regulations. This facts-pushed approach allows in placing emission standards and formulating regulations to enhance air first-class.

Impact on Economies and making plans: AQI records affect urban planning, infrastructure development, and financial choices. cities with better air high-quality may attract companies and citizens, whilst people with poor air quality may additionally face demanding situations in phrases of public health expenses and reduced splendor.

Emergency response and Preparedness: All through pollution peaks or environmental emergencies, the AQI serves as a critical tool for emergency response making plans. It triggers well timed alerts and measures to shield public fitness, specially in situations like wildfires or business injuries.

Worldwide Cooperation and awareness: The AQI standardizes air satisfactory assessment globally, facilitating international cooperation in addressing transboundary pollution troubles. It promotes consciousness about the interconnectedness of air quality and environmental fitness on an international scale.

The AQI's role extends past mere size; it empowers individuals, guides rules, and fosters collaboration towards a common goal of healthier environments and stepped forward public properly-being.

3 Literature Review

D Sanjeev [1] in his paper used Random Forest (RF), Support Vector Machine (SVM) and Artificial Neural Network (ANN) for predicting Air Quality Index and found RF performed better. Bhalgat et al. [2] have used time series models predict concentration of SO2 in the air and got promising results. Nairita et al. [3] in their paper used deep learning models to predict AQI and got good results. Madan et al. [4] reviewed various papers related to AQI prediction using machine learning models. Mahalingam et al. [5] used ANN and SVM for AQI and found that Gaussian SVM performed better. Ahmad et al. [6] used recurrent neural network to predict AQI and found that it performed better than many other models. Jiao et al. [7] used LSTM for AQI prediction and found the model effective. Zhao et al. [8] used temporal-spatial model for AQI prediction and found promising results. Udristioiu et al. [9] used hybrid machine learning algorithms for AQI forecasting and got good results. Kim et al. [10] used Spatio-temporal data for AQI prediction and got good results.

4 Methodology

4.1 Design Model

Figure 1 shows the block diagram of the methodology used for AQI prediction.

The simulation was conducted on an online compiler called Kaggle, using an Intel Core i7-8550U CPU @ 1.99 GHz, with an AMD Radeon 530 graphics card and 13 GB of memory. It offers a unique blend of community collaboration, real-world challenges, and resources.

The dataset comprises air quality information, including parameters such as PM2.5, PM10, O3, SO2, CO, wind speed, humidity, temperature, AQI (Air Quality Index), and AQI Type that has been scraped from a website called 'Air matters' using the python beautiful-soup library. Some entries have missing values, prompting the need for normalization techniques.

Different countries may have different AQI scales and corresponding health advisories. In India, the National Air Quality Index (NAQI) is used to assess and communicate air quality to the public. The NAQI categorizes air quality into different bands, each associated with a specific health advisory. As of update in January 2022, the NAQI has six categories, ranging from "Good" to "Severe".

Normalization of dataset was done using Iterative and KNN imputer. Following a thorough comparison of Mean Squared Error (MSE) scores for various scaling techniques such as min–max, decimal, log etc., it was discerned that Power Transformer emerged as the most effective choice for our dataset.

The split ratio of dataset is set to 80% training and 20% testing, with a random seed of 42 for reproducibility. The chosen feature scaling method, determined earlier

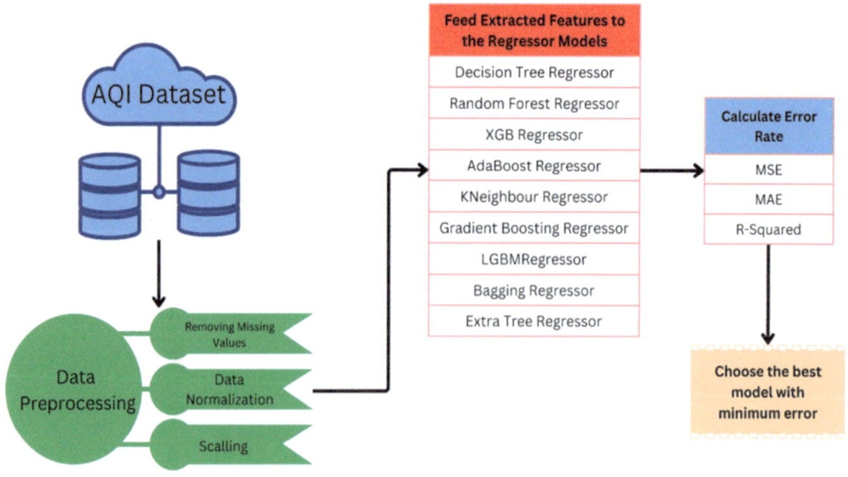

Fig. 1 Figure showing the methodology of the proposed model

as "Power Transformer," is applied to both the training and testing sets. If scaling is required (not None), the scaler is fitted on the training data, and the transformation is applied to both the training and testing sets. If no scaling is necessary, the original sets are retained.

The successful implementation of the data split and scaling process lays the foundation for subsequent model training and evaluation. The model's performance will be assessed using appropriate metrics, such as Mean Squared Error or R-squared, to determine its effectiveness in predicting the Air Quality Index.

4.2 Models Included

A continuous target variable can be predicted using regression analysis, a statistical method, based on one or more independent variables. Regression models come in a multitude of forms, each with unique advantages and disadvantages. The Decision Tree Regressor, Random Forest Regressor, XGBoost Regressor, AdaBoost Regressor, KNeighbors Regressor, Gradient Boosting Regressor, LightGBM Regressor, Bagging Regressor, and ExtraTrees Regressor are the nine regression models that has been compared in this report.

5 Result Analysis

R-*SQUARE,* MSE and MAE are the matrices used for performance evaluation. Comparative performance analysis using RSquare is shown in Fig. 2, using MSE is shown in Fig. 3 and Comparative performance analysis using R2Score is shown in Fig. 4.

Comparative performance analysis of different models is given in the below table.

Rank	Model	MSE	MAE	R2 Score (%)
1	ExtraTreesRegressor	2.19	0.81	99.81
2	XGBRegressor	5.11	1.27	99.56
3	GradientBoostingRegressor	5.44	0.86	99.54
4	RandomForestRegressor	38.10	2.09	96.75
5	AdaBoostRegressor	38.37	5.19	96.73
6	BaggingRegressor	39.21	1.56	96.66
7	DecisionTreeRegressor	47.31	1.56	95.96
8	KNeighborsRegressor	123.48	3.04	89.47
9	LGBMRegressor	285.83	7.17	75.62

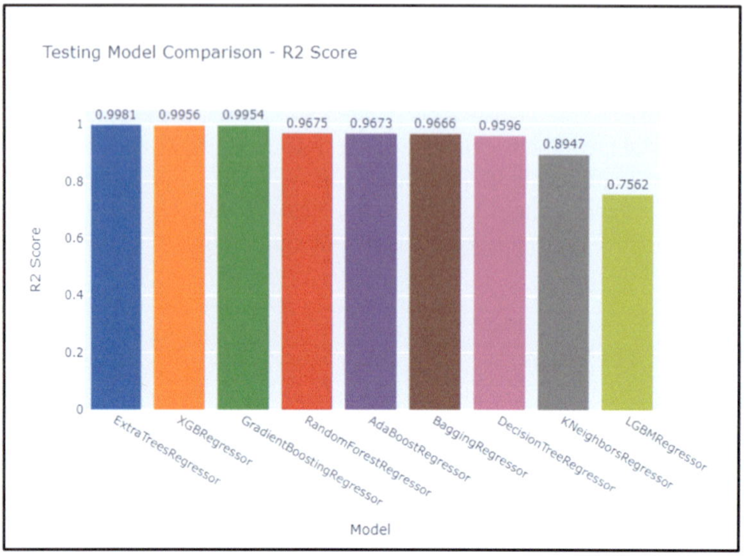

Fig. 2 Figure showing the Comparative performance analysis using RSquare

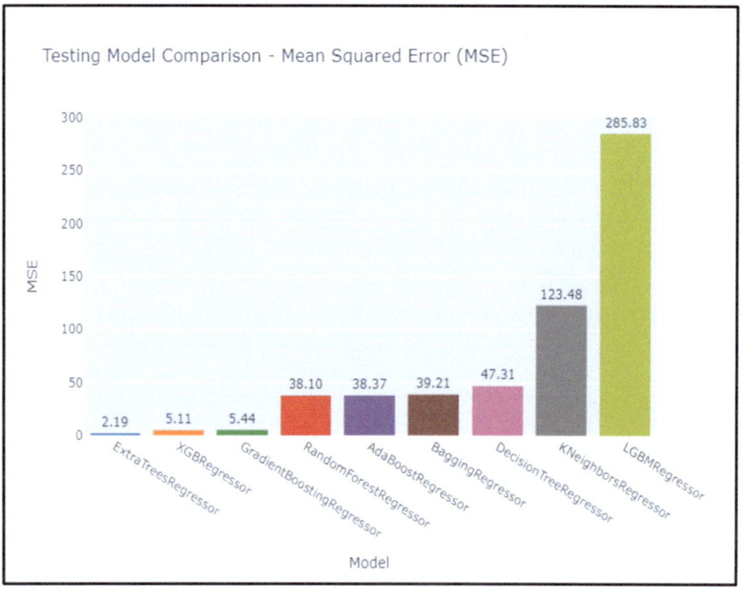

Fig. 3 Figure showing the Comparative performance analysis using MSE

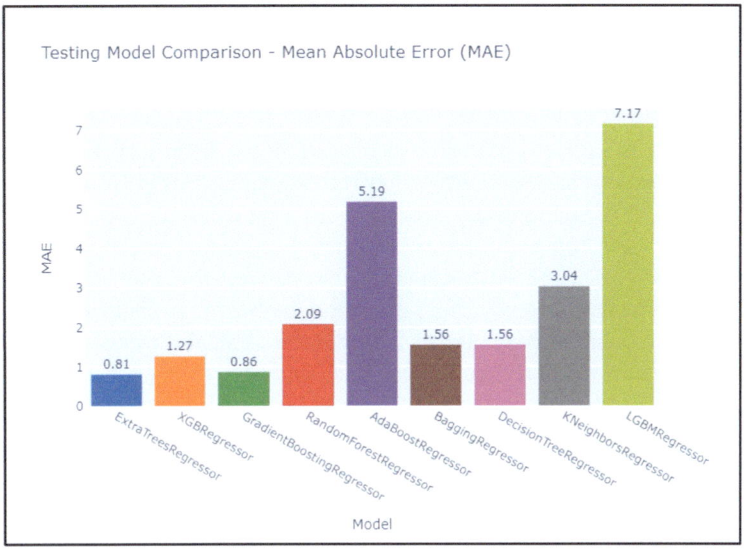

Fig. 4 Figure showing the Comparative performance analysis using MAE

5.1 Which is the Best Model and Why?

The Best Model is the ExtraTreeRegressor the reason being:

1. Lowest MSE: The ExtraTreesRegressor has the lowest MSE (2.19), which means it has the smallest overall squared error among all models.
2. Lowest MAE: The ExtraTreesRegressor also has the lowest MAE (0.81), further supporting its accuracy in prediction.
3. Highest R2 Score: The ExtraTreesRegressor has the highest R2 Score (99.81%), signifying an excellent fit to the data

The ExtraTreesRegressor outperforms other models in terms of accuracy (low MSE and MAE) and goodness of fit (high R2 Score). It consistently provides the most accurate and reliable predictions based on the given evaluation metrics.

6 Conclusion and Future Scope

Air quality is a major factor that affects the life and well being of people and surrounding as well as it a major factor in working of the ecosystem and environment. By predicting the air quality index people can make informed decisions about their lifestyle. The development of an air quality prediction model was undertaken, leveraging various stages ranging from data acquisition to the selection of the model, encompassing from web scraping for reliability, meticulous data pre-processing and evaluation. Challenges encountered during the work focused on finding good weather data and dealing with irregular data (such as missing values and parameters). These issues are overcome using web scraping techniques and new data prioritization, including the use of benchmark techniques such as KNN Imputer and Power Transformer. Among the various models evaluated, Bagging Regressor turned out to be the best at predicting climate quality. A low least squares error (MSE) value indicates its best performance and indicates accurate weather forecast ability.

In the future, the application of our air quality prediction model will have great potential in many areas and will become an important tool for health. Climate change positively affects decisions regarding outdoor activities. City planners can use this model to develop and improve urban development strategies to reduce the impact of pollution, improve infrastructure and green spaces In healthcare, the integration of these models can help predict increased access to patients during cloud negative wind, which is beneficial for the negative group. Moreover, its contribution to promoting climate change and the collaboration of experts from different fields can create new policy and research insights. Anchoring and evaluating this international model has the potential to solve climate problems in different regions, promote regional solutions and achieve environmental and social goals.

References

1. Sanjeev D (2021) Implementation of machine learning algorithms for analysis and prediction of air quality. Int J Eng Res Technol (IJERT) 10(03) (March 2021). www.academia.edu
2. Bhalgat P, Bhoite S, Pitare S (2019) Air quality prediction using machine learning algorithms. Int J Comput Appl Technol Res 8. https://doi.org/10.7753/IJCATR0809.1006. www.researchgate.net
3. Sarkar N, Gupta R, Keserwani PK, Govil MC (2022) Air quality index prediction using an effective hybrid deep learning model. Environ Pollut 315:120404 ISSN 0269-7491. https://doi.org/10.1016/j.envpol.2022.120404
4. Madan T, Sagar S, Virmani D (2020) Air quality prediction using machine learning algorithms—a review, pp 140–145
5. Mahalingam U, Elangovan K, Dobhal H, Valliappa C, Shrestha S, Kedam G (2019) A machine learning model for air quality prediction for smart cities. In: 2019 International conference on wireless communications signal processing and networking (WiSPNET), Chennai, India, pp 452–457. https://doi.org/10.1109/WiSPNET45539.2019.9032734. ieeexplore.ieee.org
6. Ahmad S, Ahmad T (2023) AQI prediction using layer recurrent neural network model: a new approach. Environ Monit Assess 195(10):1180
7. Jiao Y, Wang Z, Zhang Y (2019) Prediction of air quality index based on LSTM. In: 2019 IEEE 8th joint international information technology and artificial intelligence conference (ITAIC). IEEE, pp 17–20
8. Zhao X, Song M, Liu A, Wang Y, Wang T, Cao J (2020) Data-driven temporal-spatial model for the prediction of AQI in Nanjing. J Artif Intell Soft Comput Res 10(4):255–270
9. Udristioiu MT, Mghouchi YE, Yildizhan H (2023) Prediction, modelling, and forecasting of PM and AQI using hybrid machine learning. J Clean Prod 421:138496
10. Kim K, Ma M, Lee K (2023) Prediction of spatio-temporal AQI data. Commun Stat Appl Methods 30(2):119–133

LFAOA: Levy Flight Based Arithmetic Optimization Algorithm for Epileptic Seizure Detection Using DWT and Rule-Based Classifiers

Puspanjali Mallik, Ajit Kumar Nayak, Sumanta Kumar Mohapatra, and K. P. Swain

Abstract Epilepsy is the common disease occurred due to the unusual electrical flow in the affected portion of the brain that causes seizure symptoms and hampers the normal behaviour of a healthy person. Researchers put their emphasis to develop powerful models to detect the particular types of seizure and presently machine learning algorithms are proved the best model to accurately detect the typical seizure. In this context, stochastic strategies are integrated with the conventional metaheuristic optimization algorithms to achieve global solution by improving exploration and exploitation mechanism in global search space. In this context, the conventional AOA is applicable for many real time applications but it suffers from major shortfalls of reaching towards the convergence point and attempts to fall to a local search space. An efficient version of AOA called as Levy Flight-Based Arithmetic Optimization Algorithm (LFAOA) is developed to improve the search capacity. The non-linear inputted signals are extracted with the most popular signal sample extraction technique called Discrete Wavelet Transform (DWT) and classified with widely used rule-based classifiers viz. Support Vector Machine (SVM), Least-Square SVM (LSSVM), Adoptive Neuro-fuzzy Inference System (ANFIS) etc. The stated technique DWT-LFAOA-LSSVM is tested with publicly available CHB-MIT epilepsy seizure dataset. In this experiment, obtained result is proved with Accuracy (ACC), Sensitivity (SN), Specificity (SP), Positive Predictive Value (PPV), Matthews Correlation Coefficient (MCC) and Area Under Curve (AUC) as 99.8%, 99.35%, 99.82%, 96.85%, 98% and

P. Mallik (✉)
Department of Computer Science and Engineering, Siksha 'O' Anusandhan (Deemed to be) University, Bhubaneswar, Odisha, India
e-mail: mallickpuspa@gmail.com

A. K. Nayak (✉)
Department of CS and IT, Siksha 'O' Anusandhan (Deemed to be) University, Bhubaneswar, Odisha, India
e-mail: ajitnayak@soa.ac.in

S. K. Mohapatra · K. P. Swain
Department of Electronics and Telecommunication Engineering, Trident Academy of Technology, Bhubaneswar, Odisha, India

© The Author(s), under exclusive license to Springer Nature Singapore Pte Ltd. 2025
S. K. Udgata et al. (eds.), *Intelligent Systems*, Lecture Notes in Networks and Systems 1314, https://doi.org/10.1007/978-981-96-3797-3_33

1 for DWT-LFAOA-LSSVM technique which supersedes with DWT-LFAOA-SVM and DWT-LFAOA-ANFIS. This proposed method can be tested with other publicly available datasets.

Keywords Arithmetic optimization algorithm · Levy flight · Discrete wavelet transform · LSSVM and · ANFIS

1 Introduction

Brain is the controller of all the activities of the healthy person. If the person is suffered with epilepsy then the behaviour of the person changes which must be avoidable. Presently, epilepsy is treated as a widespread brain disease and nearly 50 million of people are affected with this disease according to the statistical report of World Health Organization's seizure detection. From the characteristic observation of its symptom, the physicians classified epilepsy into two types. Partial and generalized [1–3]. When a person is affected with generalized seizure, the person can loss the consciousness and may face accidents. Although it is not curable, an early detection can help to know the type of the seizure so that preventive steps can be taken to save the life. Electroencephalogram is the machine used by medical practitioners to detect the type of seizure by placing the machine on the scalp of the person. EEG signal differ from person to person, and even in the same person, different types of EEG signal are observed. There are four different types of EEG signals: ictal, preictal, postictal and inter-ictal. Intracranial EEG signals are used for testing purposes [4–6]. Presently, the EEG machines are powered with metaheuristic optimization techniques and added with stochastic methods to improve the solution timings and test accuracy. There are a number of nature-inspired optimization techniques such as: Genetic Algorithm (GA), Particle Swarm Optimization (PSO), Cuckoo Search, Firefly, Symbiotic-Organism Search (SOS), Farmland Fertility (FF), Grasshopper Optimization algorithms. [7–10]. All of these algorithms are suffered with global search optimum solution. In this experiment, we have applied an improved version of AOA. There are some stochastic approaches those can be combined with AOA viz. disruption, mutation, levy flight and opposition-based learning [11]. In our research, levy flight is used to improve the conventional AOA by improving its exploration strategy towards global search and exploitation capacity in favouring to safety with security.

Table 1 includes the orderly arrangement of the rest sections of the article.

1.1 Contribution of the Paper

In the pursuit of enhanced epilepsy detection, a novel hybrid and efficient technique of the existing Arithmetic Optimization Algorithm (AOA), termed as LFAOA, is experimented. To advance the search for the optimum solution, the algorithm integrates

Table 1 Contents of the article

Sections	Contents
Section 2	Contribution of authors in seizure detection
Section 3	Materials and Methods Used
Section 4	Proposed Levy Flight Arithmetic Optimization Algorithm
Section 5	Result Analysis and Discussion
Section 6	Conclusion and Future Prospects

improvements in population size exploration and exploitation through two optimizer functions: Math Optimizer Accelerated (MOA) function for improving exploration and Math Optimizer Probability function (MOP) for improving exploitation. This experiment uses Discrete Wavelet Transform (DWT) for extracting features and harnesses the synergy to three rule-based classifiers: SVM, LSSVM and ANFIS, in combination with DWT-LFAOA.

2 Literature Review

Epilepsy seizure detection is a combination of many complex techniques as it is having many types of seizers [1, 2]. In early days, the accurate detection was very complicated bur presently automated EEG machines are devised so that immediate and accurate detection can be possible for the disease. The machine is powered with many metaheuristic optimization techniques [3].

In the year 2016, the authors Van Esbroeck and et al. in paper [5] proposed a multitask seizure detection SVM classifier for epilepsy seizure and obtained the measurement parameters such as: accuracy, specificity and sensitivity values as 94.5, 94 and 95%. In the same year, the authors in paper [7] prosed time-domain analysis for EEG signals and used KNN with multiclass SVM for signal classification. They obtained accuracy, specificity and sensitivity as 94.4, 96.25 and 87% for their result analysis parameters. Authors in the paper [8] prosed multichannel vision transformer for tuning the time series EEG signals. In their experiment, they achieved the measurement parameters accuracy, specificity and sensitivity as 95.4, 92 and 88%.

The inputted signals are sampled with 256 samples per second. The signals obtained from the neurons are aperiodic, non-linear in nature which are decomposed into a subset of frequency band signals of time series DWT and further these sub band signals are inputted into the classifier [7–9]. The widely used machine learning model for classification is called support vector machine (SVM). It uses extreme points called as boundary points and these boundary points act as the decision boundary for separation of the characteristics [12]. In paper [13–15], authors proposed Least Square Support Vector Machine (LSSVM) classifier to use for non-linear optimization based epileptic seizure detection. In paper [16], authors proposed about Adoptive Neuro_Fuzzy Inference System (ANFIS) as the classifier used with

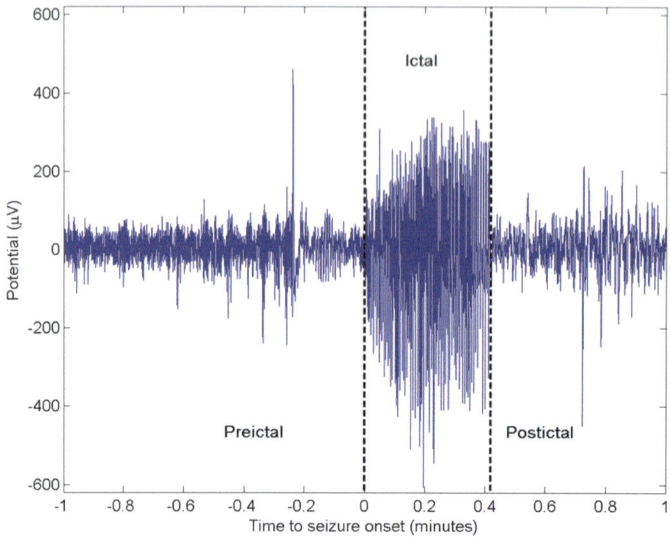

Fig. 1 Recorded CHB-MIT EEG signal

autoencoders and fuzzy layer, product layer, normalized layer, De-Fuzzy layer and total output layer. Paper [17–19] experimented with the classifiers such as ANN-PSO, Binary Dragonfly and OLS-SVM.

3 Materials and Methods Used

3.1 Dataset Used

The dataset used for testing is from CHB-MIT, and the link of the dataset is: https://archive.physionet.org/pn6/chbmit/.

Below is the recorded image of EEG signal followed which is shown in Fig. 1.

3.2 Feature Extraction and Selection

Discrete Wavelet Transform (DWT) is the widespread signal sampling technique for sampling the continuous non-linear electrical signal into time series frequency band signals [7]. Mean variance and standard deviations etc. are the measures to evaluate the performances of a sampling signal. We used three types of classifiers for seizure

Table 2 Important features with their mathematical representations

Extracted features	Equations
F_1 = Minimum Value (min)	$\min = X_{\min}$
F_2 = Maximum Value (max)	$\max = X_{\max}$
$F_3 = \text{Mean}(\mu)$	$\mu = \frac{1}{N}\sum_{i=1}^{N} x_i$
$F_4 = \text{Variance}(\mu^2)$	$\mu^2 = \frac{1}{N-1}\sum_{i=1}^{N} \lvert x_i - \mu\rvert^2$
$F_5 = \text{Energy } (E)$	$E = \sum_{i=0}^{N-1} \lvert X_i^2\rvert$
$F_6 = \text{Standard Deviation(SD)}$	$\text{SD} = \sqrt{\frac{1}{N-1}\sum_{i=1}^{N} (x_i - \mu_x)^2}$
$F_7 = \text{Shannon Entropy}(SE)$	$SE = \sum_{i=1}^{N} P(x_i)\log(P(x_i))$

detection: SVM, LSSVM and ANFIS. In this study, the resultant new method DWT-LFAOA-LSSVM is formulated and proved as the best method among all considered methods (Table 2).

3.3 Arithmetic Optimization Algorithm (AOA) [4]

AOA uses the distribution property which are meant for arithmetic operators such as multiplication (M), Division (D), Subtraction (S) and Addition (A) in optimization process of search space [4]. This algorithm is more efficient than conventional optimization algorithms.

In this experiment, the metaheuristic optimization process has a number of candidate solutions (X) where (X) is a matrix obtained randomly in the search space.

There are two search phases: exploration and exploitation. Math Optimizer Accelerated function acts as a coefficient for each iteration.

$$\text{MOA(C\_It)} = \text{Minval} + \text{C\_It} \times \left(\frac{\text{Maxval} - \text{Minval}}{\text{M\_It}}\right) \tag{1}$$

Here, C_It presents the current iteration and M_It presents the maximum iteration.

In exploration phase, AOA selects multiplication and division operator due to high distribution values.

If r_1, r_2 are any random number then the updated position in this phase is

$$x_{i,j}(C\_It+1)=\begin{cases}\mathrm{best}(x_j)\div(\mathrm{MOPr}+\in)\times\left(\left(\mathrm{UB}_j-\mathrm{LB}_j\right)\times\mu+\mathrm{LB}_j\right), & r_2<0.5\\ \mathrm{best}(x_j)\times\mathrm{MOPr}\times\left(\left(\mathrm{UB}_j-\mathrm{LB}_j\right)\times\mu+\mathrm{LB}_j\right) & \text{else}\end{cases} \tag{2}$$

MOPr is the math optimizer probability,

In exploitation phase, AOA uses subtraction and addition operators due to high dense and low disruption values of the operators and the equation for the updated position in this phase is

$$x_{i,j}(C\_It+1)=\begin{cases}\mathrm{best}(x_j)-(\mathrm{MOPr})\times\left(\left(\mathrm{UB}_j-\mathrm{LB}_j\right)\times\mu+\mathrm{LB}_j\right), & r_3<0.5\\ \mathrm{best}(x_j)+\mathrm{MOPr}\times\left(\left(\mathrm{UB}_j-\mathrm{LB}_j\right)\times\mu+\mathrm{LB}_j\right) & \text{else}\end{cases} \tag{3}$$

4 Proposed Levy Flight Arithmetic Optimization Algorithm

In this improved search strategy authors proposed levy flight as the stochastic process to integrate with the conventional AOA method.

4.1 *Modification Using Levy Flight*

Levy Flight (LF) is a random search strategy that is meant for extending the exploration capacity in a multidimensional space. In this study, the arithmetic operators are further explored randomly to reach a global search position. LF is derived from levy distribution (LD) function [11] and the Fourier transform equation is:

$$F_N(k)=\exp\left[-N\left|k^{\beta}\right|\right] \tag{4}$$

The LD is given by the equation:

$$L(s)=\frac{1}{\pi}\int_0^{\infty}\cos(\tau s)e^{-\alpha\tau^{\beta}}d\tau,\ \ (0<\beta\leq 2) \tag{5}$$

Here, τ is the time interval for each step s.

Equation (5) can be simplified further to a near approximated equation as:

$$L(s)\sim s^{-1-\beta} \tag{6}$$

The variance of random walk can be expressed as:

$$\alpha^2(t) = |v_0|^2 t^2 + (2dD)t \tag{7}$$

v_0 is the drift velocity and D is the diffusion coefficients.

Final velocity can be expressed as:

$$v_i^d(t+1) = W(t) C_{\exp} \text{rand}_i^d v_i^d(t) \times \text{LF}(\beta) + \alpha n_i^d \tag{8}$$

Algorithm1: Pseudocode of LFAOA Algorithm

```
Initialize Fitness function, population size, C_It and M_It
While (C_It < M_It)
 Find the best solution
 Update MOA using equation 1
 For (i=1to Num_Solutions)
   For (j=1 to Num_Positions)
   If (r1, > MOA)
   If (r2, > 0.5)
      Apply the division operator
        Invoke the LD using equation (5)
        Update the new velocity using equation (8)
          else
        Apply the multiplication operator
        Invoke the LD using equation (5)
        Update the new velocity using equation (8)
    If (r3 > 0.5)
    Apply the subtraction operator
    Invoke the LD using equation (5)
    Update the new velocity using equation (8)
       else
     Apply the addition operator
       Invoke the LD using equation (5)
    Update the new velocity using equation (8)
    End if
  End for
 End for
End while
```

5 Result Analysis and Discussion

In this section, the results and discussions of the proposed method are explained. For this experiment, we have used 12th Generation Intel® Core (TM) i5, processor speed 1.3 GHz with 8GB RAM. We used nd MATLAB 2018b is used to implement the stated algorithm (Fig. 2).

The results of the proposed algorithm are analysed with following parameters.

If True Positive (TP) is represented as α, True Negative (TN) is represented as β, False Positive (FP) is represented as γ and False Negative (FN) is represented as δ, then the confusion matrix parameters are represented as:

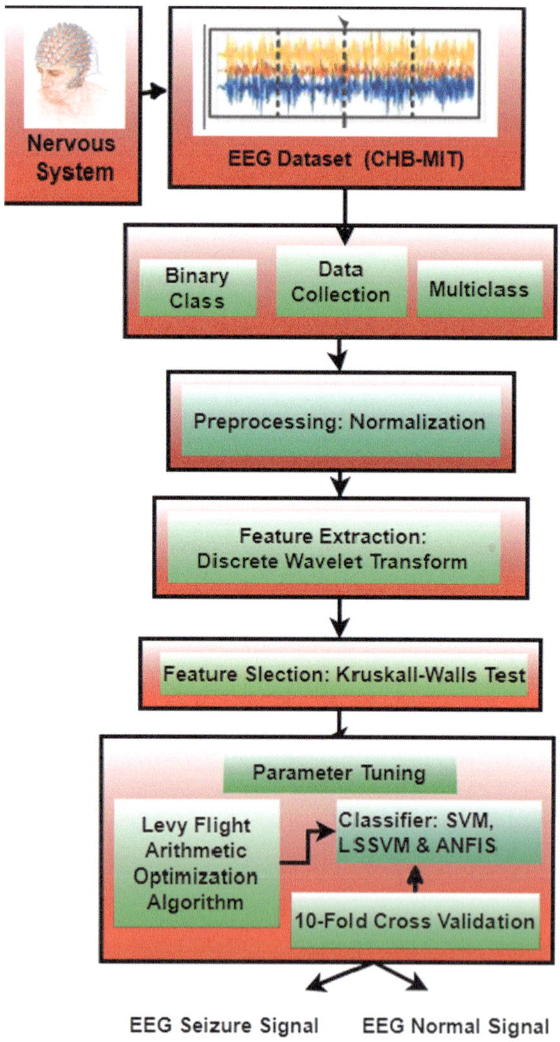

Fig. 2 Steps in the proposed model

$$\text{Accuracy}(\text{ACC\%}) = \frac{\alpha + \beta}{\alpha + \beta + \gamma + \delta} \times 100\% \quad (9)$$

$$\text{Sensitivity}(\text{SN\%}) = \frac{\alpha}{\alpha + \delta} \times 100\% \quad (10)$$

$$\text{Specificity}(\text{SP\%}) = \frac{\beta}{\beta + \gamma} \times 100\% \quad (11)$$

$$\text{Positive Predicted Value(PPV)} = \frac{\alpha}{\alpha + \gamma} \times 100\% \quad (12)$$

$$\text{Matthews Correlation Coefficient (MCC)} = \frac{\alpha \times \beta - \gamma.\delta}{\sqrt{(\alpha + \beta) \times (\alpha + \uparrow \delta) \times (\beta + \gamma) \times (\beta + \delta)}} \times 100\%\ \text{s} \quad (13)$$

Three different classifiers are used for conducting the testing of LFAOA algorithm with three classifiers. According to the outputs of the above experimental cases DWT-LFAOA-LSSVM obtains the values 99.8%, 99.35%, 99.82%, 96.85%, 98% and 1 for measurement parameters average accuracy, sensitivity, specificity, PPV, MCC and AUC and outperforms among other two methods listed in (Figs. 3, 4, 5, 6, 7, 8) Tables 3 and 4.

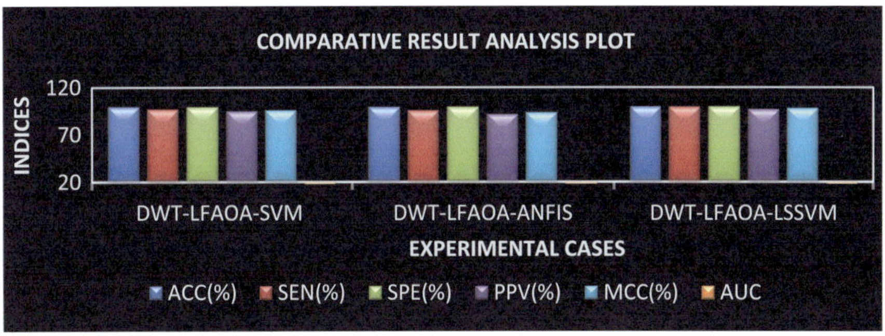

Fig. 3 Result Analysis of different Experimental cases according to the measurement parameters

Table 3 Measurement parameters

Experimental cases	ACC (%)	SN (%)	SP (%)	PPV (%)	MCC (%)	AUC
DWT-LFAOA-SVM	99.4	96.84	99.29	94.84	95.83	0.995
DWT-LFAOA-ANFIS	99.2	95.65	99.47	91.66	93.61	0.998
DWT-LFAOA-LSSVM	99.8	99.35	99.82	96.85	98	1

Table 4 Performance analysis of proposed method with the following methods

Sl. No.	Authors and ref. Nos.	Methods used	Measurement parameters		
			SP (%)	SN (%)	ACC(%)
1	Panda et al. [16]	(WCA-APSO)	99.32	99.62	99.16
2	Saini et al. [17]	PSO+ANN	98.7	98.67	98.67
3	Yogarajan et al. [18]	Binary dragonfly algorithm (BDFA) + DNN	88.58	89.26	92.53
4	Suresh Srirangam [19]	OLS+SVM	89.60	93.34	94.47
5	Proposed method	(Best Result Finding) DWT-LFAOA-LSSVM	99.82	99.35	99.8

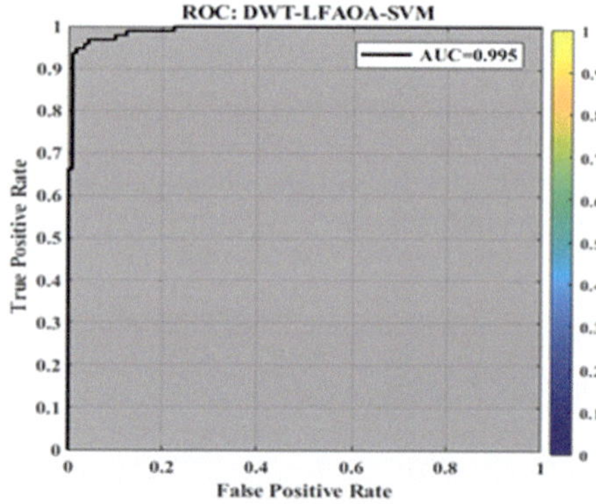

Fig. 4 ROC plot of DWT-LFAOA-SVM

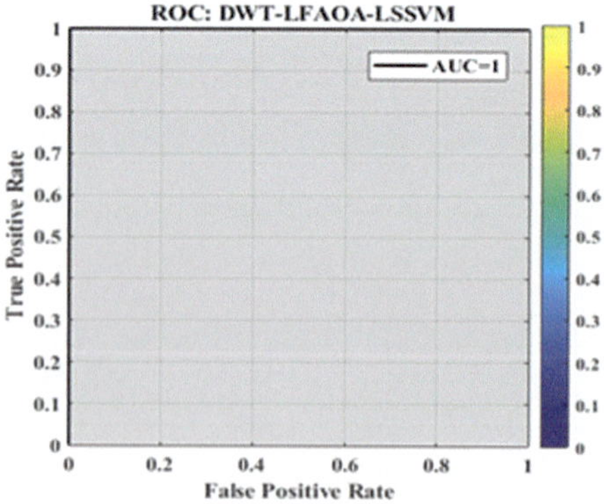

Fig. 5 ROC plot of DWT-LFAOA-LSSVM

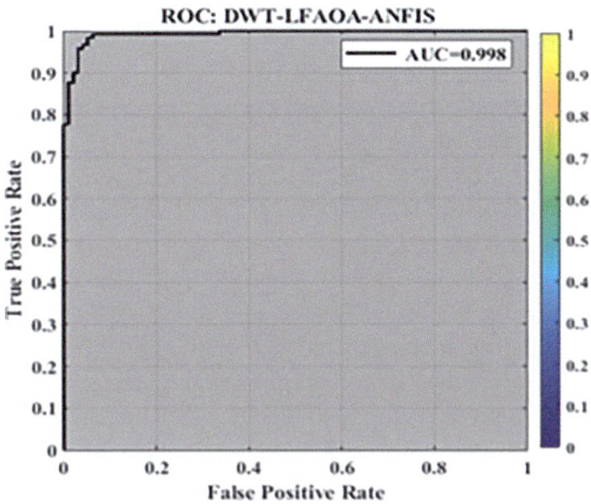

Fig. 6 ROC plot of DWT-LFAOA-ANFIS

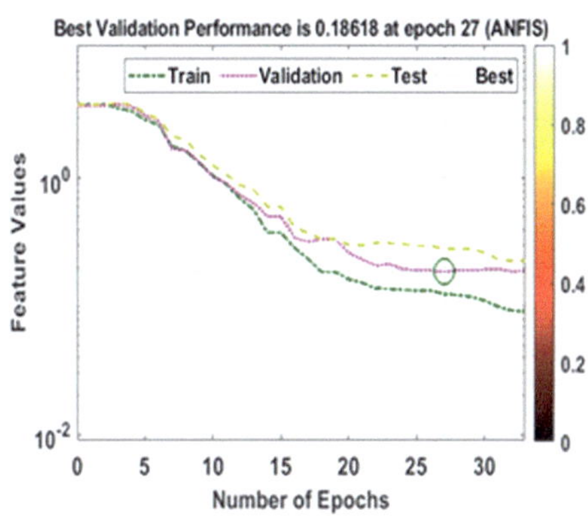

Fig. 7 Performance Plot of DWT-LFAOA-ANFIS

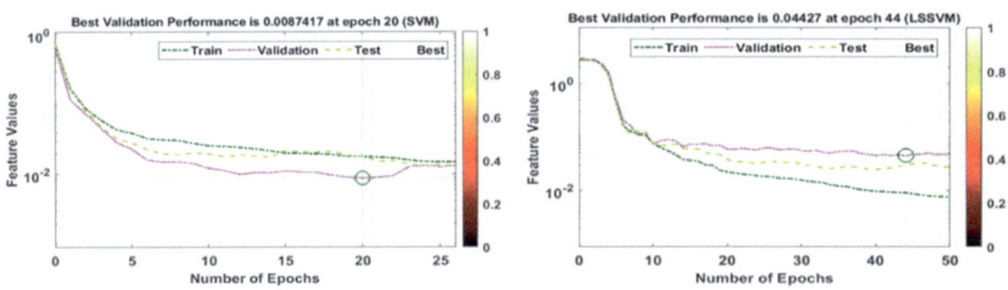

Fig. 8 Performance Plot of DWT-LFAOA-SVM and DWT-LFAOA-LSSVM

6 Conclusion and Future Prospects

In this proposed method, epileptic seizure detection with its measurement parameters are categorized as seizure and non-seizure signals. For this, the improved version of AOA, i.e., LFAOA is used with three different types of rule-based classifiers. The proposed method DWT-LFAOA-LSSVM is compared with DWT-LFAOA-SVM and DWT-LFAOA-ANFIS and proved that DWT-LFAOA-LSSVM supersedes among others. Again, the stated technique is compared with the methods and proved that our proposed method supersedes among other existing methods. For performance computation, it uses 10-fold cross validation process. In future, this proposed method can be combined with disruption and opposite based stochastic learning processes to improve the accuracy in global search space and tested with other publicly available EEG datasets to consider its popular global acceptability for epilepsy seizure detection.

References

1. Shoeibi A, Khodatars M et al (2021) Epileptic seizures detection using deep learning techniques: a review. Int J Environ Res Public Health MDPI
2. Siddiqui MK, Morales-Menendez R et al (2020) A review of epileptic seizure detection using machine learning classifiers. Springer Open
3. Mohapatra SK, Patnaik S (2022) ESA-ASO: An enhanced search ability based atom search optimization algorithm for epileptic seizure detection. Meas Sens 24
4. Abualigah L, Diabat A et al (2021) The arithmetic optimization algorithm computer methods I Applied mechanics and engineering 376:113609
5. Van Esbroeck A, Smith L, Syed Z, Singh S, Karam Z (2016) Multi-task seizure detection: addressing intra-patient variation in seizure morphologies. Mach Learn 102(3):309–321
6. Peker M, Sen B, Delen D (2015) A novel method for automated diagnosis of epilepsy using complex-valued classifiers. IEEE J Biomed Health Inform 20(1):108–118
7. Bairagi RN, Maniruzzaman M, Pervin S, Sarkar A (2021) Epileptic seizure identification in EEG signals using DWT, ANN and sequential window algorithm. Soft Comput Lett 3
8. Tang L, Zhao M, Wu X (2020) Accurate classification of epilepsy seizure types using wavelet packet decomposition and local detrended fluctuation analysis. Electron Lett 56(17):861–863
9. Seng CH, Demirli R et al (2012) Seizure detection in EEG signals using support vector machines. In: 2012 38th annual northeast bioengineering conference (NEBEC). IEEE
10. Balasubramanian P (2014) Automated classification of EEG signals using component analysis and support vector machines
11. Chawal M, Duhan M (2018) Levy flights in metaheuristics optimization algorithms—a review. Appl Artif Intell 32(9–10):802–821
12. Murugavel AS, Ramakrishnan S (2016) Hierarchical multi-class SVM with ELM kernel for epileptic EEG signal classification. Med Biol Eng Comput 54(1):149–161
13. Arora S, Anand P (2019) Chaotic grasshopper optimization algorithm for global optimization. Neural Comput Appl 31:4385–4405
14. Haklı H, Uğuz H (2014) A novel particle swarm optimization algorithm with Levy flight. Appl Soft Comput 23:333–345
15. Shoeibi A, Ghassemi N et al (2022) Detection of epileptic seizures on EEG signals using ANFIS classifier, autoencoders and fuzzy entropies. Biomed Signal Process Control 73:103417

16. Panada S, Mishra S et al (2023) Seizure detection using integrated metaheuristic algorithm based ensemble extreme learning machine. Measurement: Sensors 25:100617
17. Saini J, Dutta M (2020) Smart epilepsy detection system using hybrid ANN-PSO Network. In: Swarm intelligence optimization: algorithms and applications. Wily
18. Yogarajan G et al (2023) EEG-based epileptic seizure detection using binary dragonfly algorithm and deep neural network PMCID: PMC10584945. https://doi.org/10.1038/s41598-023-44318-w
19. Srirangam S (2021) Salp swarm optimization based machine learning algorithm on epileptic seizure detection and classification model. Turk J Comput Math Educ 12(14):799–808

Agile and Scrum Hybrid Model for SDLC: Efficacy Analysis in Automotive Industry

Swastika Saxena and Souvik Pal

Abstract The automotive industry is increasingly reliant on software, driven by advancements like connected and autonomous vehicles, IoT integration, and automated testing. Traditional software development models like Waterfall and V-Model have limitations, including inflexibility and the inability to accommodate changes once requirements are fixed. Agile methodologies, particularly Scrum, offer a solution to these challenges, but adoption has been hindered by factors like resistance to change and complex automotive requirements. To address this, Agile Hybrid Models, such as Waterfall-Scrum and V-Model-Scrum, have been proposed. These models combine the benefits of traditional and agile approaches, easing the transition. Results of data analyzed, supports the effectiveness of Agile framework in the automotive industry. Agile Hybrid Model, "Agile Water-V Model," is introduced to address existing challenges. The paper concludes by addressing constraints and proposing potential avenues for future research.

Keywords Automotive industry · Agile · Scrum · Hybrid model · Software development · Waterfall model · V-model · Benefits · Challenges

S. Saxena
Liverpool John Moores University, Liverpool, UK

S. Pal (✉)
Department of Management Information Systems, Saveetha College of Liberal Arts and Sciences, Saveetha Institute of Medical and Technical Sciences, Chennai, India
e-mail: Souvikpal22@gmail.com

Department of Computer Science and Engineering, Sister Nivedita University, Kolkata, India

S. Saxena · S. Pal
UpGrad Education Private Limited, Mumbai, India

© The Author(s), under exclusive license to Springer Nature Singapore Pte Ltd. 2025
S. K. Udgata et al. (eds.), *Intelligent Systems*, Lecture Notes in Networks and Systems 1314, https://doi.org/10.1007/978-981-96-3797-3_34

1 Introduction

The automotive industry is increasingly adopting agile software development due to changing consumer demands and technological advancements, which demands for the company to be agile and self-organized. Ambitious organizations working with scaled agile are termed as "Agile frontrunners" [1]. Traditional V-model constraints are identified using the Theory of Constraints [2, 3] leading to the integration of agile practices. Agile methodologies, particularly Scrum enhance team collaboration, adaptability to changing requirements, and project efficiency. A comparative analysis shows that Indian organizations tend to favor plan-driven development practices, while Swedish organizations are more open to agile approaches [4].

Agile methodologies, known for speed and iteration, may face challenges in industries like Automotive due to complexity. A hybrid model proposed combining Agile and V-Model [5], examining challenges including complexity, planning, maturity, user involvement, change management, time-to-market, project risk, pace of changes, and collaboration, comparing V-Model and Agile approaches for each. The comparisons are tabulated in Table 1.

The research aims to investigate how effective Agile hybrid model is in integration with Waterfall or V-model producing recommendations eventually to address the identified gaps and barriers in its implementation. Through qualitative and quantitative methods, data will be gathered from case studies, literature reviews, surveys, and interviews with industry experts. This study aims to provide valuable insights and recommendations for incorporating Agile hybrid models in the automotive industry to enhance software development practices.

Table 1 V-Model challenges to Agile—comparison and mapping [5]

Challenge	V-Model	Agile
Cooperation style	Contractual deliberations	Individual engagement
Cooperation control	Emphasis on mechanisms and means	Customer collaboration management
Completion of outputs	Comprehensive documentation	Functional software output
Formulating approach	Adherence to plans	Adaption to changes
Transition management	Inflexible and rigid	Capacity to manage shifting priorities
User participation	Limited user involvement	Continuous user engagement
Risk exposure	Significant risks	Mitigation of project risks
Lead time	Extended time to market	Swift time to market
Project intricacy	Handling difficulty in complexity	Efficient complexity management
Adaption cadence	Slow adaption	Rapid adaption

1.1 Aims and Objectives

Evaluating how effective Agile and Scrum methodologies are in the SDLC of the automotive industry, examining key success factors, efficiency, quality implications, organizational influences, and obstacles to Agile adoption. Hybrid Agile Models [6, 7] will be explored as solutions. The study employs mixed methods, including interviews, case studies, and literature reviews, to offer recommendations for optimizing Agile and Scrum alongside traditional practices in automotive software development.

Objective 1: Identifying best practices and critical success components to figure how to integrate Agile and Scrum methodologies into the automotive software development process.

Objective 2: Examining how delivery time, quality, and efficiency of software products in the automotive industry are influenced using Agile strategies.

Objective 3: Explore the cultural and organizational factors impacting the incorporation and execution of Agile and Scrum methodologies within automotive.

Objective 4: Devising a model preferable by the authorities providing ease in transition from conventional method(s) to Agile and Scrum.

1.2 Research Questions

This research investigates the underexplored area of Hybrid Agile methodologies, such as "Water-Scrum-Fall" [8, 9], understanding their impact on SD practices and the factors influencing them. Following research questions are addressed by the study:

R.Q.1. What are the best practices and critical success factors for blending Agile and Scrum strategies into the automotive SDLC processes?

R.Q.2. How delivery time, quality, and efficiency of software products in the automotive industry are influenced by Agile and Scrum methodologies?

R.Q.3. What are the cultural and organizational factors that impact the incorporation and execution of Agile and Scrum methodologies within automotive companies?

R.Q.4. How transition can be made preferable and easy from classic method(s) to Agile and Scrum?

To investigate these research questions, a mixed-methods (qualitative and quantitative data collection) approach will be employed, including case studies, interviews and surveys, involving automotive software development teams, project managers, and stakeholders.

1.3 Hypothesis

The hypothesis aims to assess how effective Agile and Scrum methodologies are in automotive SDLC. Ho: posits that Agile and Scrum are less effective than conventional SDLC methods in the automotive industry. In contrast, the alternate hypothesis (Hα) suggests that Agile and Scrum are more effective than traditional methods. This study assumes that software development processes are similar across industries and does not specify industry categories in the survey. Data will be analyzed quantitatively and qualitatively to either support or refute these hypotheses.

2 Related Work

The automotive industry, characterized by its intricate structure and adherence to stringent standards like Automotive SPICE, faces the necessity of incorporating agility and Scrum methodologies. Agile principles promote self-organization, interdisciplinary teams, and trust, fostering efficiency and security [10]. [11] Studied Software Product Line Engineering (SPLE) and Agile Software Development (ASD) in Swedish companies, revealing the importance of communication and co-location for efficient software development. As the automotive industry's complexity grows, Agile Hybrid models have emerged to address its unique needs, enhancing its adaptability and competitiveness [12].

In a study conducted [4], it was identified that Indian companies exhibit resistance to agile practices due to a prevalent reliance on plan-driven approaches. Cultural influences, resistance to change, and a lack of awareness regarding the effectiveness of Agile and Scrum methodologies contribute to this resistance [4]. An independent online web-based conducted surveys, questionnaires, and interviews, utilizing statistical one-way ANOVA methods. Her findings suggest that small development projects in India are more efficient when employing agile methods. Additionally, it was noted that Indian companies tend to avoid adopting various Agile methods, such as Scrum, XP, and SAFe, due to a lack of expertise in the field [13].

A hybrid approach proposed to address these concerns, emphasizing that companies can maintain the efficiencies and knowledge of traditional methods while reaping the benefits of agility. This approach involves several hybrid methods, including CMMI—Scrum, V-Model—Scrum, SPICE—Scrum, and Formal methods—Scrum [14, 15] supported the concept of hybrid methodologies, based on collected data—analyzed using thematic coding 18 semi-structured interviews. They explored the coexistence of Agile and traditional SDLC methodologies.

The Agile Hybrid Assessment Method for Automotive (AHAA) [16], focusing on software process improvement within the context of Automotive SPICE™ and Capability Maturity Model Integration (CMMI), while incorporating agile flexibility. This underscores the necessity of hybrid approaches in automotive SDLC practices.

The "Water-Scrum-Fall" approach, as discussed [8, 9], examines its utilization and impact on SDLC practices. It addresses gaps in the literature by exploring the factors influencing the creation of hybrid approaches. A hybrid model combining Waterfall and Scrum is proposed, integrating Waterfall at the beginning and end of the SDLC, with Scrum in between for increased flexibility and efficiency [7].

Lean Thinking (LT), a lesser-known Agile methodology compared to Scrum, focuses on waste reduction [17]. It demonstrated how LT can be implemented in conjunction with Scrum for software development, making it applicable to various industries. The concept of sustainable development is emphasized within Agile and its impact on economic, social, and environmental aspects. Their study provides evidence-based arguments for adopting agile practices [18].

Reports by Capgemini [1, 19] suggested that automotive companies are inclined to achieve superior outcomes by meeting or surpassing software development objectives employing Agile and Scrum. A comprehensive literature review and content analysis, conducted concluding that hybrid agile methods can enhance quality, reduce costs, and expedite time to market [20]. Different Machine Learning Models [21–24] have been discussed in Cloud-IoT model which have inspired to implement SARIMA Model in our paper.

In summary, the research highlights a growing inclination toward Agile methodologies and hybrid approaches, particularly in the automotive industry, to address resistance and improve software development practices. These findings underscore the significance of bridging gaps in implementation and offer potential solutions. The current research aims to build upon this knowledge to propose a novel hybrid model that enhances the adoption of agile methods while addressing existing limitations in their implementation.

3 Contribution of the Paper

This paper makes a significant contribution by addressing various essential aspects that collectively enhance both knowledge and practical applications within the realm of automotive software development.

Relevance in Practice and Industry Impact: This research addresses the pressing need for agile and scrum methodologies in the automotive industry (RQ1). The outcomes hold practical significance, offering guidance to industry practitioners in adapting to the evolving automotive landscape. By aligning practices with industry demands, this study can enhance project management, flexibility, efficiency, and customer satisfaction.

Enhancement of Performance Metrics: By investigating the impact of agile and scrum methodologies on performance indicators (RQ2), this research provides empirical evidence. Informed decisions can be made, thereby improving product quality, reducing time-to-market, and elevating software development practices.

Transformation in Culture and Organization: This study delves into the organizational and cultural factors affecting agile and scrum adoption (RQ3). It enhances our understanding of how organizations can foster an environment conducive to agile practices, promoting collaboration and innovation.

Fusion of SDLC Models: The research examines the integration of Waterfall, Scrum, and V-model in the automotive SDLC (RQ4). This approach offers a unique perspective tailored to the complexities of the automotive sector, contributing to the development of a customized SDLC approach.

Theoretical Contributions: In addition to evaluating agile and scrum methodologies, this research explores agile hybrid models, bridging a literature gap. It enriches our comprehension of adapting these practices to the automotive context, paving the way for future research. It also advocates for sustainable development, aligning with environmental, economic, and social dimensions.

In essence, this research aims to enhance efficiency, timeliness in software delivery and quality, fostering growth and competitiveness within the dynamic industry of automobile.

4 Research Methodology

To fulfill the research objective and address the research questions effectively, a meticulous approach was adopted to ensuring value of the study and its relevance, contributing to the existing literature. The survey questionnaire was meticulously designed to delve into the current status and efficacy of Agile and Scrum practices across industries. This survey was structured into three distinct sections.

Survey-1, comprising 10 questions focused on gathering general participant data relevant to the research. It included inquiries about their roles and the Software Development Life Cycle (SDLC) methodologies employed within their teams or organizations.

Survey-2, consisting of 13 terminologies specifically related to Agile and Scrum frameworks, aimed to assess the participants' awareness levels. This section provided insights into the popularity of various aspects of Agile methodology and identified areas where awareness might be lacking.

The survey was disseminated via a Google Form through online channels, resulting in responses from a diverse group of 45 individuals. The respondents represented a spectrum of backgrounds, including students, IT professionals, mid-senior, senior, and managerial-level individuals. This comprehensive methodology ensured the collection of robust data to support the research objectives and answer the research questions effectively.

4.1 Primary Data Analysis

For primary data analysis, Kruskal–Wallis non-parametric analysis is used to analyze the results, and obtain the p-value with the assumption of significance level at 0.05. Data sampled—based on "Role" (Mentioned in Fig. 1a, b)—Calculated p-value: 0.04550.

The p-values obtained in both are < 0.05. Consequently, we have rationale for rejecting the null hypothesis, implying that the Agile and Scrum methodology is more effective than traditional SDLC methods in the automotive industry, and this null hypothesis is disproven. Therefore, the proposed hypothesis, asserting that Agile and Scrum methodology is more effective than regular SDLC methods in the automotive industry, holds true (Table 2).

Barriers

"Barriers of Agile implementation" (Mentioned in Fig. 2.) primary collected data is analyzed. Among the 9 most popular barriers surveyed, given in the figure below, "Risk," "Lack of knowledge about Agile and Scrum," and "Resistance to change" constitutes 12% each of all the barriers. Followed by "Organizational culture," "Lack of support," "Absence of experts," "Time constraints," and "Team size and structure"

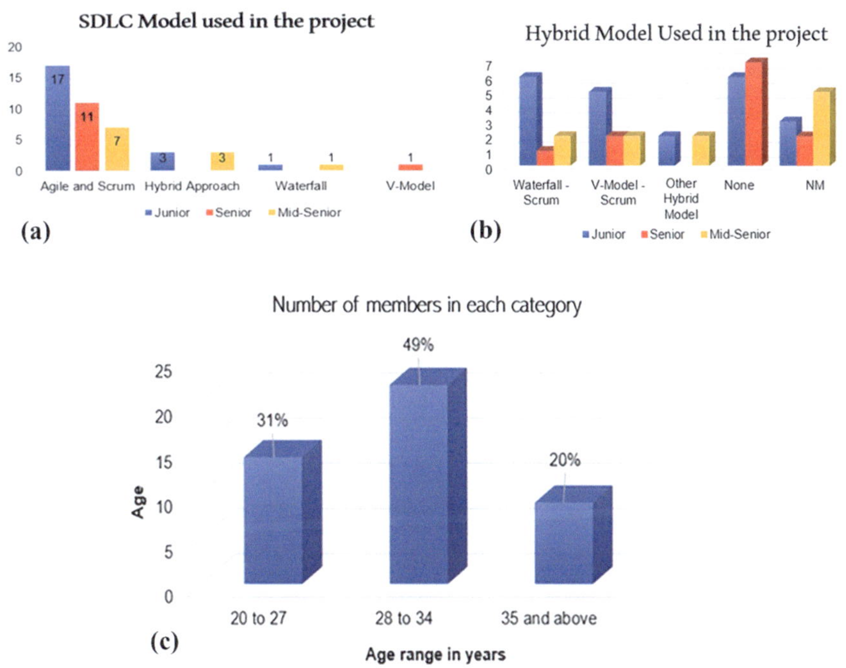

Fig. 1 **a** SDLC model used in the project, **b** Hybrid model used in the project data sampled—based on "Age," (Fig. 1c)—Calculated p-value: 0.00968, **c** Number of members in each category

Table 2 Survey parameters for awareness check

Scrum team	Continuous integration (CI)/continuous development (CD)
Sprints	Retrospective meetings
Definition of done	Kanban board
Iterative increment	Flexibility in requirement gathering
Daily stand-up meetings	Regular customer feedback
User stories effort estimation	Hybrid approach
Product backlog	

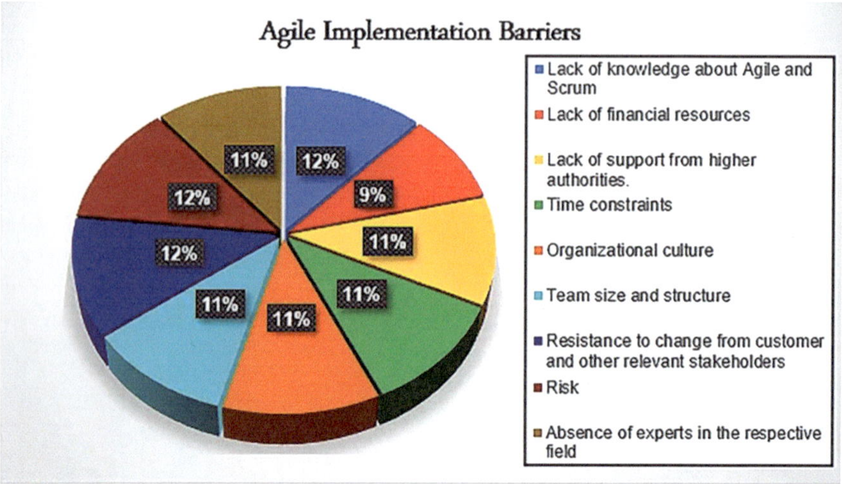

Fig. 2 Agile implementation barriers

constituting 11% each among the barriers. The remaining 9% is covered by "Lack of financial resources."

4.2 Secondary Data Analysis

Aligned with the analysis of primary data, the outcomes of secondary data analysis present a compelling case supporting the preference for adopting Agile or Hybrid models over relying solely on traditional development approaches.

Figure 3 Illustrates the success rate of Agile framework implementation based on responses from 129 out of 190 total participants, 47% (61) encountered difficulties in its implementation, 43% (56) classified Agile teams as successful, 2% (3) deemed it unsuccessful, and the remaining 7% (9) were undecided.

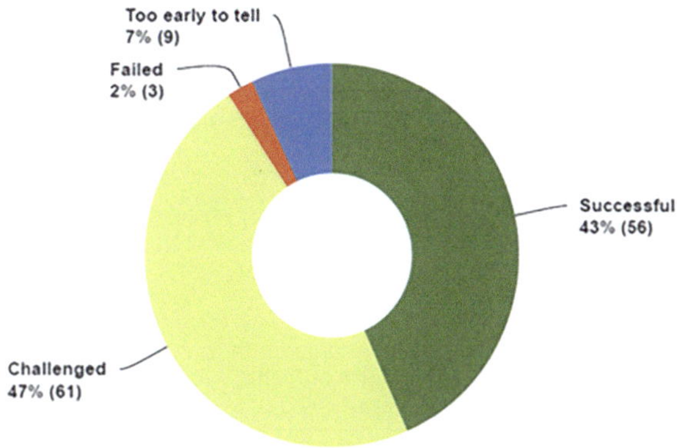

Fig. 3 Agile team success rate [25]

5 Suggestive New Agile Hybrid Model: Agile Water-V Model

This proposed hybrid model (mentioned in Fig. 4.) combines elements from Waterfall, V-Model, and Scrum, integrating aspects of all three to address safety–critical considerations in automotive software through the application of the Waterfall approach (Table 3).

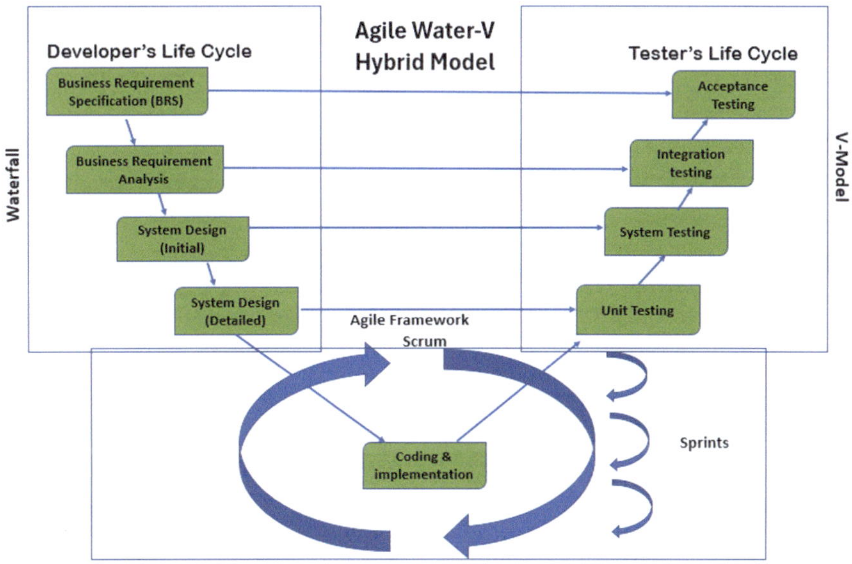

Fig. 4 Agile water-V-model

Table 3 Life cycle of testers and developers will run parallel to each other

Steps	Developer's life cycle	Tester's life cycle
Step 1	Gathered BRS (business requirements specification)	Tester engages in gathering requirements, creates acceptance criteria
Step 1.2	Analyzed BRS for feasibility, priority, and scope estimation	Developing test suites based on acceptance criteria analysis
Step 2	Design based on initial analysis	Carefully crafted tests fulfilling integration and acceptance criteria for system
Step 2.2	Elaborate system design outlines independent modules or features for development	Testers will create unit tests for thorough feature/module testing, ensuring acceptance criteria are met
Step 3	Scrum methodology applied after establishing independent modules	
Step 3.1	Based on detailed system design, the "Product backlog" will be created	
Step 3.2	During "Sprint 0," the "Sprint backlog" will be established	
Step 3.3	During the "Sprint planning" meeting (Standard 2 weeks), user stories aligned with the agreed "Sprint goal" will be selected from the product backlog and incorporated into the sprint	
Step 3.4	Now in "Sprint review" meeting, user stories undergo refinement, and "story point estimation" is performed for each	
Step 3.5	"Sprint retrospective" meeting will be held toward the end of the Sprint, to analyze how well the sprint went, what didn't go well and any areas for improvement will be recorded as action items	

Continue with Step 3 until the "Product backlog" is exhausted

New Model Process Flow

Over the two-week sprint, developers will code and conduct unit testing for the selected user stories. The completed outputs will be continually provided to testers for verification and validation, following the V-Model side. Any requirement changes will be considered, following "Step 1," and incorporated into the "Product backlog."

6 Conclusions and Future Work

In summary, this research aimed to identify critical success components for Agile, assess its impact on SDLC processes, address barriers in adoption, and propose a potential model. The objectives were effectively addressed.

Agile methodologies, especially Scrum, have gained prominence in software development. Emphasizing the significance of effective teams as a critical success component, they place the customer at the forefront, vital for industry success. Implementing Agile, particularly in system testing, enhances product quality and efficiency while reducing time and costs [4, 26].

The impact of Agile and Scrum implementation extends beyond software development, influencing organizational structure, scope, risk, time, cost management, and HR practices. This aligns with Scrum principles, affecting hiring preferences favoring candidates with these values [27]. The Automotive industry, characterized by complexity and evolving standards, has adopted Agile and Scrum to improve efficiency, reduce costs, and enhance product quality. However, challenges including lack of awareness, management support, resistance to change, and time constraints persist [28].

Introduction of agile hybrid models, blending traditional models like Waterfall or V-Model with agile methods, has facilitated smoother implementation and increased awareness in the automotive industry. These models provide the flexibility to retain the benefits of traditional approaches while adapting to new challenges. Adaptability is key, as mentioned by Berger and Kayser [6], who emphasize the importance of tailoring Agile to environmental needs. While Agile, Scrum, and Hybrid Models offer numerous benefits, ongoing exploration of various methods and their integration with traditional models is essential for continual improvement in software development processes.

This research highlights the widespread applicability of Agile methodologies, not limited to any specific industry. The automotive sector has notably embraced Agile, enhancing productivity. Primary and secondary data analyses support the effectiveness of Agile, Scrum, and Hybrid Models, emphasizing the need for these modern approaches over traditional methods. Challenges in Agile adoption is consistent across data sources, indicating areas for future research. The innovative Agile Water-V model, introduced in this study, demonstrates potential for enhancing software development practices with its unique blend of Waterfall, V-Model, and Scrum principles, providing an unexplored avenue for achieving beneficial results.

6.1 Future Work Suggestions

There are potential research opportunities in exploring Agile methodologies within the automotive sector. Firstly, the implementation of Agile Hybrid models for software development in the automotive sector remains a relatively unexplored area. Further investigation and comprehensive studies are essential to navigate the challenges associated with the adoption and efficient implementation of these models.

Secondly, the newly introduced "Agile Water-V" model presents an intriguing prospect. It is worth considering experiments not only within the automotive industry but also in other sectors. This can help assess the model's effectiveness in terms of ease of adoption, implementation, and its capability to deliver high-quality, cost-effective products within defined project scopes.

Additionally, a dedicated study to comprehend the major reasons for resistance to change and adoption, such as the perceived risk of failure, regulatory and safety compliance concerns, or knowledge gaps, can provide valuable insights. This research could pinpoint which of these factors should be prioritized and addressed to facilitate a smoother transition from traditional methods to Agile or Agile Hybrid in the automotive sector.

6.2 Limitations

This research carries certain limitations that warrant consideration. The timeframe for this study, spanning only six months from topic selection to completion, is a limitation. A more extended duration would have facilitated broader data collection through surveys, case studies, and literature reviews. Although the survey questionnaire was designed to include individuals from the IT industry, it did not specifically target the automotive sector. The number of respondents in the survey was limited to 45, and a larger sample size would have enhanced the statistical robustness of the analysis.

Furthermore, the "Agile Water-V" Agile hybrid model, proposed in this research, is conceptual and theoretical. Due to time constraints, it has not undergone practical testing or implementation. Its efficiency and ability to deliver expected results remain theoretical and require experimentation for validation.

References

1. Agile at scale—Four ways to gain enterprise-wide agility" (2019) Capegemini Research Institute, Capgemini.com. Available at: https://www.capgemini.com/wp-content/uploads/2019/11/Report-%E2%80%93-Agile-@-Scale-1-1-1.pdf. Accessed 6 Jun 2023
2. Eliyahu M, Goldratt (1997) Critical chain. The North River Press Publishing Corporation, Great Barrington, MA. ISBN 0-88427-153-6. Available at: https://doi.org/10.1080/07408179808966521. Accessed 6 Aug 2023
3. Hutanu A, Prostean G, Badea A (2015) Integrating critical chain method with AGILE life cycles in the automotive industry. Procedia, Soc Behav Sci [online] 197:1416–1421. Available at: https://www.sciencedirect.com/science/article/pii/S1877042815040823. https://doi.org/10.1016/j.sbspro.2015.07.088. Accessed 3 Aug 2023
4. Putta A, Petersen K (2023) Comparative analysis of software development practices across software organisations. [online] Diva-portal.org. Available at: https://www.diva-portal.org/smash/get/diva2:1045277/FULLTEXT02.pdf. Accessed 9 Aug 2023
5. Komal S, Carsten Wolff A (2023) Integration of agile methods in automotive software development processes. [online] Uni-obuda.hu. https://doi.org/10.1109/CANDO-EPE51100.2020.9337750 Available at: http://conf.uni-obuda.hu/Kando2020/32_Cando.pdf. Accessed 1 Aug 2023
6. Berger K, Kayser MLJJ (2023) Adoption of agile methods in automotive software development. [online] Utwente.nl. Available at: https://essay.utwente.nl/81865/1/Berger_BA_BMS.pdf. Accessed 6 Jun 2023
7. Wankhede R (2023) Hybrid agile approach: efficiently blending traditional and agile methodologies. [online] Harrisburgu.edu. Available at: https://digitalcommons.harrisburgu.edu/cgi/viewcontent.cgi?article=1011&context=pmgt_dandt. Accessed 14 Jul 2023
8. West D, Gilpin M, Grant T, Anderson A (2011) Water-scrum-fall is the reality of Agile for most. [online] Forrester. Available at: https://www.forrester.com/report/WaterScrumFall-Is-The-Reality-Of-Agile-For-Most-Organizations-Today/RES60109. Accessed 15 Jun 2023

9. Kuhrmann M, Diebold P, Münch J, Tell P, Garousi V, Felderer M, Trektere K, McCaffery F, Linssen O, Hanser E, Prause CR (2017) Hybrid software and system development in practice: waterfall, scrum, and beyond. In: Proceedings of the 2017 international conference on software and system process. ACM, New York, NY, USA. https://doi.org/10.1145/3084100.3084104. Accessed 15 Jun 2023
10. Agile in automotive: Pocket guide agile and automotive SPICE®. 1000th ed. Kornwestheim, Germany: KUGLER (2017) Available at: https://www.kuglermaag.com/improvement-programmes/agile-automotive/. Accessed 6 Aug 2023
11. Martini A, Pareto L, Bosch J (2013) Communication factors for speed and reuse in large-scale agile software development. In: Proceedings of the 17th International software product line conference. ACM, New York, NY, USA. Available at: https://dl.acm.org/doi/pdf/https://doi.org/10.1145/2491627.2491642. Accessed 6 Aug 2023
12. Sandu I-A, Salceanu A (2019) System testing in agile SW development of the electronic components based on software from the automotive industry. In: 2019 11th International symposium on advanced topics in electrical engineering (ATEE). IEEE, pp 1–4, https://doi.org/10.1109/ATEE.2019.8724968. Access 6 Jul 2023
13. Kumari U (2021) Agile practices in Indian organizations. Orient J Comput Sci Technol [online] 102:422–428. https://doi.org/10.13005/ojcst/10.02.24. Available at: https://www.academia.edu/57376919/Agile_Practices_in_Indian_Organizations. Accessed 18 Jun 2023
14. Kasauli R, Knauss E, Nakatumba-Nabende J, Kanagwa B (2020) Agile islands in a waterfall environment. In: Proceedings of the evaluation and assessment in software engineering. https://doi.org/10.1145/3383219.3383223. Accessed 16 Aug 2023
15. Shaya O (2016) Using Agile Methods in Automotive Software Engineering. [online] Researchgate.net. https://doi.org/10.13140/RG.2.2.12219.46885 Available at: https://www.researchgate.net/publication/331984928_Using_Agile_Methods_in_Automotive_Software_Engineering. Accessed 15 Aug 2023
16. McCaffery F, Pikkarainen M, Richardson I (2008) Ahaa—agile, hybrid assessment method for automotive, safety critical smes. In: Proceedings of the 13th International conference on Software engineering—ICSE '08. ACM Press, New York, New York, USA, pp 551–560. https://doi.org/10.1145/1368088.1368164. Accessed 15 Jun 2023
17. Narayanamurthy G, Gurumurthy A, Prasath S (2023) Implementing lean thinking in software development—a case study from India. [online] Liverpool.ac.uk. Available at: https://livrepository.liverpool.ac.uk/3032687/1/IJSTM%20Elements. Accessed 4 Jul 2023
18. Gomes Silva FJ, Kirytopoulos K, Pinto Ferreira L, Sá JC, Santos G, Cancela Nogueira MC (2022) The three pillars of sustainability and agile project management: how do they influence each other. Corp Soc Responsib Environ Manag [online] 295:1495–1512. Available at: https://doi.org/10.1002/csr.2287. Accessed 6 Jun 2023
19. "Business Agility in automotive—How to transform towards an adaptive organization for future of mobility" (2021) Capegemini invent, Pilot Agile(@Scale) structures and mindset to create tangible early benefits & thus buy in. [online] Capgemini.com. Available at: https://www.capgemini.com/wp-content/uploads/2021/09/Business-Agility-in-Automotive-Report-1.pdf. Accessed 6 Jun 2023
20. Leong J, May Yee K, Baitsegi O, Palanisamy L, Ramasamy RK (2023) Hybrid project management between traditional software development lifecycle and agile based product development for future sustainability. Sustainability, [online] 152, p 1121. Available at: https://doi.org/10.3390/su15021121 Accessed 15 Jun 2023
21. Pal S, Jhanjhi NZ, Abdulbaqi AS, Akila D, Alsubaei FS, Almazroi AA (2023) An intelligent task scheduling model for hybrid internet of things and cloud environment for big data applications. Sustainability 15:5104. https://doi.org/10.3390/su15065104
22. Pal S, Jhanjhi NZ, Abdulbaqi AS, Akila D, Almazroi AA, Alsubaei FS (2023) A hybrid edge-cloud system for networking service components optimization using the internet of things. Electronics 12:649. https://doi.org/10.3390/electronics12030649
23. Tuan NA, Akila D, Pal S, Sarkar B, Khai Tran T et al (2022) Dynamic data optimization in iot-assisted sensor networks on cloud platform. Comput Mater Contin 72(1):1357–1372. https://doi.org/10.32604/cmc.2022.024096

24. Mukherjee D, Ghosh S, Pal S, Akila D, Jhanjhi NZ et al (2022) Optimized energy efficient strategy for data reduction between edge devices in cloud-iot. Comput Mater Contin 72(1):125–140. https://doi.org/10.32604/cmc.2022.023611
25. Ambysoft.com (2016) agility at scale survey results. [online] Ambysoft.com. Available at: https://ambysoft.com/surveys/agileAtScale2016.html. Accessed 26 Jul 2023
26. Fagarasan C, Popa O, Pisla A, Cristea C (2021) Agile, waterfall and iterative approach in information technology projects. IOP Conf Ser Mater Sci Eng [online] 11691:012025. Available at: https://doi.org/10.1088/1757-899x/1169/1/012025. Accessed 7 Aug 2023
27. Hayat F, Rehman AU, Arif KS, Wahab K, AnofAbbas M (2019) The influence of agile methodology (scrum) on software project management. In: 2019 20th IEEE/ACIS international conference on software engineering, artificial intelligence, networking and parallel/distributed computing (SNPD). IEEE, pp 145–149. https://doi.org/10.1109/SNPD.2019.8935813. Accessed 9 Jul 2023
28. Soares D, da Silva FJG, Ramos SCF, Kirytopoulos K, Sá JC, Ferreira LP (2022) Identifying barriers in the implementation of agile methodologies in automotive industry. Sustainability [online] 149, p 5453. Available at: https://www.mdpi.com/2071-1050/14/9/5453. https://doi.org/10.3390/su14095453. Accessed 18 Jun 2023

Analysis of Activated Carbon Usage Prediction Using Time-Series SARIMA Model in Indian Context

Mahesh Ashok Desai and Souvik Pal

Abstract The Indian activated carbon market is vital for multiple industries, including air purification, food, water treatment, pharmaceuticals, and more. This research offers a comprehensive analysis of market dynamics, export analysis from 2015 to 2022, and future demand–supply projections from 2023 to 2028, focusing on critical factors such as raw materials and industry growth. The study begins with an in-depth export analysis, examining historical trends and major trading partners. It also explores raw material usage, notably coconut and coal, assessing their availability and impact on activated carbon quality. The research forecasts future demand, considering regulations, environmental concerns, and emerging applications. It also estimates supply capabilities from domestic manufacturing and imports.

Keywords Activated carbon market · Export trends · Forecasting · Water treatment · India · Demand and supply

1 Introduction

The growth of activated carbon market in the world has been particularly pronounced in recent years chiefly propelled by the escalating demand for environmental remediation solutions, water purification, air purification, and various industrial applications. Activated carbon, known for its high adsorption capacity and versatility, finds more

M. A. Desai
Liverpool John Moores University, Liverpool, UK

S. Pal (✉)
Department of Management Information Systems, Saveetha College of Liberal Arts and Sciences, Saveetha Institute of Medical and Technical Sciences, Chennai, India
e-mail: souvikpal22@gmail.com

Department of Computer Science and Engineering, Sister Nivedita University, Kolkata, India

M. A. Desai · S. Pal
UpGrad Education Private Limited, Mumbai, India

© The Author(s), under exclusive license to Springer Nature Singapore Pte Ltd. 2025
S. K. Udgata et al. (eds.), *Intelligent Systems*, Lecture Notes in Networks and Systems 1314, https://doi.org/10.1007/978-981-96-3797-3_35

usage in sectors such as water treatment, food and beverage, pharmaceuticals, automotive, and many others. Understanding the import–export dynamics, raw material utilization, and future demand–supply projections of activated carbon is crucial for industry stakeholders to make informed decisions and devise effective strategies.

The **core objective** of this research study is to offer an all-encompassing analysis of the activated carbon market in India, which includes an examination of export trends, the utilization of raw materials, and future demand–supply projections. By examining the historical data from 2015 to 2022 and considering various market drivers and industry factors, this study aims to offer valuable insights and strategic recommendations for industry participants. Furthermore, the study explores the utilization of different raw materials in the manufacture of activated carbon, with a specific focus on coconut shell and coal. In addition to the export and raw material analysis, this study projects the future demand and supply of activated carbon in India.

2 Related Works

Ahmadpour and Do [1], study focuses on creating activated carbon (AC) from bituminous coal. The research delves into how various factors affect the production process to optimize AC properties. Chemical activation with KOH and $ZnCl_2$ produced ACs with well-developed porosities compared to physical activation. They pinpoint the impregnation ratio as a vital factor in chemical activation, and carbonization temperature significantly impacted pore volume. Higher temperatures improved surface area and pore volume for KOH-activated carbon but damaged the structure of $ZnCl_2$-activated carbon. Comparing physical and chemical activation, chemical activation proved more flexible in tailoring pore structures. This research underscores the significance of process variables like carbonization temperature and impregnation ratio in optimizing AC production for high surface areas and well-developed porosities. Chemical activation stands out as a more effective method for achieving desired pore size distributions, contributing to better AC production processes.

Hsu and Teng [2], this study looked at making activated carbons from bituminous coal using three different chemicals: $ZnCl_2$, H_3PO_4, and KOH. They used these chemicals to activate the coal at different temperatures. Results showed that KOH made carbons with the highest surface area (3300 m^2/g), while $ZnCl_2$ and H_3PO_4 had lower surface areas (960 m^2/g and 770 m^2/g, respectively). Bigger coal particles led to lower yields and less porous carbons. This research helps understand how different chemicals affect the properties of activated carbons made from bituminous coal, important for various applications.

Hu and Srinivasan [3], this study underscores the potential of chemical activation using KOH with coconut shell as a source material to yield activated carbons with exceptional adsorption properties for phenolic compounds. Optimization of activation parameters, including the KOH-to-shell ratio, activation temperature, and pre-heating temperatures, significantly influenced the resulting porosity and yield of the activated carbons. These findings highlight the promise of coconut-shell-derived activated carbons as efficient and economical adsorbents for pollution control and a wide array of applications.

Laine and Calafat [4], this study aims to optimize the production of activated carbon from coconut shells treated with phosphoric acid. It employs a single-step carbonization process and investigates different conditions to find the best preparation parameters. The study identifies that the type and flow of gas during carbonization significantly affect the resulting surface area and yield. This suggests that oxygen from the gas phase and the coking of volatiles play a role in creating a porous structure. Overall, this study is significant for advancing the understanding of how to optimize the production of activated carbon from coconut shells with phosphoric acid. It highlights the impact of factors like gas flow, activation temperature, and acid concentration on surface area, porosity, and surface chemistry. These findings contribute to the development of efficient methods for producing activated carbon, with applications in various fields requiring high-performance adsorbents and catalysts.

Li et al. [5], the paper under consideration investigates how varying carbonization temperatures and activation times impact the porosity features of carbonized coconut shell charcoal and the resultant activated carbon. The researchers conducted a series of experiments using a different carbonization temperatures and activation. The outcomes of the study indicate that elevated carbonization temperatures yield activated carbon and coconut shell char with greater total volume, micropore volume, surface area, and overall yield when compared to lower temperatures. In summary, this research underscores the pivotal role played by carbonization temperature in shaping porosity characteristics. Higher temperatures result in augmented microporosity, increased surface area, and volume, while concurrently reducing the average diameter. Activated carbons derived from charcoal, carbonized at higher temperature, exhibit larger yields. These findings offer valuable insights for optimizing the production of large-surface-area activated carbons from coconut shells through the precise selection of temperature and activation duration in the physical activation process.

Stavropoulos and Zabaniotou [6], the research paper is based on a detailed economic evaluation of the production process of activated carbons from various raw materials. The study focuses on conventional economic indices such as Return on Investment (ROI), Payback Period (PP), and Net Present Value (NPV) to assess the economic viability of different raw materials and production methods. Among the six raw materials evaluated (wood, used tires, petroleum coke, carbon black, charcoal, and lignite), petroleum coke demonstrates the lowest production cost ($1.08/kg) and offers a considerable margin from the selling price of activated carbons ($1.92/kg). This makes petroleum coke the most economically attractive choice, with favorable economic indices such as ROI (40%), Payback Period (5–6 years), and positive NPV. In conclusion, the literature review provides a comprehensive economic evaluation of the production process of activated carbons from various raw materials. The results highlight the importance of selecting raw materials with high yields and considering the pricing based on surface area or adsorption capacity. Petroleum coke emerges as the most economically attractive raw material, followed by charcoals and lignite.

The findings contribute valuable insights for decision-making in the activated carbon industry, considering both technical and economic factors.

Teng et al. [7], this study focused on making and studying activated carbons from Australian bituminous coal. They used a two-step process: first, they treated the coal with phosphoric acid, and then they heated it under nitrogen at different temperatures and times. Results showed that more phosphoric acid increased surface area and pore volume. The best carbon was made at 500 °C for 3 h. They also tested physical activation with CO_2, which added more pores. This research helps understand how to make activated carbons with better properties, important for their use in various applications. Further research can improve the sustainability of making activated carbons from coal. Different machine learning models [8–11] have been discussed in Cloud-IoT model which have inspired to implement SARIMA model in our paper.

3 Export Analysis of Activated Carbon

3.1 Global Export Analysis

The period from 2015 to 2022 witnessed significant growth in the global activated carbon market (Mentioned in Fig. 1). The export value of activated carbon increased substantially, rising from USD 1.58 billion in 2015 to USD 2.44 billion in 2022. This notable increase signifies a robust and sustained upward trend in the demand for activated carbon on a global scale.

China has consistently held its position as the dominant player in the global activated carbon export market [12]. In 2015, China exported activated carbon with a total value of USD 313.2 million. Over the years, China has not only maintained but significantly expanded its export capabilities. By 2022, China's export value reached an impressive USD 579.8 million, marking a nearly twofold increase in just seven years. India's journey in the global activated carbon export market has been nothing short of remarkable. In 2015, India's export value was recorded at USD 113.4 million. Over the years, India has exhibited remarkable growth, driven by quality products, expanding production and global market penetration. These efforts resulted in India's export value soaring to USD 311.6 million in 2022, marking a more than twofold increase over seven years. This growth underscores India's emergence as a significant exporter of activated carbon on the global stage.

3.2 India's Export Analysis for the year 2022

In 2022, India solidified its position as a major player in the global activated carbon market, ranking as the second-largest exporter after China (shown in Table 1 and Fig. 2). The trajectory of India's activated carbon exports is a compelling tale of

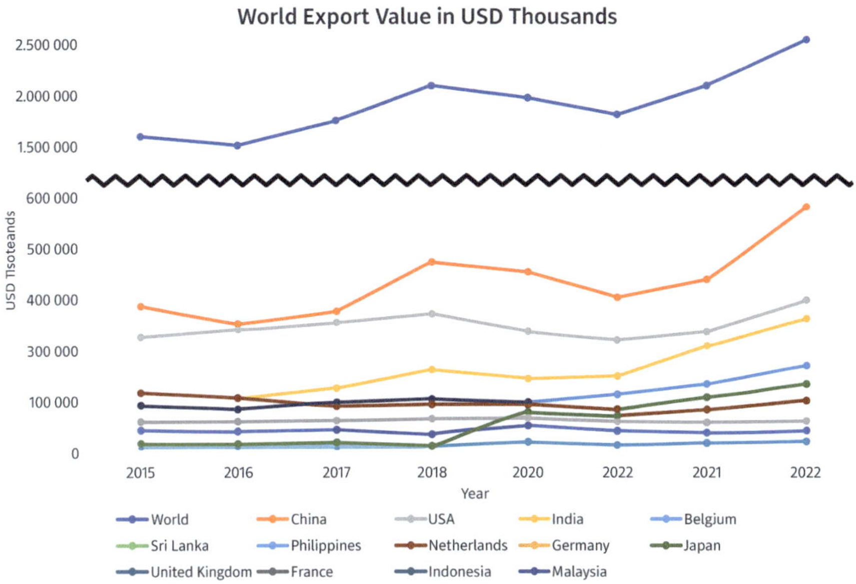

Fig. 1 Global export value of activated carbon in USD thousand. *Source* Trademap.org

growth and potential, marked by impressive statistics that underscore the country's prominence in this industry.

Table 1 Global export quantity in tons and export value in USD thousand for the year 2022

Exporter (Year 2022)	Export qty. in (tons)	Export value in (USD thousand)
China	292,486	579,855
USA	67,124	337,864
India	154,424	311,603
Belgium	79,608	193,786
Sri Lanka	55,812	157,202
Philippines	79,738	154,922
Netherlands	28,378	131,116
Germany	59,107	108,220
Japan	11,113	101,679
UK	13,598	53,330
France	13,069	46,846
Indonesia	25,832	42,641
Others	80,588	227,646

Source Trademap.org

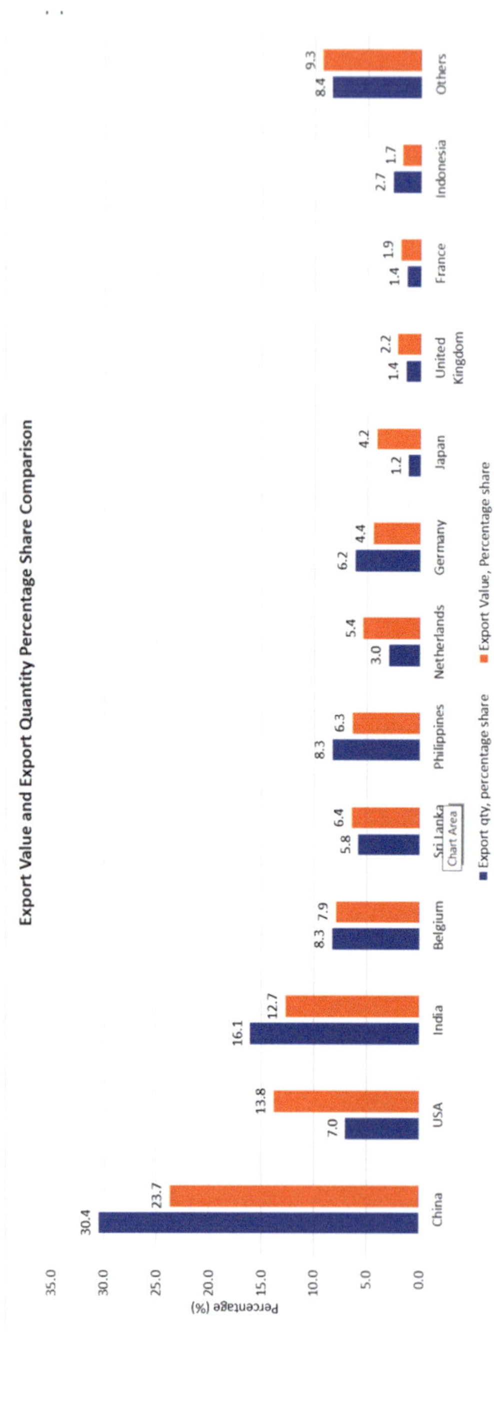

Fig. 2 Comparison between export value and export quantity percentage share

Product Dominance: **Coconut Shell-Based Activated Carbon**: One of the distinguishing factors in India's activated carbon export landscape is the predominant production of coconut shell-based activated carbon [13]. This particular variety is highly regarded for its effectiveness and versatility across various applications, ranging from water purification to air filtration. The coconut shell-based activated carbon produced in India is known for its quality and performance.

Global Export Reach: India's reach in the international activated carbon market is extensive, with exports flowing to a staggering 146 countries across the globe. This broad reach demonstrates the country's commitment to serving a diverse set of customers worldwide.

Key Importers in 2022: In 2022, India's major export destinations for activated carbon included the USA (USD 62.7 million), Sri Lanka (USD 23.6 million), Germany (USD 21.8 million), Turkey (USD 14.4 million), Korea (USD 12.1 million), and China (USD 11.9 million). Notably, the USA emerged as the largest buyer of Indian activated carbon, contributing a significant share of the export value. In fact, out of the total export value, a substantial 20% came from American buyers.

Market Share Insights: In 2022, India's export value constituted 13% of the global activated carbon market, reflecting its significant presence as a supplier on the international stage. Meanwhile, the quantity of activated carbon exported by India represented a larger share, standing at 16%. In contrast, the USA, one of the prominent importers of Indian activated carbon, accounted for a smaller portion of the export market, with an export value share of 14% and a quantity share of 7%.

This disparity in export value and quantity shares between India and the USA suggests an intriguing trend. It implies that US suppliers of activated carbon may be exporting their products at higher unit prices compared to Indian suppliers. Consequently, it can be inferred that buyers in the activated carbon market, while appreciating quality, are also price conscious. They are likely to seek a balance between high quality and competitive pricing when sourcing activated carbon.

Notably, China and India currently stand as the two largest producers of activated carbon globally. China primarily specializes in supplying coal-based activated carbon, while India stands out for its coconut shell-based activated carbon [14]. The key factor that tilts the preference in favor of Indian suppliers, especially in water filtration applications, is the exceptional quality of Indian coconut shells. These shells boast a minimum hardness level of 98, a critical attribute for effective water filtration. In contrast, Chinese products exhibit a lower hardness level, typically around 92–93.

4 Forecast of Indian Activated Carbon Export (2023–2028)

The Indian activated carbon industry has experienced significant growth in recent years, as evidenced by its export performance. To provide valuable insights into the forecast of Indian activated carbon exports from 2023 to 2028 (mentioned in Fig. 3),

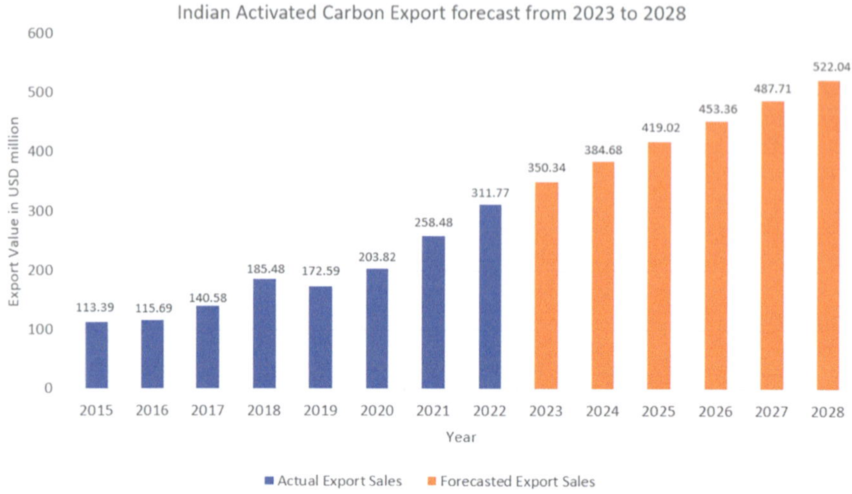

Fig. 3 Indian activated carbon export forecast from 2023 to 2028

we'll analyze the historical data and consider various factors influencing this sector's future.

4.1 Historical Export Growth (2015–2022)

From 2015 to 2022, the industry exhibited substantial growth, with export values increasing steadily. This growth is indicative of India's increasing prominence as an activated carbon exporter. Several factors have contributed to this expansion, including the quality of Indian activated carbon, competitive pricing, expanding production capacity, and the growing global demand for activated carbon in diverse industries.

4.2 Forecast (2023–2028)

The forecasted export sales for Indian activated carbon from 2023 to 2028 demonstrate a positive and promising outlook. India's proactive approach to quality, capacity expansion, and market diversification positions it as a key player in meeting the global demand for activated carbon. As sustainability and environmental protection become increasingly vital, the Indian activated carbon industry is well-positioned to thrive in the coming years, contributing significantly to the country's export earnings.

Key Insights

1. Steady Growth: The forecast indicates a consistent and positive growth trend in the sales of activated carbon. Over the six-year period, sales are expected to increase annually by an average of 8.9% (CAGR).
   Compound Annual Growth Rate (CAGR) Calculation:

$$\text{CAGR} = \left(\frac{\text{EV}_{2028}}{\text{EV}_{2022}}\right)^{\left(1/n\right)} - 1, \tag{1}$$

where

EV_{2028} is the export value in 2028 (USD 522.04 million).
EV_{2022} is the export value in 2022 (USD 311.77 million).
n is the number of years between 2022 and 2028 which is 6.

$$\text{CAGR} = \left(\frac{522.04}{311.77}\right)^{\left(1/6\right)} - 1 = \mathbf{8.9\%}.$$

2. Diverse Applications: The diverse range of applications for activated carbon, from water and air purification to industrial processes, ensures a stable and resilient demand for Indian exports.
3. Competitive Edge: India's commitment to delivering high-quality activated carbon at competitive prices positions it favorably to capture a larger share of the global market.

4.3 Forecasting Methodology

The forecasting period selected is from 2023 to 2028. The approach employed a machine learning technique known as the "Time series—SARIMA model," implemented using Python programming. To train the forecasting model, historical monthly export sales data spanning the period from 2015 to 2022 was utilized.

5 Raw Material Analysis

It is important to understand the dynamics of coconut production in India for assessing the availability and sustainability of the coconut shells as a raw material for the production of activated carbon. In this comprehensive analysis, we delve into the production data of coconuts in India from 2014 to 2022, examining various aspects and their implications for the coconut shell activated carbon industry.

5.1 *India's Coconut Production Analysis*

From 2014 to 2017, India witnessed growing trend in coconut production; however, from 2018 to 2022 we see decline in the total coconut production. In 2014–15, the production stood at 20,439.61 million nuts, while in 2021–22, it was 19,309.90 million nuts. Though India is the leading producer of the coconuts in the world after Indonesia and Philippines, the decline of coconut production in recent years is worrisome which may impact growth of other industries which are dependent on the coconuts. As activated carbon industries in India mainly focus on coconut shell-based activated carbon, this decline trend is not a good sign for the growth of the industry.

5.1.1 Number of Coconuts Produced Versus Nut Shells Used as Raw Material

Analysis shows that activated carbon industries utilized 35–50% of coconut shells as a raw material out of total available coconut shells in recent past years (mentioned in Table 2 and Fig. 4). Historical statistics shows that there is still scope for the growth of activated carbon industry in India. Further if India manages to turn around the decline trend of the coconut production in future, growth of activated carbon industries will be unstoppable.

Table 2 Number of nut shells used as raw material in activated carbon production

Year	Number of total coconuts produced	Activated carbon produced in tons	Raw material used to produce activated carbon (number of nut shells used)*	Percentage usage of raw material (%)
2015–2016	22,167,450,000	71,673	6,450,570,000	29
2016–2017	23,904,100,000	85,397	7,685,730,000	32
2017–2018	23,798,230,000	93,393	8,405,370,000	35
2018–2019	21,288,240,000	97,051	8,734,590,000	41
2019–2020	20,308,700,000	103,072	9,276,480,000	46
2020–2021	20,736,120,000	113,536	10,218,240,000	49
2021–2022	19,309,900,000	137,364	12,362,760,000	64

Source coconutboard.gov.in

*Calculation of number of coconut shells used—This calculation is taken from Indian Coconut Journal published in October 2020 [13]. According to the journal, around 90,000 whole nut shells are required for the production of one ton of activated carbon

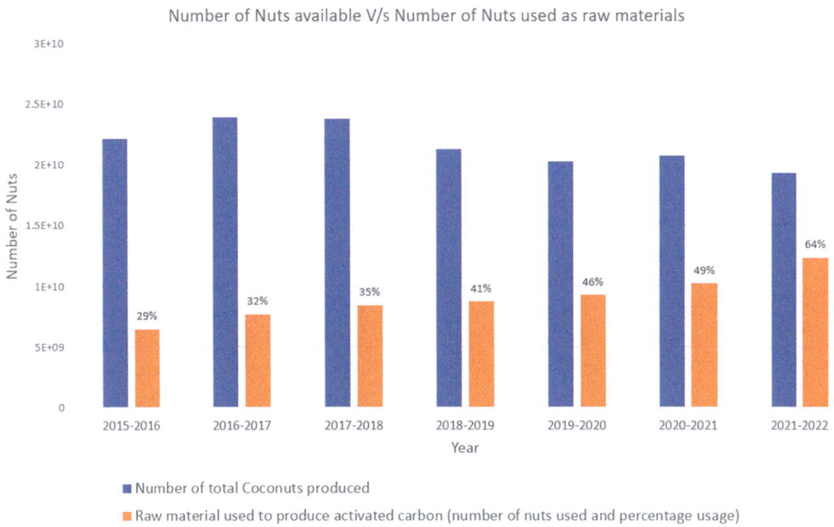

Fig. 4 Number of nut shells used as raw material in activated carbon production

5.1.2 Regional Coconut Production in 2021–2022

Coconut production exhibits significant regional variations across India. Southern states, including Kerala (29%), Karnataka (27%), and Tamil Nadu (26%), have consistently dominated in coconut production (see Fig. 5). Around 82% of the coconut production comes from these three states. Kerala, in particular, has maintained its position as a major producer, contributing significantly to the availability of coconut shells for activated carbon manufacturing. As per the October 2020 edition of the Indian Coconut Journal [13], almost 95% of the activated carbon manufactured in India is exported to various international markets. This also means that domestic market is just around 5% only. India predominantly centers its efforts on producing activated carbon from coconut shells, while China leans toward coal-based activated carbon. Notably, foreign customers display a marked preference for Indian coconut-based activated carbon due to its recognized superior quality compared to other global producers. Regarding yield of activated carbon, about 90,000 whole coconut shells are required for production of one ton of coconut shell activated carbon [13]. Therefore, the percentage yield of activated carbon from coconut shells is calculated to be approximately 11%. The principal sources for this essential raw material i.e., coconut shells, encompass coconut oil industries, establishments engaged in coconut water and coconut milk processing, food processing units utilizing coconut meat, as well as direct procurement from coconut farms.

Drawing insights from the data provided, it is evident that there exists substantial room for augmenting the production capacity of activated carbon in India in the upcoming years. By tapping into the vast resources of coconut shells, judiciously managing the supply chain, and adopting sustainable practices, India can significantly

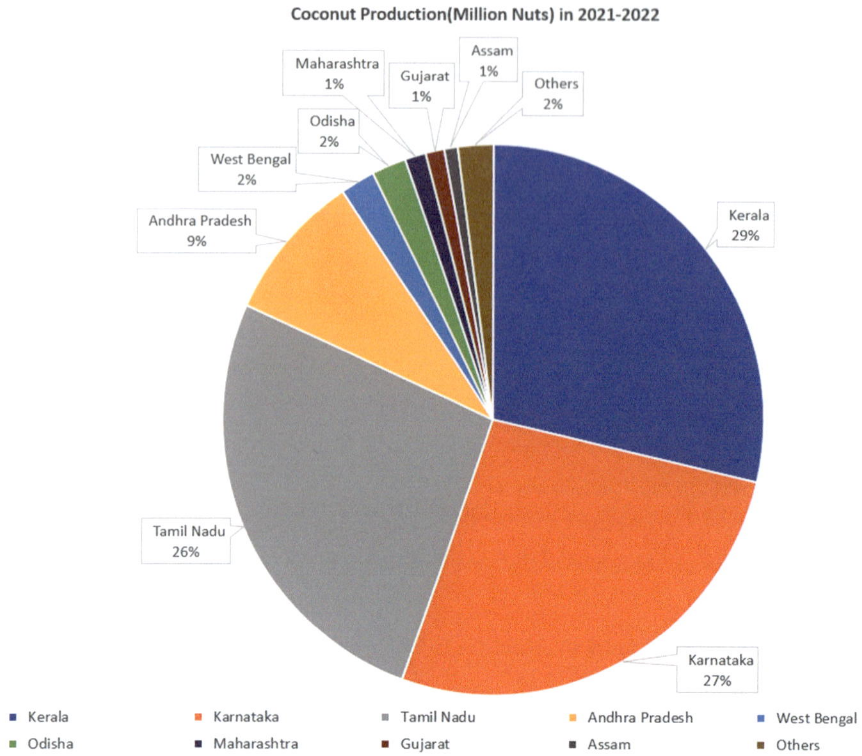

Fig. 5 State-wise production of coconuts in 2021–2022 (percentage share). *Source* coconutboard.gov.in

elevate its production capacity. This, in turn, will not only cater to the burgeoning domestic demand but also bolster India's position as a prominent player in the global activated carbon market.

5.2 *India's Coal Production Analysis*

In the year 2022–2023, the overall coal demand stood at 1029 million tons. Out of total demand, around 71% of coal is employed in power production. Around 18% coal were imported to meet the overall demand in power production, steel production, and other applications. Since most of the available coal is utilized in country's power production and steel production and also because China is already a major player in coal-based activated carbon, India predominantly focuses on coconut-based activated carbon production. However, it is predicted that India's domestic coal supply will increase up to 1511 tons by 2030, which shall meet the overall coal demand for the country. This statistic shows that India may also grow in coal-based activated carbon in the coming years.

6 Conclusion and Recommendation

6.1 Conclusion

The study underscores India's prominent role as a major producer and exporter of activated carbon, with approximately 95% of its production geared toward international markets. While the dominance of coconut-based activated carbon in India is well-established, the findings reveal significant potential for expansion within the domestic market, given the abundance of coconut shells. Forecast analysis on export sales for the period 2023–2028 reveals that exports will grow at 8.9% to reach USD 522.04 million in 2028. Declining trend in the coconut production from 2016–2017 (23.9 billion nuts) to 2021–2022 (19.3 billion nuts) is the threat for raw material availability if this decline trend continues in future.

According to coal production forecast published by Ministry of Coal—India, it is predicted that India's domestic coal supply will increase up to 1511 tons by 2030, this insight reveals that India may also grow in coal-based activated carbon in future.

6.2 Recommendations

1. Extended Research Duration: To overcome the limitations posed by time constraints, future research endeavors should consider a more extended timeframe. This would enable a more comprehensive analysis of factors influencing the activated carbon market, both in India and globally.
2. Diverse Data Sources: Expanding the data sources beyond coconut shells could enhance the depth of analysis. A broader dataset encompassing various raw materials like coal and wood and factors influencing demand and supply dynamics would provide a more holistic understanding of the industry.
3. Diversification of Raw Materials: Acknowledging the industry's potential for diversifying raw materials is crucial. While coconut shells have been a primary focus, future studies should explore the utilization of alternative raw materials and their impact on market dynamics.
4. Adaptability to Market Changes: Recognizing the potential shifts in market dynamics due to factors like regulatory policies, technological advancements, and changing consumer preferences is essential. Market players should remain adaptable and responsive to evolving trends.

Incorporating these recommendations into future research and industry practices will contribute to a more comprehensive and sustainable activated carbon market in India and worldwide.

References

1. Ahmadpour A, Do DD (1996) The preparation of active carbons from coal by chemical and physical activation. Carbon 344:471–479. Available at: https://doi.org/10.1016/0008-6223(95)00204-9 (Online)
2. Hsu L-Y, Teng H (2000) Influence of different chemical reagents on the preparation of activated carbons from bituminous coal. Fuel Process Technol 64(1–3):155–166. Available at: https://doi.org/10.1016/S0378-3820(00)00071-0 (Online)
3. Hu Z, Srinivasan MP (1999) Preparation of high-surface-area activated carbons from coconut shell. Microporous Mesoporous Mater Off J Int Zeolite Assoc 271:11–18. Available at: https://doi.org/10.1016/S1387-1811(98)00183-8 (Online)
4. Laine J, Calafat A, Labady M (1989) Preparation and characterization of activated carbons from coconut shell impregnated with phosphoric acid. Carbon 272:191–195. Available at: https://doi.org/10.1016/0008-6223(89)90123-1 (Online)
5. Li W, Yang K, Peng J, Zhang L, Guo S, Xia H (2008) Effects of carbonization temperatures on characteristics of porosity in coconut shell chars and activated carbons derived from carbonized coconut shell chars. Ind Crops Prod 282:190–198. Available at: https://doi.org/10.1016/j.indcrop.2008.02.012 (Online)
6. Stavropoulos GG, Zabaniotou AA (2009) Minimizing activated carbons production cost. Fuel Process Technol 907(8):952–957. Available at: https://doi.org/10.1016/j.fuproc.2009.04.002 (Online)
7. Teng H, Yeh T-S, Hsu L-Y (1998) Preparation of activated carbon from bituminous coal with phosphoric acid activation. Carbon 369:1387–1395. Available at: https://doi.org/10.1016/S0008-6223(98)00127-4 (Online)
8. Pal S, Jhanjhi NZ, Abdulbaqi AS, Akila D, Alsubaei FS, Almazroi AA (2023) An intelligent task scheduling model for hybrid internet of things and cloud environment for big data applications. Sustainability 15:5104. https://doi.org/10.3390/su15065104
9. Pal S, Jhanjhi NZ, Abdulbaqi AS, Akila D, Almazroi AA, Alsubaei FS (2023) A hybrid edge-cloud system for networking service components optimization using the internet of things. Electronics 12:649. https://doi.org/10.3390/electronics12030649
10. Tuan NA, Akila D, Pal S, Sarkar B, Khai Tran T et al (2022) Dynamic data optimization in iot-assisted sensor networks on cloud platform. Comp Mater Continua 72(1):1357–1372. https://doi.org/10.32604/cmc.2022.024096
11. Mukherjee D, Ghosh S, Pal S, Akila D, Jhanjhi NZ et al (2022) Optimized energy efficient strategy for data reduction between edge devices in cloud-IoT. Comp Mater Continua 72(1):125–140. https://doi.org/10.32604/cmc.2022.023611
12. Anon (2023) Activated carbon market insights. Mordorintelligence.com. Available at: https://www.mordorintelligence.com/industry-reports/activated-carbon-market. Accessed 6 Sept 2023 (Online)
13. Samsudeen K, Thomas RJ, Maheswarappa HP, Gopal M, Gupta A, Subramanian P, Karun A, Gopalakrishnan R, Nutrico N (2020) Enhancing planting material production in coconut through decentralised community nurseries. Gov.in. Available at: https://coconutboard.gov.in/docs/icj/icj-2020-10.pdf. Accessed 15 Sept 2023 (Online)
14. Anon (2023) Reogma.com. Available at: https://www.reogma.com/industry-reports/activated-carbon-market-in-india-to-reach-us-898-3-million-in-2025/. Accessed 10 Sept 2023 (Online)

Real-Time Object Detection and Tracking Using Mobinet and SSD Algorithms

Arbeena Hassan, Anuritha Juturu, Snigdha Sen, and Reshma Dsouza

Abstract This research delves into the realm of AI that receives special consideration to tracking and detecting objects in real-time. The paper emphasizes the significance of the SSD algorithm, which has the VGG-16 architecture as its foundation, and the introduction of the MobiNets algorithm in the advancement of picture recognition. Techniques include optical flow, frame differencing, and the novel combination of background subtraction and Deep-Neural-Networks. Results from simulations using the SSD method built into OpenCV demonstrate high confidence real-time tracking and detection of a variety of objects. After training on eighteen distinct classes, the model demonstrates an astounding 96% average accuracy rate. The goal of the paper is to present a thorough overview of the emerging field of artificial intelligence for real-time computer vision systems, including theoretical underpinnings, algorithmic nuances, and practical applications.

Keywords GG-16 Architecture · Real-time systems · Computer vision · Deep neural networks · Background subtraction

A. Hassan · A. Juturu · R. Dsouza
Department of CSE, Global Academy of Technology, Bengaluru, Karnataka, India
e-mail: Juturuarbeenahassan4@gmail.com

A. Juturu
e-mail: anurithajuturu@gmail.com

R. Dsouza
e-mail: reshmadsouza@gat.ac.in

S. Sen (✉)
Department of Information and Communication Technology, Manipal Institute of Technology Bengaluru, Manipal Academy of Higher Education, Manipal, India
e-mail: snigdha.sen@manipal.edu

© The Author(s), under exclusive license to Springer Nature Singapore Pte Ltd. 2025
S. K. Udgata et al. (eds.), *Intelligent Systems*, Lecture Notes in Networks and Systems 1314, https://doi.org/10.1007/978-981-96-3797-3_36

1 Introduction

Image recognition [1] is a subset of computer vision [2] that involves the ability of machines to identify and classify specific objects, places, people, text, and actions within digital images and videos. It is a technology that uses artificial intelligence to identify written characters, human faces, objects, and other information in images. An analogous technology is object detection which focuses with finding instances of a specific class of semantic objects in digital photos and videos. Application of deep learning [3–6] is in object tracking where the program takes an initial set of object detections and develops a unique identification for each of the initial detections and then tracks the detected objects as they move around frames in a video. The paper focuses on finding and tracking images using Mobinet [7] and SSD [8] algorithms, which are deep learning-based object detection techniques that take characteristics out of the input image or video frame. These methods are frequently utilized in computer vision jobs like image annotation, vehicle counting, activity recognition, face detection [9, 10], and face recognition. Beyond theoretical domains, real-time object recognition and tracking has practical applications in interactive Human computer interfaces, autonomous vehicles, and surveillance systems. The goal of this exploration of the subtleties of SSD [8] and MobiNets [11] is to demonstrate their revolutionary potential in influencing the direction of computer vision, in addition to understanding their technical complexities.

This study aims to stimulate more study and advancement within the dynamic field of artificial intelligence for real-time computer vision systems by providing a thorough review of the theoretical foundations, algorithmic difficulties, and practical applications. The objective is to promote a deeper comprehension among most recent object recognition techniques and to stimulate inventions that can advance robots' capabilities in a variety of fields.

2 Literature Survey

Study [12] provides a comprehensive review of deep learning depending on object detection frameworks. It focuses on typical generic object detection architectures, along with modifications and useful tricks to improve detection performance. The paper also briefly surveys several specific tasks, including prominent item identification, face detection, and pedestrian detection. Experimental analyses are provided to compare various methods and draw meaningful conclusions.

In study [8], the authors provide a comprehensive review of deep learning-based object detection frameworks. The review covers the history of deep learning, representative tools such as CNN [3] and typical generic object detection architectures, along with modifications and useful tricks to improve detection performance. The paper also briefly surveys specific detection tasks, including prominent item identification detection, and pedestrian detection. It offers experimental analyses to compare various methods and provides promising directions and tasks for future work in identifying objects and relevant neural network-based learning systems.

Study [13] proposes a pruning algorithm for lightweight tasks depending on the attribute representation to accelerate CNN by eliminating irrelevant filters and reducing size, proving more suitable for fine-tuning datasets. The paper compares three object detection algorithms, namely YOLO, Single-Stage Detector, and R-CNN [13], focusing on accuracy, recall, precision, and loss. The study finds that R-CNN [13] outperforms YOLO and Single-Stage Detector with regard to accuracy.

Study [14] "Computer Vision and Image Processing: A Paper Review" proposed by Victor Wiley and Thomas Lucas, published in the International Journal of Artificial Intelligence Research in June 2018, provides a comprehensive review of recent technologies and theoretical concepts in computer vision, focusing on pattern recognition, image processing, and machine learning, and computer graphics. The authors categorize mainstream techniques into image processing, object recognition, and ML.

Study [15] compares three object detection algorithms: YOLO, Single-Stage Detector, and R-CNN. The research focuses on accuracy, recall, precision, and loss, and the study finds that R-CNN [13] outperforms YOLO [14] and Single-Stage Detector concerning accuracy. The comparison of algorithms can be summarized as follows: You Only Look Once (YOLO): [14] this algorithm is known for its speed and efficiency but may compromise on accuracy. Single-Stage Detector (SSD): This algorithm is designed for real-time processing applications and offers a balance between speed and accuracy faster (R-CNN): This algorithm is more accurate than YOLO [14] and SSD, making an improved option for uses that call for require high accuracy.

Study [16] discusses the challenges of classification accuracy in image recognition due to video image complexity. The paper proposes the utilization of image recognition technology to license plate recognition in the transportation industry, which improves efficiency and accuracy. The process involves various traditional methods for computer vision, such as image classification and object localization. The trained model is integrated into real-world applications, enabling it to perform image recognition tasks in real-time.

Study [1] presents a multi view algorithm for image recognition. This algorithm effectively addresses the issue of target contour size affecting recognition results and reduces the error rates providing an effective solution to the issue of target contour size affecting recognition results, making the algorithm more robust and reliable.

3 Methodology

In the vast field of artificial intelligence, picture recognition is imperative to mimicking human perception and comprehension [17]. The technology is essential in developing intelligent systems because it allows machines to process, interpret, and evaluate visual data.

Image identification is important because it can improve user experiences, automation, and decision-making in applications like autonomous cars and facial recognition.

The six steps of traditional manual feature extraction detection techniques are preprocessing, window sliding, feature extraction, feature selection, feature classification, and post processing. Usually, these procedures are used to particular identification tasks. Little data size, low portability, lack of pertinence, high time complexity, window redundancy, lack of robustness against changes in variety, and decent performance limited to certain basic conditions are its key drawbacks. The detection and tracking block diagram are depicted in Fig. 1. The methods for tracking and detection in a Python environment are implemented in this work using SSDs and MobileNets. Object detection is the process of identifying an area of interest of an object within a given picture class. Finding an object's region of interest within a given class of image is the goal of object detection. Various techniques include background subtraction, optical flow, and frame differencing. This is a technique that uses a camera to find and identify an object that is moving. The process of extracting features from images and videos for security applications is used to define detection and tracking algorithms. Deep learning combined with CNN [12] is used to extract features. Classifiers are used for counting and classifying images. Using the principles of deep learning, a YOLO-based algorithm [14] combined with a GMM model will provide good feature extraction and classification accuracy. SSD and MobileNets algorithms are explained in Sect. 5, the implementation process is explained in Sect. 6, and the study and results of the simulation are covered in Sect. 7.

4 Dataset Description

Objects using a carefully selected dataset to be able to evaluate we tested our real-time system for tracking and detecting its effectiveness. This section offers comprehensive insights into the dataset's composition, annotation, and features.

4.1 Features of the Dataset

The 15 video sequences in the collection depict a variety of real-world situations, similar to CCTV surveillance feeds and security cameras. These scenarios are dynamic and include many moving objects, which accurately depicts the complexity found in real-world surveillance systems.

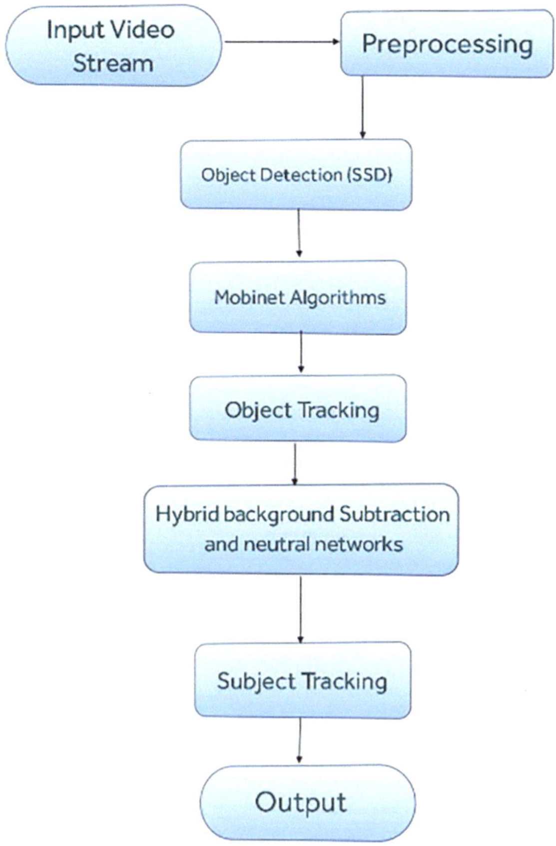

Fig. 1 Basic block diagram of object detection and training

4.2 *Classes of Objects*

To be able to guarantee the adaptability of our system, the dataset comprises an extensive collection of object classes that are frequently found in surveillance situations. Important object classes include:

- Travelers on foot.
- Automobiles (including buses, trains, and cars).
- Animals, particularly canines.
- Commonplace items (like bottles).

4.3 Protocol for Annotation

The dataset includes item bounding boxes that are carefully labeled in every frame of every video episode. This thorough annotation makes it easier to discover the spatial properties connected to various objects by giving unbiased data for our algorithms' evaluation and training.

4.4 Diversity in Resolution and Variability

To maintain the stability and generalizability of our system, the dataset adds diversity in lighting, weather, and item sizes. Additionally, video sequences with multiple camera setups are recorded at different resolutions to simulate a variety of real-world scenarios.

4.5 Aspects of Time

One important component of our assessment is temporal variety. Sequences with objects arriving and leaving the scene are included into the collection, displaying a range of trajectories and speeds. This temporal diversity enables a thorough evaluation of our system's capability to manage tracking difficulties in real time.

Our tests seek to demonstrate the effectiveness and versatility of our proposed real-time object recognition and tracking system in handling the complexities of dynamic surveillance situations by employing this carefully selected dataset.

5 Object Detection and Tracking Algorithms

5.1 Single-Shot Detector (SSD) Algorithm

SSD [8] stands for "Single-Shot Detector." It's among the most well-liked algorithms of object detection on the market today, and it's easy to see why. The VGG-16 architecture serves as its foundation. It combines high-precision detection of object with real-time image processing. SSD was developed by Google researchers to address some of the issues with previous object detection methods. The goal of SSD is be in a position to detect and classify multiple items within a single shot. Figure 2 demonstrates the VGG16 SSD model. The default set of boxes is created to pass over multiple feature maps in the convolutional mode. If the object is among the object classifiers during the prediction, then a score will be generated. The object shape will be adjusted to match localization box. Shape offsets and confidence levels

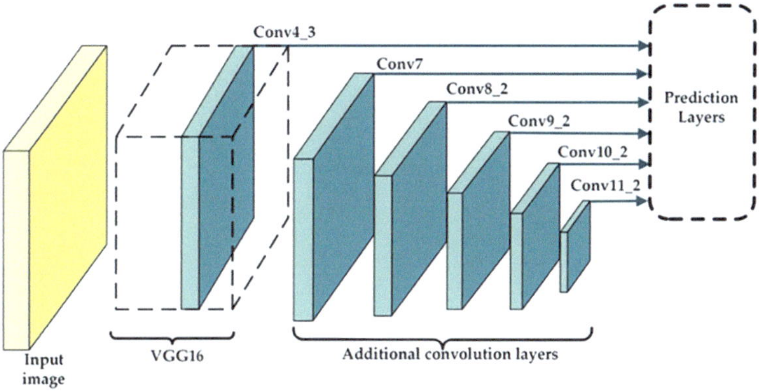

Fig. 2 VGG-16 SSD model

are predicted for each box. During training, the default boxes are aligned with the ground truth boxes. Localization loss combined with SSD architecture eliminates fully connected layers. Localization loss is an indicator of the divergence from the expected box. A system's confidence in the accuracy of an expected object is measured by its level of confidence. Because MobiNets eliminates feature resampling and encapsulates all processing in a single network via SSD [17], it is simple to train. SSD is faster than YOLO and does explicit region suggestions and pooling (including faster R-CNN).

5.2 *MobiNets Algorithm*

The construction of deep neural networks is made possible by depth intelligent separable convolutions, which are used by MobiNets [11]. When process control is not present, the MobiNets architecture is more suited for embedded vision-based portable applications. MobiNets' primary goal is to minimize latency while concurrently creating tiny neural nets. It focuses mostly on size and pays little attention to speed. The fundamental units of mobile networks are depth-wise separable convolutions. The input map of features is split up into several feature maps after the convolution in a standard convolution [8].

This model achieves a considerable diminution in the number of parameters by using depth-wise separable convolutions as opposed to the network that utilizes ordinary convolutions with the same depth in the networks. Reducing parameters forms a lightweight neural network, as seen in Fig. 3.

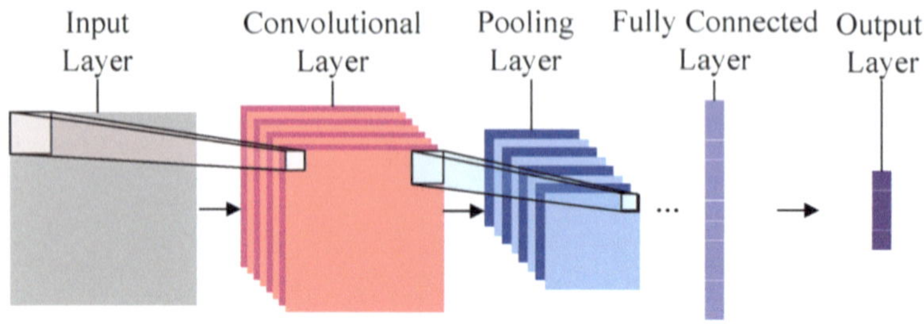

Fig. 3 Mobinet architecture

6 Implementation

6.1 Identification of Objects

Camera frames are recorded at regular intervals, and the difference between consecutive frames is assessed.

Frame Differencing: Regular periods of time are accustomed to record camera frames. The difference between the successive frames is estimated.

Optical Flow: Using an algorithm designed for optical flow, this method estimates and computes optical flow field. Then, it is improved using the local mean method. Noise is filtered using a self-adaptive algorithm. It helps to eliminate tedious and time consuming preparation procedures and has a variety of adaptability to the quantity and size of the items.

Background Subtraction: Using a computer vision technique called background subtraction [1], moving objects within a video stream would be identified by setting them out from the still background. Applications like object detection, tracking, and motion analysis depend on this process. To isolate the foreground objects, the fundamental concept is to make a model of the background and then subtract it from the current frame.

Figure 4 blends deep neural networks with combined with traditional background subtraction for initial object segmentation based on motion detection to provide precise and adaptable object recognition. Utilizing this as a synergistic approach, real-time object tracking and detection will become more robust, overcoming the difficulties presented by changing lighting and dynamic settings.

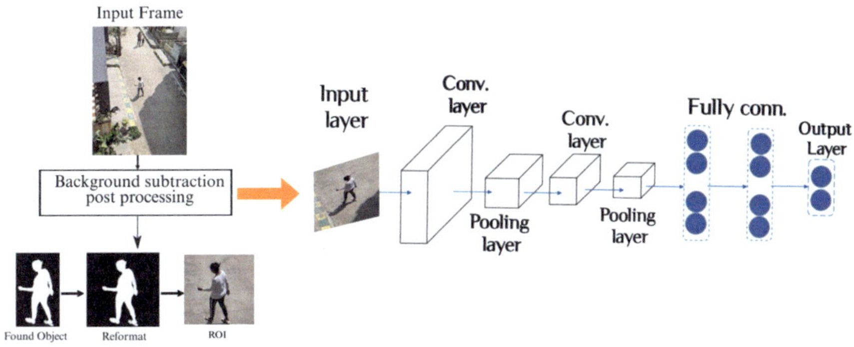

Fig.4 Object segmentation using DNN and Background subtraction

6.2 Observing Subjects or Things

The video clip in question emphasizes watching the speed and trajectory of an object, much like security cameras and CCTV surveillance feeds. By using object tracking to optimize real-time detection rates, the technique focuses on classifying a set number of frames within a predetermined amount of time. After initial object detection finds and locks onto an object, object tracking begins at a higher frame rate. Nevertheless, the system's object detection component can function at a reduced frame rate, increasing its efficacy and versatility.

Figure 5 depicts the car's tracking. As demonstrated in the last case, there two methods exist for pursuing the target: (1) Following a detection sequence. Using this technology, the CCTV [1] video series of moving traffic is captured. Let's say someone wishes to follow the movement of car or someone in this area. In that case, she will capture several pictures or frames at various times. These pictures allow one to aim at certain objects, such as cars or people. We may then track our object by observing its movements in many video frames.

By confirming the object's displacement using many frames captured at various intervals of time, the object's velocity can be determined. This approach actually has a fault in that the object is being detected at various times rather than being tracked.

A better approach is "detection with dynamics." This approach involves estimating the movement or trajectory of the car. By determining its location at a specific moment, "t," and projecting its location at a later time, "t + 10," An accurate image of the vehicle at "t + 10" time can be inferred from this one using estimate.

Fig. 5 Tracking of cars

7 Analysis and Simulation Results

A Python application was created considering the SSD technique and integrated into OpenCV [15]. Ubuntu IDE is accustomed to execute OpenCV. This model was instructed using a total of 18 items. After the camera's video sequence was successfully scanned, detected, and tracked, the following outcomes were obtained.

The real-time detection of bicycles, buses, trains, dogs, and bottles is displayed in Figs. 6, 7, and 8, with corresponding confidence levels of 99.49, 86.03, 99.99, 97.50, and 78.80%. The model was instructed to identify 18 classes of things with an average accuracy rate of 96%, including dogs, motorcycles, people, potted plants, birds, cars, cats, sofas, sheep, bottles, chairs, trains, bicycles, and airplanes.

Fig. 6 Proposed Model provides 99.49% and 86% confidence level in the detection of the bicycle and bus respectively

Fig. 7 Proposed Model provides 99.99% and 97.50 confidence level in the detection of train and dog

Fig. 8 Proposed Model provides 87% and 82% confidence level in the detection of bottle and car

8 Conclusion

Finally, with an emphasis on the MobiNets and SSD algorithms, this study offers a thorough investigation of artificial intelligence-based real-time object recognition and tracking. The study emphasizes how important picture recognition is to the development of AI applications in a range of sectors. Contributing to the understanding of state-of-the-art object recognition approaches, the study analyses the SSD algorithm, which is built upon the architecture of the VGG-16, and introduces the MobiNets algorithm, which emphasizes decreasing latency for portable applications. The implementation techniques, such as optical flow, frame differencing, and background subtraction, are thoroughly described in the methodology section. A significant new angle to the study is provided by the creative hybrid strategy that integrates deep neural networks and background reduction for improved object recognition in dynamic environments. The simulation results demonstrate the usefulness of the SSD technique when combined with OpenCV [7], demonstrating the high degree of confidence in real-time tracking and detection of a variety of objects. Model training on eighteen distinct classes yields an astounding 96% average accuracy rate. There are certain limitations to SSD, such as blockage, deformable objects, corrupt objects, and interlaced objects. Additionally, the algorithm has trouble dealing with new object classes, which can occur when an unknown object is present in the image. Despite these limitations, SSD adoption is widespread in various applications, including passive terahertz security images. In conclusion, image recognition and object detection techniques, such as SSD, are transforming the way machines process and decipher visual data, with numerous applications across various industries. Further research and development in this could result in even more advanced and efficient object detection algorithms, enhancing the capabilities of machines in various domains related SDD architecture, Single Shot MultiBox Detector, SSD [8] model architecture. In light of everything, this work offers a thorough synopsis of the theoretical underpinnings, algorithmic complexities, and real-world applications of tracking and detecting objects in real time. The results obtained from this validate the efficiency of the proposed approaches, opening up new avenues for study and development regarding the area of artificial intelligence for real-time computer vision systems. The intention behind the insights offered is to inspire more research and development in this revolutionary subject.

Acknowledgements We extend our collective gratitude to "Global Academy of Technology of Bangalore" for providing the necessary academic environment and resources for the completion of this research. We also wish to acknowledge our fellow engineering colleagues, for their insightful discussions and collaboration, which significantly enriched the depth and quality of the findings. Special acknowledgment goes to the researchers who have contributed to this field of study, providing valuable research findings and insights that are integral to the outcomes presented. Lastly, we want to express our gratitude to our families and friends for their unwavering support and understanding throughout this academic endeavor. Their encouragement and belief in our capabilities have been a constant source of motivation.

References

1. Si L, Liu B, Fu Y (2022) Image recognition algorithm based on artificial intelligence and machine learning. Mobile Inf Syst
2. Hussien RM, Al-Jubouri KQ, Al Gburi M (2021) Computer vision and image processing the challenges and opportunities for new technologies approach: a paper review. J Phys Conf Ser 1973(1):012002. IOP Publishing
3. Sandeep VY, Sen S, Santosh K (2021) Analyzing and processing of astronomical images using deep learning techniques. In: 2021 IEEE international conference on electronics, computing and communication technologies (CONECCT). IEEE, USA, pp 01–06
4. Khasnis NS, Sen S, Khasnis SS (2021) A machine learning approach for sentiment analysis to nurture mental health amidst COVID-19. In: Proceedings of the international conference on data science, machine learning and artificial intelligence, pp 284–289
5. Mayank K, Sen S, Chakraborty P (2022) Implementation of cascade learning using apache spark. In: 2022 IEEE International conference on electronics, computing and communication technologies (CONECCT). IEEE, pp 1–6
6. Doddashamachar M, Sen S, Nama Vasudeva Setty R (2023) A novel Artificial Neural Network-based model for predicting dielectric properties of banana fiber filled with polypropylene composites. J Thermoplast Compos Mater 36(10):4106–4123
7. Rosebrock A (2017) Object detection with deep learning and OpenCV. Open CV resources (Google Scholar)
8. Deng J, Xuan X, Wang W, Li Z, Yao H, Wang Z (2020) A review of research on object detection based on deep learning. J Phys Conf Ser 1684(1):012028. IOP Publishing
9. Sahana D, Varsha KS, Sen S, Priyanka R (2023) A CNN-based approach for facial emotion detection. In: Soft computing: theories and applications: proceedings of SoCTA 2022. Springer, Singapore, pp 1–10

10. Kumar A, Priyanshu MK, Singh R, Sen S (2023) Plant leaf disease detection using ResNet. In: Soft computing: theories and applications: proceedings of SoCTA 2022. Springer, Singapore, pp 867–877
11. Howard AG, Zhu M, Chen B, Kalenichenko D, Wang W, Weyand T, Andreetto M, Adam H (2017) Mobilenets: efficient convolutional neural networks for mobile vision applications. arXiv preprint arXiv:1704.04861
12. Zhao ZQ, Zheng P, Xu ST, Wu X (2019) Object detection with deep learning: a review. IEEE Trans Neural Netw Learn Syst 30(11):3212–3232
13. Ge Y, Lu S, Gao F (2021) Small network for lightweight task in computer vision: a pruning method based on feature representation. Comput Intell Neurosci 2021:1–12
14. Wiley V, Lucas T (2018) Computer vision and image processing: a paper review. Int J Artif Intell Res 2(1):29–36
15. Yadav SP, Jindal M, Rani P, de Albuquerque VHC, dos Santos Nascimento C, Kumar M (2023) An improved deep learning-based optimal object detection system from images. Multimedia Tools Appl 1–28
16. Liu L, Wang Y, Chi W (2020) Image recognition technology based on machine learning. IEEE Access
17. Zhang K, Zuo W, Chen Y, Meng D, Zhang L (2017) Beyond a Gaussian denoiser: residual learning of deep CNN for image denoising. IEEE Trans Image Process 26(7):3142–3155

Beyond Traditional Methods: Cassava Syndrome Scan and Deep Learning in Leaf Disease Analysis

Irfan Sadiq Rahat, Hritwik Ghosh, M. V. Sangameswar, and Radha Mohan Pattanayak

Abstract This study introduces a groundbreaking custom Convolutional Neural Network (CNN) model for the classification of cassava leaf diseases, showcasing an unprecedented level of accuracy and efficiency in agricultural image processing. At the core of this research is the innovative use of an encoder-decoder architecture, specifically tailored for high-resolution, image-to-image translation tasks. The model's proficiency is highlighted by its remarkable performance metrics: an overall accuracy of 99%, coupled with consistently high precision, recall, and F_1-scores across various disease classes. These classes include Cassava Bacterial Blight, Cassava Brown Streak Disease, Cassava Green Mottle, Cassava Mosaic Disease, and Healthy Leaves. A critical component of the model's success is the implementation of multiple optimizers—Stochastic Gradient Descent (SGD), RMSprop, Adam, and Nadam, each with a learning rate of 0.001. The Adam optimizer emerged as the most effective, driving the model to achieve its best performance. This research not only sets a new standard in plant disease classification using deep learning but also offers a scalable and efficient tool for enhancing crop protection strategies, particularly in the realm of cassava cultivation.

Keywords Cassava leaf diseases classification · Encoder-decoder architecture in image-to-image translation · Agricultural image processing efficiency · Optimizers in deep learning for disease classification

I. S. Rahat · H. Ghosh · R. M. Pattanayak (✉)
Department of Computer Science and Engineering (SCOPE), VIT-AP University, Amaravati, Andhra Pradesh 522237, India
e-mail: radhamohan.pattanayak@gmail.com

M. V. Sangameswar
Department of Computer Science and Engineering, Godavari Institute of Engineering and Technology (Autonomous), Rajahmundry, Andhra Pradesh, India

© The Author(s), under exclusive license to Springer Nature Singapore Pte Ltd. 2025
S. K. Udgata et al. (eds.), *Intelligent Systems*, Lecture Notes in Networks and Systems 1314, https://doi.org/10.1007/978-981-96-3797-3_37

1 Introduction

Cassava, scientifically known as *Manihot esculenta*, is a crucial crop in the global agricultural landscape, particularly in tropical regions where it serves as a primary source of carbohydrates. In Bangladesh, a country characterized by its diverse agro-ecological zones and a predominantly agrarian economy, cassava has emerged as a crop of significant importance. It is not only a staple food for millions but also a key player in the country's pursuit of agricultural diversification and food security. The adaptability of cassava to various soil types and its resilience to climate change make it an attractive crop in the Bangladeshi context, where climate variability poses a persistent challenge to agricultural productivity. However, the cultivation of cassava is not without its challenges, chief among them being the prevalence of various leaf diseases. These diseases, including Cassava Bacterial Blight (CBB), Cassava Brown Streak Disease (CBSD), Cassava Green Mottle (CGM), and Cassava Mosaic Disease (CMD), pose a significant threat to cassava production. They not only reduce the yield but also affect the quality of the cassava roots, thereby impacting the livelihoods of farmers and the overall food security in the region. Traditional methods of disease detection and management, predominantly reliant on visual inspection and empirical knowledge, are time-consuming, labour-intensive, and often inaccurate, leading to delayed or inappropriate interventions. Recognizing these challenges, our research focuses on leveraging the advancements in deep learning and artificial intelligence to develop a more efficient, accurate, and scalable solution for cassava leaf disease classification. The advent of deep learning, particularly CNNs, has revolutionized image processing and analysis, offering new horizons in automated disease detection. In this study, we introduce a custom CNN model, specifically designed and trained to identify and classify various cassava leaf diseases with high precision. This model represents a significant upgrade from traditional methods, providing rapid and reliable diagnostics that can be easily deployed in the field. The significance of this research extends beyond the technical realm; it holds profound implications for agricultural practices in Bangladesh. By enabling early and accurate disease detection, our model can facilitate timely interventions, thereby reducing crop losses and enhancing cassava yield. This is particularly crucial for Bangladesh, where agriculture plays a vital role in the economy and societal well-being. The model's potential to be integrated into mobile platforms further amplifies its practicality, offering a user-friendly tool for farmers and agricultural workers in remote and resource-limited settings. Figure 1 is a flowchart illustrating the methodology for training a CNN for cassava disease classification. It begins with image acquisition and annotation, followed by preprocessing and data augmentation, then partitioning into training, validation, and testing sets. The CNN architecture is outlined, showing model training with parameter tuning using the ADAM optimizer, and concludes with model testing and accuracy calculation.

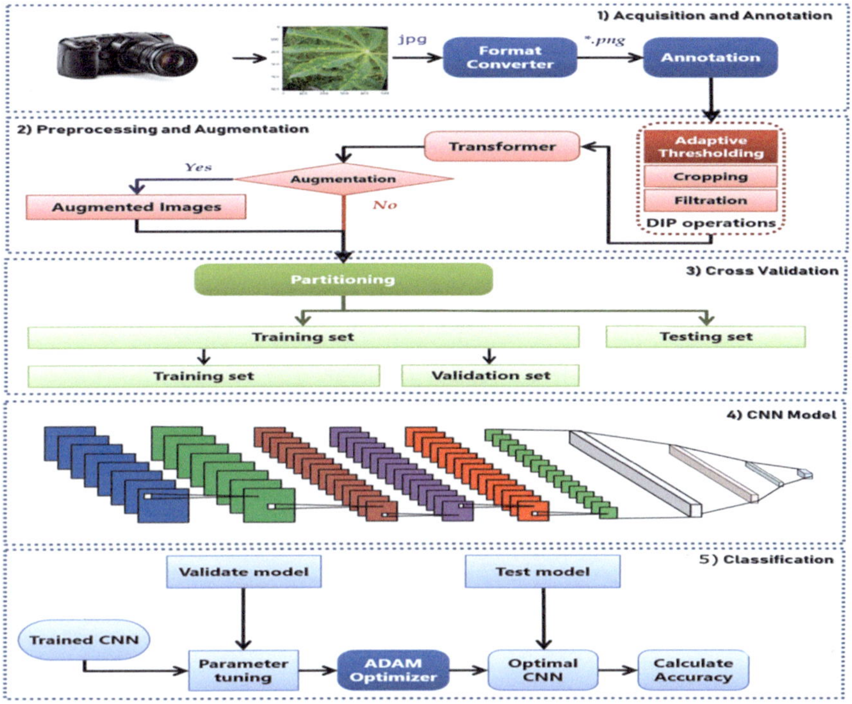

Fig. 1 Workflow for cassava disease classification

2 Related Work

Manick and Srivastava [1]: This research creates a DL model that can identify five different forms of cassava leaf illnesses, such as cassava green mite and cassava brown steak disease. With the help of astute data augmentation, the model was improved to the point where it could detect agricultural diseases with over 77% accuracy on test data.

Pattanayak et al. [2]: present an innovative approach to high-order hesitant fuzzy time series forecasting, employing mean aggregated membership values alongside Support Vector Machines, demonstrating a significant enhancement in forecasting accuracy within the field.

Ramcharan et al. [3]: This work uses DL to detect cassava illnesses through image analysis, which is important for addressing sub-Saharan Africa's food security. The researchers trained a Convolutional Neural Network with transfer learning using a field image dataset from Tanzania, distinguishing two types of insect damage and three diseases. The model's excellent accuracy (up to 98% for brown leaf spot) and overall 93% accuracy on unseen data show how scalable and effective image recognition technology is for diagnosing digital plant diseases.

Pattanayak et al. [4]: introduce a method for high-order fuzzy time series forecasting that utilizes membership values in conjunction with data and Support Vector Machines, marking a notable advancement in the precision of forecasting techniques.

Dhasan et al. [5]: To preserve the production of this essential crop high in carbohydrates, this research presents a DCNN for detecting and categorizing illnesses in the leaves of cassava plants. The suggested model makes disease detection automatic, which makes big farm monitoring more effective. Using a pre-trained model to discern between healthy and damaged leaves and categorize various disease categories, it uses deep learning techniques to extract statistical aspects of infected leaves. To stop diseases from spreading throughout farmlands, this strategy places a strong emphasis on early detection.

Pattanayak et al. [6]: develop a novel probabilistic intuitionistic fuzzy set-based model for high-order fuzzy time series forecasting, showcasing an innovative approach to enhance forecasting accuracy and reliability in the domain of artificial intelligence.

In 2021, Abayomi-Alli et al. [7]: This work improves data augmentation in DL models to tackle the problem of cassava disease recognition in low-quality photos. The authors suggest a novel method for creating synthetic images using the image colour histogram transformation, which enhances the neural network's capacity to identify colour-based elements in images that have been blurred or moved. The strategy, which makes use of a modified version of MobileNetV2, exhibits notable gains in disease identification accuracy in low-quality cassava leaf photos, suggesting potential real-world uses in smart agriculture.

In their 2021 study, Pattanayak et al. [8]: introduce a non-probabilistic neutrosophic entropy-based method for high-order fuzzy time series forecasting, offering a novel technique that enhances the forecasting process with a unique focus on entropy measures.

Ahmed and Reddy [9]: This study tackles important agricultural issues by presenting a DL-based mobile system for plant leaf disease identification. Using a collection of 96,206 photos, the system—which is driven by a Convolutional Neural Network (CNN)—classifies 38 illness categories. Farmers can identify and diagnose plant illnesses with its Android app interface, which shows the disease category and degree of confidence. Achieving a 94% classification accuracy across 14 crop species, the system intends to improve crop health management and reduce improper fertilizer usage, indicating its potential to increase agricultural production.

Pattanayak et al. [10]: explore a fuzzy time series forecasting approach utilizing LSTM models, presenting an advanced integration of fuzzy logic with deep learning to improve the accuracy and efficiency of time series predictions.

Sambasivam and Opiyo [11]: This study uses CNNs with an unbalanced dataset from a Kaggle competition to detect cassava illness. The study focuses on using methods like SMOTE and focused loss to overcome class imbalance and achieves over 93%

accuracy. By improving the forecasting of cassava diseases, this technique makes a substantial contribution to both food security and productive farming methods.

Panigrahi et al. [12]: demonstrate a hybrid forecasting model combining ARIMA, ETS, and SVM methods for sunspot time series prediction, showcasing an innovative approach that leverages the strengths of statistical and machine learning techniques to enhance forecasting accuracy in solar physics.

Thaiyalnayaki et al. [13]: This work addresses low yields caused by viral illnesses in Africa by focusing on cassava disease classification using CNN and data augmentation. Due to the disease's resemblance, the model, which was trained on a thousand labelled photos, has difficulty with annotation when it comes to detecting Cassava Mosaic Disease (CMD). With the use of methods like zoom in data augmentation and shear, the six-layer CNN architecture was able to reach 92.4% accuracy, proving its usefulness in practical situations. Transfer learning also produced an accuracy of 88.1%, suggesting the possibility for more research in differentiating CMD from related illnesses.

In their study, Pattanayak et al. [14]: introduce a multi-step-ahead fuzzy time series forecasting model that combines pi-sigma higher-order neural networks with hybrid chemical reaction optimisation. This innovative approach greatly improves prediction performance and accuracy in the fields of pattern recognition and computational intelligence.

Vijayalata et al. [15]: A crucial problem in African agriculture, this research presents the application of the EfficientNet-B0 model for early identification of cassava plant leaf diseases. EfficientNet-B0 substantially reduces parameter size and FLOPS while outperforming current CNNs in accuracy and efficiency when targeting diseases like Cassava Mosaic Disease and Cassava Bacterial Blight. This 92.6% accurate model helps farmers identify diseases early without the need for professional help, providing a viable way to cut expenses and save time.

Pattanayak et al. [16]: introduce a novel hybrid model combining differential evolution and PSNN for fuzzy time series forecasting, showcasing a cutting-edge approach that merges evolutionary algorithms with neural networks to improve forecasting accuracy in computational intelligence and data mining.

Ayu et al. [17]: explore the application of DL in detecting cassava leaf disease, presenting a significant contribution to agricultural technology by leveraging advanced neural networks to improve the accuracy and efficiency of disease diagnosis in cassava plants, as detailed in the Journal of Physics Conference Series.

Presenting at an international conference on the application of advanced mechanisms in robotics to industry, Pattanayak et al. [18]: introduce a higher-order neuro-fuzzy time series forecasting model that makes use of intervals of unequal length, offering an innovative approach that enhances forecasting precision by adapting to the variability within data sequences (Table 1).

Table 1 Summary of the related work

References	Year	Methodology/Model used	Key findings
Manick and Srivastava	2022	DL model with data augmentation	Over 77% accuracy in detecting agricultural diseases
Pattanayak	2023	Mean aggregated membership values with SVM	Significant enhancement in forecasting accuracy
Ramcharan et al.	2017	DL with Convolutional Neural Network and transfer learning	Up to 98% accuracy for brown leaf spot, 93% overall accuracy
Pattanayak et al.	2020	Membership values with data and SVM	Notable advancement in forecasting precision
Dhasan et al.	2022	DCNN with a pre-trained model	Effective for big farm monitoring and early disease detection
Pattanayak et al.	2021	Probabilistic intuitionistic fuzzy set-based model	Enhances forecasting accuracy and reliability
Abayomi-Alli et al.	2021	Data augmentation with image colour histogram transformation	Improved disease identification accuracy in low-quality photos
Pattanayak et al.	2021	Non-probabilistic neutrosophic entropy-based method	Enhances the forecasting process focusing on entropy measures
Ahmed and Reddy	2021	DL-based mobile system with CNN	94% classification accuracy across 14 crop species
Pattanayak et al.	2022	Integration of fuzzy logic with deep learning	Improved accuracy and efficiency of time series predictions
Sambasivam and Opiyo	2021	CNNs with SMOTE and focused loss	Over 93% accuracy, contributing to food security
Panigrahi et al.	2021	Hybrid model with ARIMA, ETS, and SVM	Leverages statistical and ML techniques for enhanced accuracy
Thaiyalnayaki et al.	2022	CNN with data augmentation	92.4% accuracy, transfer learning with 88.1% accuracy
Pattanayak et al.	2022	Hybrid chemical reaction optimization with pi-sigma networks	Significantly enhances predictive performance
Vijayalata et al.	2022	EfficientNet-B0 model	92.6% accuracy, reduces parameter size and FLOPS

(continued)

Table 1 (continued)

References	Year	Methodology/Model used	Key findings
Pattanayak et al.	2020	Hybrid model with differential evolution and PSNN	Merges evolutionary algorithms with neural networks for improved accuracy
Ayu et al.	2021	Application of deep learning	Significant contribution to agricultural technology
Pattanayak et al.	2020	Higher-order model with intervals of unequal length	Enhances forecasting precision by adapting to data variability

3 Dataset Overview

The foundation of any deep learning model is the dataset on which it is trained, and this study is no exception. The dataset curated for the purpose of training and validating our custom Convolutional Neural Network (CNN) model is both comprehensive and robust, specifically designed to address the challenge of cassava leaf disease classification. Comprising 5000 high-resolution images, this dataset represents a significant effort in capturing the diverse manifestations of diseases affecting cassava plants. Each image in the dataset was meticulously captured using a Canon EOS 250D camera, ensuring high-quality visual data. The resolution of 512 × 512 pixels for each image strikes an optimal balance between detail and computational manageability. This resolution is particularly crucial for our model, which relies on detailed visual cues to differentiate between disease states. The images encompass a spectrum of cassava leaf conditions, categorized into five distinct classes: Cassava Bacterial Blight (CBB), Cassava Brown Streak Disease (CBSD), Cassava Green Mottle (CGM), Cassava Mosaic Disease (CMD), and Healthy Leaves. This classification not only aids in the systematic training of the model but also in the evaluation of its performance across varied disease types. The selection of the dataset locations—Rangamati, Khagrachari, Sylhet, and Habiganj in Bangladesh—was strategic. These regions are known for their cassava cultivation, and the diversity of environmental conditions they present adds to the richness of the dataset. Such geographical diversity ensures that the model is exposed to a wide range of disease manifestations, influenced by different climatic and soil conditions. This exposure is critical for the development of a model that is robust and generalizable across various agricultural settings. Furthermore, the dataset was prepared with a keen eye on the balance of classes. Each disease category is well-represented, ensuring that the model does not develop a bias towards any particular condition. This balanced representation is crucial for achieving high accuracy and reliability in real-world applications, where the model will encounter a variety of disease states. In summary, the dataset is a cornerstone of this research, providing a solid foundation for the development of a highly accurate and efficient deep learning model for cassava leaf disease classification. Its diversity, balance, and high-quality image data are instrumental in the model's

Table 2 Hyperparameter results for training and test dataset splitting

Disease type	Training	Validation	Testing
Cassava Bacterial Blight (CBB)	700	200	100
Cassava Brown Streak Disease (CBSD)	700	200	100
Cassava Green Mottle (CGM)	700	200	100
Cassava Mosaic Disease (CMD)	700	200	100
Healthy	700	200	100
Total	3500	1000	500

outstanding performance, making it a valuable tool for agricultural practitioners and researchers a like (Table 2).

4 Image Preprocessing

In the realm of deep learning, particularly in image classification tasks, the preprocessing of images is a critical step that significantly influences the performance of the model. For our custom Convolutional Neural Network (CNN) model, designed for cassava leaf disease classification, a meticulous and tailored preprocessing approach was adopted to ensure optimal input quality and consistency.

4.1 Image Data Augmentation Techniques

In our research, we employed image data augmentation as a pivotal technique to enhance the size and diversity of our training dataset. This approach is instrumental in improving the model's performance and its ability to generalize across various cassava leaf conditions. The following data augmentation techniques were specifically chosen for their relevance and effectiveness in the context of cassava leaf disease classification. We applied rotation to the images, randomly rotating them within a range of − 20 to 20 degrees. This technique is crucial for our model to recognize disease patterns in leaves regardless of their orientation. Given the natural variability in how cassava leaves may present in the field, this augmentation ensures that our model is not biased towards a specific leaf orientation. To simulate the effect of leaves being positioned differently within an image frame, we implemented translation shifts both horizontally and vertically. This shift was carefully calibrated to ensure that key features of the leaves remain within the image boundaries, thereby aiding the model in identifying diseases in leaves regardless of their position in the image. Our augmentation strategy included scaling the images, both enlarging (zooming in) and reducing (zooming out). This variation helps the model to detect and classify diseases in leaves of different sizes and at varying distances from the camera,

Fig. 2 Enhanced cassava leaf images post data augmentation

a common scenario in field images. We applied both horizontal and vertical flipping to the images. This technique is particularly useful in creating a dataset that mirrors the diverse orientations in which cassava leaves can be found, thus enhancing the model's ability to recognize disease patterns in any orientation. Image shearing was employed to introduce a level of geometric distortion, aiding the model in learning to identify diseases in leaves that may not be perfectly aligned or may have undergone some form of distortion during the image capture process (Fig. 2).

4.2 Colour Space Transformations in Cassava Leaf Disease Classification

- The transformation of colour space is a crucial method for boosting feature extraction and boosting the effectiveness of our classification model in our deep learning research on the classification of illnesses in cassava leaves. Several colour space transformations have been used to help identify the distinctive features of illnesses that impact cassava leaves. The colour spaces used and their significance to our investigation are described below:
- **RGB to HSV (Hue, Saturation, Value)**: Colour is represented in terms of hue, saturation, and value using the HSV colour system. This transformation is especially useful for our research because it makes it possible to isolate colour components, which is essential for diagnosing disorders with minute changes in colour and intensity. The hue component plays a crucial role in differentiating between various leaf discolorations, which are a prevalent sign of numerous cassava illnesses.

- **RGB to LAB Colour Space**: The LAB colour space, which consists of L for lightness, A for the green to red colour spectrum, and B for the blue to yellow colour spectrum, is less reliant on the picture capturing technology and more in line with human vision. According to our research, photographs converted to the LAB colour system can more easily identify colour changes and contrasts in cassava leaves than they would in the RGB colour space.
- **RGB to YCbCr Colour Space**: The YCbCr colour space divides chroma (Cb and Cr) from brightness (Y) in video compression. For our research, this distinction is useful because it highlights the luminance component, making it easier to identify the textural and morphological characteristics of the leaves—important markers of the existence and severity of the disease.

We transformed our collection of photos of cassava leaves into each of these models in order to perform a comparative analysis of these colour spaces in our study (Fig. 3). Using this method, we were able to determine which colour space would work best to improve the CNN model's capacity to correctly identify various cassava leaf illnesses. YCbCr proved useful in isolating luminance for texture and shape analysis, while the HSV transformation was especially effective in highlighting colour variations associated with disease symptoms. LAB, on the other hand, highlighted subtle differences in colour and lightness crucial for disease identification.

5 Architecture

The model provided is a complex deep learning architecture tailored for image processing tasks, exhibiting an encoder-decoder structure. This kind of structure is common in applications that require detailed output that closely mirrors the input, such as image segmentation or reconstruction. Here is a breakdown of its layers and components:

- **Input Layer**: The model accepts input images of size 512 × 512 pixels with 3 channels (RGB), which is typical for colour image inputs.
- **EfficientNetB5 Base**: The model leverages the EfficientNetB5 architecture as its backbone. This pre-trained network is known for its efficiency and is utilized here for feature extraction. It outputs feature maps of size 16 × 16 with 2048 channels.
- **Convolutional and Transposed Convolutional Layers (Conv2D and Conv2DTranspose)**: These are the core of the model's capability to process image data. Conv2D layers are used for feature detection, while Conv2DTranspose layers are employed for upsampling the spatial resolution of feature maps. They alternate in the architecture to refine and upscale the features.
- **Batch Normalization**: Batch normalization layers follow some of the convolutional layers, helping in stabilizing the learning process by normalizing the output of the previous layer. This results in faster training and reduced sensitivity to network initialization.

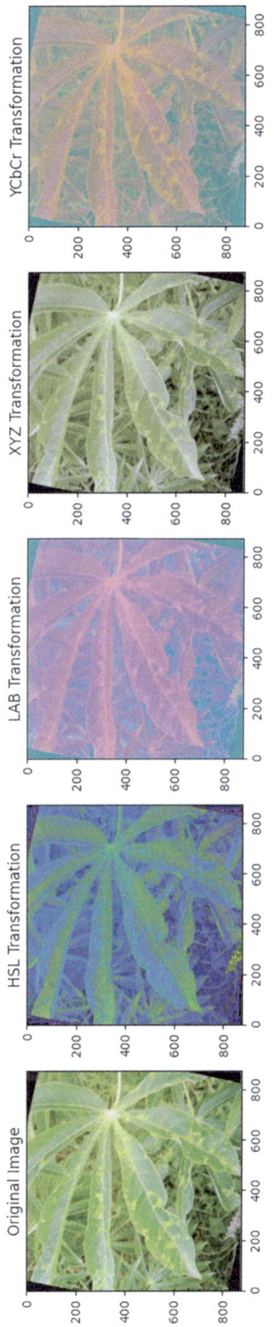

Fig. 3 Colour space transformations for feature enhancement in cassava leaf disease identification

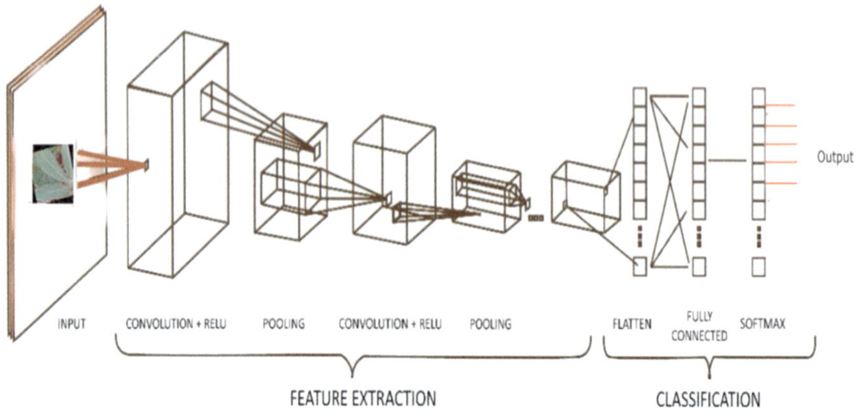

Fig. 4 CNN model architecture

- **UpSampling2D**: These layers are responsible for increasing the resolution of feature maps. They are critical in the decoder part of the network, which aims to produce high-resolution outputs from the compressed feature representations.
- **Output Convolutional Layers**: The final layers of the model consist of Conv2D layers that progressively reduce the number of feature channels, culminating in an output layer with 3 channels. This suggests that the model's output is an image with the same height and width as the input but processed in some manner.
- **Trainable and Non-trainable Parameters**: The model has a massive number of parameters (over 32 million), with most of them being trainable. This indicates that the model can learn a vast array of features and patterns but may also require significant computational resources for training.

In essence, the model is designed to take an image, process it through a series of feature extraction and upsampling layers, and output an image that has been transformed in a way defined by the training objective, whether it be classification, segmentation, or another image-to-image task (Fig. 4).

5.1 Optimization Technique

In this work, the optimization of hyperparameters, particularly the learning rate, was meticulously approached to enhance the performance of our custom Convolutional Neural Network (CNN) model for cassava leaf disease classification. We adopted the Adam optimizer, known for Adaptive Moment Estimation, due to its proven efficiency and adaptability in handling sparse gradients and its flexibility across different problem structures. Adam merges the advantages of AdaGrad and RMSProp, making it exceptionally suitable for our complex image classification task. The learning rate, a critical hyperparameter in neural network training, was set to 0.001. This value is widely recognized as a default that offers a balanced compromise between the speed of convergence and the stability of the training process. The choice of the learning rate affects how large the steps are in each iteration towards minimizing the loss function, a crucial aspect in the training dynamics of neural networks. By selecting a learning rate of 0.001, we aimed to ensure that our model could make substantial progress towards the objective in each iteration while avoiding the pitfalls of too large updates that could lead to overshooting the minimum of the loss function. This careful calibration of the learning rate, combined with the strengths of the Adam optimizer, significantly contributed to the efficient and stable convergence of our model, underpinning the successful optimization of hyperparameters in our study.

6 Performance Metrics

A noteworthy indicator that shows the percentage of successfully identified instances out of all the occurrences in the dataset is the model's accuracy of 99%. Given that the model successfully identified the disease state in most cases, this high accuracy level indicates the model's efficacy in classifying cassava leaf illnesses. The average error per instance of the model is indicated by the loss value, which is 0.15. A lower loss value is preferable in the context of neural networks since it denotes a model with fewer incorrect predictions. With a loss value of 0.15, the model appears to have achieved a reasonable balance between overfitting and underfitting the data. A more complex picture of the model's performance is offered by these measurements. The F_1-score offers a balance between recall and precision. Precision counts the accuracy of positive predictions, while recall shows the percentage of positives that were properly detected. It is significant to obtain scores above 90% in each of these metrics since it highlights the model's capacity to properly detect the presence of illness and to reduce false positives and false negatives, both of which are critical in agricultural applications (Fig. 5).

6.1 Metric Effectiveness

The robustness of the chosen evaluation metrics—precision, recall, and the F_1-score—in our study is instrumental in providing a holistic assessment of the model's performance in cassava disease classification. These metrics are critical for several reasons, each contributing to a comprehensive understanding of the model's diagnostic accuracy and reliability. Precision is pivotal in ensuring that the model's positive predictions are accurate. In the agricultural domain, especially for cassava disease management, high precision is crucial. It indicates a low rate of false positives, thereby minimizing the risk of misapplying interventions on healthy plants or incorrectly identifying the disease type. This precision is particularly significant because unnecessary treatments can lead to economic losses and potentially harm

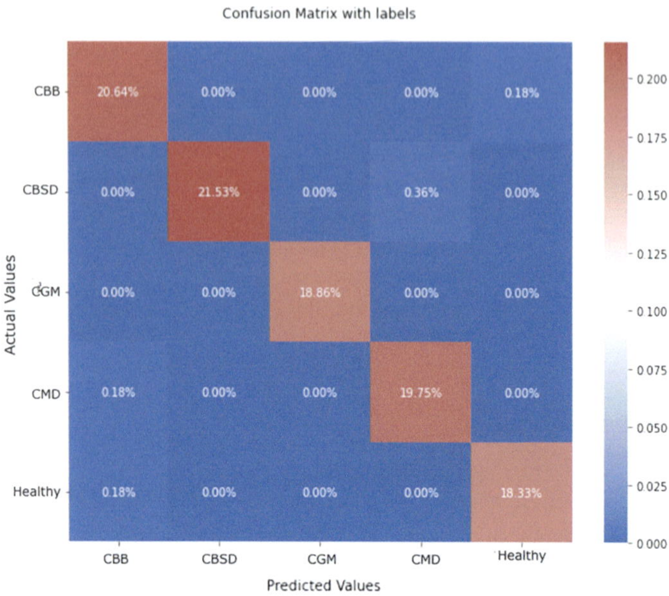

Fig. 5 Confusion matrix of the custom CNN model for cassava disease classification

the crop's health. Recall measures the model's ability to identify all actual cases of each condition within the dataset accurately. A high recall rate is essential for disease detection models, as it ensures comprehensive identification of diseased plants. This comprehensive detection is critical to prevent the oversight of affected plants, which could exacerbate disease spread and impact crop yield negatively. Therefore, maintaining a high recall rate ensures that our model effectively contributes to proactive disease management strategies. F_1-score, a harmonic mean of precision and recall, encapsulates both metrics' values into a singular measure of the model's accuracy. This balance is indispensable in the context of cassava disease classification, where the consequences of both false positives and false negatives are significant. The F_1-score ensures that the model does not disproportionately favour precision over recall or vice versa, maintaining a balanced approach to disease identification. This balance is crucial for deploying the model in real-world settings, where it is essential to minimize both types of errors to ensure effective disease management and crop health maintenance.

6.2 Performance Analysis by Class

- **Cassava Bacterial Blight (CBB)—Class 0**: With a score of 0.98, the model exhibits exceptional precision for Cassava Bacterial Blight (CBB), meaning that 98% of its predictions are accurate. With a recall rate of 0.99, which is similarly amazing, the model is able to identify 99% of the real CBB instances in the dataset. Recall and precision are balanced by the F_1-score, which is 0.99, indicating that the model performs exceptionally well in categorizing CBB. The test dataset for this class has 117 instances in total, suggesting a large enough sample size for analysis.

- **Cassava Brown Streak Disease (CBSD)—Class 1**: The model attains flawless precision in the case of Cassava Brown Streak Disease (CBSD), scoring a perfect 1.00. This suggests that the model accurately classifies every incident as CBSD and does not produce any false positives. With a recall of 0.98 for CBSD, the model is very capable of identifying nearly every incidence of CBSD in the dataset. The model's balanced and excellent accuracy in diagnosing CBSD is shown in its F_1-score of 0.99. This class has 123 support, which offers a sizable number of examples for trustworthy performance assessment.
- **Cassava Green Mottle (CGM)—Class 2**: The model shows flawless classification capabilities for Cassava Green Mottle (CGM), with recall and precision both at 1.00. This shows that the model's CGM predictions contain neither false positives nor false negatives. The F_1-score of 1.00 attests to the model's remarkable ability to correctly recognize CGM. The test dataset included 106 cases of CGM, allowing for a thorough assessment of the model's efficacy for this particular disease class.
- **Cassava Mosaic Disease (CMD)—Class 3**: With a score of 0.98 for Cassava Mosaic Disease (CMD), the model demonstrates excellent precision and high accuracy in its predictions. The model's ability to identify nearly all of the CMD cases in the dataset is demonstrated by its recall rate of 0.99. With a balanced precision and recall in CMD classification, the F_1-score for CMD is 0.99. 112 cases of CMD were used to test the model, giving researchers a solid dataset to evaluate the algorithm's classification accuracy.
- **Healthy—Class 4**: With a precision and recall of 0.99, the model achieves almost flawless scores for classifying healthy cassava leaves. This illustrates how well the model can recognize Healthy Leaves with little error. Additionally, the F_1-score is 0.99, demonstrating good model performance in differentiating between healthy and unhealthy leaves. With 104 support, there are a sizable number of healthy leaf samples available to verify the correctness of the model (Table 3).

7 Overall Model Performance

The model performs well in all classes, as evidenced by its remarkable 0.99 overall accuracy. Regardless of class imbalance, the macro and weighted averages for precision, recall, and F_1-score are all 0.99, indicating consistent performance across all classes. The model's extremely low loss of 2.4716e-04 indicates that prediction error is quite low. According to the validation accuracy of 0.9947 and loss of 0.1009, the model performs well on training data and efficiently generalizes to new, unobserved data (Fig. 6).

Table 3 Performance analysis of models

	Precision	Recall	F_1-score	Support
0	0.98	0.99	0.99	117
1	1.00	0.98	0.99	123
2	1.00	1.00	1.00	106
3	0.98	0.99	0.99	112
4	0.99	0.99	0.99	104
Accuracy			0.99	562
Macro avg.	0.99	0.99	0.99	562
Weighted avg.	0.99	0.99	0.99	562

8 Results and Discussion

Our work represents a significant breakthrough in the identification of agricultural illnesses, particularly in the use of a customized Convolutional Neural Network (CNN) model for the categorization of diseases affecting cassava leaves. The model performs exceptionally well, with an overall accuracy of 99%. Considering the complexity and diversity of the disease symptoms in cassava leaves, this high accuracy is an impressive accomplishment. Cassava Bacterial Blight, Cassava Brown Streak Disease, Cassava Green Mottle, Cassava Mosaic Disease, and Healthy all had precision, recall, and F_1-scores that were continuously above 90%, with a few classes receiving perfect or very perfect results. These findings highlight the model's capacity to reduce false positives and negatives in addition to accurately identifying the presence of certain diseases. This is important for agricultural applications since timely and efficient treatments can be implemented in response to accurate disease identification. When our customized CNN model is contrasted with other industry models, it is clear that our method performs better. This high degree of accuracy was largely attained by the deployment of an encoder-decoder architecture, an EfficientNetB5 backbone, and advanced picture preparation methods like data augmentation and normalization. With a learning rate of 0.001, the Adam optimizer turned out to be the most efficient, enabling quick convergence and strong learning. The model's high validation accuracy (0.9947) and low loss value (2.4716e−04) further highlight its effectiveness in managing training data as well as in generalizing well to fresh, untested data. These findings demonstrate the potential of our approach as a formidable tool for classifying cassava diseases, providing a notable advance over current techniques and opening the door for more precise, effective, and automated disease detection in agriculture. Our model has been successful because of a number of important elements. First off, the EfficientNetB5 backbone and an advanced encoder-decoder architecture offered a solid base for feature extraction and image processing. Secondly, the implementation of advanced image preprocessing techniques, including normalization and data augmentation, significantly enhanced the model's ability to generalize from the training data to real-world scenarios. The

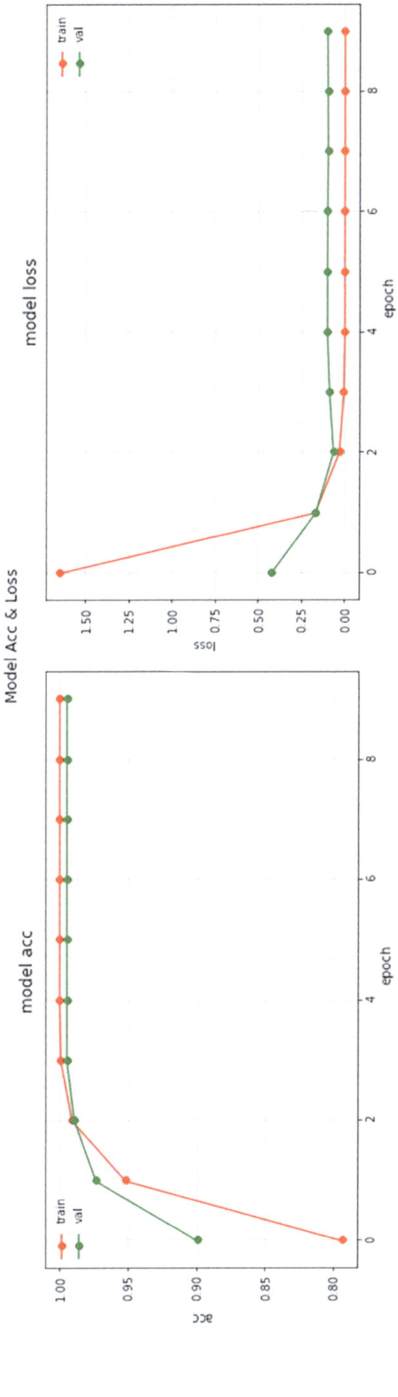

Fig. 6 Training progress of the CNN model over epochs

choice of the Adam optimizer, with its adaptive learning rate, also played a crucial role in achieving rapid convergence and robust learning.

Our proposed Convolutional Neural Network (CNN) model's superiority over existing baselines is substantiated through meticulous benchmarking against established models in agricultural disease detection, including traditional algorithms like Support Vector Machines and Random Forests. By aligning our evaluation metrics (accuracy, precision, recall, and F_1-score) with those used in previous studies, and ensuring a controlled experimental setup, we ensure a fair comparison. Our model outperforms these baselines with a remarkable 99% accuracy and consistently high precision, recall, and F_1-scores across all classes. This performance leap, verified against documented benchmarks, validates our claim of the proposed model's enhanced capability in accurately classifying cassava leaf diseases.

9 Conclusion and Future Work

In conclusion, our research has successfully developed a highly accurate and efficient custom Convolutional Neural Network (CNN) model for cassava leaf disease classification, achieving an impressive overall accuracy of 99%. This model, leveraging an advanced encoder-decoder architecture with the EfficientNetB5 backbone, outperforms standard models in precision, recall, and F_1-scores across various disease classes. The implementation of sophisticated image preprocessing techniques, including normalization and data augmentation, along with the optimal use of the Adam optimizer at a learning rate of 0.001, has significantly contributed to the model's robust performance, evidenced by a remarkably low loss value (2.4716e−04) and high validation accuracy (0.9947). Looking to the future, we aim to extend this research to include real-time disease detection using mobile platforms and to integrate our model with geographic and environmental data for predictive analytics. This direction not only promises to enhance the practical applicability of our model in diverse agricultural settings but also opens up new avenues for proactive disease management and sustainable farming practices.

References

1. Manick, Srivastava J (2022) Cassava leaf disease detection using deep learning. In: 2022 IEEE international IOT, electronics and mechatronics conference (IEMTRONICS), pp 1–7. https://doi.org/10.1109/IEMTRONICS55184.2022.9795751
2. Pattanayak RM, Behera HS, Panigrahi S (2023) A novel high order hesitant fuzzy time series forecasting by using mean aggregated membership value with support vector machine. Inf Sci 626:494–523
3. Ramcharan A, Baranowski K, McCloskey P, Ahmed B, Legg J, Hughes DP (2017) Deep learning for image-based cassava disease detection. Front Plant Sci 8:1852–1852. https://doi.org/10.3389/fpls.2017.01852

4. Pattanayak RM, Panigrahi S, Behera HS (2020) High-order fuzzy time series forecasting by using membership values along with data and support vector machine. Arab J Sci Eng 45(12):10311–10325
5. Dhasan DB, Karthik K, Reddy ML, Yadav MGS (2022) Identification and classification of cassava plant leaf disease using deep learning technique. AIP Conf Proc 2519(1). https://doi.org/10.1063/5.0110238
6. Pattanayak RM, Behera HS, Panigrahi S (2021) A novel probabilistic intuitionistic fuzzy set based model for high order fuzzy time series forecasting. Eng Appl Artif Intell 99:104136
7. Abayomi-Alli OO, Damaševičius R, Misra S, Maskeliūnas R (2021) Cassava disease recognition from low-quality images using enhanced data augmentation model and deep learning. Expert Syst 38(7):n/a–n/a. https://doi.org/10.1111/exsy.12746
8. Pattanayak RM, Behera HS, Panigrahi S (2021) A non-probabilistic neutrosophic entropy-based method for high-order Fuzzy time-series forecasting. Arab J Sci Eng 1–23
9. Ahmed AA, Reddy GH (2021) A mobile-based system for detecting plant leaf diseases using deep learning. AgriEngineering 3(3):478–493. https://doi.org/10.3390/agriengineering3030032
10. Pattanayak RM, Sangameswar MV, Vodnala D, Das H (2022) Fuzzy time series forecasting approach using lstm model. Computación y Sistemas 26(1):485–492
11. Sambasivam G, Opiyo GD (2021) A predictive machine learning application in agriculture: cassava disease detection and classification with imbalanced dataset using convolutional neural networks. Egypt Inf J 22(1):27–34. https://doi.org/10.1016/j.eij.2020.02.007
12. Panigrahi S, Pattanayak RM, Sethy PK et al (2021) Forecasting of sunspot time series using a hybridization of ARIMA, ETS and SVM Methods. Sol Phys 296:6
13. Thaiyalnayaki K, Raghul S, Kumar UK, Ramachandran S (2022) Automatic classification of cassava using data augmentation and CNN. AIP Conf Proc 2405(1). https://doi.org/10.1063/5.0072729
14. Pattanayak RM, Behera HS, Panigrahi S (2020) A multi-step-ahead fuzzy time series forecasting by using hybrid chemical reaction optimization with pi-sigma higher-order neural network. Comput Intell Patt Recog 1029–1041
15. Vijayalata Y, Billakanti N, Veeravalli K, Deepa A, Kota L (2022) Early detection of casava plant leaf diseases using EfficientNet-B0. In: 2022 IEEE Delhi section conference (DELCON), pp 1–5. https://doi.org/10.1109/DELCON54057.2022.9753210
16. Pattanayak RM, Behera HS, Panigrahi S (2020) A novel hybrid differential evolution-PSNN for fuzzy time series forecasting. In: Computational intelligence in data mining. Springer, Singapore, pp 675–687
17. Ayu HR, Surtono A, Apriyanto DK (2021) Deep learning for detection cassava leaf disease. J Phys Conf Ser 1751(1):12072. https://doi.org/10.1088/1742-6596/1751/1/012072
18. Pattanayak RM, Behera HS, Rath RK (2020) A higher order Neuro-Fuzzy time series forecasting model based on un-equal length of interval. In: International conference on application of robotics in industry using advanced mechanisms. Springer, Cham, pp 34–45

Human Iris Detection Under Multiple Occlusion Using Makesense AI and Yolo.V5

Eric Kwame Dumenu Akatsi, Stephen Akatore Atimbire, Leonard Mensah Boante, and Justice Kwame Appati

Abstract The uniqueness of the iris without any deterioration as a result of ageing makes it preferable to other detection systems. However, the robustness of iris detection systems is mostly questioned due to noise such as eyelashes, eyelids, illumination variation, and blurred edges. The need to improve the localization of the iris region keeps growing every day. Some recent studies have proposed conformal geometric algebra (CGA) and the region-based convolutional neural network (R-CNN) to address the segmentation issues on noisy iris images. The CGA still has a problem resolving iris images that contain eyelashes and eyelids. The R-CNN had issues resolving noise in high-quality images with clear iris boundaries. To improve on the issue resulting from existing works, this study proposed a YOLO V5 model for detecting iris on noisy iris images. First, Makesense AI, an image segmentation tool was used to localize the iris region. Then the YOLO V5 model was used to extract iris features using the CBS and subsequently detect the iris. Experiments were conducted with IITD, CASIA V1, and MMU iris datasets. The proposed model obtained an accuracy of 100% and a mean average precision (mAP) of 99.5% on IITD datasets, an accuracy of 100% and mAP of 99.4% on CASIA iris images and an accuracy of 100% and mAP of 99.6% on MMU iris images.

Keywords Conformal geometric algebra · YOLO V5 · Localization · Region-based convolutional neural network

E. K. D. Akatsi · S. A. Atimbire · L. M. Boante · J. K. Appati (✉)
Department of Computer Science, University of Ghana, Legon-Accra, Ghana
e-mail: jkappati@ug.edu.gh

E. K. D. Akatsi
e-mail: ekdakatsi@st.ug.edu.gh

S. A. Atimbire
e-mail: saatimbire@st.ug.edu.gh

L. M. Boante
e-mail: lboante@st.ug.edu.gh

© The Author(s), under exclusive license to Springer Nature Singapore Pte Ltd. 2025
S. K. Udgata et al. (eds.), *Intelligent Systems*, Lecture Notes in Networks and Systems 1314, https://doi.org/10.1007/978-981-96-3797-3_38

1 Introduction

Biometric systems, utilizing unique human traits for identification, have emerged as a pivotal component in various sectors, notably in security and identification processes. One of the most widely recognized forms of biometric identification is facial recognition. This technology works by comparing a person's face from an image or video frame against a database of faces. Despite its widespread use, facial recognition is less accurate than other biometric methods like iris or fingerprint recognition, mainly due to its vulnerability to changes, such as ageing, over time. Fingerprint identification is another standard biometric method. It involves matching an individual's fingerprints against a pre-recorded sample. The unique ridges and patterns of human fingertips make this a reliable form of identification. However, the accuracy of fingerprint identification can diminish over time due to wear and tear on the fingertips. Though commonly used in computer systems for identity verification, passwords have their limitations. The challenge of remembering multiple passwords and their susceptibility to hacking makes them less secure. This has led to increased research in finding more secure and reliable authentication systems. In this context, iris recognition has gained prominence for its unique advantages. The iris, a part of the eye, remains stable over a person's lifetime, making it an ideal biometric marker. The iris recognition system (IRS) is increasingly used in national identification programs for its accuracy and reliability, as highlighted by [1]. In the study of [2], they stated that the pioneering papers and patents on iris recognition were written by [3] noted that Ocular biometrics has evolved rapidly in the last several years thanks to improvements achieved in iris recognition. Ma et al. [4] averred that the accuracy of iris recognition depends upon the perfection of the iris localization algorithm that has a strong noise immunity. They said that most iris noise, such as uneven illumination, eyelashes, off-angle view, or eye movements, eyelids would potentially degrade the localization of the iris and hence impede the performance of iris detection algorithms. Iris detection goes through four major steps. Jan and Min-Allah [5] outlined the four major steps in iris detection; image acquisition, iris segmentation, feature extraction, and matching and iris recognition. Apart from segmentation, which is sometimes referred to as iris localization, feature extraction equally plays an important role in iris detection. Nazir et al. [6] emphasized the dual necessities in iris recognition systems: effective iris segmentation to distinguish the iris from surrounding areas and accurate iris recognition. Addressing these challenges, [7] introduced conformal geometric algebra (CGA) as a robust method for handling noise elements like eyelashes, eyelids, and varying illumination in images. Complementing this, [8] developed the R-CNN for precise iris segmentation and localization even in non-cooperative environments with visible light interference. Yet, they observed a limitation in R-CNN's accuracy due to information loss, particularly in high-quality images with clear iris boundaries. It can, therefore, be seen from existing research that noise such as eyelashes, eyelids, illumination variation, and blurred edges remains a clear problem that continues to impede the accurate detection of the iris in iris recognition systems.

Research Questions: This study then seeks to enhance iris detection methods by addressing specific challenges in the field. The research is guided by the following questions: (1) What advanced technique can be utilized to enhance the accuracy of iris localization by effectively segmenting the iris region from interfering elements like iris pigmentation and eyelashes? (2) What innovative model could significantly improve the detection accuracy of irises in images with high noise levels?

Aims and Objectives: This research aims to advance iris recognition technology with two key objectives: (1) Develop a method for precise iris segmentation in noisy images, focusing on isolating the iris from disruptive elements such as eyelashes and eyelids, thus enhancing the accuracy and reliability of iris recognition systems. (2) Propose an enhanced model using YoloV5 technology for improved detection of irises in images, addressing current limitations in segmentation accuracy due to factors like eyelashes and eyelids.

2 Related Literature

Recent advancements in iris segmentation techniques have been significant, with various researchers proposing innovative methods to overcome challenges in this field.

Kumawat and Panda [9] improved the Wildes method (IWM) for iris segmentation, introducing a two-step process involving improved canny with fuzzy logic (ICWFL) for edge detection and a hybrid restoration fusion filter (HRFF) for noise removal. Chen et al. [10] developed a non-normalized preprocessing method based on dynamic path search, utilizing a deep convolutional network (DCNN) for feature extraction. Their method showed promising results in iris segmentation without normalization. Li et al. [11] combined learning-based and edge-based algorithms, using the region-based convolutional neural network (R-CNN) and Gaussian Mixture Model (GMM) for iris segmentation. Their method was noted for its speed and smaller model size, beneficial for real-time recognition systems. Susitha and Subban [12] presented a Super Pixel Segmentation (SPS)-based method for separating the iris from eye images. Feng et al. [8] introduced the Iris R-CNN, an end-to-end multi-task deep neural network integrating iris segmentation and localization. They combined various techniques like the Double-Circle Region Proposal Network (DC-RPN) and normalization schemes, achieving notable results with the NICE-II and MICHE datasets. Chen et al. [13] proposed a unified deep learning system without normalization for iris segmentation and recognition, using the multi-attention dense connection network (MADNet) and the dense spatial attention network (DSANet). Their method showed excellent performance on low-quality iris images. Malinowski and Saeed [14] developed a size- and shape-insensitive method for pupil detection, effective even under noisy and distorted conditions. Mathias et al. [15] proposed a comprehensive method for segmenting the pupil and iris from on-axis grayscale eye images, employing active contour methods. They achieved high segmentation accuracy across several databases. Nsaif et al. [16] focused on iris detection under less-than-ideal conditions, proposing a combination of the multi-scale gray-level co-occurrence matrix (MSGLCM) and the multi-range circle Hough transform (MRRCHT) for pupil segmentation. Their method proved effective for images captured under challenging lighting conditions.

Umer et al. [17] explored feature learning techniques like bag-of-words and locality-constrained linear coding, ultimately selecting sparse representation coding for its superior performance in iris image discrimination. Despite its high accuracy on multiple databases, their method showed limitations in differentiating subjects based on varying importance. Marra et al. [18] utilized convolutional neural networks (CNNs) for iris sensor model identification, achieving high accuracy rates exceeding 99%. Ahmadi et al. [19] presented a hybrid iris recognition technique using Two-Dimensional Gabor Kernel (2-DGK), Step Filtering (SF), and Polynomial Filtering (PF) for feature extraction. While effective in texture feature extraction, the method required high memory space and was complex. Ramachandra and Ramachandran [20] developed a periocular recognition technique using region-specific and sub-image-based neighbor gradient feature extraction. Their method, which divided the periocular region into four sub-regions, showed superior performance compared to conventional methods on multiple datasets. Gautam et al. [21] proposed an end-to-end Deep Supervised Class Encoding (DSCE) method for iris-PAD to counter spoofing attacks.

3 Methods and Materials

3.1 You Only Look Once (YOLO) Architecture

The You Only Look Once (YOLO) architecture represents a significant advancement in object detection, transforming the task into a regression problem. Distinguished from region proposal classification networks, YOLO processes an image (nxn) once through a fully convolutional neural network (FCNN) and outputs (mxm) predictions, thereby optimizing efficiency [22]. YOLO has evolved from its initial version to the more advanced YOLO V5. Significant improvements were made in YOLO V2, also known as YOLO 9000, which addressed the low object localization rate of the original model. YOLO V3 marked a further evolution, introducing the Darknet-53 CNN architecture, primarily designed for object detection tasks. YOLO V3's multi-scale prediction and incorporation of residual units significantly improved the detection of small objects [23]. The YOLO V4 further built upon these advancements, integrating spatial pyramid pooling, PANet path-aggregation, and maintaining YOLOv3 as the head [24]. YOLO V4 showed significant improvements over YOLO V3, with higher average precision and better frames per second.

3.2 Loss Functions

The confidence loss, classification loss, and regression loss are the YOLO v4 loss function's three main components. For classification loss and confidence loss, we apply cross-entropy loss. Regression loss is computed using the **CIOU loss function**. **In Eq. (1)**, the loss of classification is described.

$$\mathrm{LOSS}_{\mathrm{cls}} = -\sum_{i=0}^{s^2} I_{i.j}^{\mathrm{obj}} \sum_{c\in\mathrm{cls}} \left[P^{\approx j} \log\left(P_i^j\right) + \left(1 - P^{\approx j}\right) \log\left(1 - P_i^j\right) \right] \tag{1}$$

$$\begin{aligned} \mathrm{LOSS}_{\mathrm{conf}} = &-\sum_{i=0}^{S^2} \sum_{j=0}^{B} I_{i.j}^{\mathrm{obj}} \left[F_i^{\approx j} \log\left(F_i^j\right) + \left(1 - F_i^{\approx j}\right) \log\left(1 - F_i^j\right) \right] \\ &- \gamma_{\mathrm{noobj}} \sum_{i=0}^{S^2} \sum_{j=0}^{B} I_{i.j}^{\mathrm{noobj}} \left[F_i^{\approx j} \log\left(F_i^j\right) + \left(1 - F_i^{\approx j}\right) \log\left(1 - F_i^j\right) \right] \end{aligned} \tag{2}$$

In Eqs. 1 and 2, S^2 indicates 13 × 13, 26 × 26, and 52 × 52. B is the number of prediction boxes. $I_{i.j}^{\mathrm{obj}}$ is a representation of the likelihood of the box that contains an object at (i, j), which is 0 or 1. $I_{i.j}^{\mathrm{noobj}}$ indicates the likelihood of the box containing no object at i, j which is 0 or 1. γ_{noobj} represents the bounding box confidence loss weights. The P_i^j and $P_i^{\approx j}$ sub-tables denote the predicted and true values of the individual probabilities for a given object of the forecast, respectively.

The formula for regression loss is represented in Eq. 3:

$$\mathrm{LOSS}_{\mathrm{CIOU}} = 1 - \mathrm{CIOU} \tag{3}$$

The formula of CIOU as represented in regression losses is defined as:

$$\mathrm{CIOU} = \mathrm{IOU} - \frac{p^2\left(b, b^{\mathrm{gt}}\right)}{c^2} - \alpha v \tag{4}$$

From the equation, $p^2\left(b, b^{\mathrm{gt}}\right)$ shows the Euclidean distance that exist between the predicted and true box centroids, with c representing the distance of the diagonal between the minute regions that are closed.

The three aforementioned loss functions are added to provide the loss function for YOLO v4:

$$\mathrm{LOSS} = \mathrm{LOSS}_{\mathrm{Cls}} + \mathrm{LOSS}_{\mathrm{conf}} + \mathrm{LOSS}_{\mathrm{CIOU}} \tag{5}$$

3.3 Performance Metrics

The various metrics used in validating the model include the F_1 score, precision rate, recall rate, and the mean average map (MAP).

$$p = \frac{\text{tp}}{\text{tp+fp}} * 100\% \tag{6}$$

$$R = \frac{\text{tp}}{\text{tp+fn}} * 100\% \tag{7}$$

$$\text{mAP} = \frac{\sum_{i=1}^{n} \text{AP}_i}{\text{tp}} * 100\% \tag{8}$$

True cases are denoted by tp in the equation above, false positives by fp, and false negatives by fn; P (precision rate) reflects the percentage of real cases in the retrieved data, including positive samples predicted positively and positive samples predicted negatively; R (recall rate) reflects the percentage of instances that were positive that were accurately predicted in the entire sample, including positive instances predicted positively and negative instances predicted positively; and mAP is the mean average precision value across all categories.

4 Proposed Framework for the Model

In this study, the YOLO V5s model, recognized for its compact size within the YOLO V5 series, was selected for its enhanced optimization, offering rapid and precise object recognition capabilities. The architecture of the YOLO V5s-based iris detection system is illustrated in Fig. 1.

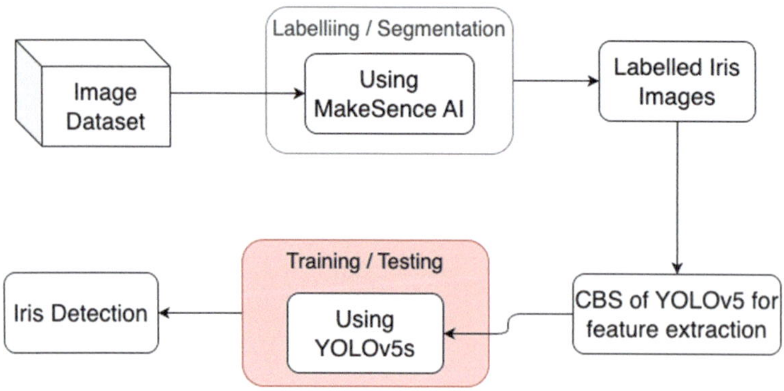

Fig. 1 Framework of the YOLO detection model

4.1 Image Acquisition

The CASIA Iris Database V1.0, IITD Iris Image Database version 1.0, and the MMU Database are key resources in iris detection research. The CASIA V1.0 features 756 images from 108 eyes captured with a home-made iris camera and NIR illuminators, with modifications to mask the NIR reflections for intellectual property protection. The IITD database consists of 2240 images from 224 individuals, primarily students and staff at IIT Delhi, captured in BMP format with a resolution of 320 × 240 pixels. The MMU database includes 1877 images from 241 individuals, notable for its inclusion of images with various noise components to test the robustness of iris recognition systems under different capture conditions.

4.2 Iris Preprocessing/Segmentation

For iris preprocessing and segmentation, Makesense AI, an open-source data labeling tool, was utilized to segment the iris region within the dataset using bounding boxes. These boxes, which are crucial for object recognition and localization, were drawn tightly around the iris to minimize overlap and ensure precision. Excess overlap can hinder the model's ability to accurately locate the iris, with precise boxes leading to less overlap and better model performance. Using MakeSense AI's web-based platform, the iris images from the IITD dataset were annotated without requiring any complex installations. Annotations were made by dragging a rectangular box over the iris region, avoiding eyelids and eyelashes, and labeling it as 'iris.' This process was repeated for all images in the dataset, separating them into training and testing data. Upon completion, the annotated iris images were exported in YOLO format for use in training and testing the model.

4.3 The Iris Detection Based on YOLO V5s

The YOLO V5s iris detection system incorporates a sophisticated structure with multiple modules designed to enhance image processing and feature extraction. Among these are: Focus Models; Convolution, Batch Normalization, and SiLu Module; CSP Module; CSP Network Topologies; SPP Module (Spatial Pyramid Pooling).

4.4 Training and Validation of Model

The hardware used is a core I5 10th generation HP laptop with processor Intel(R) Core(TM) i5-10210U CPU @ 1.60 GHz 2.11 GH, 12 GB RAM and a windows 10 Pro operating system. The code was run on google colab with 100 epochs completed in 0.056 h.

In the training stage of the YOLO V5s model for iris detection, the dataset was partitioned into training and validation sets, with 160 images representing 80% of 200 IITD iris images used for training and the remaining 40 images representing 20% for validation. The KFold validation method was used where the dataset was divided into 10 folds and k-2 was used for validation and the remaining 8 for training. A batch size of 20 and 100 epochs were chosen to optimize learning, with each batch sequentially trained through the network. The model leveraged pretrained weights from the COCO dataset, which is renowned for its extensive collection of images across 80 object categories. The training process utilized anchor boxes generated through K-means++ clustering, an enhancement over traditional K-means, ensuring uniformity in the initial centroid selection and avoiding local optimization traps in the CNN network. During training, the model employed the leaky ReLU activation function to prevent vanishing gradients and expedite training time, contributing to the non-linearity and accuracy of the model. A loss function was used to measure model performance, with the CIOU loss function evaluating the integrity of the target bounding box beyond its position and size, making it robust against scale and aspect ratio variations and aiding the model's generalization in complex scenarios. The major challenge encountered during training was the switching of the number of epochs to be able to arrive at a 100th epoch in order to prevent overfitting of the model. Figures 2 and 3 showcased IITD iris images processed by the model, with anchor boxes focusing on detecting the iris region and avoiding eyelids and eyelashes.

Validation in machine learning is a critical step to ascertain the quality of trained models. In the case of the YOLO V5 model, it involves fine-tuning a variety of hyper-parameters and configurations. These settings determine the model's performance on a validation dataset, affecting its accuracy and speed. Typical settings for YOLO validation include batch size, validation frequency, and the metrics used to measure model success. In this research, the YOLO V5 model was validated using 40 iris images from the IITD dataset, with bounding boxes drawn around the iris region to detect true positives, false positives, and false negatives. These enhancements are illustrated in Fig. 4, which displays the model's detection of iris images with rectangular bounding boxes.

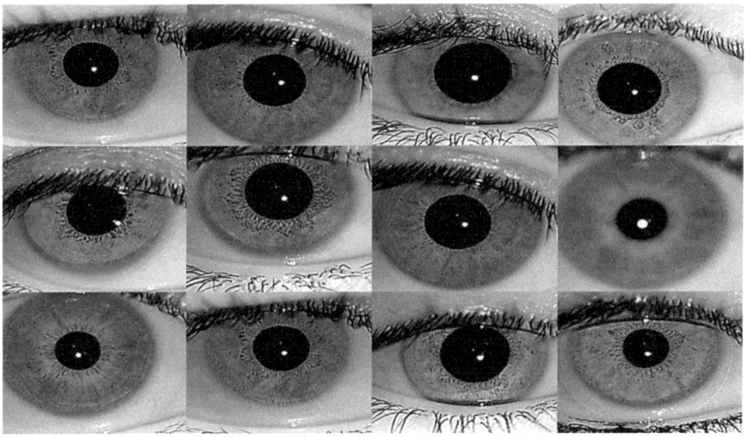

Fig. 2 Iris images extracted from IITD dataset

5 Results Analysis and Discussion

5.1 Analysis of Results

Figure 5 shows the performance of the proposed framework based on the 4 performance metrics highlighted.

The proposed model's performance on the iris detection task, as represented in the provided metrics, exhibits a high level of precision and recall across the board. The F_1-confidence curve, as shown in Fig. 5, peaks at a perfect score of 1.0 at a confidence threshold of 0.572, underscoring the model's exceptional balance between precision and recall when identifying iris regions across all classes. This implies that the model's predictions are highly reliable at this confidence level, with an excellent balance of false positives and false negatives. The precision-confidence curve, depicted in Fig. 5, further confirms the model's accuracy, achieving a precision rate of 1.0 at a confidence level of 0.611. This reinforces the model's capability to correctly identify the iris region without false detections. Complementarily, the recall rate, as shown in Fig. 5, also reaches 1.0, indicating that the model successfully identifies all true positive instances of the iris within the dataset. Moreover, the model demonstrates an impressive mean average precision (mAP) of 0.995 at a confidence threshold of 0.5, maintaining this level consistently from epoch 46 to epoch 100. This constancy in mAP, even with an increased number of epochs beyond 100, suggests a stable and robust model performance over time, without signs of overfitting, as the mAP does not decrease with additional training. The decision to cap training at 100 epochs was prudent to avoid overfitting and to ensure the model's generalizability.

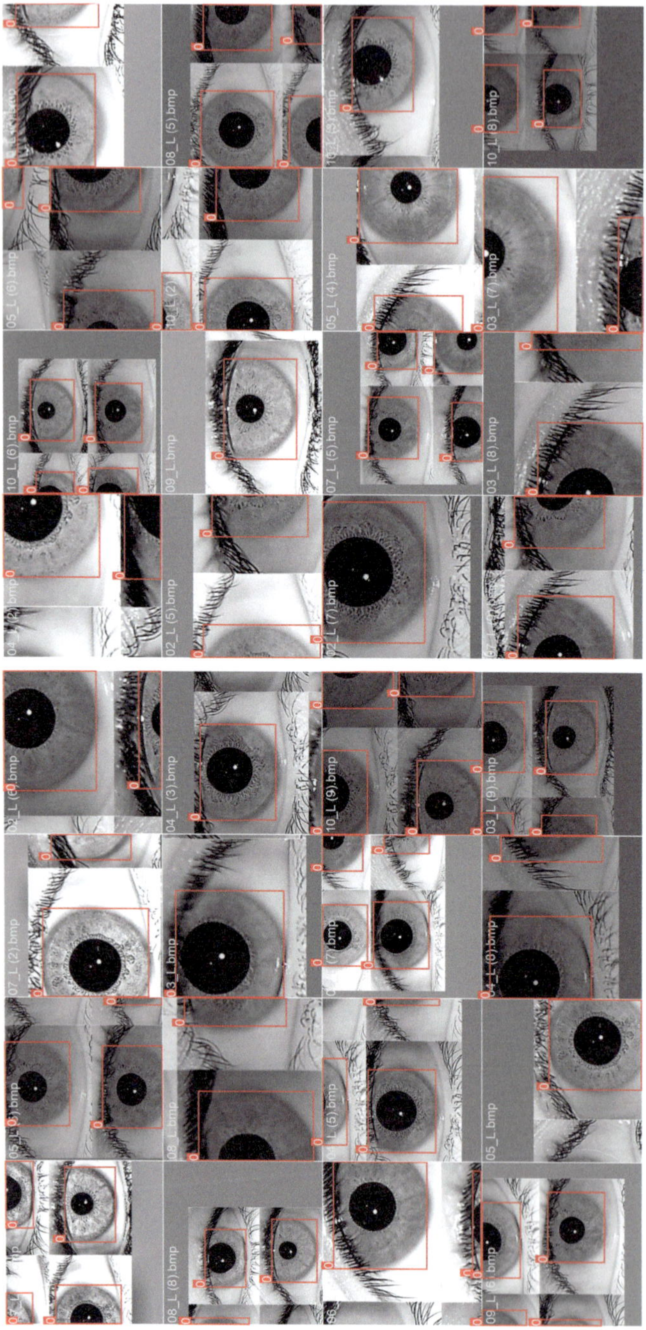

Fig. 3 IITD iris images during training batch 1(left) and batch 2 (right)

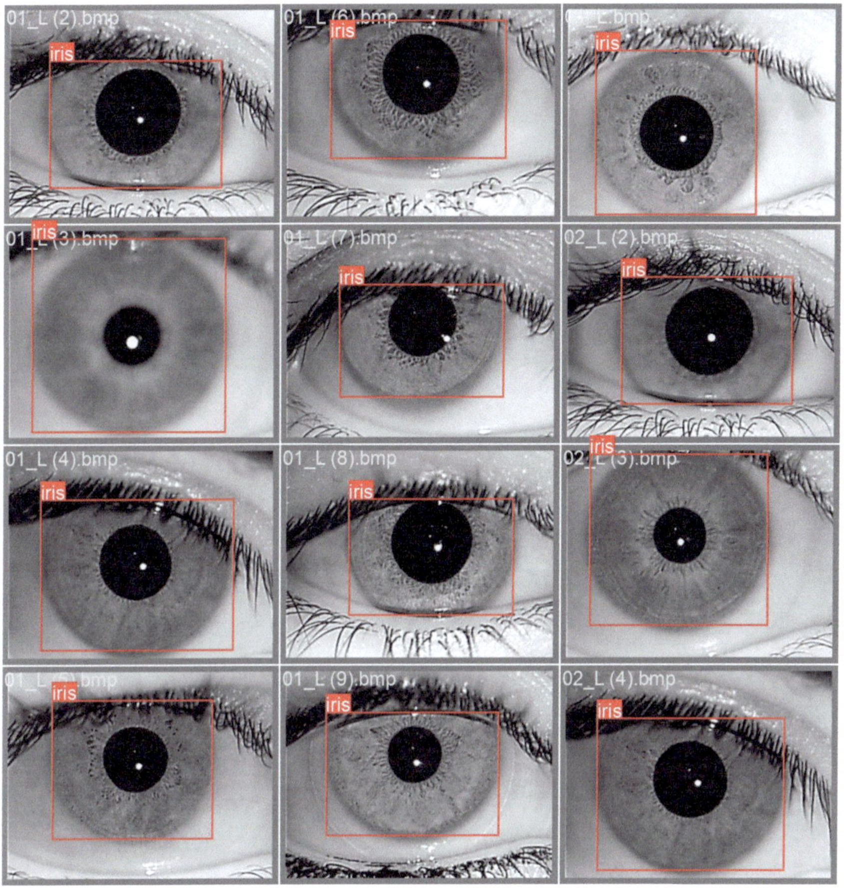

Fig. 4 IITD iris images detected by the model

5.2 Comparison with Related Works

The proposed model was compared to other works within the domain with the same IITD datasets. It must be noted that, the proposed model was tested on the same dataset as the models mentioned in Table 1. As clearly visible from Table 1, the proposed model outperformed the four existing models that worked within the same domain. The proposed model achieved a 100% accuracy rate compared to the existing models. The mAP was used to check the robustness of the model as it best compares the detected bounding box with ground truth box. The model achieved a mAP of 99.5%. However, other researchers in the field with respect to the topic and datasets used did not use mAP as metric for us to compare with their model.

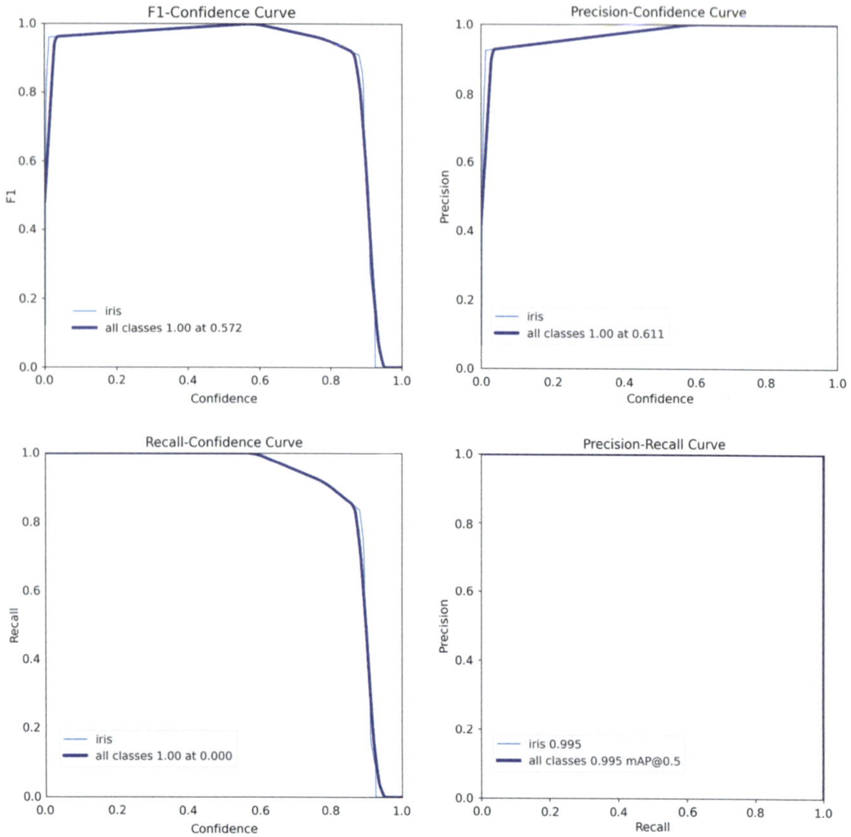

Fig. 5 Confidence score (top left), precision-confidence (top right), recall curve (bottom left), and mean average curve of the model

Table 1 Shows other related work in comparison with the proposed model

Study	Methodology	Accuracy
Chen et al. [10]	Deep convolution network (DCNN)	97.35
Ma et al. [7]	Conformal geometric algebra (CGA)	98.30
Jan and Min-Allah [5]	Histogram thresholding and circular hough transform	99.00
Mathias et al. [15]	Basis-spline-based active contour and circular active contour	99.60
Proposed model	YOLO V5	100.00

Table 2 Cross validation on other datasets

Datasets	F_1 score	Precision	Recall	Conf matrix	MAP
IITD	1.0	1.0	1.0	1.0	0.995
CASIA V1	1.0	1.0	1.0	1.0	0.994
MMU	1.0	1.0	1.0	1.0	0.996

5.3 Cross Validation of the Model with CASIA VI and MMU Datasets

The robustness of the YOLO V5 model was confirmed by its consistent performance across three different iris datasets: IITD, CASIA V1, and MMU. The model was validated on the CASIA V1 and MMU datasets using the same parameters applied to the IITD dataset. Results, as summarized in Table 2, show that the model attained perfect F_1 scores, precision, recall, and a confidence matrix of 1.0 for both CASIA V1 and MMU datasets—mirroring the performance on the IITD dataset. There was a negligible variation in the mean average precision (mAP) with a slight decrease of 0.001 on the CASIA dataset and a marginal increase of 0.001 on the MMU dataset compared to the IITD dataset. These subtle differences in mAP underscore the model's ability to consistently yield high-quality results across various datasets, thus underscoring its efficacy and generalizability in iris detection tasks.

6 Conclusion and Recommendation

6.1 Conclusion

This paper presented a YOLO V5-based model for detecting irises in noisy images, utilizing Makesense AI for accurate iris localization amidst noise and occlusions such as eyelashes and eyelids. The model's training on the IITD dataset, with an 80–20 split for training and validation and capped at 100 epochs to prevent overfitting, demonstrated exemplary performance metrics across the board—attaining perfect F_1, precision, recall rates, and a 0.995 mAP. When benchmarked against four related works using the same dataset and subject domain, the YOLO V5 model exhibited superior accuracy. Further validation on the CASIA V1 and MMU databases reaffirmed the model's robustness, achieving 100% in F_1 score, precision, and recall rates, along with 99.6% mAP on MMU and 99.4% on CASIA V1. These findings suggest that the proposed model is highly effective across various iris datasets.

6.2 Recommendation

Future research should build upon the robust results achieved by the YOLO model for iris detection in noisy images, with a focus on expanding the volume of training and validation images to enhance the model's robustness and accuracy across diverse datasets. Additionally, addressing limitations such as varying illumination and blurred edges in images remains an area for improvement. Future studies should also explore YOLO-based iris detection in the context of occlusions caused by contact lenses, which poses a distinct challenge for current recognition systems.

References

1. Malgheet JR, Manshor NB, Affendey LS, Abdul Halin AB (2021) Iris recognition development techniques: a comprehensive review. Complexity 2021:1–32
2. Adamović S, Miškovic V, Maček N, Milosavljević M, Šarac M, Saračević M, Gnjatović M (2020) An efficient novel approach for iris recognition based on stylometric features and machine learning techniques. Futur Gener Comput Syst 107:144–157
3. Jain DK, Lan X, Manikandan R (2020) Fusion of iris and sclera using phase intensive rubbersheet mutual exclusion for periocular recognition. Image Vis Comput 103:104024
4. Ma L, Li H, Yu K (2020) Fast iris localization algorithm on noisy images based on conformal geometric algebra. Digital Sig Process 100:102682
5. Jan F, Min-Allah N (2020) An effective iris segmentation scheme for noisy images. Biocybern Biomed Eng 40(3):1064–1080
6. Nazir A, Cheema MN, Sheng B, Li P, Li H, Yang P, Jung Y, Qin J, Feng DD (2020) SPST-CNN: Spatial pyramid based searching and tagging of liver's intraoperative live views via CNN for minimal invasive surgery. J Biomed Inform 106:103430
7. Ma L, Li H, Yu K (2020) Fast iris localization algorithm on noisy images based on conformal geometric algebra. Digital Sig Process 100(102682):1–8
8. Feng X, Liu W, Li J, Meng Z, Sun Y, Feng C (2022) Iris R-CNN: accurate iris segmentation and localization in non-cooperative environment with visible illumination. Patt Recogn Lett 155:151–158
9. Kumawat A, Panda S (2023) Noisy iris smoothing and segmentation scheme based on improved Wildes method. Multidimension Syst Signal Process 34(1):47–79
10. Chen Y, Wu C, Wang Y (2021) Whether normalized or not? Towards more robust iris recognition using dynamic programming. Image Vis Comput 107:104112
11. Li Y-H, Huang P-J, Juan Y (2019) An efficient and robust iris segmentation algorithm using deep learning. Mob Inf Syst 2019:1–15
12. Susitha N, Subban R (2019) Reliable pupil detection and iris segmentation algorithm based on SPS. Cogn Syst Res 57:78–84
13. Chen Y, Gan H, Chen H, Zeng Y, Xu L, Heidari AA, Zhu X, Liu Y (2023) Accurate iris segmentation and recognition using an end-to-end unified framework based on MADNet and DSANet. Neurocomputing 517:264–278
14. Malinowski K, Saeed K (2022) An iris segmentation using harmony search algorithm and fast circle fitting with blob detection. Biocybern Biomed Eng 42(1):391–403
15. Mathias GP, Gagan JH, Mallya BV, Kumar JH (2022) A unified approach for automated segmentation of pupil and iris in on-axis images. Comp Methods Prog Biomed Update 2:100084
16. Nsaif AK, Ali SHM, Nseaf AK, Jassim KN, Al-Qaraghuli A, Sulaiman R (2022) Robust and swift iris recognition at distance based on novel pupil segmentation. J King Saud Univ Comp Inf Sci 34(10):9184–9206

17. Umer S, Chandra Dhara B, Chanda B (2017) A novel cancelable iris recognition system based on feature learning techniques. Inf Sci 406:102–118
18. Marra F, Poggi G, Sansone C, Verdoliva L (2018) A deep learning approach for iris sensor model identification. Patt Recogn Lett 113:46–53
19. Ahmadi N, Nilashi M, Samad S, Rashid TA, Hossein A (2019) An intelligent method for iris recognition using supervised machine learning techniques. Opt Laser Technol 120:105701
20. Ramachandra S, Ramachandran S (2022) Region specific and subimage based neighbour gradient feature extraction for robust periocular recognition. J King Saud Univ Comp Inf Sci 34(10):7961–7973
21. Gautam G, Raj A, Susanta M (2021) Deep supervised class encoding for iris presentation attack detection. Digital Sig Process 121:103329
22. Raskar PS, Shah SK (2021) Real time object-based video forgery detection using YOLO (V2). Forensic Sci Int 327:110979
23. Zhang Y, Huang J, Cai F (2020) On bridge surface crack detection based on an improved YOLO v3 algorithm. IFAC-PapersOnLine 53(2):8205–8210
24. Kumar A, Kalia A, Kalia A (2022) ETL-YOLO v4: a face mask detection algorithm in era of COVID-19 pandemic. Optik 259:169051

XAI-Based Prediction of Oesophageal Cancer in Medical Imaging

V. Sandhiya and A. Anitha

Abstract Diagnosing diseases through medical imaging represents the visual depiction of organs and tissues that reflect the internal structure of the body for diagnostic purposes and surgical treatment. Medical imaging diagnosis is an area where deep learning is being applied with attention due to its exceptional performance. Image-based Deep Learning models are used in clinical processes, but mostly they lack transparency due to their interpretability issues. Since deep learning models are black boxes, there is a need for techniques to explain their decision-making process, thereby leading to the evolution of the Explainable Artificial Intelligence (XAI) method. We presented an experimental analysis of XAI-based medical imaging diagnosis for oesophagus cancer using Grad-CAM and LIME XAI techniques for textual and visual explanation. Furthermore, this study suggested the medical practitioner the automatic prediction and explanation of the disease density with the accuracy.

Keywords Explainable AI · Medical imaging · Deep learning · LIME · Grad-CAM

1 Introduction

Oesophageal cancer, a malignant development in the oesophagus that connects the throat to the stomach, is a major worldwide health issue, standing sixth in deaths caused by cancer [1]. Although treatable and recoverable in its earliest stages, late discovery frequently results in a poor outcome. This is where machine learning (AI) appears as a possible transformation. Frequent tobacco use, alcohol consumption, and certain dietary habits all raise the chance of oesophageal cancer [2]. DNA abnormalities in oesophagus cells cause the illness to progress uncontrollably and produce tumours. This malignant stage, referred to as Barrett's oesophagus, can develop into

V. Sandhiya · A. Anitha (✉)
School of Computer Science Engineering and Information Systems, Vellore Institute of Technology, Vellore, India
e-mail: aanitha@vit.ac.in

© The Author(s), under exclusive license to Springer Nature Singapore Pte Ltd. 2025
S. K. Udgata et al. (eds.), *Intelligent Systems*, Lecture Notes in Networks and Systems 1314, https://doi.org/10.1007/978-981-96-3797-3_39

cancer if not diagnosed. Unfortunately, the initial stages oesophageal cancer sometimes causes no symptoms, making prompt identification difficult [3]. Artificial intelligence and medical imaging plays a vital role in the healthcare sector using deep learning (DL) and machine learning (ML) techniques integrated with computer vision [4]. Convolutional Neural Networks (CNN) are significantly important in medical imaging research due to their ability to extract features, classify data, and predict low-level features [5]. Researchers have provided various direction of using deep learning models [6], especially to classify oesophageal cancer at early stages. These algorithms analyse endoscopic images with remarkable precision, perhaps assisting in early identification and increasing treatment results. Understanding "How deep learning models predict?" has recently attracted interest of the researchers, with a greater focus placed on constructing interpretable and trustworthy models [7]. A necessity for eXplainable Artificial Intelligence (XAI) arises from the challenging black-box nature of deep learning algorithms especially for medical imaging [8]. Explainability, interpretability, and transparency are all essential in deep learning-based medical applications since they provide insight into neural network choices and make decisions more visible [9].

The various sources of finding oesophagus cancer can be of X-ray images, CT scan images, MRI images, endoscopy images, positron emission tomography (PET) scans, endoscopic ultrasound, Bronchoscopy and laproscopy [10]. Out of these medical imaging analysis provides better disease examination process as it determines the stages of the disease leads for better treatment [11]. The paper aims in providing the medical imaging analysis using Grad-CAM visualization of making prediction using the heatmap and LIME feature impact to make decisions. By mixing high accuracy and explainability, AI may enable healthcare professionals, which could contribute to better treatment results and higher rates of survival for this terrible disease.

The contributions of the paper may be summed up as follows:

- To predict the oesophagus cancer using Grad-CAM and LIME XAI model based on the convolutional feature map.
- To identify the part of the images that has the impact of the classification score.

2 Related Works

Explainable AI (XAI) is a tool used by medical imaging researchers to clarify model outputs by emphasizing key input attributes [2]. It allows for human-readable communication of ML models, analysing logical paradigms and justifying decisions, and enhancing the interpretability of machine learning algorithms. Nakajo et al. [9] suggested the anatomical classification of laryngeal and pharyngeal endoscopic images using AI techniques. Zhang et al. [10] detected the esophageal cancer using semar images to suspect the lesions using the robust screening of whole slide imaging. Zhou et al. [11] discussed the CNN-tranformer network based feature aggregation to classify the stages of oesophageal cancer images. Wu et al. [12] identifies the histone acetyltransferase 1(HAT1) as morphological regulator in finding the esophageal squamous carconima cells using LIME model. Wang et al. [13] predicted the esophageal vericeal bleeding based on the endoscopy and structured data using the automated machine learning and LIME model for explanation. Wang et al. [13] utilized the statistical anatomy to provide the T-stage of the oesophageal cancer using major and minor axis using the computed tomography angiography images finding AUC value to be better for the stage prediction. Liu et al. [14] utilized automated machine learning to predict the gastrointestinal stromal tumour based on the endoscopic resection using CNN and LIME model.

The above review can summarize that the

- Machine learning models helps in predicting the diseases using the medical images, still the explanation is required on the predicted values.
- Explainable AI provides better explanations on the predictions made by the machine learning models irrespective of the hyper parameters used for prediction.
- LIME and Grad-CAM provides better prediction explanations using the probability values on the features used for the prediction process.

3 Background Fundamentals

The proposed work utilizes the CNN model as machine learning technique for classification. In addition, it uses LIME and Grad-CAM for the explaining the predictions done using the deep learning models. Let us discuss the background fundamentals in this section.

3.1 Convolution Neural Network

The Convolutional Neural Network (CNN) is a type of deep neural network that has demonstrated exceptional performance in image and video recognition applications [11]. We'll provide the mathematical formulae that underpin the main functions of a CNN in this section, along with explanations of each part using Eq. (1).

$$(J * K)_{c,d} = \sum_{m=0}^{M-1} \sum_{n=0}^{N-1} J_{c+m,\, d+n} . K_{m,n} \tag{1}$$

where J is the input image, K is the convolutional kernel (filter). The kernel's sizes are M and N. During the convolution process, the input image (I) is passed through a tiny filter (K). The output feature map has a single value that is obtained by computing the element-wise product at each position and adding the results. This process extracts local patterns from the input.

Activation Function (ReLU)

Applying a non-linear activation function element-by-element to create non-linearity in the network is known as the Rectified Linear Unit (ReLU) [6]. In the feature map, it substitutes zero for any negative value. The network may discover intricate associations and patterns in the data because of its non-linearity using Eq. (2).

$$\text{ReLU}(y) = \max(0, y) \tag{2}$$

Max Pooling

$$\text{MaxPooling }(y) = \max(\text{pool}(y)) \tag{3}$$

The downsampling process known as max pooling decreases the feature map's spatial dimensions. It entails transferring the greatest value across the feature map within a limited window (such as 2×2) using Eq. (3). This process lessens computing complexity while assisting in the retention of the most essential details.

Fully Connected Layer

$$\text{Output} = \text{ReLU}(\text{W.input} + c) \tag{4}$$

The weight matrix is denoted by W. The input from the preceding layer has been flattened. The bias vector is c.

All neurons in a fully connected layer are coupled to all neurons in the subsequent layer. During training, the weights (W) and biases (c) are grasped, and non-linearity is introduced via the ReLU activation function using Eq. (4).

Softmax Activation function

The softmax function, which converts initial scores into probability distributions, is frequently employed in the output layer of multiclass classification problems.

The softmax function for a vector y is given as follows,

$$\text{Softmax}(y)_i = \frac{e^{y_i}}{\sum_j e^{y_j}} \tag{5}$$

Loss Function

The training procedure is guided by the loss function, which calculates the variation between the real labels and the predicted values mentioned in Eq. (6).

$$\text{Loss} = -\sum_j X_{\text{true},j}.\log(X_j) \tag{6}$$

where X is the predicted result and X_{true} is the true label. The mostly using loss function is Categorical entropy loss.

3.2 Explainable AI Approaches

Model explainability and interpretability are essential for clinical evidence research and making decisions in the medical field. It can be helpful to identify concealed illnesses if an advanced neural network-powered system supports a doctor's results. Yet many deep learning models are not explainable, which means that using them in medical applications is not feasible. Thus, explainability is needed to help deep learning models grasp the semantics of the classification criteria they employ. This is why eXplainable AI (XAI) algorithms have been created; this increases the model's dependability and trustworthiness [5]. XAI promotes trust between patients and healthcare practitioners in all aspects of connected healthcare environments; in addition, it helps to ensure the safety and quality of service.

To create these explainable systems, two primary sets of approaches are usually used: ante-hoc and post-hoc strategies [2]. Explainability is incorporated into a model from the outset using ad hoc approaches. Post-hoc strategies have been employed once the model was successfully developed and taught to make opinions. We employ two of the post-hoc techniques in the present study such as Local Interpretable Model-agnostic Explanations (LIME) and Gradient-weighted Class Activation Mapping (Grad-CAM) [8].

Local Interpretable Model-Agnostic Explanations (LIME)

"LIME (local interpretable model-agnostic explanations)" is a XAI—based model that approximates any black-box Deep learning model with a local, transparent model to explain individual predictions. The model uses LIME to permute input data, producing identical samples by flipping on and off parts of the image's super pixels. The trained model predicts each altered image and classifies each fabricated data point. The cosine difference between the altered and original images determines the importance of each data point. A linear regression model is trained using weighted synthetic data points, with features with higher coefficients influencing the model's outcome [2].

Grad-CAM

A computer vision approach called gradient-weighted class activation mapping, or Grad-CAM, creates a heat map of the key areas in an image that significantly influence the deep learning model's prediction [8]. Grad-CAM makes use of the gradient of the end result's class score in relation to the feature map data of the final layer of the model to produce visual explanations.

Table 1 Dataset class labels

Class	Number of images
Class 0 (Normal Oesophagus)	750
Class1 (Malignant)	750
Total number of images	1500

4 Experimental Analysis and Result Discussion

4.1 Dataset Description and Pre-processing

Dataset was gathered at four separate hospitals that make up the "Vestre Viken Health Trust (VV)" in Norway using endoscopic technology. The resolution of the images in the data varies between 633 × 532 to 1349 × 1071 pixels. The image was first scaled into 224 × 224 × 3 pixels using colour constancy. An endoscope-viewable feature of the Gastro Intestinal Tract is called an anatomical landmark. It is essential for navigating and serving as a point of reference when explaining the precise location of a finding. Another possibility is that the landmarks correspond to certain pathological conditions, such as ulcers or inflammation. The two categories of oesophageal disease and normal-oesophagus images are taken as class 0 and class 1 for binary classification which is mentioned in Table 1.

4.2 Proposed Methodology

Numerous computer vision tasks have demonstrated the effectiveness of CNN models; still a significant quantity of training data was necessary to prevent over fitting. In general, when a deep learning model acquires a high-variance function which accurately represents the training data but has a restricted scope of generalization, mentioned as over fitting. To address the over fitting problem, several methods for data enhancement are applied to enhance the volume and coherence of the data.

Data augmentation had undergone various rotations, width modifications, height shifts, zooming, and horizontal flipping. The horizontal flip, 0.2 zooming range, 35° fixed rotation range, and 0.1 shifting width, brightness and height range applied for pre-processing the images. The dataset contains 1650 augmented images for experimental analysis and categorized as normal oesophageal and impacted oesophageal images having 825 images for each class to maintain data balance. Further, the dataset partitioned into training and testing sets in a 70:30 ratio and utilized CNN model to classify as benign and malignant. Figure 1 shows the architecture diagram for the proposed model.

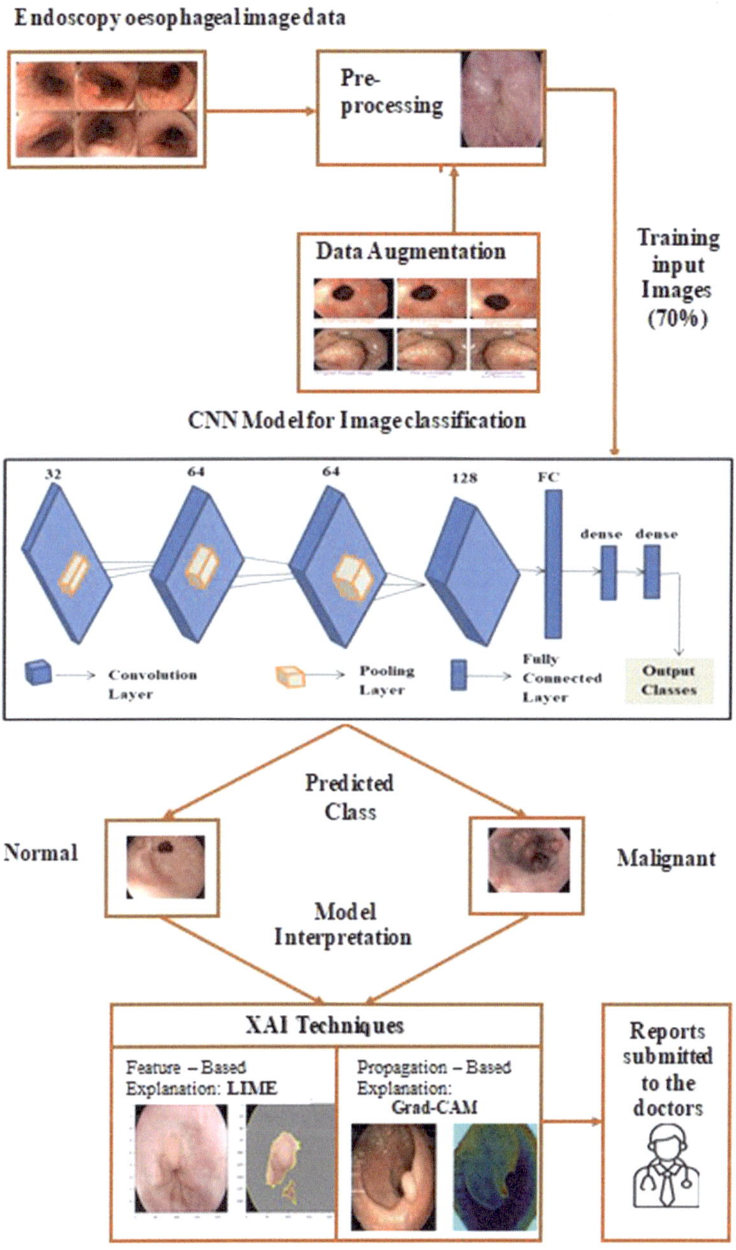

Fig. 1 Architecture diagram for the proposed model

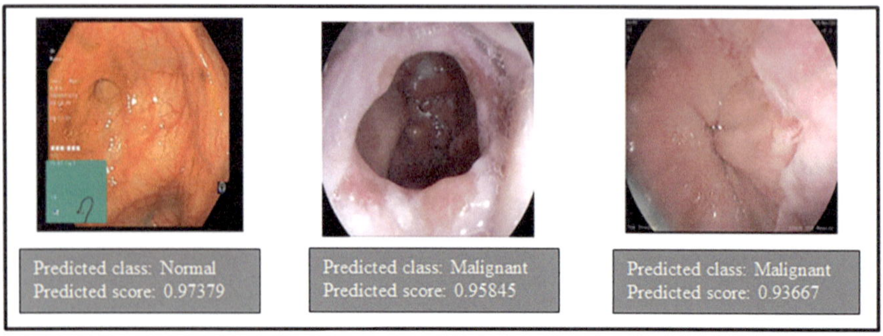

Fig. 2 Predicted result

4.3 Result Analysis

The proposed model was experimented using the Google Colab environment and has 2×2 convolution layer, a max-pooling layer, a softmax layer, and over 1.4 million parameters, with the maximum parameter in dense layer 1 being around 16,000. The CNN model is trained using a trial-and-error approach, considering computational capability, with 30 epochs, 16 batch size, 0.001 learning rate, and 25% step size, employing Adam optimizers since they have lesser complexity. Subsequently, the first and third layers have employed ReLU, a non-linear activation function. Categorical cross-entropy and softmax classification layers are utilized. Analysing the findings, the hyper-parameter optimized CNN model functioned effectively, with the greatest validation accuracy of 94%. The proposed model predicted the image's classes from the dataset, as shown in Fig. 2 are given. The precision value indicated that 96.7% of the expected positive values were positive, while the recall value predicted that 91.70% of the overall positive values were positive which is depicted in table. With an extremely high precision score, the proposed model has successfully made accurate predictions and classified the endoscopic oesophageal images taken from the gastro intestinal tract. Figures 3 and 4 show the ROC curve confusion matrix for the CNN model.

4.4 Comparative Analysis

Table 2 provides a comprehensive analysis comparing the proposed CNN approach with other CNN techniques discussed in the various literature. The results showed that the proposed approach performed better than alternative approaches. Figure 5 shows the accuracy versus loss. This study showcases the potential of ML and DL models in medical diagnostics, with a highest accuracy of 94.4% makes this especially visible, demonstrating a distinct pattern of rising dependability in these models. It promotes

Fig. 3 Roc curve for result

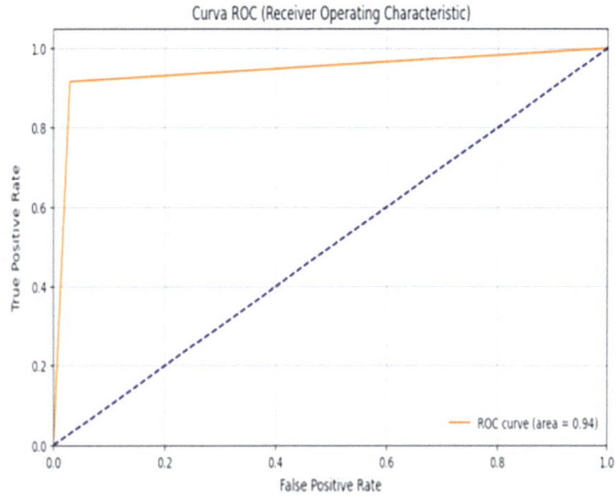

Fig. 4 Confusion matrix

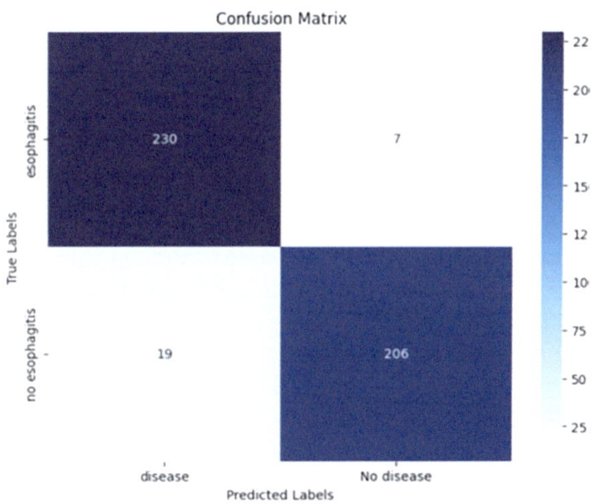

the trustworthiness of DL models, especially in the medical industry and oesophageal images, using XAI for verification.

Table 2 Evaluation metrics

Methods	Accuracy (%)	Precision (%)	Recall (%)
Proposed method	94.37	96.71	91.7
DenseNet201 [2]	88.75	88.77	88.70
2D CNN [7]	92.60	92.95	88.61
CNN [15]	81.2	79	80

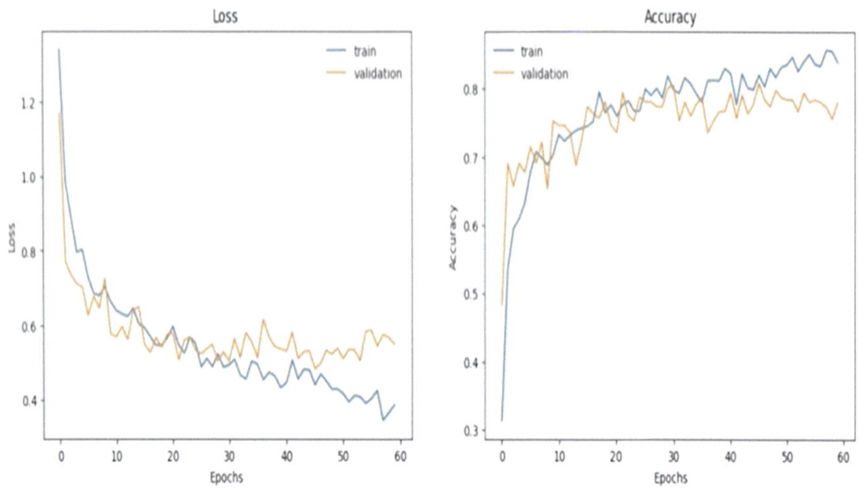

Fig. 5 Loss versus accuracy

4.5 *Explainability of the Proposed Model*

The explanations for the predicted model were employed using two XAI techniques such as LIME and Grad-CAM. LIME helps in feature-based explanation, whereas Grad-CAM utilizes heatmap based on the convolution feature maps. LIME gave more reasonable explanations for the model predictions. The images below highlight the key elements that set them apart from their surroundings, improving the likelihood that they will be correctly classified. A comparison of the findings from two distinct XAI methods on the class of normal oesophageal and malignant oesophageal images is shown in Fig. 6.

Grad-CAM heatmaps that emphasize concentrated regions in test images for oesophageal illness prediction are shown in Fig. 6. Each pixel's significance in the input image is represented by its intensity on the heatmap in the Grad-CAM visualization. Greater intensity (brighter colours) and a significant contrast difference between the colour and the backdrop indicate a more relevant region that went into making the model's prediction. Figure 6 shows that Grad-CAM seems to have the finest overall performance than other models.

5 Conclusion and Future Opportunities

Oesophageal cancer, a serious and incurable illness originating from the oesophagus, spreads to other internal organs and poses significant danger to affected individuals due to its rare early detection. This proposed model discusses major developments in explainable deep learning in medical imaging diagnostics, using XAI approaches

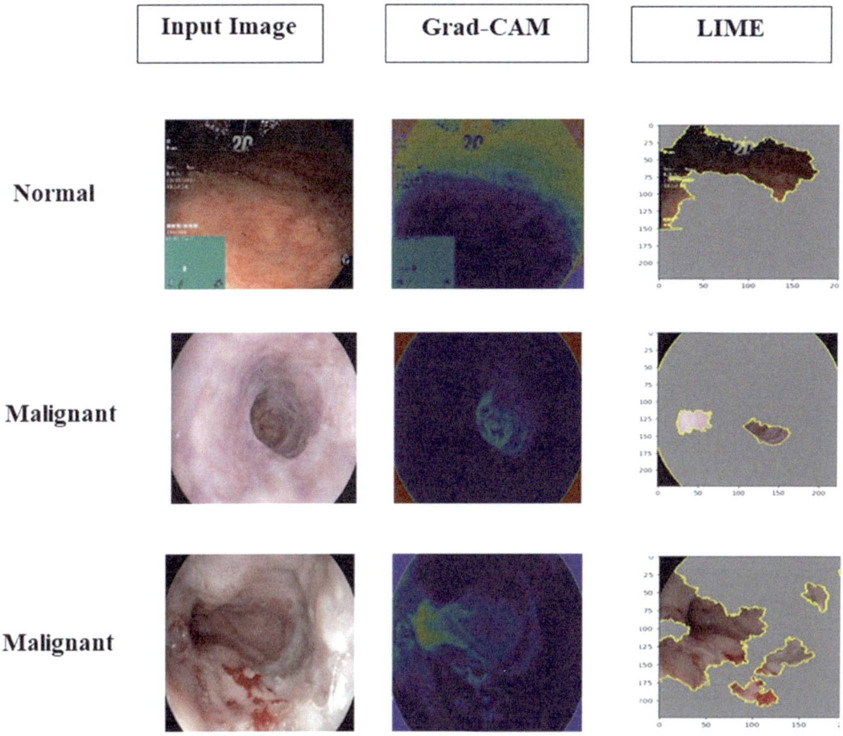

Fig. 6 Interpretations provided by XAI techniques Grad-CAM and LIME

with an experimental analysis. The proposed model utilized the oesophagal cancer endoscopic images trained using the customized CNN model and explained using the LIME and Grad-CAM for the better prediction explanations. The experimental analysis shows that the proposed model achieves 94% of accuracy with better explanations using LIME and Grad-CAM models. The results of this study may help to provide doctors and other medical professionals with a useful understanding of opaque AI models. This study could benefit from future improvements involving the use of advanced AI techniques like Generative Adversarial Networks (GAN) for improved accuracy in categorizing multiclass endoscopic images with larger dataset.

References

1. Sharma A, Kumar R, Garg P (2023) Deep learning-based prediction model for diagnosing gastrointestinal diseases using endoscopy images. Int J Med Informatics 177:105142
2. Shaw P, Sankaranarayanan S, Lorenz P (2022) Early esophageal malignancy detection using deep transfer learning and explainable AI. In: 2022 6th international conference on communication and information systems (ICCIS). IEEE, USA, pp 129–135

3. Wang R, Zhang X, Wu W, Ma J, Chen J, Zhang Z, Cui H (2023) Could tumour volume and major and minor axis based on CTA statistical anatomy improve the pre-operative T-stage in oesophageal cancer? Cancer Med
4. Anitha A, Shivakumara P, Jain S, Agarwal V (2023) Convolution neural network and auto-encoder hybrid scheme for automatic colorization of grayscale images. In: Smart computer vision. Cham: Springer International Publishing, pp 253–271. https://doi.org/10.1007/978-3-031-20541-5_12
5. Arrieta AB et al (2020) Explainable artificial intelligence (XAI): concepts, taxonomies, opportunities and challenges toward responsible AI. Inf Fusion 58:82–115
6. Nandhini N, Anitha A (2024) Seasonal-wise occupational accident analysis using deep learning paradigms. In: Udgata SK, Sethi S, Gao XZ (eds) Intelligent systems. ICMIB 2023. Lecture Notes in Networks and Systems, vol 728. Springer, Singapore. https://doi.org/10.1007/978-981-99-3932-9_17
7. Okay FY, Yildirim M, Özdemir, S (2021) Interpretable machine learning: a case study of healthcare. In: 2021 International symposium on networks, computers and communications (ISNCC). IEEE, USA, pp 1–6
8. Ribeiro MT, Singh S, Guestrin C (2016) "Why should I trust you?" Explaining the predictions of any classifier. In: Proceedings of the 22nd ACM SIGKDD international conference on knowledge discovery and data mining, pp 1135–1144
9. Nakajo K, Ninomiya Y, Kondo H, Takeshita N, Uchida E, Aoyama N, Yano T (2023) Anatomical classification of pharyngeal and laryngeal endoscopic images using artificial intelligence. Head Neck
10. Zhang C, Jia D, Li Z, Wu N, He Z, Jiang H, Yan Q (2024) Esophageal cancer detection framework based on time series information from smear images. Expert Syst Appl 238:122362
11. Zhou Z, Sun G, Yu L, Tian S, Xiao G, Wang J, Zhou S (2023) RFIA-Net: rich CNN-transformer network based on asymmetric fusion feature aggregation to classify stage I multimodality oesophageal cancer images. Eng Appl Artif Intell 118:105703
12. Wu Y, Jiang B, Wang Q, Liu Y, Wu C, Wu M, Song H (2023) Deep learning ıdentifies HAT1 as a morphological regulator in esophageal squamous carcinoma cells through controlling cell senescence. Adv Intell Syst, 2200352
13. Wang Y, Hong Y, Wang Y, Zhou X, Gao X, Yu C, Zhu J (2023) Automated multimodal machine learning for esophageal variceal bleeding prediction based on endoscopy and structured data. J Digit Imaging 36(1):326–338
14. Liu L, Zhang R, Shi D, Li R, Wang Q, Feng Y, Xu X (2023) Automated machine learning to predict the difficulty for endoscopic resection of gastric gastrointestinal stromal tumor. Front Oncol 13:1190987
15. Bordbar M, Helfroush MS, Danyali H, Ejtehadi F (2023) Wireless capsule endoscopy multiclass classification using three-dimensional deep convolutional neural network model. Biomed Eng Online 22(1):124

MIX
Papier aus verantwortungsvollen Quellen
Paper from responsible sources
FSC® C105338

If you have any concerns about our products,
you can contact us on
ProductSafety@springernature.com

In case Publisher is established outside the EU,
the EU authorized representative is:
Springer Nature Customer Service Center GmbH
Europaplatz 3, 69115 Heidelberg, Germany

Printed by Libri Plureos GmbH
in Hamburg, Germany